HIGH TEMPERATURE MATERIALS and MECHANISMS

HIGH TEMPERATURE MATERIALS and MECHANISMS

Edited by
Yoseph Bar-Cohen

CRC Press
Taylor & Francis Group
Boca Raton London New York

CRC Press is an imprint of the
Taylor & Francis Group, an **informa** business

CRC Press
Taylor & Francis Group
6000 Broken Sound Parkway NW, Suite 300
Boca Raton, FL 33487-2742

First issued in paperback 2017

© 2014 by Taylor & Francis Group, LLC
CRC Press is an imprint of Taylor & Francis Group, an Informa business

Version Date: 20131101

ISBN 13: 978-1-4665-6645-3 (hbk)
ISBN 13: 978-1-138-07154-4 (pbk)

Library of Congress Cataloging-in-Publication Data

High temperature materials and mechanisms / edited by Yoseph Bar-Cohen.
    pages cm
  Includes bibliographical references and index.
  ISBN 978-1-4665-6645-3 (hardback)
  1. Heat resistant materials. I. Bar-Cohen, Yoseph.

TA418.26.H527 2014
620.1'1296--dc23
                                                     2013041092

Visit the Taylor & Francis Web site at
http://www.taylorandfrancis.com

and the CRC Press Web site at
http://www.crcpress.com

Contents

Preface

High-temperature materials and mechanisms are critical technologies for pushing the boundaries of human capabilities. The use of high-temperature materials is dated as far back as the beginning of human civilization. As early as humans started working with fire and heat, they have used materials that can sustain high temperatures. Initially, the primitives used rocks to handle fire that they applied for cooking and heating. With time, the applications were expanded to making tools that were processed and used at high temperatures. It is interesting to note that the human ability to process and handle materials has improved considerably with their advancement in melting high-temperature materials. During the Stone Age, humans were not able to smelt any ore, whereas during the Bronze and Iron Ages they were able to smelt bronze and iron, respectively, as well as produce artifacts from them. The Industrial Revolution led to an enormous rise in applications and requirements for increased capability and operational safety.

At the end of the nineteenth century, some of the important applications that emerged include the steam turbine, while in the early part of the twentieth century the filament lamp and the combustion engine were developed, and these developments have been followed by jet engines around the 1940s and the space shuttle engine in the 1980s. Steel alloys were introduced at the end of the nineteenth century, followed by stainless steel and nickel-chrome alloys in the beginning of the twentieth century, and these alloys had a major impact on the applications of high-temperature materials. Today, the selection of high-temperature materials has been widened to such metal alloys as refractory metals, super-alloys, and titanium alloys. Other materials that are being used include certain ceramic materials, carbon/carbon composites, metal matrix composites, and many others.

Generally, in materials science the term *high temperature* is defined as the temperature that is equal to, or greater than, about two-thirds of the melting point of a solid. However, there are other definitions too, including application-based definitions such as heat resistance (e.g., strength and corrosion resistance) above 500°C. Since increasing the temperature decreases the material strength, they need to have high strength at the required operating temperatures with a safety margin in order to make these materials effective and economical. High-temperature materials need to be resistant to such causes of damage as oxidation and corrosion, which are accelerated with the increase in temperature. Generally, extrapolating the material properties and chemical behavior do not correlate with the high-temperature data. Specifically, as the temperature rises, chemical reactions become pronounced and thermodynamic properties determine the reactivity rather than kinetics. Also, there are various effects that take place, including disorder of the material structure.

It is interesting to follow the parallel development of materials and the resulting industrial applications. Specifically, the driver of engineering requirements at the end of the nineteenth century has been the availability of materials and their processing, including powder metallurgy, casting, and vacuum melting. The availability of compatible materials is critical to the ability to develop high-temperature-related technologies such as high-speed airplanes and rockets that can reach 5 Mach and beyond having surface temperature that rises above 500°C. Besides the capability of the materials and structures to sustain high temperatures without irreversible changes and degradation of the properties, it is also necessary to have effective structural designs that allow for efficient heat dissipation, transfer of the associated high heat fluxes, and the ability to sustain the related thermal conditions.

This book includes Chapters 1 through 5 that cover subjects of high-temperature materials and mechanisms from many angles, including the chemistry and thermodynamics, overview of the various materials, including refractory metals, ceramics and composites for high-temperature structural and functional applications, and adhesives and bonding and their failure causes. Chapters 6 through 8 cover the topics of processes, materials characterization methods, and their nondestructive evaluation and health monitoring. The application of high-temperature materials to actuators and sensors is described in Chapters 9 through 15, including electromechanical materials, such as piezoelectric, thermo-acoustics materials, shape memory and super-elastic alloys, thermoelectric materials drilling mechanisms, as well as electronics. Further, Chapters 16 and 17 cover sensor design challenges and various high-temperature materials and mechanisms applications and challenges.

Acknowledgments

The editor thanks the chapters' reviewers for their very valuable technical comments and suggestions and they were

Chapter 1

Ted Iskenderian, Jet Propulsion Laboratory (JPL)/Caltech, Pasadena, California

Yiannis Pontikes, Department of Metallurgy and Materials Engineering, KU Leuven, Belgium

Chapter 2

Elizabeth J. Opila, Department of Materials Science and Engineering, University of Virginia, Charlottesville, Virginia

William T. Petuskey, Arizona State University, Tempe, Arizona

Chapter 3

James Gordon Hemrick, Oak Ridge National Laboratory, Oak Ridge, Tennessee

Vilupanur A. Ravi, California State Polytech University, Pomona, California

Mark L. Weaver, University of Alabama, Tuscaloosa, Alabama

Eric Wuchina, Office of Naval Research (ONR), Arlington, Virginia

Chapter 4

John Bishopp, Star Adhesion Limited, Cambridge, UK

Lucas F. M. da Silva, Editor in Chief of *The Journal of Adhesion*, University of Porto (FEUP), Portugal

Linda Del Castillo, Jet Propulsion Lab (JPL)/Caltech, Pasadena, California

Erol Sancaktar, University of Akron, Ohio

Chapter 5

Sebastien Dryeopondt, Oak Ridge National Laboratory, Oak Ridge, Tennessee

Paul Gannon, Montana State University, Bozeman, Montana

Brian Gleeson, University of Pittsburgh, Pittsburgh, Pennsylvania

Bryan Harder, NASA's Glenn Research Center, Cleveland, Ohio

Chapter 6

William Fahrenholtz, Missouri University of Science and Technology, Rolla, Missouri

Yiannis Pontikes, Department of Metallurgy and Materials Engineering, KU Leuven, Belgium

Eric Wuchina, Office of Naval Research, NSWCCD, W. Bethesda, Maryland

Chapter 7
Vince Crist, Nanolab Technologies, Milpitas, California
Doreen D. Edwards, North Carolina State University, Raleigh, North Carolina
Kip O. Findley, Colorado School of Mines, Golden, Colorado

Chapter 8
Ali Abdul-Aziz, NASA Glenn Research Center and Cleveland State University, Cleveland, Ohio
Edward Ecock, Consolidated Edison Company, New York, New York
Darby B. Makel, Makel Engineering, Inc., Chico, California
Subhasish (Subh) Mohanty, Argonne National Laboratory, Lemont, Illinois
Philip G. Neudeck, NASA Glenn Research Center, Cleveland, Ohio
Don J. Roth, NASA Glenn Research Center, Cleveland, Ohio

Chapter 9
Iqbal Husain, North Carolina State University, Raleigh, North Carolina
Ted Iskenderian, Jet Propulsion Laboratory (JPL)/Caltech, Pasadena, California
Robert Troy, Jet Propulsion Laboratory (JPL)/Caltech, Pasadena, California

Chapter 10
Troy Y. Ansell, Oregon State University, Corvallis, Oregon
David Cann, Oregon State University, Corvallis, Oregon
Dragan Damjanovic, Institute of Materials, School of Engineering, EPFL, Lausanne, Switzerland
Xiaoning Jiang, North Carolina State University, Raleigh, North Carolina

Chapter 11
Wael Akl, Faculty of Engineering, Ain Shams University, Cairo, Egypt
Hyeong Jae Lee, Jet Propulsion Laboratory (JPL)/Caltech, Pasadena, California
Ji Su, NASA Langley Research Center (LARC), Langley, Virginia

Chapter 12
Alexander Czechowicz, Ruhr-University, Bochum, Germany
Eugenio Dragoni, University of Modena and Reggio Emilia, Italy
Osman E. Ozbulut, University of Virginia, Charlottesville, Virginia

Chapter 13
Sabah K. Bux, Jet Propulsion Laboratory (JPL), Pasadena, California
Jeff Sakamoto, Michigan State University, East Lansing, Michigan
Eric Toberer, Colorado School of Mines, Golden, Colorado

Rama Venkatasubramanian, RTI International, Research Triangle Park, North Carolina

Chapter 14

George Cooper, University of California at Berkeley, Berkeley, California

Timothy J. Szwarc, Department of Aeronautics and Astronautics, Stanford University, California

Chapter 15

Michael Pecht, University of Maryland, College Park, Maryland

Leon M. Tolbert, University of Tennessee, Knoxville, Tennessee

Fred (Fei) Wang, University of Tennessee, Knoxville, Tennessee

Chapter 16

Carl W. Chang, NASA Glenn Research Center, Cleveland, Ohio

Liang-Yu Chen, NASA Glenn Research Center, Cleveland, Ohio

Michael Iten, Marmota Engineering AG, Zürich, Switzerland

David Parks, Idaho National Laboratory, Idaho Falls, Idaho

Bernhard R. Tittmann, Pennsylvania State University, University Park, Pennsylvania

Chapter 17

Josephine Aromando, Consolidated Edison Company, New York, New York

Robert D. Cormia, Foothill College, Los Altos Hills, California

Somrah, Dowlatram, Consolidated Edison Company, New York, New York

Edward Ecock, Consolidated Edison Company, New York, New York

Jeff Hall, Jet Propulsion Laboratory (JPL)/Caltech, Pasadena, California

Editor

Dr. Yoseph Bar-Cohen is a supervisor of the Advanced Technologies Group (http://ndeaa.jpl.nasa.gov/) and a senior research scientist at the Jet Propulsion Lab/California Institute of Technology, Pasadena, California. In 1979, he earned a PhD in physics from the Hebrew University, Jerusalem, Israel.

His research is focused on electro-mechanics, including planetary sample-handling mechanisms, novel actuators that are driven by such materials as piezoelectric and EAP (also known as artificial muscles), and biomimetics. Using ultrasonic waves in composite materials, he discovered the polar backscattering (1979) and leaky lamb waves (1983) phenomena. He (co)edited and (co)authored 8 books, coauthored over 360 publications, co-chaired 44 conferences, and has 22 registered patents. His notable initiatives include challenging engineers and scientists worldwide to develop a robotic arm driven by artificial muscles to wrestle with humans, and he held contests in 2005 and 2006.

For his contributions to the field of artificial muscles, *Business Week* named him in April 2003 one of five technology gurus who are "Pushing Tech's Boundaries." His accomplishments earned him two NASA Honor Award Medals, two SPIE's Lifetime Achievement Awards, Fellow of two technical societies—ASNT and SPIE, as well as many other honors and awards.

Contributors

Grigory Adamovsky
NASA Glenn Research Center
Cleveland, Ohio

Osama Aldraihem
Department of Mechanical Engineering
King Saud University
Riyadh, Saudi Arabia

Masood Taheri Andani
Dynamic and Smart Systems Laboratory
University of Toledo
Toledo, Ohio

Mircea Badescu
Jet Propulsion Laboratory (JPL)/Caltech
Pasadena, California

Xiaoqi Bao
Jet Propulsion Laboratory (JPL)/Caltech
Pasadena, California

Yoseph Bar-Cohen
Jet Propulsion Laboratory (JPL)/Caltech
Pasadena, California

Amr Baz
Department of Mechanical Engineering
University of Maryland
College Park, Maryland

Linfeng Chen
Department of Electrical Engineering
University of Arkansas
Fayetteville, Arkansas

Sang H. Choi
NASA Langley Research Center
Hampton, Virginia

Robert D. Cormia
Department of Engineering
Foothill College
Los Altos Hills, California

R. Peter Dillon
Jet Propulsion Laboratory (JPL)/Caltech
Pasadena, California

Mohammad Elahinia
Dynamic and Smart Systems
    Laboratory
University of Toledo
Toledo, Ohio

Jeffrey W. Fergus
Wilmore Laboratory
Auburn University
Auburn, Alabama

Olivia A. Graeve
Department of Mechanical and
    Aerospace Engineering
University of California, San Diego
La Jolla, California

and

Kazuo Inamori School of Engineering
Alfred University
Alfred, New York

Andrew Gyekenyesi
Ohio Aerospace Institute
Cleveland, Ohio

Christoph Haberland
Dynamic and Smart Systems
    Laboratory
University of Toledo
Toledo, Ohio

Wesley P. Hoffmann
Air Force Research Laboratory
Edwards, California

Gary Hunter
NASA Glenn Research Center
Cleveland, Ohio

Nathan S. Jacobson
NASA Glenn Research Center
Cleveland, Ohio

James P. Kelly
Kazuo Inamori School of Engineering
Alfred University
Alfred, New York

Hyun Jung Kim
National Institute of Aerospace
Hampton, Virginia

Matthew M. Kropf
The Energy Institute
University of Pittsburgh at Bradford
Bradford, Pennsylvania

Nishant Kumar
Honeybee Robotics Spacecraft Mechanisms
    Corporation
Pasadena, California

Hyeong Jae Lee
Jet Propulsion Laboratory (JPL)/Caltech
Pasadena, California

Jungmin Lee
Department of Electrical Engineering
University of Arkansas
Fayetteville, Arkansas

John D. Lekki
NASA Glenn Research Center
Cleveland, Ohio

Zhenxian Liang
Oak Ridge National Laboratory
Knoxville, Tennessee

Shyh-Shiuh Lih
Jet Propulsion Laboratory (JPL)/Caltech
Pasadena, California

Gyanesh N. Mathur
Department of Electrical Engineering
University of Arkansas
Fayetteville, Arkansas

Alexandra Navrotsky
Peter A. Rock Thermochemistry
    Laboratory
University of California, Davis
Davis, California

Mostafa Nouh
Department of Mechanical
    Engineering
University of Maryland
College Park, Maryland

Sulata Kumari Sahu
Peter A. Rock Thermochemistry
    Laboratory
University of California, Davis
Davis, California

Stewart Sherrit
Jet Propulsion Laboratory (JPL)/Caltech
Pasadena, California

Thomas R. Shrout
Materials Research Institute
Pennsylvania State University
University Park, Pennsylvania

James L. Smialek
NASA Glenn Research Center
Cleveland, Ohio

Vijay K. Varadan
Department of Electrical Engineering
University of Arkansas
Fayetteville, Arkansas

Mark Woike
NASA Glenn Research Center
Cleveland, Ohio

Kris Zacny
Honeybee Robotics Spacecraft Mechanisms
    Corporation
Pasadena, California

Shujun Zhang
Materials Research Institute
Pennsylvania State University
University Park, Pennsylvania

1
Introduction

Yoseph Bar-Cohen

CONTENTS

1.1 Introduction

As early as humans started working with fire and heat, they have been using materials that can sustain high temperatures. Initially, the primitives used rocks, which they picked from their neighborhood. As they became more capable in controlling the heat that they produced and the related applications, increasingly efforts were made to control the properties of the produced materials and the performance of the structures that were made. The applications have grown from preparing food and heating their living areas to sophisticated processing methods at higher temperatures as well as applications that involve extreme thermal conditions. Today, the selection of high-temperature materials (Chapter 3) has been widened to such metal alloys as refractory metals, stainless steels, superalloys, and titanium alloys. Other materials that are being used include certain ceramic materials, carbon–carbon composites, metal matrix composites (MMC) and many others. Applications of high-temperature materials include aircraft jet engines (see example of an aircraft fighter jet engine in Figure 1.1), nuclear reactors, and industrial

FIGURE 1.1
A Pratt & Whitney F100 turbofan engine for the F-15 Eagle being tested at an Air National Guard base in Florida. The tunnel behind the engine has the functions of allowing the exhaust gases to escape and muffling noise. (From Wikimedia Commons, http://en.wikipedia.org/wiki/File:Engine.f15.arp.750pix.jpg; http://upload.wikimedia.org/wikipedia/commons/2/2e/Engine.f15.arp.750pix.jpg.)

gas turbines. Also, such materials are used in furnaces and ducts, as well as electronic and lighting devices.

Generally, in materials science the term high temperature is defined as the temperature that is equal to, or greater than, about two-thirds of the melting point of a solid (Spear et al., 2006). However, there are other definitions, too, including application-based ones as well as indicating materials that are used for their heat resistance (e.g., strength and corrosion resistance) above 500°C as described by Meetham and Van de Voorde (2000). Since increasing the temperature decreases the material strength, in order to make them effective and economical, they need to have high strength at the required operating temperatures with a safety margin. High-temperature materials need to be resistant to the related causes of damage including oxidation and corrosion, which are accelerated with the increase in temperature (Chapter 5). Generally, extrapolating the material properties and chemical behavior do not correlate with the high-temperature data (Chapter 2). Specifically, as the temperature of materials rise, chemical reactions become pronounced and thermodynamic properties determine the reactivity rather than kinetics. As the temperature increases, there are various effects that take place including disorder of the material structure.

1.2 Historical Perspective

The use of high-temperature materials is dated as far back as the very ancient historic years of the start of humans' civilization. Initially, heat was used for cooking and overcoming cold weather but with time the applications expanded to processing extremely high-temperature materials and their use at harsh conditions. It is interesting to note that the eras of technologies that are used to date humans' ability to handle materials is based on their evolved ability to melt high-temperature ones. This includes the Stone Age where humans were not able to smelt any ore, whereas the Bronze and Iron Ages where humans were able to smelt bronze and iron, respectively, as well as produce artifacts from them. The industrial revolution led to enormous rise in applications and requirements for increased capability and operation safety of both the fabrication of the materials and their usage. Further, new materials were introduced and improved processes were developed for their fabrication and treatment to enhance their strength and durability (Chapters 2 and 6).

Some of the applications that emerged include the development of the steam turbine at the end of the nineteenth century, while filament lamps, combustion engine, and petrochemical industry in the early twentieth century were followed by the development of the jet engine around the 1940s and the Space Shuttle engine in the 1980s (see photo of the space shuttle Atlantis and its three main engines in Figure 1.2). The development of the steel alloys at the end of the nineteenth century as well as the stainless-steel (Brearley, 1913) and the nickel–chrome alloys (Crawford, 1959) in the beginning of the twentieth century had major impact on the application of high-temperature materials. The carbon–carbon composites (Savage, 1993) were developed around 1970.

It is interesting to follow the parallel development of materials and the resulting industrial applications. Specifically, the driver of requirements at the end of the nineteenth century has been the powering of ships and it led to the development and advancements in steam turbines/engines. The development of the gas turbine has benefited greatly from the use of nickel superalloys (Sims, 1987), which were developed around the 1940s. These

FIGURE 1.2
The Space Shuttle engine emitted significant level of heat and the temperature rose significantly during launch. This figure shows close-up of the three main engines of the space shuttle Atlantis that launched the shuttle into space. Because the engine burns hydrogen and oxygen, it emits steam through its exhaust. (From NASA/Rusty Backer and Michael Gayle. http://www.nasa.gov/mission_pages/shuttle/flyout/ssme.html; high resolution: http://www.nasa.gov/images/content/502050main_ssme_firing.jpg.)

turbines were used to drive electric power-generating plants and later, jet aircraft. The use of metal alloys with various additives led to substantial improvement of the properties of the available high-temperature materials, including the corrosion resistance of nickel-chrome that was achieved by adding aluminum and titanium. Various manufacturing processes were introduced to improve the available high-temperature materials and these include the use of powder metallurgy, casting, and vacuum melting.

The availability of compatible materials can be critical to the ability to develop high-temperature-related technologies such as high-speed airplanes and rockets that reach speeds of several Mach (Martelluci and Harris, 1991). Specifically, the issue for reaching 5 Mach and beyond is the rise in surface temperature above 500°C, which initially causes large stresses and vaporization of the surface material (fast ablation). Besides the capability of the materials and structures to sustain the high temperatures without irreversible changes and degradation of the properties, it is also necessary to have effective structural design that allows for efficient heat dissipation, transfer of the associated high heat fluxes and the ability to sustain the related thermal conditions. Generally, MMC and titanium alloys are used in structures that need to operate in the range from 300 to about 1000°C and they are selected for their being light weight and easy to manufacture (Tenny et al., 1988). Such alloys as the 6Al–4 V titanium and AM-350 CRES steel have high strength properties, however they tend to fail above 400°C (Jenkins and Landis, 2003). For temperatures above 1000°C, such materials as ceramics and carbon–carbon composites are used.

1.3 Need for High-Strength High-Temperature Materials

Generally, there is a need for materials that are capable of being applicable at higher temperatures than available and their development can benefit many industries. For example, significant increase can be reached in the efficiency of pulverized coal power plants that operate under ultra-supercritical conditions (Meetham and Van de Voorde, 2000). Since the material strength degrades with the rise in temperature, the applicability of high strength materials is limited to the requirements of the developed structure. Titanium metal matrix unidirectional composite with SiC reinforcement has strengths as high as over 1400 MPa, but the application of this material is limited to the range of about 500°C. On the other hand, for applications in the range of 1500°C, one can use refractory metals with strength that is less than about 100 MPa or carbon/carbon with strength above 350 MPa. Besides the loss of strength as a function of temperature, one needs to take into account the ability of the material to sustain accelerated oxidation and other degradation effects that may be involved with the exposure to high temperatures.

Continually, new high-temperature materials and alloys are reported (Chapter 3). When new materials emerge from research and development with highly attractive and desirable properties, they are not immediately used in engineering applications. It is essential to identify their disadvantages and to minimize them with proper precaution and attention to the critical issues that are identified. Examples include the recognition by aerospace engineers that titanium is a lightweight, high-temperature material with potential applications to supersonic aircraft structures. However, attention is needed to developing effective processing, thermal treatment, adhesion, welding, and others as well as methods of preventing premature cracking before titanium alloys became materials of choice for integrity critical structures in supersonic aircraft and others.

1.4 HT Materials

Various high-temperature materials are in use and an extensive review is given in Chapter 3. This chapter includes discussions related to some of the properties of concern to users of high-temperature materials. Generally, these properties include friction and wear, tensile strength, compressive strength, bearing strength, shock, electrical properties, thermal CTE, and heat capacity. Herein, a brief review is given to provide a broad background. When referring to composite materials the discussion covers materials that consist of two or more bonded constituents having significantly different physical or chemical properties. Composite materials consist of filler in the form of fibers or powder and a matrix of a monolithic material that is reinforced by the embedded filling component(s). Other high-temperature materials include refractory materials. These materials retain their strength at high temperatures and, according to ASTM C71, they have chemical and physical properties that make them applicable at temperatures above 538°C (1000°F).

1.4.1 Carbon–Carbon Composites

Carbon fiber-reinforced carbon, also known as carbon–carbon composites, is an effective high-temperature material for structural applications that require thermal shock resistance, a low coefficient of thermal expansion, thermal protection, lightweight, and good strength retention up to about 1400°C (Tenny et al., 1988). However, they have low impact resistance and, in the pure form of carbon–carbon composites, they are susceptible to significant oxidation at temperatures above about 500°C. To address the oxidation issue, overcoat sealers are used; however, in applications to high-speed flight vehicles, such overcoats are not an effective solution due to the fact that long duration of heat exposure can rapidly erode the coating if it is based on low viscosity sealants. Applying coating layers that consist of SiC and hafnium carbide (Ohlhorst et al., 2006) was determined to be effective in preventing oxidation at temperatures as high as 2200°C (Ohlhorst et al., 2006). Carbon–carbon composites are used for making structures of high-speed flight vehicles, such as the nose leading edge, horizontal control surfaces, and the tail leading edge (Ohlhorst et al., 2006). Other applications include aircraft brakes, high-temperature bearings and clutches, nozzles, exit cones, satellite structures, and heat shields. Also, for protection against the reentry temperature that exceeds 1260°C, the Space Shuttle nose cap (see Figure 1.3), its chin area between the nose cap and nosewheel doors, and the wing leading edges were also made of carbon–carbon.

1.4.2 Carbon–Silicon Carbide Ceramic Matrix Composite

Carbon fiber-reinforced silicon carbide has a relatively constant strength at temperatures up to 1600°C and the use of carbon fiber makes this composite material lightweight (Schmidt et al., 2004). It has a relatively low load-bearing properties resulting from micro-cracking that are produced during its fabrication and the formation of empty spaces that are microstructural voids that are produced during the carbon fiber weaving (Glass et al., 2002).

1.4.3 Ceramics

Ceramics are inorganic nonmetallic solids that have mostly crystalline structures but, as in the case of glass, they may have amorphous structure. Certain ceramic materials, including silicon borides, carbides, and nitrites, can sustain extreme temperatures of more than

FIGURE 1.3
The nose cap of the Space Shuttle was made of carbon–carbon. (From Wikimedia Commons, http://
en.wikipedia.org/wiki/Space_Shuttle_thermal_protection_system#cite_note-tech-0; high resolution image:
http://upload.wikimedia.org/wikipedia/commons/2/2f/Thermal_protection_system_inspections_from_
ISS_-_Shuttle_nose.jpg.)

2000°C (Wuchina et al., 2007). Ceramics are used for handling molten metals, and for aerospace applications such as hypersonic flight, scramjet propulsion, rocket propulsion, and thermal shield for atmospheric reentry. For these applications, the materials are used to produce nozzles of gas burners, nuclear fuel uranium oxide pellets, coatings of jet engine turbine blades, ceramic disk brakes, nose cones of missile, as well as refractory applications in furnaces of materials production. Also, they were used to produce the protection tiles of the Space Shuttle.

1.4.4 Ceramic Composites

For thermal protection of structures, increasingly ceramic composites such as hafnium diboride and zirconium diboride are being used (Opeka et al., 2004). This is the result of their very high melting point of about 3200°C, high oxidation resistance up to 2000°C (Opeka et al., 2004), and their low ablation rates at high temperatures (Malone, 2000). One of the most effective ceramic composites that have been used in heat shields is the phenolic impregnated carbon ablator (PICA).

 PICA is a lightweight ceramic ablator that is designed to burn away slowly and in a controlled manner. This is done in order to carry heat away from the spacecraft by the gases generated in the ablative process while the remaining solid material insulates the craft from superheated gases. PICA was developed in the 1990s at the NASA's Ames Research Center. It has a very low density (weighing about 20% of conventional heat shields), and can withstand temperatures as high as 1930°C. The material consists of carbon fibers coated with a thin layer of phenolic polymeric resin. While the resin provides bonding, it also creates a light, durable, and heat-resistant shield. In January 2006, PICA was used on the NASA's Stardust Sample Return Capsule (SRC) and it entered the Earth's atmosphere at the fastest recorded entry speed. Another well-known application of PICA for a heat shield was on the Mars Science Lab (MSL) that landed on Mars in August 2012 (see a photo of this heat shield in Figure 1.4). Due to the large size of the Curiosity Rover of the Mars Science Laboratory mission, a single-piece heat shield could not be used (Beck et al., 2010) and therefore a total

FIGURE 1.4
(**See color insert.**) PICA was used in tiles form to produce the heat shield (bottom cone) of the MSL mission that landed on Mars in Aug. 2012. (Courtesy of JPL/NASA, Reference Figure No. MSL-2011-05-26-143545-IMG_0959.JPG.)

of 113 tiles (3.2 cm thick) were employed with adhesive that filled the gap between them. The produced heat shield was capable of sustaining thermal loading of 197 W/cm².

1.4.5 Cermets

These are ceramic–metallic composites that combine the high-temperature resistance and hardness properties of ceramics with the ability of metals to deform plastically. The metal, which may be cobalt molybdenum or nickel with less than 20% in volume, serves as a binder of the ceramic constituent that may be a boride, carbide, or oxide. Other metallic constituents may be a nickel-based superalloy with such elements as columbium, molybdenum, and tantalum (Jenkins and Landis 2003) in a matrix of high-temperature ceramic material such as silicon carbide (SiC) (Jenkins and Landis, 2003). Cermets are used in high-temperature electronics to produce resistors and capacitors, as well as other components such as seals of fuel cells.

1.4.6 Metal Matrix Composites

These materials consist of at least two constituents and with at least one made of metal, while the other one(s) can be a different metal, ceramic, or other material(s). These materials are also called hybrid composites when the MMC consists of three or more constituents. Generally, at elevated temperatures the creep and yield strengths of MMC are relatively higher compared to most metal alloys. Increasingly, MMC are used in high-performance systems and example includes a structural component of the landing gear F-16 Fighting Falcon jet aircraft. This component is made of monofilament silicon carbide fibers in a titanium matrix (a photo of the landing gear of the F-16 can be seen in Figure 1.5). An example of an effective MMC is the titanium MMC. It is capable of sustaining extensive cyclic loading in corrosive environments (Stephens, 1987) and structures made of this composite can endure temperatures as high as 1650°C (Stephens, 1987).

FIGURE 1.5
A structural component of the landing gear F-16 Fighting Falcon jet aircraft is made of metal matrix composite, http://en.wikipedia.org/wiki/Metal_matrix_composite. (From Wikimedia Commons, high resolution: http://upload.wikimedia.org/wikipedia/commons/0/05/F-16_CJ_Fighting_Falcon.jpg.)

Examples of titanium MMC include

Titanium–Aluminum alloy: This MMC is a high strength alloy with good air oxidation resistance for temperatures up to 650°C (Draper et al., 2007), where above 650°C it sustains oxidation and embrittlement (Tobin, 1997). This problem can be overcome by producing a stable oxide layer during the stage of fabrication. But, this process can also produce undesirable characteristics such as very low ductility, low fracture toughness and poor creep properties (Tobin, 1997). To prevent the oxidation, Ti–Al alloy substrates are suspended in vapor of aluminum to create a uniform coating layer of TiAl3 (Draper et al., 2007).

Titanium–Zirconium–Molybdenum alloy: This alloy has high strength and, with a dispersion of TiC and ZrC in the molybdenum matrix, its high strength at elevated temperatures is well preserved. The carbide complexes (TiC and ZnC) provide benefit in making this composite highly weldable. Its low coefficient of thermal expansion allows designers to integrate this composite material at areas that are exposed to high temperatures without the need to prevent potential buckling from thermal stresses (Paull, 2006).

1.4.7 Refractory Metals

These are metals that have extremely high melting points, have significant resistant to heat and wear, and they are very stable against creep deformation. The elements that are included in this category are molybdenum, niobium, rhenium, tantalum, and tungsten. These elements have melting point above 2000°C and they are very hard at room temperature. Applications of refractory metals include casting molds and wire filaments (see example in Figure 1.6) and they are widely used in powder metallurgy. There are also silicon-based refractory compounds, including SiC, Si_3N_4, and $MoSi_2$,

FIGURE 1.6
The filament of a 200 Watt incandescent light bulb, http://en.wikipedia.org/wiki/Refractory_metals. (From Wikipedia, http://upload.wikimedia.org/wikipedia/commons/0/08/Filament.jpg.)

which have excellent oxidation resistance at temperatures as high as 1700°C (Wuchina et al., 2007).

One of the widely used refractory materials is silica. The need for lightweight insulation and low heat conductivity has been addressed in the Space Shuttle thermal protection system by covering it with tiles made of Li-900. This material was developed and manufactured by Lockheed Missiles and Space Company in Sunnyvale, California. Li-900 consists of 10% in volume of silica fibers that are made of pure quartz and the rest is air and this combination leads to 0.144 g/cm³ bulk density (Jenkins, 2007). Due to the fact that these tiles are not flexible, the Space Shuttle Orbiter (the orbital spacecraft of the Space Shuttles) was covered with thousands of tiles that were glued to its surface (see an example of a tile in Figure 1.7). The low impact resistance of the tiles on the Space Shuttle Orbiter was the cause of the Columbia destruction in 2003 during its return to Earth. This took place after piece of insulation foam fell from the external tank and impacted the tiles. The debris struck the leading edge of the left wing and damaged the Shuttle's thermal protection system that shields the vehicle from the intense heat generated from atmospheric compression during reentry (http://www.nasa.gov/columbia/home/CAIB_Vol1.html).

1.4.8 Superalloys

Superalloys, also known as high-performance alloys, are materials with excellent mechanical strength and creep, good surface stability, and corrosion and oxidation resistance at high temperatures. The base elements of most superalloys consist of nickel, cobalt, or nickel-iron. Examples of superalloys include

Hastelloy: This group of superalloys consists of 22 different nickel-based, highly corrosion-resistant metal alloys. The name is a registered trademark name of Haynes International, Inc. Besides nickel, the following elements are used: molybdenum, chromium, cobalt, iron, copper, manganese, titanium, zirconium, aluminum, carbon, and tungsten. Hastelloys are applicable to corrosive and erosive environments and they are used in pressure vessels of chemical reactors, distillation equipment, nuclear reactors, as well as chemistry pipes and valves.

Inconel: This superalloy is based on nickel–chromium. Its strength gradually decreases as the temperature reaches about 650°C (Jenkins and Landis, 2003) and

FIGURE 1.7
A photo from the backside of one of the Space Shuttle tiles. (From Wikipedia, http://en.wikipedia.org/wiki/File:Shuttle_tile.jpg; high resolution image: http://upload.wikimedia.org/wikipedia/commons/6/6f/Shuttle_tile.jpg.)

it is relative easy to create complex shapes with this superalloy. To avoid stress cracking in welded structures, which may result from thermal stresses, softer weld materials are used.

Rene alloys: This superalloy is also nickel-based and it maintains its room temperature strength to as high temperatures as 980°C (Smith, 2001; Hebeler, 1963). These alloys are used to produce jet engine and missile components that require very high strength at high temperatures. The outer shell of the NASA Mercury capsule was made of Rene 41 alloy.

Waspalloy is an age hardened austenitic nickel-based superalloy that maintains excellent strength and good corrosion resistance up to about 980°C. The name of this alloy is a registered trademark of United Technologies Corp. and it is used for high-temperature applications such as gas turbines.

1.5 HT Processes

Effective processing methods are critical to the availability of high-temperature materials for practical applications. Understanding the chemical kinetics, fluid dynamics, thermodynamics, and transport phenomena are essential to controlling and optimizing the processes of fabricating high-temperature materials. Besides the ability to produce such materials there is increasing need to address their environmental impact as well as the ability to recycle the produced materials. Furthermore, there are major developments in the area of processing of nano-scale materials and they are leading to important material capabilities.

Owing to the challenges that are involved with the processing and fabrication of high-temperature materials, there are many different methods that have emerged to address the needs of the specific materials and related properties. The higher the temperature that is applied in the processes, the greater the acceleration of the related kinetics and the reaction rate tends to be exponential with the temperature. Processing at high temperatures increases the solubility of various oxides when forming dissolutions, as well as enabling certain reactions and phase transformations that are essential to some of the produced materials. Rapid cooling is also used in some of the processes and it prevents molecules in the molten phase from rearranging into crystallographic structures. Thus, amorphous solids can be formed allowing the fabrication of metastable phases.

Some of the processing methods include smelting, powder sintering, and roasting, and they require sufficient understanding of the related kinematic and thermodynamic processes, effective process modeling, simulation, and optimization capabilities, as well as the ability to measure physical properties at high temperatures (Chapter 7). The processing methods include vacuum deposition methods, such as physical vapor deposition (PVD), where thin films are deposited by condensation of desired film material in vaporized form onto various part surfaces. This coating method involves high-temperature vacuum evaporation and condensation, or plasma sputter bombardment. Another deposition process is the chemical vapor deposition (CVD), which is a chemical process that is used to produce high-purity, high-performance solid materials. Generally, the modeling of processes involves both theoretical and experimental studies and includes multi-phase equilibrium modeling to allow predicting reaction equilibrium and transport properties of complex multicomponent and multiphase systems. The models involve using computational fluid dynamics (CFD) that are also used to simulate fluid flow in high-temperature systems. The performance of high-temperature metallurgical systems can be affected by a number of factors including: liquid and solid chemistries; reaction kinetics; such physical properties as viscosity and surface tension; transport properties including electrical and thermal conductivity; interfacial properties between liquids, solids, and gases. Some of the properties that need to be measured include the viscosity, as well as the surface and interfacial tension at high temperatures, electrical properties, thermal conductivity, and the reaction rates and order.

Increasingly, producers of high-temperature materials need to reduce greenhouse gas emissions, improve product quality and productivity, and improve the energy efficiency of the production processes.

1.6 Actuators, Devices, Mechanisms, and Jet Engine Turbines

Making devices, mechanisms, and actuators that can operate at high temperatures is a great challenge and it increases as the temperature is raised (Chapters 9 through 17). The related issues that need attention include material compatibility, chemical reactions, alloying, annealing, and diffusion characteristics. These may affect the chemical and physical nature of the components that are used. One of the key mechanical design issues is a mismatch of the thermal expansion coefficients, and this can be catastrophic to components that need to fit precisely inside a structure.

Driving mechanisms at high temperatures requires adequate capability of the actuators and their drive and control electronics. Due to technology limitations, most electronic devices are limited to the maximum range of 250–300°C, whereas the actuators that are driven by coils or electroactive materials are limited to applications in the range of up to about 500°C (Chapters 9 through 12 and 17; Bar-Cohen and Zacny, 2009).

Turbojet engines are operated at much higher temperatures and the specific temperature is bounded by the durability of the blades and the nozzle materials (Stone, 1999). The combustion chamber of these engines burns fuel at large quantities and it is supplied via spray nozzles. The work, which is done by the compressor, raises the air temperature to the range of 200–550°C, while the combustion process raises the temperatures to the range of 650–1150°C. After combustion, the gas temperature reaches the range of about 1800–2000°C, which may be too high for the turbine nozzle guide vanes. To lower the temperature inside the combustion chamber, the emitted gas is used for cooling.

1.7 NDE and Characterization Methods for HT Materials and Mechanism

A critical part of assuring the ability of high-temperature materials to sustain the operating conditions for which they are designed is their integrity, material quality, and the properties consistency with the design requirements (Bar-Cohen et al., 2003; Chapter 8). The detection of defect and the characterization of properties without affecting the integrity of the material or structure require the use of nondestructive testing (NDT) or nondestructive evaluation (NDE) methods. Increasingly, in-service nondestructive health monitoring methods are used to monitor the integrity of structures and materials properties. The term health monitoring is inspired by the medical term of monitoring the health of our body and similarly it is essential to monitor and control the material integrity and properties at all stages of the structure's life, from cradle to retirement. A typical health monitoring system consists of a sensor, data acquisition, and setup for processing and controlling the system using a microprocessor that acquires and analyzes the data, providing real-time information. The methods that are used at room temperature to inspect high-temperature materials include the common ones such as ultrasonics, magnetic-particles, liquid penetrants, radiography, thermography, visual inspection, eddy-current testing, and laser interferometry. While most NDE methods are applied at room temperatures, certain methods of health monitoring of high-temperature materials and mechanisms can be applied at the elevated temperatures, which may reach several hundreds of degree Celsius. Besides the use of NDE methods, characterization methods are used to examine the materials, structures, and mechanisms at room temperatures. These methods are also used to test representative samples of actual parts and structures that are subjected to service conditions and tested afterward at room temperature (Chapter 7). These methods are used to perform surface, bulk, and structural analysis using various tools that include chemical and mechanical analysis.

1.8 Summary/Conclusions

Development of high-temperature materials that can maintain their properties and integrity are critical to applications where extended exposure to hot conditions is required. Generally, even primitives used such materials in the earlier days of civilization but, with time, materials and processes were developed to allow more control over the properties of the produced materials leading to commercial quantities. Despite the significant efforts, the materials that need to operate at extreme temperature environments possess certain related limitations. These require engineering their applications to work within their acceptable limits.

Many advances in developing high-temperature materials have been made in the past 50 years, particularly for such applications as high-speed flight vehicles such as the Space Shuttle and hypersonic vehicles. Generally, the required materials need to be lightweight, have high tolerance to cyclical loading, and the ability to form into complex shapes. High heating rates and temperatures that are reached during service pose many engineering challenges that led to the development of many effective materials, as well as methods of thermal management. The developed materials include high-temperature metal alloys, superalloys, MMC, ceramic composites, and carbon–carbon composites. Key factors in choosing materials for high-temperature applications include the cost and the complexity of fabrication. To support the capability of high-temperature materials to endure the related harsh conditions effective coating methods are applied to prevent oxidation and surface cracking.

Acknowledgments

Some of the research reported in this chapter was conducted at the Jet Propulsion Laboratory (JPL), California Institute of Technology, under a contract with the National Aeronautics and Space Administration (NASA). The author thanks Yiannis Pontikes, Department of Metallurgy and Materials Engineering, KU Leuven, Belgium; and Ted Iskenderian, Jet Propulsion Laboratory (JPL)/Caltech, for reviewing this chapter and for providing valuable technical comments and suggestions. Also, the author expresses his appreciation of Ted Iskenderian, JPL/Caltech for his assistance in getting photos related to the MSL mission.

References

Bar-Cohen Y., *High Temperature Technologies for Sample Acquisition and In-Situ Analysis*, JPL Technical Report, Document No. D-31090, Sept. 8, 2003.

Bar-Cohen Y. and K. Zacny (Eds.), *Drilling in Extreme Environments—Penetration and Sampling on Earth and Other Planets*, Wiley-VCH, Hoboken, NJ, ISBN-10: 3527408525, ISBN-13: 9783527408528, 2009 pp. 1–827.

Beck R. A. S., H. H. Hwang, M. J. Wright, D. M. Driver, and E. M. Slimko, The evolution of the Mars Science Laboratory heatshield, *7th International Planetary Probe Workshop*, Barcelona, Spain, June 16, 2010.

Crawford C. A., *Age-Hardenable Nickel-Chromium Alloys*, ASIN: B005I7UKZ0, International Nickel Company, NY, 1959, pp. 1–222.

Draper S. L., D. Krause, B. Lerch, I. E. Locci, B. Doehnert, R. Nigam, G. Das, P. Sickles, B. Tabernig, N. Reger, and K. Rissbacher, Development and evaluation of TiAl sheet for hypersonic applications, *Materials Science and Engineering* 2007, 1–13.

Glass D. E., N. R. Merski, and C. E. Glass, Airframe research and technology for hypersonic air-breathing vehicles, *NASA Technical Memorandum*, 2002-211752, 2002, 1–15.

Hebeler H. K., Hypersonic cooling system, United States Patent 3,089,318, 1963, 1–4.

Jenkins D. R., and T. R. Landis, *Hypersonics: The Story of the North American X-15*, Specialty Press, USA, 2003.

Jenkins D. R., *Space Shuttle: The History of the National Space Transportation System*, Voyageur Press, ISBN 0-9633974-5-1 2007, pp. 1–524.

Malone J. E., Materials may allow spacecraft design change, *Aerospace Technology Innovation*, 8, (6), 2000, 1.

Martelluci A., and T. B. Harris, Assessment of key aerothermal issues for the structural design of high speed vehicles, *Thermal Structures and Materials for High Speed Flight AIAA Journal*, 140, 1991, 59–91.

Meetham G.M., and M.H. Van de Voorde, *Materials for High Temperature Engineering Applications*, Springer, New York, ISBN 3-540-66861-6, 2000.

Ohlhorst C. W., Development of X-43A Mach 10 leading edges, International Astronautical Congress Report 05-D2.5.06, 2006, pp. 1–9

Opeka M. M., I. G. Talmy, and J. A. Zaykoski, Oxidation-based materials selection for 2000°C+ hypersonic aerosurfaces: Theoretical considerations and historical experience, *Journal of Materials Science*, 39, 2004, 5587–5904.

Paull A, M. K. Smart, and E. H. Neal, Flight data analysis of the HyShot 2 scramjet flight experiment, *AIAA Journal*, 44, (10), 2006, 2366–2375.

Savage E., *Carbon–Carbon Composites*, ISBN-10: 0412361507, ISBN-13: 978-0412361500, Springer, NY, 1993, pp. 1–400.

Schmidt S., S. Beyer, H. Knabe, H. Immich, R. Meistring, and A. Gessler, Advanced ceramic matrix composite materials for current and future propulsion technology applications, *Acta Astronautica*, NY, 55, 2004, 409–420.

Sims C. T., N. S. Stoloff, W. C. Hagel, *Superalloys II*, ISBN-10: 0471011479, ISBN-13: 978-0471011477, Wiley-Interscience; 2nd Edition, 1987, pp. 1–615.

Smith T. Dyna Soar X-20: A look at hardware and technology, *Quest Magazine* 2001, 1–5.

Spear K. E., S. Visco, E. J. Wuchina, and E. D. Wachsman, High temperature materials, Vol. 15, No. 1, *Interface Journal*, the Electrochemical Society Conference, Spring 2006, pp. 48–51.

Stephens J. R, High temperature metal matrix composites for future aerospace systems, *NASA Technical Memorandum* 100212, 1987, 1–17.

Stone R., *Introduction to Internal Combustion Engines*, Society of Automotive Engineers Inc., Warrendale, PA, 3rd Edition, ISBN-10: 0768004950, ISBN-13: 978-0768004953, 1999, pp. 1–641.

Tobin A. G., Oxidation protection method for titanium, United States Patent 5,672,436, 1997, pp. 1–3.

Wuchina E., E. Opila, M. Opeka, W. Fahrenholtz, and I. Talmy, UHTCs: Ultra-high temperature ceramic materials for extreme environment applications, *Proceedings of the Winter 2007 Electrochemical Society Interface*, 2007, pp. 30–36.

Internet Resources

High Temperature Materials for Hypersonic Flight Vehicles, viewed on August 05 2012, http://seit.unsw.adfa.edu.au/ojs/index.php/Hypersonics/article/viewFile/21/12

Meetham G.W., and M.H. Van de Voorde, Materials for High Temperature Engineering Applications, Section 1.2 "High Temperature Materials" at Google Books: http://books.google.com/books?id=cxl1I9llVLcC&pg=PA3&source=gbs_toc_r&cad=4#v=onepage&q&f=false

MSL—Aeroshell and Heat Shield, viewed on August 10, 2012, http://www.spaceflight101.com/msl-aeroshell-and-heat-shield.html

Principles of Jet Engine Operation, viewed on August 10, 2012, http://www.century-of-flight.net/Aviation%20history/evolution%20of%20technology/jet%20engines.htm

Refractory Metals: Niobium, Molybdenum and Rhenium, viewed on August 05, 2012, http://www.key-to-metals.com/Article123.htm

Report of Columbia Accident Investigation Board http://www.nasa.gov/columbia/home/CAIB_Vol1.html

Thermographic nondestructive evaluation of the space shuttle main engine nozzle, viewed on August 10, 2012, http://ntrs.nasa.gov/archive/nasa/casi.ntrs.nasa.gov/20000096499_2000138014.pdf

UHTCs: Ultra-High Temperature Ceramic Materials for Extreme Environment Applications, viewed on August 05, 2012, http://www.electrochem.org/dl/interface/wtr/wtr07/wtr07_p30–36.pdf

2

High-Temperature Materials Chemistry and Thermodynamics

Sulata Kumari Sahu and Alexandra Navrotsky

CONTENTS

2.1 Introduction

Many of the uses of materials at high temperature require their persistence, with minimal change in chemical composition and microstructure, over long periods of time (hours to years). Chemical reactions at high temperature often occur fast and thus the thermodynamic equilibrium state, rather than persistence of metastable states, often governs the suitability of a given material or set of materials for a proposed application. The term "high temperature" cannot be defined uniquely; perhaps the most useful practical definition is one that says that for a given material, high temperature is the temperature above which the material is strongly chemically reactive in use. From the point of view of solid state chemistry and thermodynamics, there is vast empirical knowledge about such reactivity and moderately adequate thermodynamic data bases for common alloys and ceramics (Sundman et al., 1985; Schenck and Dennis, 1989; Pankratz, 1982; Ondik and Messina, 1989; Liu et al., 2003; Levin and McMurdie, 1975; Kubaschewski and Alcock, 1979; Gisby et al., 2002; Eriksson and Hack, 1984; Davies et al., 2002; Chase et al., 1985a,b; Chart, 1978; Barry et al., 1979). Nevertheless, an inexperienced or only moderately knowledgeable scientist or engineer may have a hard time selecting materials for high temperature use and avoiding pitfalls of unanticipated reactivity and degradation. The purpose of this article is to organize such chemical knowledge into a short and useful guide on materials use and compatibility, taking advantage of general trends governed by the regularity of chemical bonding and thermodynamics among groups of elements in the periodic table.

2.2 How Materials Fail at High Temperature

This section briefly summarizes failure modes (Walter, 1999; Tillack, 2000; Priest, 1992; Kerans, 2000; Jarvis, 2006; Danzer, 1993; Blachnio, 2009; Johnson, 1950; Ashby and Jones, 2005). More than one process can occur simultaneously. The following sections give specific examples and general trends.

2.2.1 Melting and Softening

Obviously melting destroys the integrity of a solid and the melt or partial melt (melt/crystal "mush") is generally chemically very reactive with its environment. Thus, one needs high melting compounds for high-temperature applications. Impurities usually lower the melting point, and deep eutectics can be disastrous. Glasses do not melt but they soften on going through their glass transition. Glasses can crystallize, sometimes transforming a dense solid material to a porous powdery one. Even crystalline solids become soft and deformable at high temperature and brittle materials become ductile (Petty et al., 1968). Such effects occur typically above about two-thirds of the melting point in degrees Kelvin.

2.2.2 Vaporization

Many solids vaporize (sublime) rather than melt. Others melt and then boil. At reduced pressure vaporization is more extensive. The vaporization behavior of metals and their oxides can be quite different (Tietz and Wilson, 1965). For example, metals such as tungsten, molybdenum, rhenium, and osmium are refractory and useable to temperatures above 2000°C, while their oxides are low melting and volatile. Their ready oxidation and the poor high-temperature behavior of their oxides limit the use of such metals to non-oxidizing environments (inert gas or vacuum).

2.2.3 Corrosion and Chemical Reaction with the Atmosphere

Reactions include oxidation, nitridation, reaction with H_2O and CO_2 to form hydrates and carbonates, and reaction with traces of HCl, SO_2, and other corrosive gases. Oxidation, especially of small particles and dusts, can be explosive and reaction of active metals such as zirconium with water to form the metal oxide plus hydrogen can lead to hydrogen explosions if air is also present. Such hydrogen explosions, with H_2 generated by the reaction of zirconium-containing fuel rod cladding (zircalloy) with water or steam (Grosse et al., 2010; Kim et al., 2010; Tanabe, 2011) are the major source of breach of containment in nuclear reactor accidents such as those at Chernobyl and Fukushima (Armstrong et al., 2012; Burns et al., 2012). One should also note that nitrogen is not always a suitably inert gas for metals that readily form stable nitrides at high temperature, for example, titanium, zirconium, and tantalum (Loremez and Woolcock, 1928; Gulbransen and Andrew, 1950; Ono et al., 1996; Fromm, 1970; Wilkinson, 1969).

2.2.4 Diffusion and Solid-State Reaction

When two different materials are in contact, they will diffuse into each other or form a layer of reacted material if such reaction is thermodynamically favorable. Grains will grow and the material may become porous and change in other physical properties. In composite structures, such as thermal barrier coatings, different layers may delaminate or material may spall off. On the other hand, adherent protective coatings of reacted (oxidized) metal or intentionally added other protective layers can greatly extend the use and lifetime of materials at high temperature, for example, in protective oxide layers and thermal barrier coatings (Stowell et al., 2001; Schmitz and Stamm, 2009; Igolkin, 2003; Beele, 1999).

2.2.5 Solid–Solid Phase Transformations

If there are phase transitions among different crystal structures, the properties of the high-temperature phase may be quite different from those of the low-temperature phase (Lopato et al., 1974; Navrotsky and Ushakov, 2005; Wang et al., 1992; Eyring, 1979; Adachi and Imanaka, 1998; Barnal et al., 2004). If the phase transition involves a large volume change, cracking or disintegration may occur (Navrotsky and Ushakov, 2005; Wang et al., 1992; Scott, 1975).

Various elements, alloys, and oxides are used in high-temperature applications as structural elements, resistive heaters, and containers. Table 2.1 lists the more common ones and their approximate upper temperature for continuous use in a given atmosphere and common mode of failure. The reader is warned that there is no such thing as a totally inert material at temperatures above about 1500°C, and often one has to choose the least deleterious scenario rather than the perfect solution for a given application.

2.3 Noble Metals

Platinum and platinum–rhodium alloys are used for thermocouples, electrodes, and sample crucibles at high temperature (Rajan, 2004; Pollock, 1985; Swindells, 1968; Groza et al., 2007; Soszko et al., 2011). Rhodium raises operating temperature and adds strength (Ochiai, 1993; Winkler, 2000). Iridium is also sometimes used for crucible material; it is higher melting and strong but even more expensive than platinum (Cardarelli, 2000). Common thermocouple types, with wires commercially available, are given in Table 2.2, which also gives data on use temperatures and melting points of various noble metals. Though platinum group metals are considered to be noble metals, they in fact chemically reactive under a variety of conditions, sometimes with disastrous consequences to the user. There are a number of binary and ternary noble metal oxides (see Table 2.3), whose stability is generally limited to temperatures below about 1150°C (Jehn, 1984; Krier and Jaffe, 1963; Raub, 1959; Raub and Plate, 1957; Schwartz and Prewitt, 1984; Jacob et al., 2009; Chaston, 1964, 1969, Bettahar et al., 1987; Tancret et al., 1996) and is of course enhanced by strongly oxidizing conditions such as a pure oxygen atmosphere. Platinum

TABLE 2.1

Various Elements, Alloys, and Oxides Used in High Temperature Applications

Materials	Approximate Upper Use Temperature (°C)	Atmosphere	Common Mode of Failure
Nickel–chromium–iron alloys (American Foundrymen's Association, 1957; Taylor, 1991; Hussain et al., 1995)	900–1100	Various	a, b, c
Iron–chromium–aluminum alloys (Prescott and Graham, 1992)	1000–1150	Various	a, b, c
Ti-Ni-Al alloys (superalloys) (Wang et al., 1995)	800–1000	Various	a, b, c
Molybdenum	<500	Oxidizing	a, b, c, d
	1600	Reducing	
Platinum (Jahn, 1984)	1600	Various	b, c, d
Iridium (Cardarelli, 2000)	2200	Various	b, c, d
Rhodium (Jahn, 1984)	1700	Oxidizing Reducing Vacuum	
Tantalum (Wilkinson, 1969)	1000	Oxidizing	a, b, c, d, g
	800	Nitrogen	
	2200	Reducing	
Tungsten (Wilkinson, 1969)	<500	Oxidizing	a, b, c, d
	2200	Reducing	
SiC (Singhal, 1976)	1700	Reducing	a, c, d
Si_3N_4 (Singhal, 1976)	1400	Reducing	a, c, d
C (graphite) (Jahnson, 1950)	2000	Reducing	a, c, d
$MoSi_2$	1700	Oxidizing, forms a protective coating	
	1350	Reducing	
ThO_2 (Jahnson, 1950)	2300	Oxidizing, reducing	b, e
MgO (Jahnson, 1950)	1600–1700	Oxidizing, reducing	b, d, e
ZrO_2 (Jahnson, 1950; Cardarelli, 2000)	2300	Oxidizing	b, e, f
Al_2O_3 (Jahnson, 1950; Cardarelli, 2000; Badkar, 1991)	1950	Oxidizing, reducing	b, e, f

Source: Geddes, B. et al., *Superalloys: Alloying and Performance*, ASM International, SBN: 9781615030408, 2010, pp. 1–176; Davis, J. R. *ASM Specialty Handbook: Heat-Resistant Materials*, ASM International. ISBN: 0871705966, 1997, pp. 1–591, 51.

[a] Oxidation.

[b] Grain growth.

[c] Temperature lowered if corrosive gases or various incompatible metals oxides are present.

[d] Vaporization.

[e] Melting.

[f] Phase transformation a problem, "stabilized zirconia avoids this but may have somewhat lower temperature limit.

[g] Forms nitrides.

TABLE 2.2

Properties of Standard Thermocouples

Type	Base Composition	Melting Point (°C)	Ambient	Temperature Range (°C)
J	Fe (+) 44Ni-55Cu(−)	1400 1210	Oxidizing or reducing	7–600[a]
K	90Ni-9Cr(+) 94Ni-Al,Mn, Si, Fe, Co(−)	1350 1400	Oxidizing	−270–1372[a]
N	84Ni-14Cr-1.4Si(+) 95Ni—4.4Si-0.15 Mg(−)	1410 1400	Oxidizing	−270–1260[a]
T	Cu(+) 44Ni-55Cu(−)	1083 1210	Oxidizing	−200–370[a]
E	90Ni-9Cr(+) 44Ni-55Cu(−)	1350 1210	Oxidizing	−200–870[a]
R	87Pt-13Rh(+) Pt(−)	1860 1769	Oxidizing or inert	−50–1768
S	90Pt-10Rh(+) Pt(-)	1769 1927	Oxidizing or inert	−50–1768
B	70Pt-30Rh(+) 94Pt-6Rh(−)	1927 1826	Oxidizing, vacuum, or inert	800–1820
C	W-5Re (+) W-26Re (−)	3350 3120	Inert	0–2760[a]

Source: Davis, J. R. *ASM Specialty Handbook: Heat-Resistant Materials*, ASM International. ISBN: 0871705966, 1997, pp. 1–591–51. From Springer Science+Business Media: *Materials Handbook: A Concise Desktop Reference*, 1st Edition, 2000, pp. 1–595, Cardarelli, F.

[a] Use in oxidizing atmospheres limited by easy oxidation of the metals, may be used somewhat higher in inert atmospheres, but wires become very brittle on heating.

is also seriously corroded by basic oxides such as Na_2O and K_2O, especially when these melt. Under reducing conditions, one must be cautious when using platinum because of its tendency to react with base metals, often to the extent of pulling them out of oxides in contact with it (Berndt and Keller, 1974). Lead precipitated in the grain boundaries of platinum appears to enhance grain growth and can weaken grain—grain contacts and

TABLE 2.3

Binary and Ternary Platinum Oxides

Oxide	Decomposition Temperature (°C)
PtO	555
OsO_4	Boils at 130
IrO_2	1100
RhO_2	1127
RhO	1120
Rh_2O_3	880
PtO_2	430
Pb_2PtO_4	735
$PbPt_2O_4$	750
$Ca_2Pt_3O_8$	720

cause leaks in crucibles (Cardarelli, 2000; Darling et al., 1970a). Similarly, hot Pt should never be placed in contact with graphite. Iron and other transition metals will alloy with platinum, embrittle it, and change the composition of oxide samples contained in Pt foil or crucibles. At high temperatures, platinum will react strongly with refractories such as alumina, zirconia, and thoria when oxygen is effectively removed from the surrounding atmosphere. Magnesia is the only refractory so far examined which resists this type of decomposition, which can occur at temperatures as low as 1200°C (Darling et al., 1970a). There are very stable Pt-Zr intermetallic compounds, so Pt in contact with zirconia under reducing conditions can spell trouble (Darling et al., 1970a). Such reactions contaminating Pt thermocouples will change their e.m.f. (often decreasing it) and could cause erroneous apparently low-temperature readings which will be compensated by control systems and cause the furnace to overheat, sometimes to the point of melting the Pt. A low eutectic in the Pt-Si system is also a potential hazard. The binary phase diagrams for Pt-Si, Pt-Pb, Pt-Zr, Pt-Al (McAlister and Kahan, 1986) are given in Figure 2.1a through d. Table 2.4 shows the eutectics between low melting point metals and noble metals (Groza et al., 2007).

2.4 Materials above 2000°C

In the absence of oxygen, tungsten, tantalum, and iridium can be used to temperatures of above 2000°C, encountering neither melting nor serious vaporization (Davis, 1997; Cardarelli, 2000). They can be used in vacuum or in an inert noble gas atmosphere but not in nitrogen because of potential nitride formation (Stowell et al., 2001; Schwarzkopf and Kieffer, 1953; Jehn and Ettmayer, 1978). They are used for containers, thermocouples (e.g W-Rh) and heating elements (Wilkinson, 1969; Davis, 1997; Cardarelli, 2000; Schwarzkopf and Kieffer, 1953). However these metals oxidize easily, and their oxides, in contrast, melt and boil in the 800–1000°C range (Revie, 2010; Kofstad, 1966; Gulbransen and Wysong, 1948; Gulbransen et al., 1963). Thus, air or oxygen leaks in systems using these materials can cause catastrophic failure. Graphite can be used to about 2200°C (Johnson, 1950; Davis, 1997). It appears to sublime rather than melt, and has significant vapor pressure and mobility (as well as transport as CO gas species) above 2000°C (Palmer, 1970; Joseph et al., 2002; Scheindlin, 1984; Brewer, 1952). Thus, it is almost impossible to avoid carbon contamination of samples in graphite containers and furnaces, or even in high vacuum systems containing traces of carbon from lubricants or other sources. The formation of carbides of zirconium, titanium, tungsten, and molybdenum is a complication when these metals and graphite are present in the same high temperature system (Marmer et al., 1971).

Maintaining oxidizing conditions and a good sample containment/measurement environment above 2000°C is challenging because of the lack of appropriate materials. Table 2.1 shows few oxides with melting points above 2000°C, which in principle make suitable crucible materials or protective coatings. However, they themselves are reactive toward other oxides, often resulting in compounds with lower melting points or systems with deep eutectics (Jarvis, 2006; Danzer, 1993; Adachi and Imanaka, 1998; Davis, 1997; Cardarelli, 2000; Marmer et al., 1971). Dense thoria is an excellent crucible material, stable with no change in composition or structure in atmospheres ranging from air down to very low oxygen fugacity. It is slow to react with other oxides. Its use has gone out of vogue because of its modest radioactivity, but it remains very useful as a container, refractory, and when

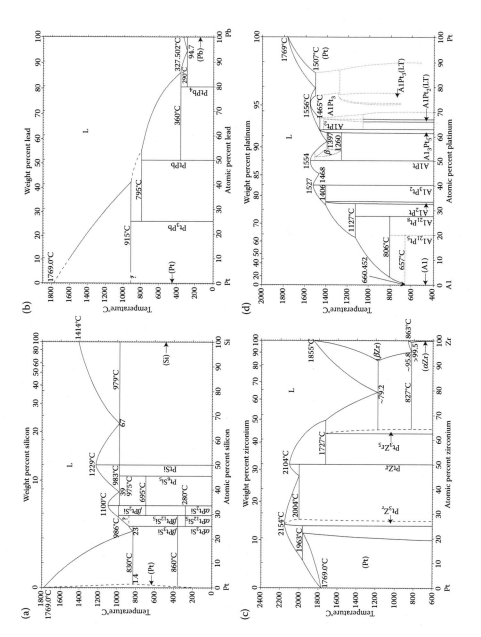

FIGURE 2.1

(a) Pt-Si Phase diagram (From Tanner, L. E. and H. Okamoto, *Journal of Phase Equilibria*, 12(5), 1991, 571–574.); (b) Pt-Pb phase diagram. (From Massalski, T. B. et al. *American Society for Metals*, 1986, 1–2224.); (c) Pt-Zr phase diagram (From Darling, A. S. et al., *Platinum Metals Review*, 14(4), 1970b, 124–30.); (d) Pt-Al phase diagram. (From McAlister, A. and D. J. Kahan, *Bulletin of Alloy Phase Diagrams*, 7(1), 1986, 47–51, 83–84.)

TABLE 2.4

Noble Metal Eutectics (°C) with Selected Elements

Element	Pt	Rh	Ir
B	825	1131	1046
P	588	1254	1262
Si	890	1389	1470
Sn	1070	–	–
Pb	327	–	–
Bi	730	–	–
Sb	633	–	–
As	597	–	–

doped with rare earths, a solid electrolyte which maintains purely ionic conductivity down to extremely low oxygen fugacity (Chaudhary et al., 1980; Subbarao, 1980; Bonnell and Hastie, 1990; Fauquier, 1961). A number of oxides, notably the rare earth oxides, zirconia, and hafnia, undergo a series of phase transitions with substantial changes in enthalpy and volume, prior to melting, which can compromise their physical integrity (Navrotsky and Ushakov, 2005; Wang et al., 1992; Adachi and Imanaka, 1998; Navrotsky, 1994; Suresh et al., 2003; Pitcher and Navrotsky, 2003; Pitcher et al., 2005; Ushakov et al., 2004). Transition points and thermodynamic properties are listed in Table 2.5. A further complication is the existence of deep eutectics in oxide systems; which means that, although specific compounds may be stable to high temperature, melting can ensue as much as 500°C lower in binary and multicomponent systems (Bowen, 1914; Alper et al., 1962; Chatterjee and

TABLE 2.5

Phase Transitions and Thermodynamic Properties of Some Refractory Oxides

Oxide	Phase Transition	Transition Temperature (°C)	Enthalpy of Transition kJ.mole^{-1}	Entropy of Transition J mol^{-1} K^{-1}
ZrO_2	m–t	1202	5.4 ± 0.3	3.7 ± 0.3
	t–c	2300	4.2 ± 0.5	1.6 ± 0.2
	m–c	Metastable	9.6	5.3
HfO_2	m–t	1650	32.5 ± 1.7	
	t–c	2700		
	m–c	Metastable		
Y_2O_3	C–c	2200 ± 50	24 ± 5	
SiO_2	Quartz-	870	1.32	
(Stevens et al., 1997;	tridymite	1473 ± 30	0.96 ± 0.25	
Roy and Roy, 1964;	tridymite-cristobalite	525	9.49 ± 0.25	
Richet et al., 1982;	cristobalite	835	10.20 ± 0.33	
Navrotsky et al.,	α-β	1427		
1980)	cristobalite- quartz	1726		
	amorphous SiO_2			
	-quartz			
	amorphous			
	SiO_2-cristobalite			

Source: Navrotsky, A., *Journal of Materials Chemistry,* 15(19), 2005, 1883–1890.
Note: M, monoclinic; t, tetragonal; c, cubic fluorite; C, cubic C-type (ordered vacancy).

Zhmoidin, 1972). Such effects are illustrated in the phase diagrams for MgO-Al$_2$O$_3$, MgO-SiO$_2$, and Al$_2$O$_3$-CaO in Figures 2.2a, b, and c, respectively.

Considering a variety of compositions and composite materials can enhance design capability and functionality. Among the rare-earth chromite refractories, lanthanum chromite is the most refractory (m.p. 2487°C). It is also the least expensive and is under consideration as a high-temperature material for magnetohydrodynamic power generation (MHD). However, the limitation to the use of lanthanum chromite at very high temperatures is its volatility in oxidizing atmospheres due to volatilization of gaseous CrO$_3$. Doping of lanthanum chromite with strontium reduces the volatility and improves the sinterability and conductivity (Khattak and Cox, 1977; Barnal et al., 2004; Wang et al., 2007; Setz et al., 2011; Acchar et al., 2012). Similarly when mullite impurity is added to the refractory material zircon (ZrSiO$_4$), it improves sintering, making the production of high-density ceramics from zircon–mullite mixtures more efficient than than from pure zircon (Rendtorff et al., 2009, 2010; Yeo et al., 2004). Also the emissive properties of rare earth hexaborides (e.g.,

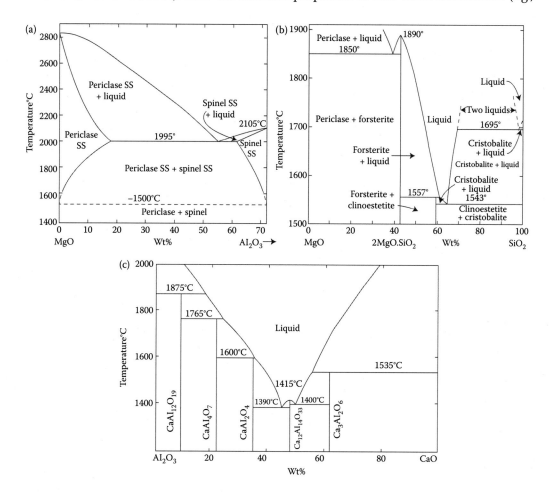

FIGURE 2.2

(a) MgO-Al$_2$O$_3$ phase diagram (Alper, 1962); (b) MgO-SiO$_2$ phase diagram (From Bowen, N. L. and O. Andersen, *American Journal of Science*, 37, 1914, 487–500.); (c) phase diagram Al$_2$O$_3$-CaO (From Chatterjee, A. K. and G. I. Zhmoidin, *Journal of Material Science*, 7(1), 1972, 93–97.).

LaB$_6$) improve by addition of refractory metals (tungsten, rhenium, titanium, and iridium) (Latini et al., 2002; Kondrashov, 1974; Bondarenko et al., 1971; Chu and Goebel, 2012).

2.5 Silicon-Based Refractories

Silicon nitride and silicon carbide are excellent structural and refractory materials under neutral atmospheres up to 1400°C and 1700°C, respectively. Si$_3$N$_4$ exists in two polymorphs, α and β, the latter stabilized by oxygen impurities (Liang et al., 1999; Jansen et al., 2002; Chen, 1993). SiC shows many complex polytypes. Rare earth oxides are used as sintering aids for Si$_3$N$_4$ and details of the mechanism of sintering and the nature and thermodynamics of intergranular crystalline, amorphous, and molten phases resulting from the interaction of nitride and oxide are still under study (Ashton, 2011; Nagano et al., 2000; Negita, 1986; Shui et al., 2011). In the past decade, the controlled pyrolysis of polymers with a silicon backbone and pendant organic groups has produced a new class of "polymer derived ceramics" (PDC) in the Si-O-C, Si-C-N, and Si-B-C-N systems (Tavakoli et al., 2012; Varga et al., 2007; Ionescu, 2012; Morcos et al., 2008a; Riedel et al., 1995, 2006; Seifert et al., 2001). Though amorphous to X-ray diffraction, they are heterogeneous at the nanoscale, with carbon rich and carbon poor (oxygen or nitrogen rich) domains. They withstand high temperature as well as or better than silicon nitride, though they are subject to crystallization, oxidation, and carbothermal reduction above 1500–1550°C. Thermochemical studies suggest considerable thermodynamic stability of PDCs relative to crystalline SiO$_2$, Si$_3$N$_4$, SiC, and C (Morcos et al., 2008a,b; Seifert et al., 2001; Durham et al., 1991).

2.6 High-Temperature Oxidation

Oxidation is an important high-temperature corrosion phenomenon, especially in metals and alloys. The rate of oxidation of metals usually decreases as oxide layer thickness increases. Very small amounts of reactive elements added to the alloy sometimes introduce a remarkable improvement in the integrity of the protective oxide, particularly in its resistance to thermal cycling (Sequeira, 2011a,b; Wright and Dooley, 2010; Boinovich et al., 2011; Del et al., 2012). The reaction product, a scale formed on the metal, acts as a physical barrier and, in favorable cases, reaction ceases after the barrier is established (Boinovich et al., 2011; Aiello et al., 2004; Fazio et al., 2001; Hata and Takahashi, 2006; Kurata and Futakawa, 2004; Zhang et al., 2005).

Thermodynamics provides an essential tool in the analysis of oxidation problems. Although not predictive of kinetics, thermodynamic analysis allows one to ascertain which reaction products are possible. It is useful to represent thermodynamic analysis in graphical form. The types of thermodynamic diagrams most often used are

1. Gibbs free energy versus composition diagrams, which are used for thermodynamics of solutions (gas, liquid, or solid) (Gaskell, 2003; Darken et al., 1953; Alper, 1970).

2. Standard Gibbs free energy of formation versus temperature diagrams, often called Ellingham diagrams, which show the thermodynamics for a given class of compounds, for example, oxides, carbides, nitrides, and sulfides (Ellingham, 1944). They allow immediate comparison of the relative stability of different materials, including solids with different oxidation states for the same metal (Ellingham, 1944; Shatynski, 1977, 1979; Olette and Ancey-Moret, 1963; Richardson and Jeffes, 1948; Kellogg, 1966; Belton and Worrell, 1970).

3. Vapor species diagrams, sometimes called Kellogg diagrams, which present the vapor pressure of compounds as a function of convenient variables such as partial pressure of gaseous components (Revie, 2010; Kellogg, 1966; Belton and Worrell, 1970).

The Ellingham diagram is a graphical representation of the standard Gibbs free energies of oxidation of pure metals vs. temperature (Gaskell, 2003). The most useful form of representation is to express the quantities in terms of one mole of the oxidizing species:

$$xM + O_2 = M_xO_2 \tag{2.1}$$

At a given temperature, T, the standard free energy ΔG_T^0 of the reaction is

$$\Delta G_T^0 = \Delta H_T^0 - T\Delta G_T^0 \tag{2.2}$$

At equilibrium

$$\Delta G_T^0 = -RT \ln\left(\frac{a_{MxO_2}}{a_M} \cdot f_{O_2}^{-1} \right) \tag{2.3}$$

where "a" stands for activity and "f" for fugacity. The stoichiometric coefficient x is 2 for an oxide MO, 1.5 for M_2O_3, and 1 for MO_2. If the metal oxide is stoichiometric, the gas is ideal, and there is negligible solubility of oxygen in the metallic phase, the activities of M and M_xO_2 can be taken as unity and pressure can be substituted for fugacity. Thus,

$$\Delta G_T^0 = RT \ln p_{O_2} \tag{2.4}$$

Figure 2.3 is an Ellingham plot for binary oxides. The ΔG_T^0 values fall on a straight line in the diagram. The change in the slope of the line corresponds to the phase change of the metal or compounds. The partial pressure at which the metal and oxide coexist is the dissociation pressure of the oxide.

The relative stability of the various oxides is easily deduced from Figure 2.3. The most stable oxide have the most negative value of ΔG_T^0 the lowest value of P_{O_2} and the highest value of (P_{H_2}/P_{H_2O}) and (P_{CO}/P_{CO_2}). Figure 2.3 shows clearly that oxides such as CaO and Al_2O_3 can be reduced only under very low oxygen fugacity, and conversely, their metals are predicted to oxidize under a wide range of P_{O_2} conditions. The role of kinetics and protective coatings is demonstrated by the very different actual behavior of aluminum and calcium at ambient conditions. Aluminum generally forms a protective adherent alumina film and persists (metastably) for years under most conditions. Calcium oxidizes rapidly (potentially explosively) to CaO which in turn reacts with H_2O and CO_2 in the atmosphere to form $Ca(OH)_2$ and $CaCO_3$.

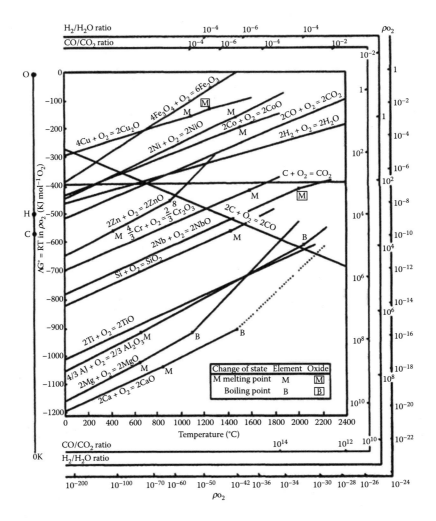

FIGURE 2.3
Ellingham plot for some common metal–metal oxides. (From Birks, N., G. H. Meier, and F. S. Pettit, *Introduction to the High Temperature Oxidation of Metals*, 2nd ed., Cambridge University Press, ISBN:9780521480420, 2006, 1–352.)

From Figure 2.3, it is clear that the oxides of Fe, Ni, and Co, which form the base of the majority of engineering alloys, are significantly less stable than the oxides of some solutes like Cr, Al, and Si. When one of these solute elements is added to Fe, Ni, Co, or their alloys, internal oxidation of the solute (the oxide remaining within the alloy as small particles) is expected to occur, if the concentration of the solute is relatively low. As the concentration of the solute increases, oxidation will change to external, resulting in an oxide scale that protects the alloy from rapid oxidation. This process is known as selective oxidation. Most of the Fe, Ni, and Co alloys rely on the selective oxidation of Cr to Cr_2O_3 scale for oxidation resistance (Wright and Dooley, 2010; Boinovich et al., 2011; Del et al., 2012; Chatha et al., 2012; Sadique et al., 2000; Stott et al., 1995; Gasik et al., 1970; Hetmanczyk et al., 1979). Some high-temperature alloys use Al to form Al_2O_3 scale for oxidation resistance (Sadique et al., 2000; Stott et al., 1995; Masalski et al., 1999; Yamamoto et al., 2007; Dunning et al., 2002).

Another important use of internal oxide particles is in the mechanical strengthening of alloys. So-called oxygen dispersion strengthened alloys, in which small (nanosized) particles of rare earth titanates and related oxide phases appear to slow the movement of defects and the propagation of cracks in alloys operating at high temperature (Wright et al., 2001; Kamise et al., 1967; Ukai et al., 2002; Chen and Dong, 2011; Oksiuta et al., 2011; Tang et al., 2012; Voevodin et al., 2011; Zhong et al., 2012; Zhou et al., 2012). The thermodynamics and kinetics of such precipitate formation reflect complex phenomena currently under active investigation.

2.7 Volatile Oxides

Some oxides exhibit high vapor pressure above 1000°C. Hence, oxide scales become less protective, when their vapor pressure is high. Cr, Mo, W, V, Pt, Rh, and Si are metals for which volatile oxygen-bearing species are important at high temperature. Oxidation of Pt and Pt group metals at high temperature is influenced by oxide volatility in that the only stable oxides are volatile. This results in continuous mass loss (Jehn, 1984). Alcock and Hopper studied the mass loss of Pt and Rh at 1400°C as a function of oxygen partial pressure (Alcock and Hooper, 1960). The gaseous species were identified as PtO_2 and RhO_2. These results have additional significance because; Pt and Pt-Rh wires are often used as thermocouples, furnace windings, electrical contacts, and sample holders to support specimens during high-temperature experiments. It is notable that although W and Mo are high melting and show low volatility in the absence of oxygen, their oxides are low melting and extremely volatile (Table 2.6).

Formation of SiO_2 on silicon-containing alloys and ceramics results in very low oxidation rates. However, a significant pressure of SiO is in equilibrium with SiO_2 and Si at oxygen pressures near the dissociation pressure of SiO_2 (Bronson and Chessa, 2008; Rocabois et al., 1996). This results in a rapid flux of SiO away from the specimen surface and the SiO_2 coating is no longer protective (Wagner, 1958, 1965). In the presence of carbon (graphite or other carbon source), the carbothermic reduction of silica by the reaction SiO_2 (solid) + C

TABLE 2.6

Melting Point and Boiling Point of Some Representative Metals and Oxides

Materials	Approximate Melting Point (°C)	Approximate Boiling or Decomposition Point (°C)
Mo	2610	5560
MoO_3	795	1155
W	3407	5727
WO_3[a]	1470	1700
Pt	1772	3800
PtO	—	d.550
PtO_2	—	d.650
IrO_2	—	d.1124
Rh	1966	3727
Rh_2O_3	—	d.1100
Si	1414	3250
SiO	1700	1880
SiO_2	1600	2230

[a] Significantly volatile at 800–1000°C.

(solid) = SiO (gas) + CO (gas) becomes thermodynamically favorable above about 1500°C (Dhage et al., 2009; Komarov et al., 2005; Shimoo et al., 1995; Tolstoguzov, 1989). In the presence of low concentrations of oxygen, the formation of SiO from silicon nitride, silicon carbide, and polymer-derived ceramics in the Si-O-C-N system (Greil, 1998) also becomes a major mechanism of corrosion. Thus, silica and carbon are basically incompatible at high-temperature and SiO as a gaseous species limits the protective effect of silica coatings, especially under low oxygen fugacity (Wagner, 1958; Opila and Jacobson, 1995). The presence of water vapor (steam) also enhances corrosion and vaporization of silica coatings (Deal and Grove, 1965; Hashimoto, 1992; Opila, 2004).

Similar volatility problems are encountered at high temperature for boron-containing materials such as the nitride, BN. Oxidation of pure BN is complex and highly dependent on the microstructure of the BN and water vapor content in the gas stream. High deposition temperature chemically vapor deposited boron nitride (CVD BN) is the most desirable, but may lead to problems in composite fabrication. As BN and SiC oxidize, there are numerous interactions between the oxidation products. It has been identified three basic mechanisms of formation of borosilicate glass, gettering of oxygen by SiC and preservation of the BN, and volatilization of the BN via B_2O_3 and $H_xB_yO_z(g)$ formation (Jacobson et al., 1999; Segal, 1984).

2.8 Summary

This overview summarizes the ranges of use and thermodynamic driving forces for failure of high-temperature materials. These constraints must be taken into account when designing systems for high-temperature use. Knowledge of the fundamental chemistry and consideration of all possible chemical interactions is essential to avoid unexpected and possibly catastrophic failure of materials at 1000°C and above, where the rates of reactions can be rapid and thermodynamic equilibrium can be established rapidly.

Acknowledgments

The authors thank Elizabeth J. Opila, Department of Materials Science and Engineering, University of Virginia, and William T. Petuskey, Associate Vice President, Office of Knowledge Enterprise Development, Arizona State University, Tempe, Arizona, for reviewing this chapter and providing valuable technical comments and suggestions. Preparation of this manuscript was supported by the U.S. Department of Energy, Office of Basic Energy Sciences, Grant DE-FG02-03ER46053.

References

Acchar, W., C. R. C. Sousa, and S. R. H. Mello-Castanho, Mechanical performance of $LaCrO_3$ doped with strontinum and cobalt for SOFC interconnect, *Materials Science and Engineering A*, 550, 2012, 76–79.
Adachi, G. and N. Imanaka, The binary rare earth oxides, *Chemical Reviews*, 98(4), 1998, 1479–1514.

Aiello, A., M. Azzati, G. Benamati, A. Gessi, B. Long, and G. Scaddozzo, Corrosion behaviour of stainless steels in flowing LBE at low and high oxygen concentration, *Journal of Nuclear Materials*, 335(2), 2004, 169–173.

Alcock, C. B. and G. W. Hooper, Thermodynamics of gaseous oxides of the platinum group metals, *Proceedings of Royal Society of London, Ser. A*, 254, 1960, 551–561.

Alper, A. M., *Phase Diagrams: Materials Science and Technology*, Academic Press, 1970, pp. 1–325.

Alper, A. M., R. N. McNally, P. H. Ribbe, and R. C. Doman, System MgO-MgAl2O4, *Journal of American Ceramic Society*, 45, 1962, 263–268.

American Foundrymen's Association, Cast Metals Handbook. 4th Edition. Des Plaines, Illinois: American Foundrymen's Society, Inc., 1957, pp. 1–756.

Armstrong, C. R., M. Nyman, T. Shvareva, G. E. Sigmon, P. C. Burns, and A. Navrotsky, Uranyl peroxide enhanced nuclear fuel corrosion in seawater, *Proceedings of the National. Academy of Sciences of the United States of America*, 109(6), 2012, 1874–1877.

Ashby, M. F. and D. R. H. Jones, *Engineering Materials 1: An Introduction to Properties, Applications and Design*, Elsevier-Butterworth-Heinemann Publications, Amsterdam, 2005.

Ashton Acton, Q., *Issues in Materials and Manufacturing Research*, Scholarly Editions, PhD. Publisher, Atlanta, Georgia, 2011.

Badkar, P. A., Alumina ceramics for high temperature applications, *Key Engineering Materials*, Trans Tech Publications, Switzerland, 56–57, 1991, 45–58.

Barnal, S., G. Blanco, J. M. Gatica, and J. Antonio, Chemical reactivity of Binary rare earth oxides, pp. 9–55; In: *Binary Rare Earth Oxides*, Editors: Adachi, G., N. Imanaka, Z. C. Kang, Kluwer Academic Publishers, 2004, pp. 1–260.

Barry, T. I., T. G. Chart, and G. P. Jones, The calculation of multicomponent alloy phase diagrams from thermodynamic data, *Phase Transformations*, 11(2), 1979, 18–21.

Beele, W., Superalloy component with a protective coating system, PatentUS5985467, 1999.

Belton, G. R. and W. L. Worrell, Heterogeneous kinetics at elevated temperatures: *Proceedings of an International Conference in Metallurgy and Materials Science*, University of Pennsylvania, September 8–10, 1969, Plenum Press, 1970, pp. 1–532.

Berndt, U., B. Erdmann, and C. Keller, Coupled reduction with platinum group metals. Preparation of high-purity rare metals, *Platinum Metals Review*, 18(1), 1974, 29–34.

Bettahar, N., P. Conflant, F. Abraham, and D. Thomas, Lead platinate (Pb_2PtO_4), a new platinum-lead oxide with edge-shared PtO_6 octahedral chains, *Journal of Solid State Chemistry*, 67(1), 1987, 85–90.

Birks, N., G. H. Meier, and F. S. Pettit, *Introduction to the High Temperature Oxidation of Metals*, Second Edition, Cambridge University Press, New York, New York, 2006, 1–352.

Blachnio, J., The effect of high temperature on the degradation of heat-resistant and high-temperature alloys, Diffusion and Defect Data, Part B, 147–149, (Mechatronic Systems and Materials III), 2009, pp. 744–751.

Boinovich, L. B., S. V. Gnedenkov, D. A. Alpysbaeva, V. S. Egorkin, A. M. Emelyanenko, S. L. Sinebryukhov, and A. K. Zaretskaya, Corrosion resistance of composite coatings on low-carbon steel containing hydrophobic and superhydrophobic layers in combination with oxide sublayers, *Corrosion Science*, 55, 2011, 238–245.

Bondarenko, V. P., V. V. Morozov, and L. V. Chernyak, Reaction of lanthanum and cerium hexaborides with refractory metals, *Poroshkovaya Metallurgiya*, 11(1), 1971, 73–78.

Bonnell, D. W., and J. W. Hastie, A predictive thermodynamic model for complex high temperature solution phases XI, In: *Materials Chemistry at High Temperatures*, Hastie, J. W. Ed., Humana Press, Gaithersburg, Maryland, 1990, pp. 313–333.95.

Bowen, N. L. and O. Andersen, The binary system: MgO-SiO2, *American Journal of Science*, 37, 1914, 487–500.

Brewer, L., Recent determinations of the vapor pressure of graphite, *Journal of Chemical Physics*, 20, 1952, 758–759.

Bronson, A. and J. Chessa, An evaluation of vaporizing rates of SiO_2 and TiO_2 as protective coatings for ultrahigh temperature ceramic composites, *Journal of the American Ceramic Society*, 91(5), 2008, 1448–1452.

Burns, P. C., R. C. Ewing, and A. Navrotsky, Nuclear fuel in a reactor accident, *Science*, 335(6073), 2012, 1184–1188.

Cardarelli, F., *Materials Handbook: A Concise Desktop Reference*, 1st Edition, Springer, ISBN: 1846286689, 2000, pp. 1–595.

Chart, T. G., The calculation of multicomponent alloy phase diagrams at the National Physical Laboratory, *NBS Special Publications (U. S.)*, 496(2), 1978, 1186–1199.

Chase, M. W., Jr., C. A. Davies, J. R. Downey, Jr., D. J. Frurip, R. A. McDonald, and A. N. Syverud, *JANAF Thermochemical Tables*. Third Edition. Part I, Aluminum-Cobalt, *Journal of Physical and Chemical Reference Data*, Supplement, 14(1), 1985a, 1–926.

Chase, M. W., Jr., C. A. Davies, J. R. Downey, Jr., D. J. Frurip, R. A. McDonald, and A. N. Syverud, *JANAF Thermochemical Tables*. Third Edition. Part II, Chromium-Zirconium, *Journal of Physical and Chemical Reference Data*, Supplement, 14(1), 1985b, 927–1856.

Chaston, J. C., Reaction of oxygen with platinum metals. I. The oxidation of platinum, *Platinum Metals Review*, 8(2), 1964, 50–54.

Chaston, J. C., Solid oxide of platinum and rhodium. Formation in high pressures of oxygen, *Platinum Metals Review*, 13(1), 1969, 28–29.

Chatha, S. S., H. S. Sidhu, and B. S. Sidhu, High temperature hot corrosion behaviour of NiCr and Cr3C2-NiCr coatings on T91 boiler steel in an aggressive environment at 750Â°C, *Surface Coating Technology*, 206(19–20), 2012, 3839–3850.

Chatterjee, A. K. and G. I. Zhmoidin, Phase equilibrium diagram of the system calcium oxide-aluminum oxide-calcium fluoride, *Journal of Material Science*, 7(1), 1972, 93–97.

Chaudhary, C. B., H. S. Maity, and E. C. Subbarao, Defect structure and transport properties, In: *Solid Electrolytes and their Applications*, Subbarao, E. C., ed., Plenum Press, New York, New York, 1980, 1–80. 94.

Chen, C. L. and Y. M. Dong, Effect of mechanical alloying and consolidation process on microstructure and hardness of nanostructured Fe-Cr-Al ODS alloys, *Materials Science and Engineering A*, 528(29–30), 2011, 8374–8380.

Chen, I. W., Silicon nitride ceramics: Scientific and technological advances, *Materials Research Society*. Edited by P. F. Becher, M. Mitomo, G. Petzow, T. S. Yen, Materials Research Society, Boston, Massachusetts, 1993, pp. 1–550.

Chu, E. and D. M. Goebel, High-current lanthanum hexaboride hollow cathode for 10-to-50-kW Hall thrusters, *IEEE Transactions of Plasma Science*, 40(9), 2012, 2133–2144.

Danzer, R., Properties of ceramics during high temperature applications. (Part one), *Ceramic Forum International*, 70(6), 1993, 280–286.

Darken, L. S., R. W. Gurry, and M. B. Bever, *Physical Chemistry of Metals*, McGraw-Hill Inc., New York, 1953, 1–535.

Darling, A. S., G. L. Selman, and R. Rushforth, Platinum and the refractory oxide. I. Compatibility and decomposition processes at high temperatures, *Platinum Metals Review*, 14(2), 1970a, 54–60.

Darling, A. S., G. L. Selman, and R. Rushforth, Platinum and the refractory oxide. III. Constitutional relations in the alloys formed, *Platinum Metals Review*, 14(4), 1970b, 124–30.

Davies, R. H., A. T. Dinsdale, J. A. Gisby, J. A. J. Robinson, and S. M. Martin, MTDATA—thermodynamic and phase equilibrium software from the National Physical Laboratory, *CALPHAD: Comput. Coupling Phase Diagrams and Thermochemistry*, 26(2), 2002, 229–271.

Davis, J. R. *ASM Specialty Handbook: Heat-Resistant Materials*, ASM International. ISBN: 0871705966, 1997, pp. 1–591, 51.

Deal, B. E. and A. S. Grove, General relation for the thermal oxidation of silicon, *Journal of Applied Physics*, 36(12), 1965, 3770–3778.

Del, G. M., A. Weisenburger, and G. Mueller, Fretting corrosion in liquid lead of structural steels for lead-cooled nuclear systems: Preliminary study of the influence of temperature and time, *Journal of Nuclear Materials*, 423(1/3), 2012, 79–86.

Dhage, S., H.-C. Lee, M. S. Hassan, M. S. Akhtar, C.-Y. Kim, J. M. Sohn, K.-J. Kim, H.-S. Shin, and O. B. Yang, Formation of SiC nanowhiskers by carbothermic reduction of silica with activated carbon, *Materials Letters*, 63(2), 2009, 174–176.

Dunning, J. S., D. E. Alman, and J. C. Rawers, Influence of silicon and aluminum additions on the oxidation resistance of a lean-chromium stainless steel, *Oxidation of Metals*, 57(5/6), 2002, 409–425.

Durham, S. J. P., K. Shanker, and R. A. L. Drew, Thermochemistry of the silicon-oxygen-nitrogen-carbon system with relation to the formation of silicon nitride, *Canadian Metallurgical. Quarterly*, 30(1), 1991, 39–43.

Ellingham, H. J. T., Reducibility of oxides and sulfides in metallurgical processes, *Journal of the Society of Chemical Industry, London*, 63, 1944, 125–133.

Eriksson, G. and K. Hack, Calculation of phase equilibria in multicomponent alloy systems using a specially adapted version of the program 'SOLGASMIX', *CALPHAD: Comput. Coupling Phase Diagrams and Thermochemistry*, 8(1), 1984, 15–24.

Eyring, L. The binary rare earth oxides, *Handbook of Physics and Chemistry of Rare Earths*, Edited by Gschneidner, Karl A., Jr.; Eyring, LeRoy North-Holland, 1979, 337–399.

Fauquier, D., The specific nature of wiikite and loranskite, *Comptes Rendus de la Congrès de la Société Savantes Paris, Section. Science*, 251, 1961, 283–287.

Fazio, C., G. Benamati, C. Martini, and G. Palombarini, Compatibility tests on steels in molten lead and lead-bismuth, *Journal of Nuclear Materials*, 296, 2001, 243–248.

Fromm, E. J., Gas-metal reactions of refractory metals at high temperature in high vacuum, *Journal of Vacuum Science and Technology*, 7(6), 1970, 100–105.

Gasik, M. I., V. I. Gubinskii, V. S. Ignat'ev, S. I. Khitrik, and Y. M. Chupakhin, Kinetics of the high-temperature oxidation of chromium and ferrochrome, *Zashchita Metallov*, 6(2), 1970, 216–218.

Gaskell, D. R., *Introduction to the Thermodynamics of Materials*, Fourth Edition, Taylor & Francis, 2003, 1–618.

Geddes, B., H. L. X. H. Blaine Geddes, H. Leon, and X. Huang, *Superalloys: Alloying and Performance*, ASM International, SBN: 9781615030408, 2010, pp. 1–176.

Gisby, J. A., A. T. Dinsdale, I. Barton-Jones, A. Gibbon, P. A. Taskinen, and R. Tinto, Predicting phase equilibria in oxide and sulphide systems, *Proceedings of 3rd International Sulfide Smelting Symposium* (Ed. R Stephens and H Sohn) TMS, Warrendale PA, 2002, 533–545.

Greil, P., Near net shape manufacturing of polymer derived ceramics, *Journal of the European Ceramic Society*, 18(13), 1998, 1905–1914.

Grosse, M., M. Steinbrueck, and J. Stuckert, Steam and air oxidation behavior of nuclear fuel claddings at severe accident conditions, *MRS Proceedings*, 1264, 1264-BB04-04 (Basic Actinide Science and Materials for Nuclear Applications), 2010 doi:10.1557/PROC-1264-BB04-04.

Groza, J. R., J. F. Shackelford, E. J. Lavernia, and M. T. Powers, *Materials Processing Handbook*, Taylor & Francis ISBN-13: 978–0849332166, 2007, pp. 1–840.

Gulbransen, E. A. and K. F. Andrew, Institute of metals division-kinetics of the reactions of columbium and tantalum with oxygen, nitrogen, and hydrogen, *AIME Transactions*, 188(3), 1950, 586–599.

Gulbransen, E. A. and W. S. Wysong, K. Andrew, Decarburization of chrome nickel alloys by their surface oxides in high vacua and at elevated temperatures, *American Institute of Mining and Metallurgical Engineers, Institute of Metals Division, Metals Technology* 15, 1948, 1–175.

Gulbransen, E. A., K. F. Andrew, and F. A. Brassart, Vapor pressure of molybdenum trioxide, *Journal of the Electrochemical. Society.*, 110(3), 1963, 242–243.

Hashimoto, A., The effect of water gas on volatilities of planet-forming major elements: I. Experimental determination of thermodynamic properties of calcium-, aluminum-, and silicon-hydroxide gas molecules and its application to the solar nebula, *Geochimica et Cosmochimca Acta*, 56(1), 1992, 511–532.

Hastie, J. W., *Materials Chemistry at High Temperatures*. Humana Press, Gaithersburg, Maryland, 1990.

Hata, K. and M. Takahashi, Corrosion study in direct cycle Pb-Bi cooled fast reactor, *Bullein of Research Laboratory for Nuclear Reactors* (Tokyo Institute of Technology), 30(1/2), 2006, 230–234.

Hetmanczyk, M., J. Laskawiec, H. Malinska, and K. Wylon, Investigation of oxidation of chromium-manganese steels of the ferrochromium type. I. Kinetics of oxidation of chromium-manganese steels, *Ochrona Przed Korozja*, 22(6), 1979, 141–144.

Hussain, N., K. A. Shahid, I. H. Khan, and S. Rahman, Oxidation of high-temperature alloys (superalloys) at elevated temperatures in air. II, *Oxidation of Metals*, 43(3/4), 1995, 363–378.

Igolkin, A. I., Thermal diffusion coatings for protection from gas corrosion, coke deposition, and carburization, *Chemical and Petroleum Engineering*, 39(5/6), 2003, 366–371.

Ionescu, E., *Polymer-Derived Ceramics*, Wiley-VCH Verlag GmbH, 2012, pp. 457–500.

Jacob, K. T., G. Rajitha, and G. M. Kale, Thermodynamic properties of Pb_2PtO_4 and $PbPt_2O_4$ and phase equilibria in the system Pb-Pt-O, *Journal of Alloys and Compounds.*, 481(1–2), 2009, 228–232.

Jacobson, N. S., G. N. Morscher, D. R. Bryant, and R. E. Tressler, High-temperature oxidation of boron nitride: II, boron nitride layers in composites, *Journal of the American Ceramic Society*, 82(6), 1999, 1473–1482.

Jansen, M., Haubner, R., B. Lux, G. Petzow, and R. Weissenbacher, *High Performance Non-Oxide Ceramics II*, Springer, ISBN-13: 978-3540431329, 2002, pp. 1–188.

Jarvis, D., Refractories in practice, *World Cement*, 37(6), 2006, 75–81.

Jehn, H., High temperature behavior of platinum group metals in oxidizing atmospheres, *Journal of Less-Common Metals*, 100, 1984, 321–339.

Jehn, H. and P. Ettmayer, The molybdenum-nitrogen phase diagram, *Journal of Less-Common Metals*, 58(1), 1978, 85–98.

Johnson, P. D., Behavior of refractory oxides and metals, alone and in combination, in vacuo at high temperatures, *Journal of the American Ceramic Society*, 33, 1950, 168–171.

Joseph, M., N. Sivakumar, and P. Manoravi, "High temperature vapour pressure studies on graphite using laser pulse heating, *Carbon*, 40(11), 2002, 2031–2034.

Kamise, T., T. Enjyo, and M. Adachi, Internal oxidation of dilute copper alloys, *Suiyokai-Shi*, 16(3), 1967, 171–174.

Kellogg, H. H., Vaporization chemistry in extractive metallurgy, *Transactions of Metallugical Society AIME*, 236(5), 1966, 602–615.

Kerans, R. J., Oxidation resistant ceramic composites: How weak interfaces became designed failure processes, *MRS Proceedings 1999*, 586 (Interfacial Engineering for Optimized Properties II), 2000, 1–293.

Khattak, C. P. and D. E. Cox, Structural studies of the lanthanum, strontium chromate ((La,Sr)CrO3) system, *Materials Research Bulletin*, 12(5), 1977, 463–471.

Kim, J.-W., M.-H. Kim, C. Lee, and H.-M. Kim, Method for forming fine protrusions on cladding surface of nuclear fuel rod containing zirconium, Postech Academy-Industry Foundation : WO2010/128691, November 11, 2010, European Patent: 2428963 A1.

Kofstad, P., *High-Temperature Oxidation of Metals*, John Wiley & Sons, 340 S, 1966, 1–240.

Komarov, S. V., D. V. Kuznetsov, V. V. Levina, and M. Hirasawa, Formation of SiO and related Si-based materials through carbothermic reduction of silica-containing slag, *Materials Transactions*, 46(4), 2005, 827–834.

Kondrashov, A. I., Interaction of lanthanum hexaboride with refractory metal carbides and borides, *Poroshkovaya Metallurgiya*, 11, 1974, 58–60.

Krier, C. A. and R. I. Jaffe, Oxidation of the platinum-group metals, *Journal of Less-Common Metals*, 5(5), 1963, 411–431.

Kubaschewski, O. and C. B. Alcock, *Metallurgical Thermochemistry*, Pergamon Press, 1979, 1–462.

Kurata, Y. and M. Futakawa, Excellent corrosion resistance of 18Cr-20Ni-5Si steel in liquid Pb-Bi, *Journal of Nuclear Materials*, 325(2–3), 2004, 217–222.

Latini, A., P. F. Di, and D. Gozzi, A new synthesis route to light lanthanide borides: Borothermic reduction of oxides enhanced by electron beam bombardment, *Journal of Alloys and Compounds*, 346(1/2), 2002, 311–313.

Levin, E. M. and H. F. McMurdie, *Phase Diagrams for Ceramists 1975 Supplement*, American Ceramic Society, Columbus, OH, 1975, pp. 1–513.

Liang, J., Topor. L, A. Navrotsky, and M. Mitomo, Heat of formation of alpha and beta Si3N4 measured using high temperature calorimetry, *Journal of Materials Research*, 14, 1999, 1959–1968.

Liu, X. J., K. Oikawa, I., Ohnuma, R. Kainuma, and K. Ishida, The use of phase diagrams and thermodynamic databases for electronic materials, *Journal of Metals*, 55(12), 2003. 53–59.

Lopato, L. M., A. V. Shevchenko, A. E. Kushchevskii, and S. G. Tresvyatskii, Polymorphic transformations of rare earth oxides at high temperatures, *Izvestiya. Akademii. Nauk SSSR, Neorganicheskie Materialy*, 10(8), 1974, 1481–1487.

Loremez, R. and J. Woolcock, Decomposition pressure of nitrides, *Zeitschriftfür Anorganischeund Allgemeine Chemie*, 176, 1928, 289–304.

Marmer, E. N., O. S. Gurvich, and L. F. Mal'tseva, *High Temperature Materials*, Freund Publishing House, 1971, pp. 1–281.

Masalski, J., J. Guszek, J. Zabrzeski, K. Nitsch, and P. Guszek, Improvement in corrosion resistance of the 316L stainless steel by means of Al_2O_3 coatings deposited by the sol-gel method, *Thin Solid Films*, 349(1/2), 1999, 186–190.

Massalski, T. B., J. L. Murray, L. H. Bennett, and H. Baker, Binary alloy phase diagrams, *American Society for Metals*, 1 and 2, 1986, 1–2224.

McAlister, A. and D. J. Kahan, The Al-Pt (aluminum-platinum) system, *Bulletin of Alloy Phase Diagrams*, 7(1), 1986, 47–51, 83–84.

Morcos, R. M., G. Mera, A. Navrotsky, T. Varga, R. Riedel, F. Poli, and K. Muller, Enthalpy of formation of carbon-rich polymer-derived amorphous SiCN ceramics, *Journal of the American Ceramic Society*, 91(10), 2008a, 3349–3354.

Morcos, R. M., A. Navrotsky, T. Varga, Y. Blum, D. Ahn, F. Poli, K. Muller, and R. Raj, Energetics of SixOyCz polymer-derived ceramics prepared under varying conditions, *Journal of the American Ceramic Society*, 91(9), 2008b, 2969–2974.

Nagano, T., K. Kaneko, G.-D. Zhan, and M. Mitomo, Effect of atmosphere on weight loss in sintered silicon carbide during heat treatment, *Journal of the American Ceramic Society*, 83(11), 2000, 2781–2787.

Navrotsky, A., Thermochemical insights into refractory ceramic materials based on oxides with large tetravalent cations, *Journal of Materials Chemistry*, 15(19), 2005, 1883–1890.

Navrotsky, A. Thermochemistry of crystalline and amorphous silica, *Reviews in Mineralogy and Geochemistry*, 29(1), 1994, 309–329.

Navrotsky, A. and S. V. Ushakov, Thermodynamics of oxide systems relevant to alternative gate dielectrics, *Materials Fundamentals of Gate Dielectrics*, Edited by A. Demkov and A, Navrotsky Springer, 2005, pp. 57–108.

Navrotsky, A., R. Hon, D. F. Weill, and D. J. Henry, Thermochemistry of glasses and liquids in the systems $CaMgSi_2O_6$-$CaAl_2Si_2O_8$-$NaAlSi_3O_8$, SiO_2-$CaAl_2Si_2O_8$-$NaAlSi_3O_8$, and SiO_2-Al_2O_3-CaO-Na_2O, *Geochimica et Cosmochimica Acta*, 44(10), 1980, 1409–1423.

Negita, K., Effective sintering aids for silicon carbide ceramics: Reactivities of silicon carbide with various additives, *Journal of the American Ceramic Society*, 69(12), 1986, 308–310.

Ochiai, S., *Mechanical Properties of Metallic Composites*, ASM International, Taylor & Francis, ISBN: 9780824791162, 1993, pp. 1–808.

Oksiuta, Z., M. Lewandowska, P. Unifantowicz, N. Baluc, and K. J. Kurzydlowski, Influence of Y_2O_3 and Fe_2Y additions on the formation of nano-scale oxide particles and the mechanical properties of an ODS RAF steel, *Fusion Engineering and Design*, 86, (9–11), 2011, 2417–2420.

Olette, M. and M. I. Ancey-Moret, Changes of the free energy of formation of oxides and nitrides with temperature. Construction of the diagrams. Examples of application, *Reviews of Metallurgy*, 60, 1963, 569–582.

Ondik, H. M. and C. G. Messina, Creating a materials data base builder and producing publications for ceramic phase diagrams, *ASTM SpecialTechnical Publications*, 1017, 1989, 304–314.

Ono, H., K. Morita, and N. Sano, Reaction rate and adsorption of nitrogen between gas and metal phases, *Nippon Tekko Kyokai*, 9, 1996, 11–12.

Opila, E. J. and N. S. Jacobson, SiO(g) formation from SiC in mixed oxidizing-reducing gases, *Oxidation of Metals*, 44(5/6), 1995, 527–544.

Opila, E. J., Volatility of common protective oxides in high-temperature water vapor: current understanding and unanswered questions, *Material Science Forum*, 461–464, 2004, 765–774.

Palmer, H. B., Equilibrium vapor pressure of graphite and the temperature of the carbon arc, *Carbon*, 8(2), 1970, 243–244.

Pankratz, L. B., Thermodynamic properties of elements and oxides, *Bulletin of United States Bureau of Mines*, 672, 1982, 1–515.

Petty, E. R., *Physical Metallurgy of Engineering Materials (Modern Metallurgical Texts)*, Elsevier, Vol. 6, 1968, 1–320.

Pitcher, M. W. and A. Navrotsky, Synthesis and stability of zirconia nanocrystals, *Abstracts of Papers of the American Chemical Society*, 225, 2003, U49–U49.

Pitcher, M. W., S. V. Ushakov, A. Navrotsky, B. F. Woodfield, G. Li, J. Boerio-Goates, and B. M. Tissue, Energy crossovers in nanocrystalline zirconia, *Journal of the American Ceramic Society*, 88(1), 2005, 160–167.

Pollock, D. D., *Thermoelectricity: Theory, Thermometry, Tool*, ASTM Special Technique Publication 852, 1985, pp. 1–15.

Prescott, R. and M. J. Graham, The formation of aluminum oxide scales on high-temperature alloys, *Oxidation of Metals*, 38(3/4), 1992, 233–254.

Priest, A. M., The high temperature performance of engineering materials, *Conference Proceedings of the Institute of Mechanical Engineering* (7, Mater. Des. Fire), 1992, pp. 85–91.

Raub, E. and W. Plate, The solid state reactions between the precious metals or their alloys and oxygen at high temperatures, *Zeitschrift für Metallkunde*, 48, 1957, 529–539.

Raub, E., Metals and alloys of the platinum group, *Journal of Less-Common Metals*, 1, 1959, 3–18.

Rendtorff, N. M., L. B. Garrido, and E. F. Aglietti, Mechanical and fracture properties of zircon-mullite composites obtained by direct sintering, *Ceramic International*, 35(7), 2009, 2907–2913.

Rendtorff, N. M., L. B. Garrido, and E. F. Aglietti, Zirconia toughening of mullite-zirconia-zircon composites obtained by direct sintering, *Ceramic International*, 36(2), 2010, 781–788.

Revie, R. W. and Editor, *UHLIG's Corrosion Handbook*, Third Edition, John Wiley & Sons, Inc., 2010, 1–1253.

Richardson, F. D. and J. H. E. Jeffes, The thermodynamics of substances of interest in iron and steel-making from 0 to 2400°C: I. Oxides, *Journal of Iron Steel Institute*, London, 160, 1948, 261–270.

Rajan, T. V., C. P. Sharma, and A. Sharma, *Heat Treatment Principles and Techniques*, PHI Learning Pvt. Ltd., 2004 1–476.59.

Richet, P., Y. Bottinga, L. Denielou, J. P. Petitet, and C. Tequi, Thermodynamic properties of quartz, cristobalite and amorphous silicon dioxide: Drop calorimetry measurements between 1000 and 1800 K and a review from 0 to 2000 K, *Geochimica et Cosmochimica Acta*, 46(12), 1982, 2639–2658.

Riedel, R., G. Mera, R. Hauser, and A. Klonczynski, Silicon-based polymer-derived ceramics: Synthesis properties and applications—a review. Dedicated to Prof. Dr. Fritz Aldinger on the occasion of his 65th birthday, *Journal of the Ceramic Society of Japan*, 114, 2006, 425–444.

Riedel, R., H.-J. Kleebe, H. Schoenfelder, and F. Aldinger, A covalent micro/nano-composite resistant to high-temperature oxidation, *Nature* (London), 374(6522), 1995, 526–528.

Rocabois, P., C. Chatillon, and C. Bernard, Thermodynamics of the Si-O-N system: II, stability of $Si_2N_2O(s)$ by high-temperature mass spectrometric vaporization, *Journal of the American Ceramic Society*, 79(5), 1996, 1361–1365.

Roy, D. M. and R. Roy, Tridymite-cristobalite relations and stable solid solutions, *American Mineralogist*, 49, 1964, 952–962.

Sadique, S. E., A. H. Mollah, M. S. Islam, M. M. Ali, M. H. H. Megat, and S. Basri, High-temperature oxidation behavior of iron-chromium-aluminum alloys, *Oxidation of Metals*, 54(5–6), 2000, 385–400.

Scheindlin, M. A., Investigation of carbon vapor pressure at very high temperatures and pressures with the aid of laser heating, *MRS Proceedings*, 22(2), 1984, 33–41.

Schenck, P. K. and J. R. Dennis, PC access to ceramic phase diagrams, *ASTM Special Technical Publication*, 1017, 1989, 292–303.

Schmitz, F. and W. Stamm, Layer system with TBC and noble metal protective layer, European Patent Application, EP2130945, 2009.

Schwartz, K. B. and C. T. Prewitt, Structural and electronic properties of binary and ternary platinum oxides, *Journal of Physics and Chemistry of Solids*, 45(1), 1984, 1–21.

Schwarzkopf, P. and R. Kieffer, *Refractory Hard Metals. Borides, Carbides, Nitrides, and Silicides. The Basic Constituents of Cemented Hard Metals and their Use as High Temperature Materials*, Macmillan 1953, pp. 1–477.

Scott, H. G., Phase relations in the zirconia-yttria system, *Journal of Material Science*, 10(9), 1975, 1527–1535.

Segal, D. L., Sol-gel processing: Routes to oxide ceramics using colloidal dispersions of hydrous oxides and alkoxide intermediates, *Journal of Non-Crystalline Solids*, 63(1/2), 1984, 183–191.

Seifert, H. J., J. Peng, H. L. Lukas, and F. Aldinger, Phase equilibria and thermal analysis of Si-C-N ceramics, *Journal of Alloys and Compounds*, 320(2), 2001, 251–261.

Sequeira, C. A. C., *High-Temperature Oxidation—Testing and Evaluation*, John Wiley & Sons, Inc., 2011a, pp. 1053–1058.

Sequeira, C. A. C., *High-Temperature Oxidation*, John Wiley & Sons, Inc., 2011b, pp. 247–280.

Setz, L. F. G., I. Santacruz, M. T. Colomer, S. R. H. Mello-Castanho, and R. Moreno, Fabrication of Sr- and Co-doped lanthanum chromite interconnectors for SOFC, *Materials Research Bulletin*, 46(7), 2011, 983–986.

Shatynski, S. R., The thermochemistry of transition metal carbides, *Oxidation of Metals*, 13(2), 1979, 105–118.

Shatynski, S. R., The thermochemistry of transition metal sulfides, *Oxidation of Metals*, 11(6), 1977, 307–320.

Shimoo, T., K. Okamura, T. Akizuki, and M. Takemura, Preparation of SiC-C composite fiber by carbothermic reduction of silica, *Journal of Materials Science*, 30(13), 1995, 3387–3394.

Shui, A., X. Xi, Y. Wang, and X. Cheng, Effect of silicon carbide additive on microstructure and properties of porcelain ceramics, *Ceramic International*, 37(5), 2011, 1557–1562.

Singhal, S. C., Thermodynamic analysis of the high-temperature stability of silicon nitride and silicon carbide, *Ceramurgia International*, 2(3), 1976, 123–130.

Soszko, M., M. Lukaszewski, Z. Mianowska, K. Drazkiewicz, H. Siwek, and A. Czerwinski, Study of palladium-platinum-rhodium alloys as electrocatalysts in direct methanol fuel cells, *Przemsyl Chemiczny*, 90(6), 2011, 1195–1200.

Stevens, S. J., R. J. Hand, and J. H. Sharp, Temperature dependence of the cristobalite Î ± -Î² inversion, *Journal of Thermal Analysis*, 49(3), 1997, 1409–1415.

Stott, F. H., G. C. Wood, and J. Stringer, The influence of alloying elements on the development and maintenance of protective scales, *Oxidation of Metals*, 44, (1–2), 1995, 113–145.

Stowell, W. R., C. P. Lee, J. F. Ackerman, G. A. Durgin, and R. W. Harris, Article with tailorable high temperature coating, US Patent Number: US 6207295 B1, March 27, 2001.

Subbarao, E. C., *Solid Electrolytes and Their Applications*. Plenum Press: New York, USA 1980.

Sundman, B., B. Jansson, and J. O. Andersson, The Thermo-Calc databank system, *Calphad*, 9(2), 1985, 153–190.

Suresh, A., M. J. Mayo, and W. D. Porter, Thermodynamics of the tetragonal-to-monoclinic phase transformation in fine and nanocrystalline yttria-stabilized zirconia powders, *Journal of Materials Research*, 18(12), 2003, 2912–2921.

Swindells, J. F., *Precision Measurement and Calibration: Selected NBS Papers on Temperature*, U.S. Government Printing Office, 1968, pp. 263–289.

Tanabe, F., Analysis of core melt accident in Fukushima Daiichi-unit 1 nuclear reactor, *Journal of Nuclear Science and Technology* (Tokyo, Japan), 48(8), 2011, 1135–1139.

Tancret, N., S. Obbade, N. Bettahar, and F. Abraham, Synthesis and ab initio structure determination from powder X-ray diffraction data of a new metallic mixed-valence platinum-lead oxide $PbPt_2O_4$, *Journal of Solid State Chemistry*, 124(2), 1996, 309–318.

Tang, Q., S. Ukai, N. Oono, S. Hayashi, B. Leng, Y. Sugino, W. Han, and T. Okuda, Oxide particle refinement in 4.5 mass% Al Ni-based ODS superalloys, *Materials Transactions*, 53(4), 2012, 645–651.

Tanner, L. E. and H. Okamoto, The Pt-Si (platinum-silicon) system, *Journal of Phase Equilibria*, 12(5), 1991, 571–574.

Tavakoli, A., J. A. Golczewski, J. Bill, and A. Navrotsky, Effect of boron on the thermodynamic stability of amorphous polymer-derived Si(B)CN ceramics, *Acta Materialia.*, 60(11), 2012, 4514–4522.

Taylor, D. F., Corrosion-resistant zirconium alloy containing copper, nickel and iron, United States Patent No. 4,986,556, 1991.

Tietz, T. E. and J. W. Wilson, *Behavior and Properties of Refractory Metals*, 1965, Stanford University Press, pp. 1–419.

Tillack, D. and D. Bagnoli, Understanding high-temperature failure modes of metals, *Chemical Engineering (N. Y.)*, 107(11), 2000, 62–68.

Tolstoguzov, N. V., Mechanism and a model of the carbothermic reduction of silica in the manufacture of silicon, *Stal'*, 5, 1989, 36–40.

Ukai, S., T. Okuda, M. Fujiwara, T. Kobayashi, S. Mizuta, and H. Nakashima, Characterization of high temperature creep properties in recrystallized 12Cr-ODS ferritic steel claddings, *Journal of Nuclear Science and Technology*, 39(8), 2002, 872–879.

Ushakov, S. V., A. Navrotsky, Y. Yang, S. Stemmer, K. Kukli, M. Ritala, M. A. Leskelae, P. Fejes, A. Demkov, C. Wang, B. Y. Nguyen, D. Triyoso, and P. Tobin, Crystallization in hafnia- and zirconia-based systems, *Physica Status Solidi B*, 241(10), 2004, 2268–2278.

Varga, T., A. Navrotsky, J. L. Moats, R. M. Morcos, F. Poli, K. Muller, A. Saha, and R. Raj, Thermodynamically stable SixOyCz polymer-like amorphous ceramics, *Journal of the American Ceramic Society*, 90(10), 2007, 3213–3219.

Voevodin, V. N., V. I. Karas, A. O. Komarov, Y. E. Kupriyanova, N. N. Pilipenko, Jr., and B. A. Shilyaev, Phase stability of ODS particles in ferritic martensitic steels, *Voprosy Atomnoj Naukii Tekhniki. Seriya:: Vakuum, Chist. Mater., Sverkhprovodn.*, 19, 2011, 157–174.

Wagner, C., Passivity and inhibition during the oxidation of metals at elevated temperatures, *Corrosion Science*, 5(11), 1965, 751–764.

Wagner, C., Passivity during the oxidation of silicon at elevated temperatures, *Journal of Applied Physics*, 29, 1958, 1295–1297.

Walter, G. F., High temperature materials. A review, Forschungsh. A, A851, (Waermedaemmstoffe im Bauwesen und in Hochtemperaturanlagen), 1999, pp. 9–19.

Wang, F., H. Lou, and W. Wu, The oxidation resistance of a sputtered, microcrystalline TiAl intermetallic compound film, *Oxidation of Metals*, 43(5/6), 1995, 395–409.

Wang, J., H. P. Li, and R. Stevens, Hafnia and hafnia-toughened ceramics, *Journal of Materials Science*, 27(20), 1992, 5397–5430.

Wang, Z, H. Zhao, Z. Tian, and W. Shen, Progress in the research of LaCrO3 connecting material for SOFC, *Dianchi*, 37(5), 2007, 401–403.

Wilkinson, W. D., *Properties of Refractory Metals*, Gordon and Breach Science Publishers, 1969 pp. 1–355, 38. Stowell, W. R., B. A. Nagaraj, C. P. Lee, J. F. Ackerman, and R. S. Israel, "Enhanced Coating System for Turbine Airfoil Applications," U.S. Patent No. 6394755, 2002

Winkler, P. J., *EUROMAT 99, Materials for Transportation Technology*, Federation of European Materials Societies, John Wiley & Sons, 2000, 1–372.

Wright, I. G. and R. B. Dooley, A review of the oxidation behaviour of structural alloys in steam, *International Materials Review*, 55(3), 2010, 129–167.

Wright, I. G., B. A. Pint, and P. F. Tortorelli, High-temperature oxidation behavior of ODS-Fe3Al, *Oxidation of Metals*, 55(3/4), 2001, 333–358.

Yamamoto, Y., M. P. Brady, Z. P. Lu, P. J. Maziasz, C. T. Liu, B. A. Pint, K. L. More, H. M. Meyer, and E. A. Payzant, Creep-resistant, Al2O3-Forming austenitic stainless steels, *Science*, 316(5823), 2007, 433–436.

Yeo, J.-G., S.-C. Choi, J.-W. Kim, J.-E. Lee, J.-H. Lee, and Y.-G. Jung, Thermal reaction behavior of $ZrSiO_4$ and $CaCO_3$ mixtures for high-temperature refractory applications, *Materials Science and Engineering A*, A368(1/2), 2004, 94–102.

Zhang, J., N. Li, Y. Chen, and A. E. Rusanov, Corrosion behaviors of US steels in flowing lead-bismuth eutectic (LBE) , *Journal of Nuclear Materials*, 336(1), 2005, 1–10.

Zhong, S. Y., J. Ribis, V. Klosek, C. Y. de, N. Lochct, V. Ji, and M. H. Mathon, Study of the thermal stability of nanoparticle distributions in an oxide dispersion strengthened (ODS) ferritic alloys, *Journal of Nuclear Materials*, 428(1–3), 2012, 154–159.

Zhou, Z., S. Yang, W. Chen, L. Liao, and Y. Xu, Processing and characterization of a hipped oxide dispersion strengthened austenitic steel, *Journal of Nuclear Materials*, 428(1–3), 2012, 31–34.

3

Refractory Metals, Ceramics, and Composites for High-Temperature Structural and Functional Applications

Jeffrey W. Fergus and Wesley P. Hoffmann

CONTENTS

3.1 Introduction

High temperatures offer advantages or are required for many applications. Materials for a specific high-temperature application are chosen based on their properties at the application temperature, the environment in which they will function, the time at temperature, and the frequency at which they will cycle between temperatures. In this chapter, rather than focusing on a particular group of materials or a particular temperature range, a brief overview of the various classes of high-temperature materials will be given and then some of the general issues associated with high-temperature applications will be discussed and illustrated with some specific examples.

The definition of what constitutes a high temperature is relative as it depends on the situation or application and is different for different materials. For example, high-temperature superconductors operate at less than –100°C, while in the rocket propulsion community a rocket nozzle is not considered to be at high temperature until it reaches 3100°C. Similarly, 550°C is a very high operating temperature for aluminum which melts at 660°C, but a low temperature for tungsten, which does not melt until 3400°C.

3.1.1 Advantages of and Needs for High Temperatures

High temperatures are generated in a variety of applications, such as in high-speed aircraft where friction generates heat. Materials processing, such as casting and heat treatment, require materials to generate the heat and to contain the materials. Energy conversion processes often involve high temperatures—nuclear fusion, for example, occurs only at very high temperatures. In other cases, high temperatures are desired because of improved performance, such as in the conversion of thermal energy into mechanical energy, the efficiency of which is limited according to the Carnot cycle. Similarly, reaction rates increase at high temperatures, so high temperatures are used for the conversion of chemical energy into electrical energy or chemicals into other chemicals.

3.1.2 Challenges Associated with High Temperatures

High temperatures create additional constraints in the design and selection of materials. The phases used are often not the equilibrium phases, in which case, phase stability can be an issue. This can include decomposition of the material into a more stable phase(s) or reaction with a gas in the surrounding environment or reaction with another condensed phase present in the system. Even if the phase is the equilibrium phase, changes in the morphology can occur. For example, high surface areas are needed for catalysts and electrodes. Similarly, fine microstructures often provide good mechanical properties. These high-interfacial-area structures are inherently unstable due to the high surface energy associated with the large interfacial area and can coarsen during high-temperature use.

Synthesis and the processing of high-temperature materials can also be a challenge. Typical material processing techniques, such as casting and sintering, are performed above the application temperature, which is clearly a challenge for high-temperature materials. In addition to synthesis, fabrication, and shaping of the individual materials, joining of high-temperature materials can be difficult.

Performance of materials can also be degraded at high temperatures. This can include mechanical properties, such as strength, or long-time deformation by creep. In addition, functional properties related to transport properties or transduction are temperature dependent.

High-temperature materials are a very broad and expansive subject and a significant number of books have been written on various individual classes of high-temperature materials. Thus, a comprehensive review of high-temperature materials is well beyond the scope of this chapter. Rather, brief descriptions of the different types of these materials, as well as some of the issues identified earlier in the introduction will be presented in the following sections.

3.2 Types of High-Temperature Materials

3.2.1 High-Temperature Metals

In general, metals have much greater ductility and higher densities than either ceramic or nonmetal matrix composites. This means that they are not subject to brittle failure and at the same time have lower specific strength and stiffness. Intrinsically, most metals have higher thermal and electrical conductivities than either ceramics or nonmetal matrix composites (except carbon–carbon composites). The three distinct types of high-temperature metals are superalloys, platinum group metals, and refractory metals.

Superalloys are usually based on group VIIIA elements (Ni, Fe, and Co). They have been developed for service up to 1100°C where they must maintain excellent mechanical strength, phase stability, resistance to creep, high surface stability, and resistance to corrosion and oxidation under relatively severe mechanical stresses. Examples of well-known superalloy families are Hastelloy, Haynes, Inconel, Rene, Monel, Incoloy, and Waspaloy. These alloys are utilized in the hot sections of industrial gas turbines, turbocharger turbines, and marine turbines, as well as for turbine blades in the hot section of jet engines (Figure 3.1).

The platinum group metals (platinum, rhodium, and iridium) with melting points ranging from 1770°C to 2450°C function well under mechanical loads and simultaneous corrosive attack since they are chemically stable, as well as being resistant to oxidation and reaction with many molten oxides. Most people know of the platinum group metals and

(a)   (b)

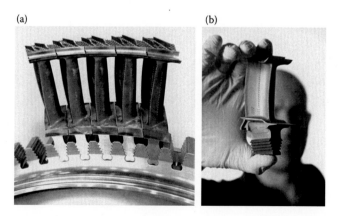

FIGURE 3.1
A jet engine turbine (a) is composed of numerous blades (b) that have internal cooling channels and a thermal barrier coating. (These photographs are reproduced with the permission of Rolls-Royce plc.)

their alloys as thermocouples, but in fact they are indispensable in many areas requiring high temperatures and resistance to chemical attack. Applications include glass melting and spinning equipment, small rocket nozzles for station-keeping, single crystal growing, as well as capsules for radioactive power sources.

The refractory metals: tantalum, niobium, tungsten, molybdenum, and rhenium all have melting points in excess of 2400°C (2477–3400°C), are extraordinarily resistant to heat, wear, and creep as well as being relatively chemically inert. However, they are not as chemically stable and oxidation resistant as the platinum group. This is why iridium is coated on top of the more ductile rhenium as an oxidation protection coating above 2300°C for carbon–carbon composites. These metals are also used, for example, in forging jet engines, rocket nozzles, and incandescent lamp filaments.

3.2.2 Ceramic Materials

By nature of their composition, virtually all ceramic materials are high-temperature materials and some are ultra-high-temperature materials. Ceramics encompass a very broad spectrum of refractory materials that makes a universal definition difficult. In general, ceramic materials are inorganic, nonmetallic materials made from compounds of a metal and a nonmetal. Ceramic materials may be crystalline, partly crystalline, or amorphous (i.e., glasses) and the vast majority are held together with ionic or covalent bonds. However it should be noted that a few ceramics (e.g., TaC) are held together by metallic bonds. In contrast to metals, the strong bonding in ceramics causes these materials to fracture before they are able to plastically deform if the temperature is below the ductile-to-brittle transition (DBT) temperature.

Ceramics enjoy numerous high-temperature applications because they offer a variety of attractive properties including high stiffness, high strength in compression, great thermal stability, low thermal expansion, and extremely high melting temperature. In addition, they are resistant to most forms of chemical (corrosion, oxidation), and physical (abrasion, wear) attack. Applications include furnace elements and insulation, catalytic converters (Figure 3.2), ball bearings, thermal barriers, fuel cells, and fuels for nuclear reactors.

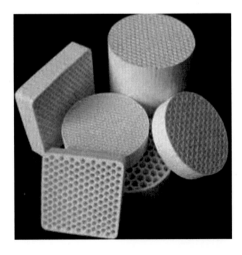

FIGURE 3.2
Catalytic converters consist of a honeycomb ceramic with platinum group catalysts on the surface.

Ultra-high-temperature ceramics (UHTC) are a class of ceramic materials with melting points in excess of 3000°C that are used in environments that require strength as well as environmental resistance to erosion and ablation at extremely high temperatures. Applications for these materials include rocket propulsion, scramjet propulsion, and hypersonic flight either as monolithic structures, composites, or coatings.

Ceramics that meet these criteria are principally carbides and borides of titanium, zirconium, niobium, tantalum, and hafnium. The carbides tend to have lower oxidation resistance than borides at intermediate temperatures due to the formation of CO gas as one of the oxidation products. However, they have higher melting points (at 3997°C, TaC has the highest melting temperature of any material) and perform well in erosive environments such as rocket nozzles. Borides on the other hand, have higher thermal conductivity than many ceramics (60–120 W m^{-1}K^{-1}) (Cutler, 1991) and are used in applications where heat needs to be moved and spread, such as the leading edges of hypersonic vehicles.

3.2.3 High-Temperature Composites

Composites comprise a large family of materials (Harris, 1991) in which a reinforcement phase is placed in a continuous matrix phase resulting in properties that are not possible to obtain with a single material. The reinforcement can be in the form of a particulate, whisker, continuous or discontinuous fiber, nanotube or nanophase particle. In high-temperature composites the reinforcement and the matrix material are limited to ceramics, metals, and carbon. The possibility of combining these various material systems results in almost unlimited variation in composition and properties because the composites are tailorable. Due to the fact that composites are more complex materials and less well known than metals and ceramics, they will be covered in greater detail.

The possible variations in composites, and thus their properties, are further increased by the fact that even after selecting the composition of the matrix and the reinforcement, as well as the type of reinforcement, there are many possibilities in how the reinforcement is placed in the composite. That is, for example, the reinforcement can be dispersed in the matrix (particles, whiskers, discontinuous fibers, etc.) leading to isotropic properties, or in the case of fibers (both continuous or discontinuous) a preform can be fabricated and subsequently filled with a matrix resulting in anisotropic properties. Preforms are fabricated to the desired shape by a variety different means. For example, 1-D preforms as well as 2-D fabric lay-up preforms can be fabricated by hand, whereas braided, 3-D to 4-D (and even up to 11-D) woven preforms, as well as felted preforms require complex looms and other machinery.

It should also be noted that the enhanced properties of composites also result from the function of their constituents. That is, the matrix and the reinforcement play complementary roles. For example, in a fiber-reinforced composite the fibers are the principal load-bearing component. The matrix binds the fibers together, holding them aligned to carry the load, and transfers the load applied to the composite through the fiber–matrix bond to the fibers enabling the composite to withstand compression, flexural and shear forces as well as tensile loads. In addition, the matrix isolates the fibers from one another so that cracks are unable to pass through all the fibers at one time which would result in brittle failure.

When pairing together a matrix material with a fiber, it is important that there be thermal compatibility and chemical compatibility. High processing or use temperatures can lead to matrix (or fiber) cracking during cooling when a thermal expansion mismatch is present. Chemical compatibility also prevents degradation at the fiber–matrix interface at elevated processing, heat-treating, and use temperatures. Degradation can be caused by chemical reactions between the materials or phase changes in either component.

Coatings are often applied to protect the fibers from chemical attack, such as when using carbon fibers in a silicon carbide matrix. Fiber coatings are also employed to tailor the interfacial bond strength, that is, the strength of the bond between the fiber and the matrix. This bond strength has a great effect on composite properties, such as decreasing toughness when it is too strong.

3.2.3.1 Carbon–Carbon Composites

Carbon–carbon (C–C) composites (Sheehan et al., 1994; Fitzer and Monocha, 1998; Savage, 1993) are a unique class of high-temperature and ultra-high-temperature materials due to their extraordinary combination of properties. These materials are stronger and stiffer than steel as well as having a lower density (1.5–1.9 g cm^{-3}) than aluminum. Their thermal shock resistance, and their excellent specific properties coupled with the fact that their mechanical properties actually increase with temperature (>3200°C) make them unsurpassed as ultra-high temperature structural materials.

Owing to their high performance and relatively high cost, C–C composites are used principally in the aerospace and astronautics industries. These materials are used to fabricate rocket nozzles, nose tips, exit cones, leading edges, and engine inlets of hypersonic vehicles (Figure 3.3) as well as high-temperature insulation, furnace hardware (Figure 3.4), and heaters for high-temperature applications. For reuseable hypersonic vehicles, the fact that carbon does not go through phase changes like some ceramics makes it a very valuable material. For satellite applications and high power laser mirrors, the high specific strength and stiffness of carbon–carbon composites as well as their near zero thermal expansion makes them an ideal material for structures that require dimensional stability as they circle the Earth.

The highest volume application of carbon–carbon composites is as stators and rotors in the braking systems of all military aircraft and most civilian jet aircraft (Figure 3.5). These composites enjoy a monopoly in this field because, in addition to the above mentioned properties, they also possess high thermal conductivity, good frictional properties, and low wear. This one application alone in which temperatures can reach 1400°C on a rejected take off makes them the highest volume high-temperature composite. For this reason, they will be used as an example and will be described in much greater detail than the other high-temperature composites.

C–C composites are fabricated through a multi-step process. Being a broad class of materials, the thermal and mechanical properties of carbon–carbon composites vary greatly depending on the type and grade of fibers used, the preform geometry, the type of process utilized for densification, as well as the ultimate graphitization (heat-treat) temperature. The carbon fibers, which carry the majority of the mechanical and thermal load, are chosen on the basis of the desired final properties of the composite. Pitch-based carbon fibers are utilized for high modulus up to (930 GPa) (Fibraplex) and high thermal conductivity (up to 1000 W m^{-1}K^{-1}) (Sullivan, 1998) applications while polyacrylonitrile (PAN)-based carbon fibers are used where high strength (up to 6.96 GPa) (HexTow) is the desired property. Rayon and cellulose-based carbon fibers are required in ablative and insulation applications. The tow size (# of fibers/tow) is specified on the basis of the desired properties and cost.

After the preform geometry has been selected and fabricated, it is then densified with a carbon matrix, which fills the space between the fibers and distributes the load among the

FIGURE 3.3
(**See color insert.**) The Space Shuttle and its launch system require carbon–carbon leading edges and a nose-tip for the shuttle as well as carbon–carbon nozzles and exit cones for the solid rocket boosters.

FIGURE 3.4
Carbon–carbon composite conveyor belt assembly for OXYNON brazing furnace. (Courtesy of Kanto Yakin Kogyo Co. Ltd.)

fibers. For other types of composites the desired final matrix material is simply used to fill the preform. However, in contrast, no carbon matrix material exists. It must be formed by conversion of some sort of hydrocarbon material to carbon by a process that involves a specific type of pyrolysis known as carbonization. Thus, matrix precursor materials are placed in the preform and then converted to carbon. This can involve a variety of processes, such as,

(a)   (b)

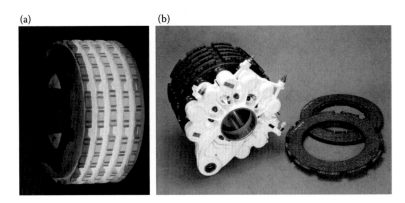

FIGURE 3.5
(See color insert.) (a) During rejected take-off the brakes turn white hot, and the tires sometimes burn. (b) Brake assembly at room temperature along with stator and rotor shown for comparison. (Courtesy of Honeywell International Inc., Aircraft Landing Systems.)

a chemical vapor deposition (CVD) process utilizing methane or propane at elevated temperatures (Benzinger and Huttinger, 1996), the charring of a polymer such as phenolic or polyimide (Ludenbach et al., 1998) as well as the polymerization and coking of a pitch material to produce a high-quality graphitizable matrix. All of these processes have been studied extensively.

The first two processes are fairly well understood, but the third process has been very difficult to elucidate until recently (Burgess and Thies, 2011) because of the large number of oligomeric species (~2000) in the pitch as well as the fact that the pitch material destroys analytical equipment. Carbon–carbon composite aircraft brakes are manufactured by the CVD process. This process requires a long period of time to densify a preform because it is necessary to keep the temperature and gas concentration as low as possible in order to diffuse the hydrocarbon gas into the preform before it deposits on the fiber surfaces. This process is limited to relatively thin (<5 cm) preforms and requires in-process machining to remove the outer dense layer in order for additional in-depth deposition to occur.

The polymer precursor route to a carbon–carbon composite matrix was developed for the Space Shuttle and has been utilized for many decades. In this process, woven carbon fiber fabric is pre-pregged with a polymer (usually phenolic) and then stacked/cured into a 2-D layered composite before carbonization and re-densification. These composites have good in-plane properties (tensile strength of 138–483 MPa and tensile modulus of 34.5–103.4 GPa) (Shih W., Allcomp Inc., private communication, 2012) with much lower inter-laminar properties (tensile strength of 1.4–13.8 MPa) (Shih W., Allcomp Inc., private communication, 2012). Processing of these composites is slow due to the need to keep gas evolution during processing to a minimum in order to avoid delamination. The processing requirements and associated costs keeps the thickness of these composites to less than 100 mm.

The most versatile processing route to densify carbon–carbon composites is to utilize a pitch-based material, which includes petroleum-based pitches, coal tar-based pitches, and various synthetic pitches to impregnate the preform. Although, in contrast to the other methods, commercial pitch processing is able to densify large thick parts, and like CVD produce a graphitizable matrix, they are not without issues. The petroleum and coal tar pitches are viscous and generally do not wet the preform surface. Thus, temperature and pressure are required to fill the preform. In addition, pressure is also needed to keep the matrix precursor in the preform during the conversion process in which pitch is converted into carbon.

To eliminate the need for pressure impregnation, alternate pitch-based precursors must be utilized. The only commercial process employing a synthetic pitch is the *in situ* densification process (Wapner et al., 2007) that is utilized by Allcomp Inc. In this process the preform is placed in a precursor such as naphthalene, which wets the carbon fiber preform and is thus drawn into the preform without the need for pressure. The precursor is then catalytically polymerized inside the preform to form a mesophase pitch that is infusible. Since this process does not require a pressure vessel or in-process machining, the densification process is less expensive and requires much less time.

After the hydrocarbon matrix is placed in the preform, it is carbonized to form the carbon matrix. Depending on the type of precursor matrix material the carbon yield can vary from ~50% (phenolic) to ~85% for a synthetic pitch. Obviously, even if the hydrocarbon matrix completely fills the preform (which it does not), several densification cycles are required. Depending on the application, either after each densification/carbonization cycle or after the last cycle, the densified preform can be graphitized at temperatures in excess of 2200°C in order to improve the properties.

Although carbon–carbon composites are excellent high-temperature structural materials, their Achilles heel is oxidation. Above 425°C, carbon starts to oxidize if unprotected limiting the use of these composites for long-term high-temperature applications. Various oxidation protection coating and inhibitor technologies have been tried, with most not being successful. A silicon carbide coating has performed well on the leading edges of the Space Shuttle where the temperature is below 1500°C and a ZrB_2-SiC coating performs well up to 1800°C. Above this temperature only rhenium coated with iridium has proved to be a successful long-term coating up to 2400°C. Another way to decrease the oxidation rate is to densify the carbon fiber preform with a ceramic matrix, such as, silicon carbide producing a ceramic matrix composite.

3.2.3.2 Ceramic Matrix Composites

Since monolithic ceramics are brittle materials, even microscopic flaws can greatly reduce the strength of the component. This makes them unsuitable for many applications. However, the inherent brittleness of ceramic materials can be overcome by the use of continuous or discontinuous ceramic or carbon fiber reinforcements resulting in a ceramic matrix composite (CMC) (Krenkle, 2008). This dispersed phase is designed to improve toughness by bridging the cracks and keeping the material intact when it fractures. The fibers can also de-bond and slide through the matrix, dissipating energy and preventing fracture.

The reinforcement consists principally of fibers of alumina, silicon carbide, carbon, titanium boride (TiB_2), aluminum nitride (AlN), zirconium oxide (ZrO_2), yttrium–aluminum garnet (YAG), and alumina-silica (mullite), with SiC and carbon fibers being the most popular due to their high strength (3 GPa - SiC) and modulus (270 GPa - SiC) (COI Ceramics). The matrix by definition is a ceramic with the most important commercial matrices being SiC and Al_2O_3. To enhance oxidation resistance oxide (fiber)-oxide (matrix) composites are utilized. Depending on the constituents, these composites are able to function well in an oxidizing environment for long periods at temperatures up to 1400°C (Keller et al., 2005; Parlier et al., 2011).

The ceramic fiber structures can be densified with the matrix material by using various processes such as slurry infiltration, polymer impregnation and pyrolysis (PIP), chemical

FIGURE 3.6
F-16 engine exhaust with 5 CMC divergent seals, identified by arrows. (U.S. Air Force photo.)

vapor infiltration (CVI), melt infiltration (MI), electrophoretic deposition of a ceramic powder, as well as sol–gel processing.

These composites are utilized in high-temperature oxidizing environments, such as in turbines, jet engines (flaps, vanes, seals, flame holders), heat shields, burner tubes, and heat exchangers (Figure 3.6).

3.2.3.3 Metal Matrix Composites

Although metal matrix composites (MMCs) do not function at as high a temperature as carbon and ceramic-based composites, they function well where ductility as well as enhanced mechanical properties are required. These composites consist of a metal matrix (Al, Mg, Ti) reinforced with a continuous or discontinuous fiber (boron or carbon) or particulate (SiC, C, Al_2O_3, B_4C). Because the matrix is metallic these composites also have good thermal conductivity. Since the vast majority of applications utilize an aluminum matrix, these composites are usually employed in temperatures below 500°C with most uses taking advantage of the stiffness and low CTE of these composites. The principal applications above 200°C have been in the automotive industry for use as cylinder liners (Hunt and Miracle, 2001), connecting rods (Chawla, 2006), as well as pistons and crank cases (Kainer, 2006) where the composite's increased strength and modulus, creep and fatigue resistance, hardness, wear and abrasion resistance, as well as thermal shock resistance are required (Figure 3.7). These composites are usually produced employing powder metallurgy techniques.

In traditional composite materials, the microstructures consists of discrete, dispersed, and isolated phases embedded in an otherwise homogeneous matrix material. Only dilute concentrations of a second phase can usually be incorporated, unless one is very clever. In the field of high-temperature composite materials there are variations in which there are no isolated phases, that is, no reinforcing fibers or particles. One such class of materials are the interpenetrating phase composites (IPCs) in which each of the discrete solid phases within the densified composite forms a completely topologically interconnected network throughout the microstructure. The easiest way to visualize these composites is by the fact that if any one of the constituent phases were removed, a self-supporting, reticulated foam would result.

(a) (b)

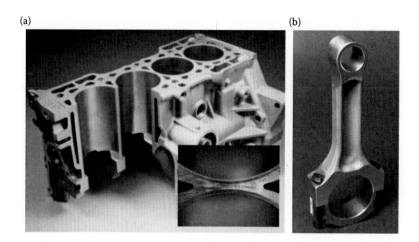

FIGURE 3.7
(a) Engine with integrally cast aluminum MMC cylinder. (b) Aluminum MMC connecting rod. (Reprinted with permission of ASM International. All rights reserved. http://www.asminternational.org.)

In a practical sense, the size of regions of each continuous phase becomes commensurate with the size of a piece of the material. These three-dimensional microstructures consist of an intertwined network of contiguous phases of the various constituents, which results in a novel class of structural or functional materials with truly multifunctional characteristics: each phase contributing its own properties to the macroscopic properties of the composite. For instance, one phase might provide strengthening to the material and the other the required transport property.

"The principal difficulty in fabricating interpenetrating phase composites is in producing the required connectivity and spatial distribution of the two or more component phases, especially on a fine scale where it is not feasible to assemble the microstructure architecturally as is done in lay-up approaches" (Clark, 1992, p. 741), such as those mentioned in Section 3.2.3. Depending upon how these ICPs are formed, their properties can be isotropic or anisotropic.

Although ICPs may consist of combinations of metals, ceramics, and polymers, the vast majority are ceramic–metal systems with Al_2O_3-Al being the most common. This material has application principally in the aeronautics and astronautics fields. Numerous techniques have been developed to fabricate these composites including self-propagating high-temperature synthesis (SHS), squeeze casting, freeform fabrication techniques, sacrificial oxide displacement reactions, directed metal oxidation, spontaneous infiltration, and so on. Just in the field of freeform fabrication techniques 3-D architectures can be formed by direct-write techniques, such as ink-jet printing, micro-pen writing, and robo-casting. In addition, these same architectures have also been formed by fused deposition of layers.

3.3 High-Temperature Stability

The importance of stability is illustrated, albeit exaggerated, in the laws of high-temperature chemistry as proposed by Alan Searcy and amended by John Margrave that (i) at high temperatures, everything reacts with everything else, (ii) the higher the temperature, the more seriously everything reacts with everything else, and (iii) the products might be anything (Spear and Dirkx, 1990).

3.3.1 Reaction with the Environment

One important high-temperature application is the development of materials used to heat other materials and objects—that is, heating elements. Refractory metals, such as tungsten and molybdenum, or graphite can be used in a vacuum, but heating in air requires materials that are stable in air. Metals that do not oxidize at high temperatures, such as platinum, are very expensive, so high-temperature heating elements are typically either oxides or nonoxide ceramics that form protective oxide scales (Zhang et al., 2011). The most common high-temperature heating elements are based on SiC or $MoSi_2$, both of which form silica scales in oxidizing environments. The growth rate of silica is low, so these heating elements can be used for long periods of time. However, the growth rate of silica is sensitive to impurities, so the lifetime can be decreased if the elements are contaminated with, for example, alkali metal ions. In addition, when $MoSi_2$ heating elements are used at temperatures below 1000°C the formation of molybdenum oxide gas species can lead to a nonprotective scale in a process referred to as pesting, which significantly decreases the lifetime. The issues with oxidation can be avoided by using an oxide, such as $LaCrO_3$, which is an electronic conductor, or ZrO_2, which is an ionic conductor. These oxides are relatively brittle and difficult to fabricate which increases cost. In addition, ZrO_2 is an insulator at room temperature, so ZrO_2-based heating elements must be preheated to about 1000°C before current can be passed through the element, which complicates the design of the furnace.

Another important application for high-temperature materials is in gas turbine engines where increasing temperature improves efficiency and enhances performance (Tejedor, 2011). The presence of oxygen and water vapor in combustion requires that materials used in gas turbine engines be oxidation resistant. The components are also under stress, so adequate mechanical strength is needed. There are tight tolerances on the dimensions of turbine engine components, so creep resistance is particularly important to maintain these tolerances during operation. Superalloys, generally described as MCrAlY (M = Fe,Ni,Co), are widely used in turbine engines (Gibbons, 2009). These alloys contain aluminum so that an aluminum oxide scale forms to provide good oxidation resistance. Chromium additions reduce the amount of aluminum required and also improve the strength, while yttrium additions improved the adherence of the alumina scale.

As shown in Figure 3.8, increasing pressure also improves conversion efficiency as in ultra-supercritical (USC) boilers (Viswanathan et al., 2009). The increase in pressure increases stress, so that mechanical properties become more important. Because of the expected long operational lifetimes, creep resistance is the critical property in materials selection (Jetter et al., 2011).

The operating temperature of superalloys can be increased with the application of a thermal barrier coating (TBC) (Wellman and Nicholls, 2007). The most common TBC coatings are based on ytrria stabilized zirconia, which has low thermal conductivity and a coefficient of thermal expansion that reasonably matches those of the alloys. As shown in Figure 3.9, the coatings typically have a columnar microstructure to accommodate thermal expansion differences between the alloy and coating by directing cracking perpendicular, rather than parallel to the alloy-coating interface, which improves the resistance to coating spallation during thermal cycling. A bond coat is applied to improve adherence between the TBC and the growing alumina scale, which is referred to as the thermally

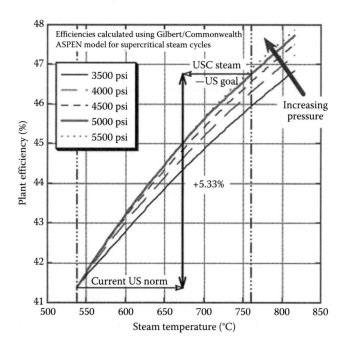

FIGURE 3.8
Effects of temperature and pressure on power plant efficiency. (From Gibbons T.B., *Mater. Sci. Tech.*, 25, (2), 2009, 129–135.)

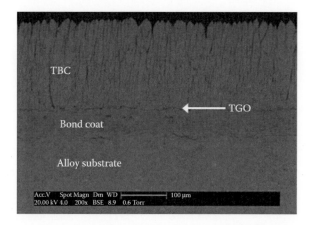

FIGURE 3.9
Scanning electron micrograph of thermal barrier coating. TBC = thermal barrier coating, TGO = thermally grown oxide. (From Wellman R.G., and Nicholls J.R., *J. Phys.* D, 40(16), 2007, R293–R305.)

grown oxide (TGO). Thus, understanding and control of the interfaces between the components of the thermal barrier systems is important to improve performance (Marino et al., 2011).

Further increases in operating temperature can be achieved with alloys of refractory metals, such as molybdenum (Heilmaier et al., 2009). In oxidizing environments, alloying additions, such as silicon, are needed to promote the formation of a protective oxide scale, as discussed above in the case of MoSi$_2$ heating elements. Ceramic compounds, such as borides, carbides, and nitrides, provide even higher operating temperatures (Wuchina et al., 2007). Such ceramic compounds are typically brittle, so improvements in the mechanical properties are needed. As discussed above, the toughness of ceramic materials can be improved by combining multiple ceramic components to inhibit crack growth through crack deflection or compressive stress generation from phase transformations (Butler and Fuller, 2004). Ceramics typically have good high-temperature stability, but can degrade in some environments. For example, Si$_3$N$_4$ is a promising ceramic material (Bocanegra-Bernal and Matovic, 2010), but still needs an environmental barrier coating (EBC) in some corrosive or abrasive applications (Klemm, 2010).

Another important energy conversion application for high-temperature materials is nuclear power generation, which includes not only high temperatures but also radiation exposures. There are materials challenges in current fission-based nuclear reactors as well as fusion-based systems that are under development (Zinkle and Busby, 2009; Allen et al., 2010; Was, 2007). The ranges of temperatures and radiation exposures for current and future nuclear power systems are summarized in Figure 3.10.

High-temperature materials applications in fission reactors include components involved directly in the energy conversion, such as the graphite moderator (Bonal et al., 2009) or uranium oxide fuels (Haertling and Hanrahan, 2007), which are exposed to high radiation levels. Radiation exposure can also affect the supporting structural materials by causing embrittlement or inducing void swelling (Chant and Murty, 2010). Ferritic–martensitic stainless steels are widely used because of their low coefficients of thermal expansion, high thermal conductivity, and resistance to radiation-induced void swelling. As with combustion-based energy conversion, increased operating temperatures increases efficiency, so very high-temperature reactors (VHTRs) have been developed that place increased demands on the corrosion resistance of the materials (Cabet and Rouillard, 2009). For example, oxide dispersion strengthened (ODS) alloys are used to improve creep resistance (Allen et al., 2009). The improved creep resistance of ODS alloys relies on the dispersion of very

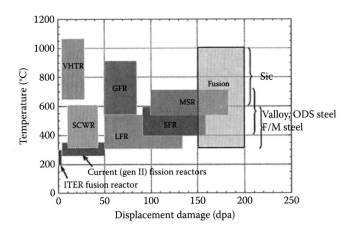

FIGURE 3.10
Overview of operating temperatures and displacement damage dose regimes for structural materials in current (Generation II) and proposed future (Generation IV) fission and fusion energy systems. The six Gen IV fission systems are very high temperature reactor (VHTR), super critical water reactor (SCWR), lead fast reactor (LFR), gas fast reactor (GFR), sodium fast reactor (SFR), and molten salt reactor (MSR). (From Zinkle S.J., and Busby J.T., *Materials Today (Oxford, UK)*, 12, (22), 2009, 12–19.)

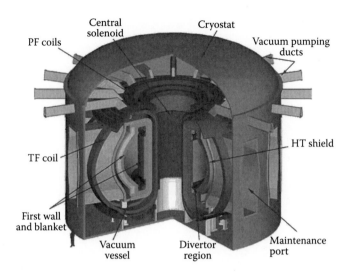

FIGURE 3.11
(**See color insert.**) The ARIES-AT fusion power core. (From Najmabadi F. et al., *Fusion Eng. Design*, 80, (1–4), 2008, 3–23.)

small oxide particles, which requires special processing to prevent particle agglomeration and thus increases cost.

The temperatures in fusion reactors are even higher and create numerous materials challenges (Baluc et al., 2007; Barabash, 2007). One challenge is the plasma-facing material (labeled as first wall and blanket in Figure 3.11), which is exposed directly to the fusion plasma at a high temperature and is under a high radiation flux. Tungsten is the primary candidate for this application due to its high-temperature strength, good thermal conductivity, and low sputtering rate (Rieth et al., 2011). SiC and SiC-composites are also attractive due to their high-temperature stability and irradiation resistance (Katoh et al., 2007). In addition, as in fission energy conversion, supporting structural materials are exposed to radiation exposure. Another materials challenge is corrosion by molten alkali metals used as the coolants (Bloom et al., 2007).

3.3.2 Vaporization

The high entropy of gases relative to condensed phases leads to vaporization at high temperatures. In addition to simple congruent vaporization, which depends simply on the vapor pressure of the relevant species, gaseous phases can form due to oxidation or reduction. This includes oxidation of metals or elements, such as the oxidation of carbon. In addition, oxides, including bulk oxides and the oxide scales formed on other materials, can be oxidized or reduced. For example, Cr_2O_3 can be oxidized to CrO_3 or $CrO_2(OH)_2$ vapor species in oxidizing atmospheres and SiO_2 can be reduced to SiO in reducing atmospheres. Volatilization can be increased if water vapor is present in the case of silica (Opila and Meyers, 2003), alumina (Opila and Meyers, 2004) and other ceramics (Nguyen, 2004). In addition, in applications with limited volumes for the gases evolved, even a small amount of vaporization can be important. For example, volatilized chromia, at levels that would be insignificant to the recession of the metal below a protective scale, can deposit at the cathode in solid oxide fuel cells and result in degradation of the fuel cell performance (Fergus, 2007).

3.3.3 Ablation

Ablation involves the removal of material from a surface encompassing processes, such as, erosion and vaporization. The high-speed flow of a gas across a surface, as in hypersonic aircraft, not only creates heat, which can accelerate oxidation, but also can lead to removal of material by ablation. For hypersonic applications the material is usually a carbon–carbon composite, a ceramic material, or a ceramic coating (Squire and Marschall, 2010). Ablation also takes place in the nozzle in a solid rocket motor burning aluminized propellant. In this case the aluminum in the propellant burns and oxidizes to molten alumina droplets that impinge on the carbon/carbon nozzle leading to erosion.

In these two situations, ablation is a negative process leading to a shorter life for the component. Ablation is also employed as a process for protection and cooling resulting in longer component life. For example, ablative materials are used in heat shields for atmospheric reentry (Figure 3.12) and in solid rocket motors to protect the entrance of the nozzle from erosion and failure. Thus, these materials usually consist of a low-density carbon-phenolic (Pulci et al., 2010) or carbon–silica–phenolic composite that acts not only as insulation from the intense heat but also removes heat by volatilizing in a controlled manner.

Another application of high-temperature ablation is in cutting tools for machining operations. In this case, very hard materials such as tungsten carbide, titanium carbide, or tantalum carbide are employed (Figure 3.13). These materials are usually coated onto a tool or alternatively mixed with a metal, such as cobalt to make a replaceable insert more durable and last longer. The object of the process is to remove material from the part being machined so the cutting tool must be very hard. Since temperatures can reach 1000°C these must also be high-temperature materials.

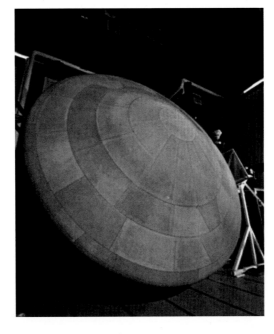

FIGURE 3.12
Ablative heat shield design used for Apollo and Orion spacecraft. (Courtesy of NASA.)

(a)　　　　　　　　(b)

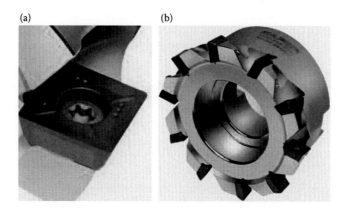

FIGURE 3.13
Metallic cutting tool with replaceable ceramic inserts.

3.3.4 Reaction among Components

In addition to decomposition, phase changes, or reaction with the surrounding atmosphere, reactions can occur between different components in a system. One example is in high-temperature sealants, which must maintain good contact with the materials being sealed, but also not form detrimental reaction products. For example, planar solid oxide fuel cells require a sealant between the electrodes, which must separate the fuel at the anode from air at the cathode (Fergus, 2005; Lessing, 2007). One approach is to use a glass or glass-ceramic that is designed with sufficient rigidity to provide a seal, but also with sufficient flow to accommodate stresses during thermal cycling. To attain the proper balance between these different requirements, the composition of the glass must be controlled carefully, since changes in the composition due to reaction with other components can lead to crystallization or other changes in the properties of the glass and degrade the seal, even if a separate reaction product phase is not formed.

3.4 High-Temperature Performance

High-temperature materials are often selected to perform certain functions in an application, so the temperature dependence of those properties must be considered in materials selection and optimization.

3.4.1 Mechanical Properties

There are, of course, many mechanical properties that are important for materials utilized in high-temperature applications. Considering that the important mechanical properties for each material are application and temperature dependent, and data for many materials is limited, only a brief comparison will be made using the tensile strength and the elastic (Young's) modulus for select materials.

TABLE 3.1

Representative Mechanical Properties of Some High-Temperature Metals

Metal	Tensile Strength (GPa)		Tensile Modulus (GPa)	
	Room Temperature	Elevated Temperature	Room Temperature	Elevated Temperature
Platinum Group				
Platinum	0.12–0.14		165	133 (900°C)
Iridium	1	0.31 (1000°C)	526	430 (900°C)
Rhodium	0.7		372	282 (900°C)
Refractory Group				
Tantalum	0.24–0.82	0.10 (1000°C)	186	152 (1000°C)
Tungsten	1.30–1.51	0.32 (1000°C)	4	3.58 (1000°C)
Molybdenum	0.48–0.69	0.26 (1000°C)	325	
Rhenium	1.13		460	
Superalloys				
Hastelloy C-276	0.79	0.61 (538°C)	205	175 (538°C)
Haynes 718	1.4	0.07 (1000°C)	200	117 (1000°C)
Inconel 600	0.66	0.55 (550°C)	207	
Inconel 718	1.24		211	
Waspaloy	1.27–1.44	0.083 (1000°C)	213	146 (1000°C)

3.4.1.1 Metals

Metals are alloyed to improve properties and in some cases (e.g., W-Re) to save cost. Since there are an extraordinary number of possible alloy compositions and very limited space in this chapter, only a survey of the properties of some pure metals and some superalloys will be presented (Table 3.1). It should be noted that, in general, the properties of metals depend on their composition and crystalline structure, as well as their heat treatment and mechanical treatment (hot working and cold working). Additionally, mechanical properties vary with temperature.

The strength and modulus of some representative high-temperature metals are presented in Table 3.1 with values given at room temperature and at the elevated temperature specified. It can be seen that tensile strength and modulus decrease with an increase in temperature.

3.4.1.2 Ceramics

For ceramics, the mechanical properties depend on the composition, crystalline structure, grain size, purity, and manufacturing process, which make it rather difficult to obtain reliable values from the literature only. One of the reasons for this variation is the strong dependence of fracture on flaw size so the distribution of flaw sizes, as typically described by the Weibull modulus, is needed to characterize the distribution of failure strengths. The surface quality of the sample is particularly important in this respect, since surface flaws can act as crack initiation sites and significantly decrease the failure stress. Thus, only ranges for a few materials will be provided for bulk materials. Aluminum oxide has a tensile strength of 240–280 MPa and a tensile modulus of ~380–430 GPa while for carbides of silicon and titanium the tensile strength is in the range 240–260 MPa with a modulus in

TABLE 3.2

Mechanical Properties of Some High-Temperature Ceramic Fibers

Fiber	Tensile Strength (GPa)	Tensile Modulus (GPa)	Reference
Alumina			
Alpha (Nextel 610)	3.1	380	Nextel
Mullite			
Nextel 550	2	193	Nextel
SiC			
Nicalon	3	210	COI ceramics
Hi-Nicalon	2.8	270	

the range of 275–495 GPa. Tantalum has a higher strength, which is in the range of 350–520 MPa but a similar modulus, which is in the range of 280–450 GPa. Titanium diboride has a lower strength than the carbides (120–130 MPa) but a higher modulus (365–575 GPa).

Ceramic fibers are used to reinforce both metal and ceramic matrices. In a fiber-reinforced composite the fibers typically have a significantly higher tensile strength and modulus than either the bulk form of the same material or the matrix because they have fewer and smaller flaws as well as a decreased gain size. However, it is usually possible to obtain only a fraction of the mechanical properties of the fibers from the composite because of the fiber matrix bond, as well as the lower properties of the bulk matrix itself. Ceramic fiber properties at room temperature are presented in Table 3.2. Properties of various composites will not be presented since they are very numerous and diverse and the values would be misleading without a complete description of the composite and its processing history, which is rarely available.

3.4.1.3 Composites

In the area of composites, there is tremendous tailorability of the mechanical properties depending on the materials chosen for the reinforcement and the matrix, their processing history (especially the heat-treatment temperature), as well as the orientation and volume fraction of the reinforcement. Of course, the measurement is also dependent on its orientation to the fiber direction.

Taking carbon–carbon composites as an example, one can see that there is a very great variability of strength and modulus in the fibers themselves as can be seen in Table 3.3. This variability is due in great part to the precursor, method of manufacture, the maximum heat-treatment temperature, and for the spun fibers, the amount of stretching at elevated temperature during manufacture.

It should be remembered that only a fraction of the properties of fibers are translated into the properties of the composite that they reinforce. Properties of composites will not be provided for the reasons given above. It is also important to know that above 400°C in an oxidizing environment carbon will oxidize. Thus, although some carbon fibers have high-temperature properties superior to those of metal and ceramic fibers and, in contrast to metals and ceramics, the properties of carbon fibers increase to 3400°C, in an oxidizing environment above 425°C these properties will decrease. A partial solution is to protect the exterior of the composite from oxidation or to utilize a ceramic matrix such as silicon carbide.

TABLE 3.3

Mechanical Properties of Carbon Fibers

Fiber	Tensile Strength (GPa)	Tensile Modulus (GPa)	Reference
Pitch-Based Fibers			
Thornel P-25	1.6	159	
Thornel P-30	1.6	207	Thornel-Pitch
Thornel P-55	1.4	414	
K-1100	3.2	930	Fibraplex, Sullivan (1998)
PAN-Based Fibers			
Thornel T 300	3.8	231	Thornel-PAN
Thornel T650	4.3	255	
Torayca T1000G	6	294	Torayca
IM10	7	303	HexTow
Vapor Grown			
Pyrograph-I	7	600	Lake (1996)
Nanotubes	30–150	800–1400	Yu (2000), Treacy (1996), Demczyk (2002)
Glassy Carbon			
Phenol-hexamine	2	7	Kawamura (1970)

3.4.2 Ionic Conduction

Ionic conduction in solids is thermally activated and thus increases with increasing temperature, which allows for their use in high-temperature electrochemical devices, such as fuel cells (Fergus, 2006). High temperatures are also good for reducing the degree of crystallographic ordering or clustering of defects, which can decrease the ion mobility (Navrotsky, 2010). High temperatures also increase the electrode kinetics, which allows for the use of a wide variety of fuels as in the solid oxide fuel cells or molten carbonate fuel cells shown schematically in Figure 3.14 (McPhail et al., 2011). This fuel flexibility expands the range of potential applications to include the use of hydrocarbon fuels, such as natural gas, which can be used after reformation or directly with internal reformation. Similarly, gaseous fuels can be produced from reformation of liquid fuels, such as diesel fuel or JP-8, or biomass sources.

Other high-temperature electrochemical devices include those using electrodes with molten sodium, such as a sodium sulfur battery (Sudworth and Tilley, 1985; Oshima et al., 2004), which is recently drawing attention for use in energy storage systems for use with renewable energy conversion technologies, such those based on wind and solar energy conversion. Molten sodium has also been used for the conversion of thermal energy based on the flow of sodium under a temperature gradient (Cole, 1983). The most widely used solid electrolyte material for these applications is beta alumina, which is a layered sodium–aluminum oxide with high sodium ion conductivity (Fergus, 2012a; Lu et al., 2010). As noted above, high-temperature materials are important for many clean energy technologies, but they are also used in reducing emission through carbon sequestration. For example, lithium-based ceramic materials can be used to absorb CO_2 produced from combustion processes and thus decrease the associated carbon footprint (Nair et al., 2009).

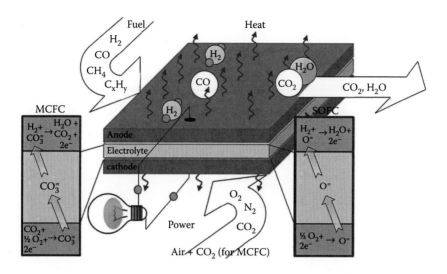

FIGURE 3.14
Schematic overview of a high-temperature fuel cell, with anode and cathode electrochemical reactions in the insets. When hydrocarbons are utilized as fuel, the anode can act as a reforming stage by $C_xH_y + xH_2O \rightarrow (1/2x + y) H_2 + xCO$. (From McPhail S.J. et al., *Int. J. Hydrogen Energy*, 36(16), 2011, 10337–10345.)

3.4.3 High-Temperature Electronics

High temperatures resulting from either the surrounding ambient environment or from high power use can affect the performance of electronic devices, so some of the materials used in conventional silicon-based electronic devices are not adequate. In particular, semiconductors with larger bandgaps and improved stability are needed for high power or high-temperature applications. High-temperature devices are typically based on SiC (Xuan et al., 2008; Chin et al., 2010) and diamond (Wort and Balmer, 2008), Diamond has the same crystal structure as silicon and SiC forms a structure with the same atomic positions as silicon, but with half of the site occupied by silicon and half by carbon. Maintaining ferromagnetic properties at high temperatures is also a challenge so alternative approaches, such as doping transition metal ions on silicon-like structures have been developed (Bonanni and Dietl, 2010).

3.4.4 Thermal Conduction

Thermal conductivity is an important parameter in many high-temperature applications. In some cases, such as heat exchangers, heat spreaders, and leading edges, high thermal conductivity is desired. Carbon-based materials can provide good thermal conduction and high thermal stability (Wang et al., 2012). In other cases, such as with thermal barrier coatings, ablatives, and insulations, low thermal conductivity is required. With the broad range of materials available there is a high-temperature material to match almost any application.

3.4.4.1 Metals

Pure metals in general have much higher thermal conductivities than ceramics, with copper and silver having the highest values of ~400 W m^{-1}K^{-1} at 25°C. However, some

TABLE 3.4

Representative Thermal Conductivity of
High-Temperature Metals at Room Temperature

Metal	Thermal Conductivity (W m⁻¹K⁻¹)
Platinum Group	
Platinum	72
Iridium	148
Rhodium	150
Refractory Group	
Tantalum	54
Tungsten	129
Molybdenum	142
Rhenium	71
Superalloys	
Hastelloy C-276	10
Haynes 718	11
Inconel 600	15
Inconel 718	12
Waspaloy	11

high-temperature metals, such as the superalloys, generally have much lower room temperature thermal conductivities as seen in Table 3.4.

3.4.4.2 Ceramics

The thermal conductivity of ceramics cover a broad range and can vary from 0.6 to 3 W m⁻¹K⁻¹ at 25°C for zirconia, ~180 W m⁻¹K⁻¹ at 25°C for titanium carbide, and >~300 W m⁻¹ K⁻¹ at 25°C for boron nitride. Thus, zirconia is used, for example, as a thermal barrier coating on turbine blades and boron nitride is used in heat sink applications. It should be noted, however, that for ceramics the thermal conductivity depends on the composition, crystalline structure, orientation within crystalline structure, grain size, purity and manufacturing process, which makes it rather difficult to obtain a reliable value from the literature only since all these details are rarely given. For this reason, only a few examples will be provided. Although there is little data for ceramic fiber composites, it should be noted that the thermal conductivity of silicon carbide fibers (Nicalon) is in the range of 3–8 W m⁻¹K⁻¹ at 25°C (COI Ceramics) and for alumina and mullite fibers (Nextel) the values are <1 W m⁻¹K⁻¹ at 25°C (Nextel). Since these values are so low, the composite value can be higher due to the matrix contribution. For all materials, the thermal conductivity varies with temperature in a nonlinear fashion.

3.4.4.3 Composites

In the area of composites, there is tremendous tailorability of the thermal conductivity depending the materials chosen for the reinforcement and the matrix, their processing history (especially the heat-treatment temperature), as well as the orientation and volume fraction of the reinforcement. Of course, the measurement is also dependent on its orientation to the fiber direction.

In a carbon fiber-reinforced composite, the fibers typically have a significantly higher thermal conductivity than the matrix because they have fewer flaws and defects. In addition, because of the fiber matrix bond as well as the properties of the bulk matrix itself, it is never possible to obtain the thermal conductivity of the fibers from the composite measurement with high accuracy.

Taking carbon–carbon composites as an example, there is a great variability of thermal conductivity in the fibers themselves as can be seen in Table 3.5. This variability is due in great part to the precursor, method of manufacture, the maximum heat-treatment temperature, and for the spun fibers the amount of stretching at elevated temperature.

By combining various fiber–matrix material pairs along with controlling fiber orientation, fiber volume fraction, as well as heat-treatment temperature, one is able to obtain a tremendous range of thermal conductivities for carbon–carbon composites. For example, utilizing a rayon-based fiber in a glassy carbon matrix, a thermal conductivity in the range of 10–20 W m^{-1}K^{-1} results. On the other end of the continuum, a uni-axial composite utilizing Pyrograph fibers (Lake, 1996) in combination with *in situ* densification with synthetic pitch (Wapner et al., 2007) resulted in a thermal conductivity of 1166 W m^{-1}K^{-1} along the fiber axes. Thus, with highly graphitic matrices, such as those from CVD or pitch, it is possible to get up to 60% of the thermal conductivity of the fiber along the fiber direction. With one exception (Ohlhorst et al., 1977) values for individual composites will not be given because material processing is proprietary and these materials have military application. Thus, although some composite properties are published, their processing histories on which these properties depend are not.

TABLE 3.5

Thermal Conductivity of Carbon Fibers

Fiber	Thermal Conductivity (W m^{-1}K^{-1})	Reference
Pitch-Based Fibers		
Thornel P-25	36	Thornel-Pitch
Thornel P-30	62	
Thornel P-55	120	
Thornel P120	650	
K-1100	950–1100	Fibraplex, Sullivan (1998)
PAN-Based Fibers		
Thornel T 300	8	Thornel-PAN
Thornel T650	14	
Torayca T1000G	32	Torayca
Rayon		
Thornel P-50	60	Kalnin (1972)
Vapor Grown		
Pyrograph-I	1950	Lake (1996)
Bezene derived	1380	Nysten (1985)
Nanotubes	~6600	Yu (2000), Treacy (1996), Demczyk (2002)

In summary, for carbon–carbon composites, low thermal conductivity composites can be employed as excellent thermal insulators for use as furnace insulation or in an ablative application whereas, in contrast, if their thermal conductivity is high, they can be used as thermal heat sinks and thermal spreaders for electronics. In all cases it should be noted that the thermal conductivity of materials varies nonlinearly with temperature.

3.4.4.4 Thermoelectric Energy Conversion

Another application in which low thermal conductivity is needed is thermoelectric energy conversion. In thermoelectric energy conversion, the flow of heat through the device directly generates an electrical current. The primary advantage of thermoelectric conversion is that there are no moving parts, which leads to good reliability. However, the efficiency is relatively low. The efficiency of the device is related to a figure of merit, which increases with increasing electrical conductivity, increasing Seebeck coefficient and decreasing thermal conductivity, so low thermal conductivity is desired for this application. The best thermoelectric materials are based on compounds of relatively uncommon elements, such as antimony, tellurium, and germanium. These compounds tend to oxidize at high temperatures, so oxide thermoelectric materials are needed. Both p-type ($Co_3O_4O_9$, Na_xCoO_2) and n-type ($CaMnO_3$, ZnO, $SrTiO_3$) oxides with good thermoelectric properties have been reported (Ohtaki, 2011; Fergus, 2012b). Although the figures of merit of these oxides are not as high as those of compounds based on antimony, tellurium, and germanium, the stabilities are much better. Thus, if the fabrication costs could be reduced sufficiently to produce cost effective device, these oxide materials could enable thermoelectric energy conversion in high-temperature applications, such as automotive exhaust systems. The maximum theoretical efficiencies for some thermoelectric oxide materials are summarized in Figure 3.15.

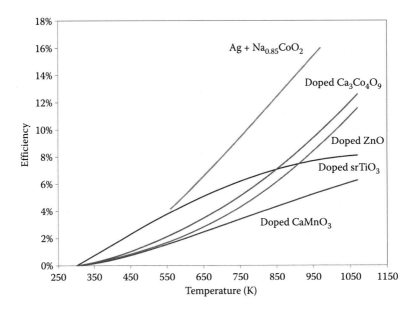

FIGURE 3.15
Maximum theoretical efficiency for oxide thermoelectric materials. (From Fergus J.W., *J. Eur. Ceram. Soc.*, 32, 2012b, 525–540.)

3.4.5 Transduction Properties

There are many high-temperature applications where high-temperature sensors or actuators are needed. For example, information on chemical concentrations is important for many high-temperature applications. Controlling combustion processes to obtain high efficiency requires information on the gas composition. The largest volume application for monitoring combustion is for automotive internal combustion engines, which is monitored by exhaust gas sensors consisting of an electrochemical cell based on a yttria stabilized zirconia solid electrolyte. While oxygen can be detected with a simple galvanic cell using an oxygen-conducting electrolyte, other species can be detected with more complex designs using alternative electrode or electrolyte materials (Schulz et al., 2010). The same approach has been used to measure dissolved gases and alloy additions in molten metals (Fergus, 2009). As noted above, the ionic conductivity of solid electrolytes increases with increasing temperature. However, the electronic conductivity also increases with increasing temperature and electronic conduction in the electrolyte can lead to an erroneous sensor signal. The most common electrolytes for chemical sensors are the oxygen ion-conducting electrolyte, yttria-stabilized zirconia, and the sodium ion-conducting electrolyte, beta alumina.

There are also needs for mechanical sensing and actuation at elevated temperatures. Piezoelectric materials provide a relation between deformation in a crystal and electric field, which can be used for sensing (deformation generates a voltage) or actuation (voltage generates a deformation). High-temperature piezoelectric materials include oxides that form the langasite (Tortissier et al., 2011) and Aurivillius (Moure et al., 2009) structures.

Another high-temperature sensor challenge is measuring temperature. Temperature can be measured with a thermocouple of from the radiation spectrum (i.e., infrared or optical pyrometer). Another approach is to use the change in the ultrasonic response with temperature and has been shown to measure temperatures up to 3000°C with a tungsten-based device (Laurie et al., 2010).

3.5 Summary/Conclusions

High temperatures are needed and provide advantages in a variety of applications, but also place additional constraints and demands on the materials used. In addition to inherent stability at high temperatures, mechanical, transport and other properties are affected by temperature. Developing materials that meet these demands and constraints is needed to enable high-temperature technologies in energy conversion, transportation, and other fields.

Acknowledgments

The authors thank James Gordon Hemrick, Oak Ridge National Laboratory, Oak Ridge, Tennessee; Vilupanur A. Ravi, California State Polytech University, Pomona, California; Mark L. Weaver, University of Alabama, Tuscaloosa, Alabama; and Eric Wuchina, Office of Naval Research (ONR), Arlington, Virginia, for reviewing this chapter and for providing valuable technical comments and suggestions. Also, the authors thank Edgar Lara-Curzio, Oak Ridge National Laboratory, Oak Ridge, Tennessee, for helping identify expert reviewers for this chapter.

References

Allen T., Burlet H., Nanstad R.K., Samaras M., and Ukai S., Advanced structural materials and cladding, *MRS Bull*, 34(1), 2009, 20–27.

Allen T., Busby J., Meyer M., and Petti D., Materials challenges for nuclear systems, *Mater. Today (Oxford, UK)*, 13(12), 2010, 14–23.

Baluc N., Abe K., Boutard J.L., Chernov V.M., Diegele E., Jitsukawa S., Kimura A. et al., Status of R&D activities on materials for fusion power reactors, *Nucl. Fusion*, 47(10), 2007, S696–S717.

Barabash V., Peacock A., Fabritsiev S., Kalinin G., Zinkle S., Rowcliffe A., Rensman J. W. et al., Materials challenges for ITER—Current status and future activities, *J. Nucl. Mater*, 367–370(A), 2007, 21–32.

Benzinger W., and Huttinger K.J., Chemical vapour infiltration of pyrocarbon: I. Some kinetic considerations, *Carbon*, 34(12), 1996, 1465–1471.

Bloom E.E., Busby J.T., Duty C.E., Maziasz P.J., McGreevy T.E., Nelson B.E., Pint B.A., Tortorelli P.F., and Zinkle S.J., Critical questions in materials science and engineering for successful development of fusion power, *J. Nucl. Mater*, 367–370(A), 2007, 1–10.

Bocanegra-Bernal M.H., and Matovic B., Mechanical properties of silicon nitride-based ceramics and its use in structural applications at high temperatures, *Mater. Sci. Eng. A*, A527(6), 2010, 1314–1338.

Bonal J.-P., Kohyama A., van der Laan J., and Snead L.L., Graphite, ceramics, and ceramic composites for high-temperature nuclear power systems, *MRS Bull.*, 34(1), 2009, 28–34.

Bonanni A., and Dietl T., A story of high-temperature ferromagnetism in semiconductors, *Chem. Soc. Rev.*, 39(2), 2010, 528–539.

Burgess, W.A., and Thies M.C., Molecular structures for the oligomeric constituents of petroleum pitch, *Carbon*, 49(2), 2011, 636–651.

Butler E.P., and Fuller E.R. Jr., Ceramic–matrix composites, *Kirk-Othmer Encyclopedia of Chemical Technology* (5th Ed.). Vol. 5, John Wiley & Sons Inc, Hoboken, NJ 2004, pp. 551–581.

Cabet C., and Rouillard F., Corrosion of high temperature metallic materials in VHTR, *J. Nucl. Mater.*, 392(2), 2009, 235–242.

Chant I., and Murty K.L., Structural materials issues for the next generation fission reactors, *JOM*, 62(9), 2010, 67–74.

Chawla N., Metal matrix composites in automotive applications, *Adv. Mater. Process.* July, 2006, 29–31.

Chin H.S., Cheong K.Y., and Ismail A.B., A review on die attach materials for SiC-based high-temperature power devices, *Metall. Mater. Trans. B*, 41B(4), 2010, 824–832.

Clark D.R., Interpenetration phase composites, *J. Am. Ceram. Soc.* 75(4), 1992, 739–59.

COI Ceramics, Hi-Nicalon Ceramic Fiber Data Sheet, COI Ceramics, Inc., www.coiceramics.com

Cole T., Thermoelectric energy conversion with solid electrolytes, *Science*, 221, 1983, 915–920.

Cutler R.A., Engineering properties of borides, 787–803 in *Ceramics and Glasses: Engineered Materials Handbook* Volume 4, ed. by S.J. Schneider, Jr., ASM International, Materials Park, OH 1991.

Demczyk B.G., Wang Y.M., Cumings J., Hetman M., Han W., Zettl A., and Ritchie R.O., Direct mechanical measurement of the tensile strength and elastic modulus of multiwalled carbon nanotubes, *Mats. Sci. Eng.*, A334, 2002, 173–178.

Fergus J.W., Sealants for solid oxide fuel cells, *J. Power Sources*, 147(1–2), 2005, 46–57.

Fergus J.W., Electrolytes for solid oxide fuel cells, *J. Power Sources*, 162(1–2), 2006, 30–40.

Fergus J.W., Effect of cathode and electrolyte transport properties on chromium poisoning in solid oxide fuel cells, *Int. J. Hydrogen Energy*, 32(16), 2007, 3664–3671.

Fergus J.W., Electrochemical sensors: Fundamentals, key materials and applications, *Handbook of Solid State Electrochemistry: Fundamentals, Methodology and Recent Advances*, V. Kharton (Ed.) (Wiley-VCH) 2009, 427–491.

Fergus J.W., Ion transport in sodium ion conducting solid electrolytes, *Solid State Ionics*, 227, 2012a, 102–112.

Fergus J.W., Oxide materials for high temperature thermoelectric energy conversion, *J. Eur. Ceram. Soc.*, 32, 2012b, 525–540.

Fibraplex, *Comparison of Carbon Fiber Manufacturer's Products*, http://www.fibraplex.com/tow.asp

Fitzer E. and Manocha L.M., *Carbon Reinforcements and Carbon/Carbon Composites*, Springer-Verlag, Berlin Heidelberg GmbH 1998.

Gibbons T.B., Superalloys in modern power generation applications, *Mater. Sci. Tech.*, 25(2), 2009, 129–135.

Haertling C., and Hanrahan R.J., Literature review of thermal and radiation performance parameters for high-temperature, uranium dioxide fueled cermet materials, *J. Nucl. Mater.*, 366(3), 2007, 317–335.

Harris B., *Engineering Composite Materials*, The Institute of Materials, London 1999.

Heilmaier M., Krueger M, Saage H, Roesler J., Mukherji D., Glatzel U., Voelkl R. et al., Metallic materials for structural applications beyond nickel-based superalloys, *JOM*, 61(7), 2009, 61–67.

HexTow Carbon Fiber Data Sheet, Hexcel Corp., www.hexcel.com

Hunt W.H., Miracle D.B., Automotive applications of metal matrix composites, in *ASM Handbook*. In: Miracle D.B., Donaldson S.L. (Eds.) *Composites*, Vol. 21. Materials Park: ASM International; 2001. pp. 1029–1032.

Jetter R.I., Sham T.-L., and Swindeman R.W., Application of negligible creep criteria to candidate materials for HTGR pressure vessels, *J. Pressure Vessel Tech.*, 133(2), 2011, 021103/1–021103/7.

Kainer K.U., Basics of metal matrix composites, in *Metal Matrix Composites. Custom-made Materials for Automotive and Aerospace Engineering*, Kainer K.U. Ed., Wiley-VCH Verlag GMBH & Co., Weinheim, Germany 2006.

Kalnin I.L., Ram N.J., and Dix R., Technical Report AFML-TR-72-151, Part 1, Air Force Materials Laboratory, 1972.

Katoh Y., Snead L.L., Henager C.H., Hasegawa A., Kohyama A., Riccardi B., and Hegeman H., Current status and critical issues for development of SiC composites for fusion applications, *J. Nucl. Mater.*, 367–370, Pt. A, 2007, 659–671.

Kawamura K., Jenkins G.M., A new glassy carbon fibre, *J. Mat. Sci.*, 5, 1970, 262–267.

Keller K.A., Jefferson, G., and Kerans R.J., Oxide–oxide composites, in *Handbook of Ceramic Composites*, Bansal N.P. ed., pp. 377–421, Springer, 2005.

Klemm H., Silicon nitride for high-temperature applications, *J. Am. Ceram. Soc.*, 93(6), 2010, 1501–1522.

Krenkle W. ed., *Ceramic Matrix Composites: Fiber Reinforced Ceramics and their Applications*, John Wiley & Sons, Weinheim, Germany 2008.

Lake M., Vapor grown carbon fiber, *Mater Tech*, 11(4), 1996, 131–44. Also see, Applied Sciences Inc., www.apsci.com

Laurie M., Magallon D., Rempe J., Wilkins C., Pierre J., Marquie C., Eymery S., and Morice R., Ultrasonic high-temperature sensors: Past experiments and prospects for future use, *Int. J. Thermophys.*, 31(8–9), 2010, 1417–1427.

Lessing P.A., A review of sealing technologies applicable to solid oxide electrolysis cells, *J. Mater. Sci.* 42(10), 2007, 3465–3476.

Lu X., Xia G., Lemmon J.P., and Yang Z., Advanced materials for sodium-beta alumina Batteries: Status, challenges and perspectives, *J. Power Sources*, 195, 2010, 2431–2442.

Ludenbach G., Peters P.W.M, Ekenhorst D., and Muller B.R., The properties and structure of the carbon fibre in carbon/carbon produced on the basis of carbon fibre reinforced phenolic resin, *J. Eur. Cer. Soc.*, 18, 1998, 1531–1538.

Marino K.A., Hinnemann B., and Carter E.A., Atomic-scale insight and design principles for turbine engine thermal barrier coatings from theory, *Proc. Natl Acad. Sci. USA*, 108(14), 2011, 5480–5487.

McPhail S.J., Aarva, A., Devianto H., Bove R., and Moreno A., SOFC and MCFC: Commonalities and opportunities for integrated research, *Int. J. Hydrogen Energy*, 36(16), 2011, 10337–10345.

Moure A., Castro A., and Pardo L., Aurivillius-type ceramics, a class of high temperature piezoelectric materials: Drawbacks, advantages and trends, *Prog. Solid State Chem.*, 37(1), 2009, 15–39.

Nair B.N., Burwood R.P., Goh V.J., Nakagawa K., and Yamaguchi T., Lithium based ceramic materials and membranes for high temperature CO_2 separation, *Prog. Mater. Sci.*, 54(5), 2009, 511–541.

Najmabadi F., Abdou A., Bromberg L., Brown T., Chan V.C., Chu M.C., Dahlgren F. et al., The ARIES-AT advanced tokamak, advanced technology fusion power plant, *Fusion Eng. Design*, 80(1–4), 2008, 3–23.

Navrotsky A., Thermodynamics of solid electrolytes and related oxide ceramics based on the fluorite structure, *J. Mater. Chem.*, 20(47), 2010, 10577–10587.

Nextel_Tech_Notebook_11.04 Data Sheet, 3M Advanced Materials.

Nguyen Q.N., Opila E.J., and Robinson R.C., Oxidation of ultrahigh temperature ceramics in water vapor, *J. Electrochem. Soc.*, 151(10), 2004, B558–B562.

Nysten B., Piraux L., and Issi J-P, Thermal conductivity of pitch-derived fibres, *J. Phys. D: Appl. Phys.* 18, 1985, 1307–1310.

Ohlhorst C.W., Vaughn W.L., Ransone P.O., and Tsou H.-T., Thermal conductivity database of various structural carbon–carbon composite materials, NASA Technical Memorandum 4787, November 1997; http://techreports.larc.nasa.gov/ltrs/ltrs.html

Ohtaki M., Recent aspects of oxide thermoelectric materials for power generation from mid-to-high temperature heat source, *J. Ceram. Soc. Japan*, 119(Nov. 2011), 770–775.

Opila E.J., and Meyers D.L., Oxidation and volatilization of silica formers in water vapor, *J. Am. Ceram. Soc.*, 86(8), 2003, 1238–1248.

Opila E.J., and Meyers D.L., Alumina volatility in water vapor at elevated temperatures, *J. Am. Ceram. Soc.*, 87(9), 2004, 1701–1705.

Oshima T., Kajita M., and Okuno A., Development of sodium-sulfur batteries, *Int. J. Appl. Ceram. Technol.*, 1, 2004, 269–276.

Parlier M., Ritti M.-H., Jankowiak A., Potential and perspectives for oxide/oxide composites, *J. Aerospace Lab*, (3) (November, 2011).

Pulci G., Tirillò J., Marra F., Fossati F., Bartuli C, Valente T., Carbon–phenolic ablative materials for re-entry space vehicles: Manufacturing and properties, *Composites: Part A*, 41, 2010, 1483–1490.

Rieth M., Boutard J.L., Dudarev S.L., Ahlgren T., Antusch S., Baluc N., Barthe M.-F. et al., Review on the EFDA programme on tungsten materials technology and science, *J. Nucl. Mater.*, 417(1–3), 2011, 463–467.

Savage G., *Carbon–Carbon Composites*, Chapman & Hall, New York, 1993.

Schulz M., Richter D., Sauerwald J., and Fritze H., Solid state sensors for selective gas detection at high temperatures-principles and challenges, *Integrated Ferroelectrics*, 115, 2010, 41–56.

Sheehan J.E., Buesking K.W., and Sullivan B.J., Carbon–carbon composites, *Annu. Rev. Mater. Sci.*, 24, 1994, 19–44.

Spear K.E., and Dirkx R.R., Role of high temperature chemistry in CVD processing, *Pure Appl. Chem.*, 62(1), 1990, 89–101.

Squire T.H., and Marschall J., Material property requirements for analysis and design of UHTC components in hypersonic applications, *J. Eur. Ceram. Soc.*, 30(11), 2010, 2239–2251.

Sudworth J.L., and Tilley A.R., *The Sodium Sulfur Battery*, Chapman & Hall, London, 1985.

Sullivan, B., Design of high thermal conductivity carbon–carbon for thermal management applications, *Proc. 22nd Annual Conference on Composites, Materials, and Structures*, Cocoa Beach, FL, Jan 25–30, 1998.

Tejedor, T.A., Gas turbine materials selection, life management and performance improvement, *Woodhead Publishing Series in Energy*, Vol. 23, *Issue Power Plant Life Management and Performance Improvement* (J.E. Oakley, Ed.), 2011, pp. 330–419.

Thornel PAN-Based Fiber Technical Data Sheet, Cytec Engineered Materials, www.Cytec.com

Thornel Pitch-Based Fiber Technical Data Sheet, Cytec Engineered Materials, www.Cytec.com

Torayca Technical Data Sheet No. CFA-008, Toray Carbon Fibers America, www.toraycfa.com

Tortissier G., Blanc L., Tetelin A., Lachaud J.-L., Benoit M., Conedera V., Dejous C., and Rebiere D., Langasite based surface acoustic wave sensors for high temperature chemical detection in harsh environment: Design of the transducers and packaging, *Sensors and Actuators* B, B156(2), 2011, 510–516.

Treacy M.M.J., Ebbesen T.W., and Gibson J.M., Exceptionally high Young's modulus observed for individual carbon nanotubes, *Nature*, 381, 1996, 678–680.

Viswanathan R., Purgert R., and Rao U., Materials for ultra supercritical coal fired power plant boilers, *Pressure Vessels and Piping*, Vol. 2, Edited by Raj, B. 2009, 1–20.

Wang Q., Han X.H., Sommers A., Park Y.T., Joen C., and Jacobi A. A review on application of carbonaceous materials and carbon matrix composites for heat exchangers and heat sinks, *Int. J. Refrig.*, 35(1), 2012, 7–26.

Wapner W.G., Hoffman W.P., and Jones S.P., U.S. Pat. Nos., 6,309,703; 706,401; 6,756,112.

Was G.S., Materials degradation in fission reactors: Lessons learned of relevance to fusion reactor systems, *J. Nucl. Mater.*, 367–370, Pt. A, 2007, 11–20.

Wellman R.G., and Nicholls, J.R., A review of the erosion of thermal barrier coatings, *J. Phys. D*, 40(16), 2007, R293–R305.

Wort C.J.H., and Balmer R.S., Diamond as an electronic material, *Materials Today (Oxford, UK)*, 11(1–2), 2008, 22–28.

Wuchina E., Opila E., Opeka M., Fahrenholtz W., and Talmy I., Ultra-high temperature ceramic materials for extreme environment applications, *Electrochem. Soc. Interface*, 16(4), 2007, 30–36.

Xuan G., Lv P.-C., Zhang X., Kolodzey J., DeSalvo G., and Powell A., Silicon carbide terahertz emitting devices, *J. Electronic Mater.*, 37(5), 2008, 726–729.

Yu M.-F., Files B.S., Arepalli S., and Ruoff R.S., Tensile loading of ropes of single wall carbon nanotubes and their mechanical properties, *Phys. Rev. Lett.* 84, 2000, 5552–5555.

Zhang Y., Li Y., Zhang K., and Tian Y., Research and development of high temperature electrothermal materials, *Adv. Mater. Res.* (Durnten-Zurich, Switzerland), 339, Advanced Manufacturing Systems, 2011, 17–22.

Zinkle S.J., and Busby J.T., Structural materials for fission & fusion energy, *Mater. Today (Oxford, UK)*, 12(22), 2009, 12–19.

4

High-Temperature Adhesives and Bonding

R. Peter Dillon

CONTENTS

4.1 Introduction

Synthetic adhesives offer a number of potential advantages when compared to more traditional joining methods (fastening, riveting, soldering, brazing, and welding) including: more uniform distribution of stress and larger stress-bearing area than conventional mechanical fasteners; joining of any combination of similar or dissimilar materials; and cure temperatures generally below those which could affect the strength of metal parts. One of the major limitations of synthetic adhesives has been degradation of these materials at service temperatures greater than 175°C. There has been growing interest, particularly in aerospace, for adhesives capable of withstanding temperatures in excess of 200°C for both short and, particularly, long term applications and industry has responded with synthetic adhesives which retain 50% of room temperature strength at 450°C (polybenzimidazole (PBI) under non-oxidative conditions). While long-term temperature resistance greater than 300°C may be best accomplished with metal, ceramic, or other inorganic materials, this chapter's focus is on the use of synthetic adhesives for elevated temperature joining applications.

4.2 Fundamentals of Adhesion

The strength of an adhesively bonded joint is limited by the weakest of the following regions of the bond (Figure 4.1): the adhesive region at the interface between the substrate (or adherend) and the adhesive; a cohesive region which has the nominal structure and exhibits the bulk properties of the adhesive; and a interphase region where the structure and properties vary from that of cohesive region to that of the adhesive region. Please note that it is possible for the bonded joint to exceed the cohesive strength of the adherends such that failure occurs in the substrate and not in any of the regions described above. It is also possible for the region of failure to change as a function of environmental exposure.

The adhesive manufacturer's technical data sheet generally provides sufficient detail to describe the properties of the cohesive region. These properties originate from the chemical bonding of the base polymer, the chemical bonding from interchain crosslinks, which result from curing thermoset adhesives, the interchain interactions in the polymer, and the mechanical entanglement of the molecular chains in the polymer. To achieve the manufacturer's stated properties often require the end user optimize cure conditions.

In circumstances where a thin bond line or thick interphase regions occur, there may be little to no cohesive region to affect the performance of the bond. The thickness of the interphase region will be determined by the degree of heterogeneity of the adhesive and

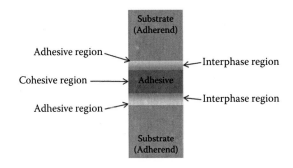

FIGURE 4.1
Anatomy of an adhesively bonded joint.

the nature of the adhesive region. For example, an adhesive may exhibit a gradient of properties across the interphase region as smaller chains diffuse into the surface of a porous substrate. Whereas, the same adhesive on a smooth, non-porous adherend would not be expected to exhibit this same type of structural gradient.

The strength of the adhesive region derives from a number of independent and interacting forces, such that the nature of the adhesion is generally complex and not simply defined. There is no unifying theory that describes the nature of all adhesive bonding. Instead there are a number of theories of adhesion to be considered singularly and in combination when evaluating adhesives and adherends for bonding. The most commonly accepted theories include mechanical, adsorption, chemisorption, electrostatic, diffusion, and weak boundary layer (Comyn 1997).

4.2.1 Mechanical

The mechanical theory of adhesion is presented first largely because much of the original thought on adhesion focused on the adhesive flowing into and interlocking with voids, such as pores and holes, in the substrate. The substrate would be mechanically fixed once the adhesive hardened. While this macroscale interlocking contributes to joint strength (Stuart and Crouch 1992), this mechanical model of joint strength has evolved as the interlocking of small adhesive molecules in the microtexture, such as the microporosity of an adherend, has been shown to be vital to joint strength. It is this microtexture that can be manipulated to enhance adhesion.

The microtexture of nonporous surfaces can be modified through roughening or abrasion of the surface. The total contact area of the roughened surface will be greater than that of the non-abraded surface, thus creating an equivalently greater surface interaction with the adhesive. Additionally, the process of abrading the adherend surface generally cleans the surface of residues or contaminants, which might otherwise impair adhesion and increases the reactivity of the surface. The forces of adhesion are thus able to develop over a greater and more reactive surface area.

Controlled grit blasting, as opposed to simple abrasion techniques, is generally the preferred methodology for surface pretreatment in metals such as mild steel, stainless steel, and, to a certain extent, titanium and fiber-reinforced composites. Very ductile materials, such as aluminum, do not form a good active surface for bonding using abrasion methods. While still macroscopically rough, the surface, instead, consists of loosely attached aluminum debris that provides weak points in the bonded joint. For additional details on surface pretreatments the reader is directed to Cognard (2005).

Mechanical interlocking with surface microtexture, particularly when the orientation of the texture opposes the stress, enhances the strength of the adhesive joint by forcing the adhesive to plastically deform. The strength is also enhanced by the crack propagation barrier, which results from the same microtexture (Figure 4.2). Cracks dissipate more energy propagating along the more tortuous path of the roughened surface.

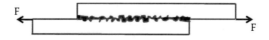

FIGURE 4.2
A tortuous crack path (exaggerated) can enhance the strength of an adhesive joint.

4.2.2 Adsorption and Wetting

Adhesion from adsorption requires intimate molecular contact between the adhesive and the adherend such that bonds may form through attractive surface forces. These forces, usually designated as secondary or van der Waals forces (2 to 4 kJ/mol), require the respective surfaces to be separated by no more than 5 angstroms in distance to develop. Thus, for the adhesive to make intimate and continuous molecular contact with the substrate surface, good wetting must occur.

When good wetting occurs, the adhesive fills any and all surface perturbations, displacing air such that any interfacial defects are avoided. Conversely, when poor wetting occurs the adhesive bridges the texture of the substrate surface, reducing the overall contact area and entrapping small air pockets resulting in reduced overall joint strength.

The degree of wetting can be measured by the wetting angle as illustrated in Figure 4.3 and defined by the balance of surface free energies (γ) between the substrate and atmosphere (γ_{SV}), the substrate and the uncured adhesive (γ_{SL}), and the uncured adhesive and the atmosphere (γ_{LV}) in Young's equation.

Young's equation:

$$\gamma_{LV} - \cos\theta = \gamma_{SV} - \gamma_{SL}$$

The work of adhesion can also be considered. The energy expended to create two new surfaces from one is the sum of the two surface energies reduced by the interfacial energy present before the new surfaces were created. The result is Dupre's equation:

$$W_A = \gamma_{LV} + \gamma_{SV} - \gamma_{SL}$$

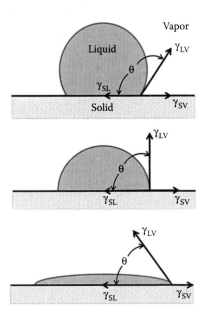

FIGURE 4.3
Wetting as related to surface energy.

This can be combined with Young's equation to make the Young-Dupre equation, which follows as

$$W_A = \gamma_{LV}(1 + \cos\theta)$$

W_A becomes the spreading coefficient, which relates to the wetting angle and the surface energy (or surface tension) between the liquid (adhesive) and the vapor (atmosphere). An adhesive with a high surface tension will help with wetting as will a spreading coefficient greater than zero. Substrates with high surface energies (ceramics, epoxy-glass laminates, metal oxides) will cause the adhesive contact angle to be less than 90°, ensuring at least partial wetting. Low-energy substrates (fluoridated surfaces or surfaces with contaminants such as silicones, oils and grease) which exhibit a contact angle greater than 90° will not wet.

To allow wetting by the adhesive, the surface energy of the adherends can be modified by coatings or surface modifications processes (e.g., chemical or plasma treatments) Wetting in and of itself is not, however, a guarantee of strong bonding of good adhesive performance. Still, the strength of the adhesive bond at the interface can be improved by these treatments such that the total number of secondary of van der Waals bonds increase, or strong bonds such as covalent bonds are introduced. The latter can be newly available bonding sites on the substrate following chemical or plasma treatments or bonding strategies formulated into coating which allow the coating to both bond to the substrate and provide high strength bonding sites to the adhesive (see da Silva et al. 2011).

4.2.3 Chemisorption

Chemisorption involves adsorption where adhesive needs to adsorb to and wet the surface of the adherend. Chemisorption goes further, however, in that the adhesive chemically reacts with the substrate to form strong covalent bonds. These bonds have bond energies between 150 and 900 kJ/mol. Adhesion promoters and coupling agents are examples of chemisorption. These molecular materials are functionalized on one end to react with the substrate and on the other end to react with the adhesive. They often enable the bonding of substrates which otherwise could not be adhered to.

4.2.4 Electrostatic

Electrostatic bonding theory considers the transfer of charge from the substrate to the adhesive to form an attractive electrical double layer. This is most easily described for a metal substrate whose electrons are transferred to the polymer (Figure 4.4). Van der Waals forces generally contribute more to adhesion than electrostatic forces; however, electrostatic forces are known to be a strong factor in biological cell adhesion. Electrostatic

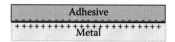

FIGURE 4.4
Adhesive metal interface showing electrical double layer.

adhesion is generally not considered a significant factor in the high-temperature systems considered in this chapter.

4.2.5 Diffusion

Diffusion-related adhesion models are considered for two polymer surfaces when there is some solubility between one or both of the polymers. Adhesion is better when both polymers have similar solubility parameters. When the temperature is above the glass transition temperature (Tg) of the polymers and if the molecules are relatively mobile (e.g., minimal to no cross-linking, minimal to no bulky side groups) the interface may eventually be eliminated due to diffusion of polymer chains, or parts of polymer chains, across the interface. The remaining interphase region is no longer discontinuous and, as such, does not exhibit any stress concentration.

Solvent welding is a joining process that leverages diffusive adhesion. By selecting the appropriate solvent for the polymer the solubility parameters can be modified and chain mobility enhanced allowing two polymer surfaces to intermix, essentially welding. This is generally done at room temperature and the interphase region is somewhat analogous to a heat-affected zone. Chain pull-out is the dominant failure mechanism for the interphase region.

As diffusion-related adhesion requires minimal-to-no cross-linking, it is generally not considered for the high-temperature thermoset adhesives in this chapter.

4.2.6 Weak Boundary Layer

The final adhesive theory considered focuses on the failure of adhesive systems due to the presence of a weak interfacial layer. This layer can be introduced to the system anytime from before or during preparation of the adhesive and adherend for joining up to and including the time in service. Some examples of weak boundary layers are residues, oxides, dirt and debris on substrates; improperly mixed adhesives, impurities in the adhesives, or other chemical constituents such as plasticizers which segregate to the substrate surface; and corrosion or weak oxide layers on metal surfaces which form from environmental exposure. When failure does occur, the weak boundary layer fails cohesively between the adherend and the adhesive. This highlights the importance of proper substrate and adhesive preparation and an understanding of the service environmental effects to prevent adhesive joint failure from weak boundary layers.

For a more in-depth discussion of the fundamentals of adhesion, refer to Comyn's "Adhesion Science," Kinloch's "Adhesion and Adhesives," Petrie's 2007 "Handbook of Adhesive and Sealants," or Licari and Swanson's 2011 "Adhesive Technology for Electronic Applications."

4.3 High-Temperature Environments and Adhesives: It Might Not Just Be a Dry Heat

Many people are familiar with the expression, "It's not the heat. It's the humidity." The additional moisture in the air on humid days causes us to feel the ambient temperature to a greater degree due to more efficient heat transfer while less effectively cooling via perspiration. Similarly, a cross-linked adhesive may also experience increased heat transfer due to increased moisture in the air and less effective cooling of the bonded structure. Excessive heat can result in thermally induced chain scission, or splitting of the polymer molecules, which reduces the molecular weight of the adhesive and results in degraded cohesive strength. For adhesive systems where excessive heat can continue, cross-linking shrinkage and bond embrittlement can result. Adams et al. (1992) give a thorough discussion of shrinkage and thermal expansion induced stresses and the effect of temperature on adhesive properties in their paper, "The effect of temperature on the strength of adhesive joints" while Schneberger (1983) examines heat and moisture as degradation mechanisms in "Adhesives in Manufacturing."

As all polymers are permeable to moisture, the effect of humidity extends beyond increasing the effectiveness of environmental heat transfer. GW Critchlow in DE Packham's 2005 Handbook of Adhesion states, "moisture ingress is thought to be responsible for many examples of premature joint failure." Water in the atmosphere can reach the adherend/adhesive interface by diffusing along the interface, through the adhesive or permeable adherends, or through defects such as cracks or voids. Increasing the temperature increases the rate of moisture uptake into and diffusion through the adhesive and along the interface. Ashcroft and Comyn in their article "Effect of Water and Mechanical Stress on Durability" in the Handbook of Adhesion Technology discuss a critical relative humidity above which joints see significant weakening. The authors report this value to be around 65% relative humidity. Hygroscopic swelling can degrade the adhesive and generate additional stress on the interface, which further deteriorates long-term durability. Water and stress may also have a combined effect, which dominates the strength of the adhesive joint through enhanced creep of the adhesive (Bowditch 1996).

At the interface, moisture can corrode the adherent or form hydroxyl ions, typically from a cathodic potential on a metallic adherend, both of which degrade adhesion. Properly selecting and applying surface pretreatments can minimize the effect of water attack at the interface. For example, in the document "Hydrophobicity, Hydrophilicity and Silane Surface Modification," Barry Arkles 2013 of Gelest, Inc. recommends two or more silanes to use as surface coatings when coupling to iron, copper, zinc, or other metals that form hydrolytically or mechanically unstable oxides. A chelating silane such as diamine, polyamine, or polycarboxylic acids would be applied first to bind with the metal surface and prevent reaction of the metal with moisture or other environmental elements. A second silane that reacts with the first and bonds with the adhesive would then be applied. The adhesive joint would be formed third.

Absorption of water into the bulk adhesive can change the physical properties of the adhesive below that required for the application. For example, absorbed water can reduce the glass transition (Tg) temperature of an adhesive by acting like a plasticizer. The adhesive may not perform as required if the application temperature is above this reduced Tg. Accelerated aging in water vapor or with water immersion can help elucidate the impact of moisture on the adhesive system.

In addition to performance, moisture may also play a role in the curing of the adhesive. Silicone and isocyanate materials require the diffusion of water vapor through the resin for chain extension and cross-linking. Low relative humidity can increase the cure times for these materials. On the contrary, excessive moisture or high relative humidity can inhibit epoxy resin systems and result in voids and porosity in the epoxy. As epoxy adhesives are often packaged "premixed-frozen," it is important to allow the material to completely thaw to room temperature prior to breaking the seal. This practice will help prevent condensation of moisture on and diffusion into the adhesive. Controlling the relative humidity where the adhesive joint is assembled is recommended for most adhesive formulations, but this is in many cases not possible.

4.4 High-Temperature Adhesives: Structures and Properties

The upper use temperature of a polymer adhesive may be limited by the softening point of the adhesive, degree of strength reduction, or degradation of some other critical material property. Elevated temperatures in combination with the environment degrade polymers through a wide range of reactions primarily related to chain scission, or the breaking of polymer bonds, both in the backbone chain and the crosslinks. Ultimately, thermal oxidation or pyrolysis can occur, at which point the material degradation is unrecoverable. Adhesives designed for high temperature exposure characteristically have rigid polymer structures with stable chemical groups and high glass transition (Tg) or softening temperatures.

The rigidity and stability of the polymer structure is enhanced by aromatic rings such as that pictured below (Figure 4.5). These structures can be monocyclic when the ring structure is made of one atom type, such as carbon, or heterocyclic when the ring structure also has non-carbon ring atoms such as nitrogen, oxygen, or sulfur. The alternating bonds help to delocalize electrons such that a highly reactive free radical is less likely to form should a bond break. These structures are often referred to as resonance stabilized ring structures.

Most stable high-temperature adhesives have ladder or semi-ladder structures (Figure 4.6). This kind of structure takes advantage of electron delocalization, also known as resonance stabilization, and adds polybonding through the additional string of atoms. Through polybonding the scission of at least two chains is required to break the ladder thereby increasing the stability of the macromolecule.

FIGURE 4.5
Typical aromatic ring structure.

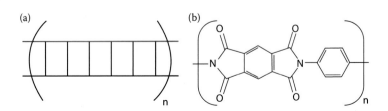

FIGURE 4.6
(a) Ladder structure concept polymer; (b) polyimide polymer, a semi-ladder structure.

In this section, we will examine polymer adhesives reportedly capable of maintaining mechanical properties and resist thermal breakdown at temperatures in excess of 175°C. The materials considered here are epoxies and epoxy phenol novolacs, polyimides, bismaleimides (BMI), PBIs, cyanate esters, and silicones.

4.4.1 Epoxy and Epoxy Phenol Novolacs

Epoxy resins are characterized by the epoxy (or oxirane) ring (Figure 4.7). Epoxies can be made either by peroxidizing olefins (the cycloaliphatics) or by reaction with epichlorohydrin (the glycidyls).

The most commonly encountered epoxy adhesive is produced from bisphenol A and epichlorohydrin to form diglycidylether of bisphenol A (Figure 4.8) or DGEBA.

DGEBA, or another epoxy resin, is then co-reacted with a curing agent such as an acid, amide, amine, anhydride, phenolic, or Lewis acid. With sufficient energy, such as heat, the curing agent acts to open the oxirane ring as part of the curing and crosslinking reactions to extend the structure. The ring is not present in the final, cured structure. The choice of co-reactant and epoxy resin ultimately determines the final structure of the polymer and thus the final properties of the cured adhesive. With such a variety of available chemistries, it is not surprising epoxy adhesives are so ubiquitous.

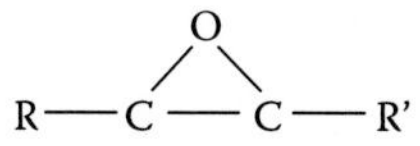

FIGURE 4.7
Generic epoxy structure showing the epoxy (or oxirane) ring comprised two carbon atoms and one oxygen atom.

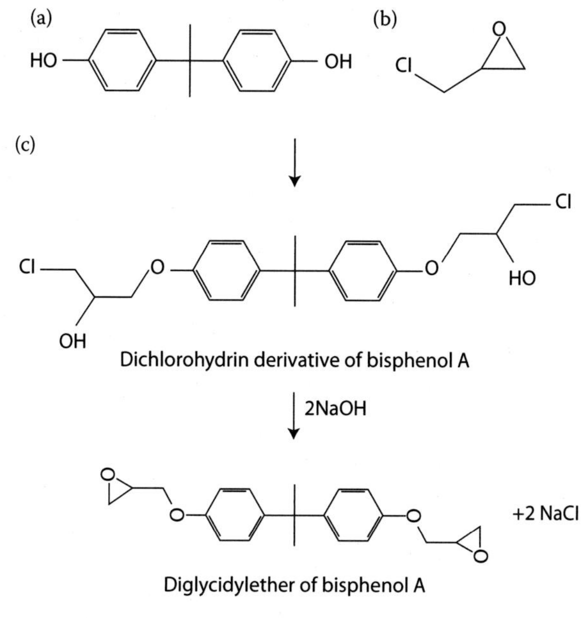

Dichlorohydrin derivative of bisphenol A

Diglycidylether of bisphenol A

FIGURE 4.8
(a) Bisphenol A and (b) epichlorohydrin molecular structures (c) react to make DGEBA.

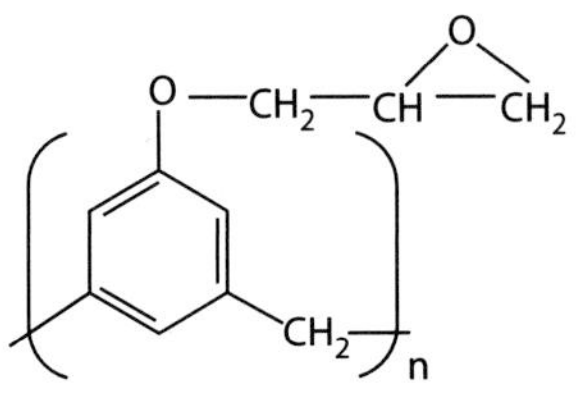

FIGURE 4.9
Epoxy phenol novolac polymer.

While epoxies can have a broad temperature use range, they are generally limited to upper use temperatures below 150°C for most formulations. That said, there are some epoxy adhesives suitable for high-temperature use (continuous exposure temperatures ≥175°C; higher short duration exposure temperatures). The most common epoxy resin system and curing agent for elevated temperature use is tetraglycidyl methylene dianiline and diaminodiphenyl sulfone, respectively. Commercially available high performance epoxy adhesives which are derivatives of this type of epoxy formulation include Araldite MT 0163, Tactix 742, and Redux 322.

Epoxy phenol novolacs are another relatively temperature tolerant adhesive. A characteristic resin repeat unit is shown in Figure 4.9. While glass transition temperatures greater than 250°C can be realized, these materials tend to be brittle and exhibit significant shrinkage when curing.

4.4.2 Polyimide

The development of polyimide is tied very closely with the aerospace industry. NASA Glenn Research Center (formerly NASA Lewis Research Center) has been particularly successful with their development of PMR-15, which is used in a variety of molding compounds and adhesives for high-temperature applications and PMR-II, which is similar to PMR-15 but provides somewhat better heat resistance (Scola, D. A. "Chapter 7: Polyimide Resins" in Vol. 21 of ASM Handbook–Composites, 2001).

Polyimide is distinguished by the repeating imide groups in the polymer chain (Figure 4.10). The typical polymer structure is along the lines of the semi-ladder polyimide polymer structure shown in Figure 4.10. The specific polyimide polymer structures, including those for PMR-II and PMR-15, are also given by Scola D. in Chapter 7: "Polyimide Resins" in Vol. 21 of ASM Handbook–Composites, 2001. Both PMR-15 and PMR-II are available as prepreg, solutions, or powders.

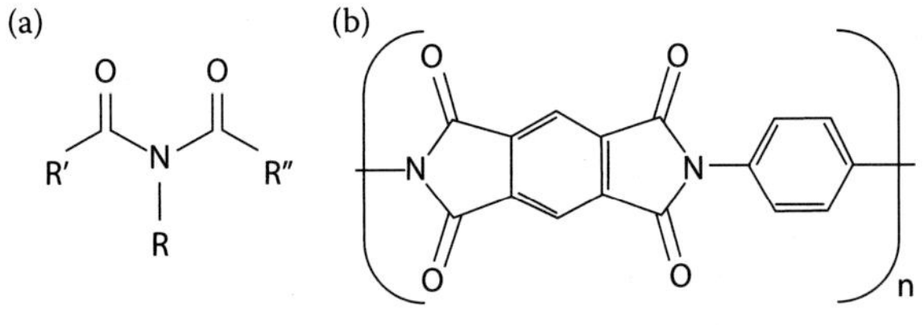

FIGURE 4.10
(a) Imide group and (b) polyimide polymer.

With regard to strength retention, polyimides are only slightly more able than epoxy phenol novolacs during short-term exposures at temperatures approaching 500°C. Where polyimides stand out over virtually all other high temperature organic adhesives is in their thermo-oxidative stability. Polyimides capable of withstanding prolonged exposure to temperatures of 250–300°C are readily available.

While there are now thermoplastic polyimides available, thermoset polyimide adhesives remain the choice for many high-temperature applications including a large number of aerospace structural composites. The two types of thermosetting polyimides available are characterized by their cure chemistry. Condensation-reaction polyimides have the highest temperature performance of the two but require additional controls during curing to prevent moisture, a by-product of the reaction, from leaving voids in the bond line and weakening the bond. Curing condensation-reaction polyimides under vacuum produces the most reliable bonds. A high temperature post-cure is used to establish the highest use temperature.

Addition-reaction polyimides are an alternative to condensation-reaction polyimides. These materials cure by addition polymerization and thus do not generate moisture during cure. These materials are often used as adhesive films, so high temperature solvents may be used to keep the films pliable (Petrie 2007). Vacuum curing such films is also beneficial. Addition-reaction polyimides do not have the same oxidative stability as the condensation type with short-term temperature exposure limited to just over 300°C and long-term exposure limited to approximately 250°C.

As previously mentioned, polyimide is available in film form. These are typically supported or reinforced. Adhesive solutions are also available and can often be found in microelectronics fabrication.

4.4.3 Bismaleimide

The high-temperature resistance of BMI originates from a molecular structure similar to polyimide. BMIs are characterized by the presence of maleimide groups in the polymer chain (Figure 4.11). A BMI polymer structure is shown in Figure 4.11. While sharing some of the same structural elements as polyimides, BMIs do not have the same high-temperature performance as polyimides. BMIs tend to bridge the gap between polyimides and epoxies and epoxy-phenolics. Short-term temperature exposure limits for BMIs tend to be around 225°C with extended duration temperature exposure limits near 200°C.

When performance and environmental requirements are such that polyimide is not required, BMI adhesives may have some desireable benefits. As BMI adhesive systems cure by addition reaction, no volatiles are given off (Figure 4.12). This will simplify the cure process to several hours under pressure followed by unloading and a postcure at

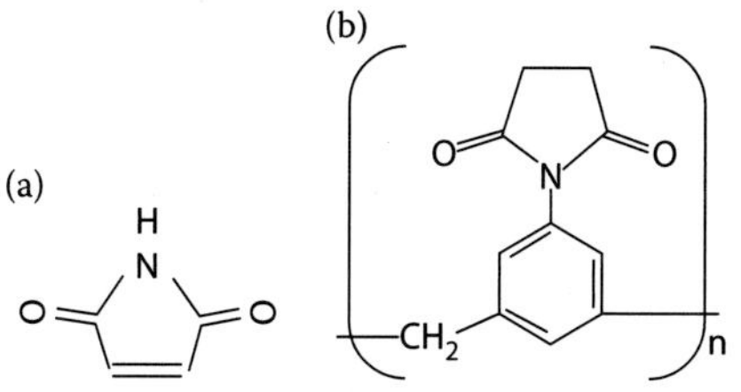

FIGURE 4.11
(a) The maleimide group characterizing BMI and (b) the generic BMI polymer structure.

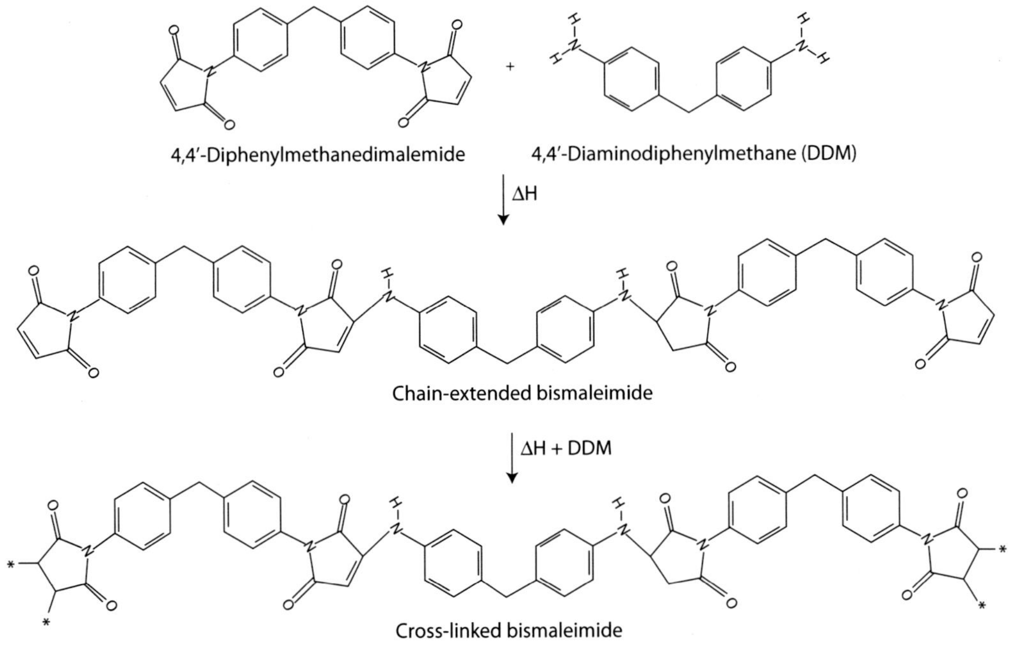

FIGURE 4.12
A BMI cure sequence.

or near the service temperature to fully develop the material's thermal stability. BMI adhesives are generally less expensive than polyimide systems and can be formulated with the same diversity as epoxies. Both resins and films are available for adhesive applications.

4.4.4 Polybenzimidazole

Before we delve into PBI, it should be noted that an off-the-shelf PBI adhesive is not readily available. A PBI adhesive manufactured under the name Imidite 850 by the Whittaker Corporation was of great interest to the aerospace industry in the 1960s and 1970s but is no longer available today. This is likely due to some of the material's drawbacks as will be presented here and the availability of alternatives such as polyimide and BMI.

PBIs are distinguished by the presence of benzimidazole groups in the polymer chain (Figure 4.13). The most common PBI polymer structure is shown in Figure 4.13.

Much of the early interest in PBI centered on its ability to retain much of its shear strength at temperatures as high as or greater than 500°C. PBI also showed good performance at low, cryogenic temperatures, making it an interesting candidate for space applications where such broad temperature ranges are readily experienced. Unfortunately, PBI had three major drawbacks, which prevented general adoption as a high performance aerospace adhesive. First, the material was found to oxidize at temperatures above ~250°C. For applications where the temperature would exceed 250°C, the exposure period would have to be limited to prevent serious degradation of the material. Second, PBI adhesives needed to be cured at high temperatures under pressure and

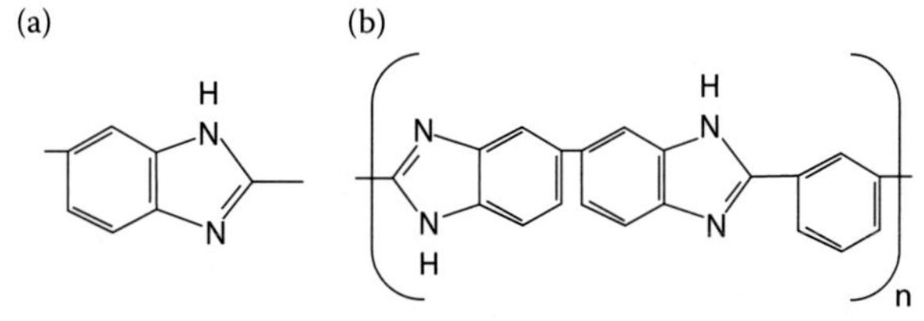

FIGURE 4.13
(a) The benzimidazole group characterizing PBI and (b) the most common polymer form.

often required post-curing above 300°C to achieve the greatest temperature resistance. Additionally, curing of PBI-released volatiles, which often embrittled the bond line and leaving it with low peel strength. Curing under partial vacuum minimized this problem, but added to the difficulty of making the adhesive joint, particularly for large structures. Finally, polyimide and BMI-based adhesives were suitable for most organic adhesive applications that were being considered for PBI and these materials were more readily available for lower cost.

Today, PBI may be supplied as a stiff, fiber reinforced film or prepreg. The material is expensive, and the raw resin is not generally available. PBI is produced by the Hoechst Celanese Corporation along with thermoplastic PBI under the name Celazole.

4.4.5 Cyanate Esters

Cyanate esters are emerging to challenge epoxies and BMIs in terms of thermal stability. In aerospace composites, cyanate esters are being selected as the matrix material for carbon reinforcement due to their greater toughness. Cyanate esters are characterized by the cyanate ester functional group (Figure 4.14a) which is usually attached to a rigid aromatic backbone. These rigid aromatic backbones are similar to the resonance-stabilized ring structures previously presented. A simple example of the cyanate ester cure process is homopolymerization of cyanate ester functional groups to the highly stable triazine ring structure (Figure 4.14b) and 3D polymer network. This process of developing this structure is also called cyclotrimerization (see Bauer et al. 1998). Polymerization is additive and the resulting cyanate esters have very low outgassing rates.

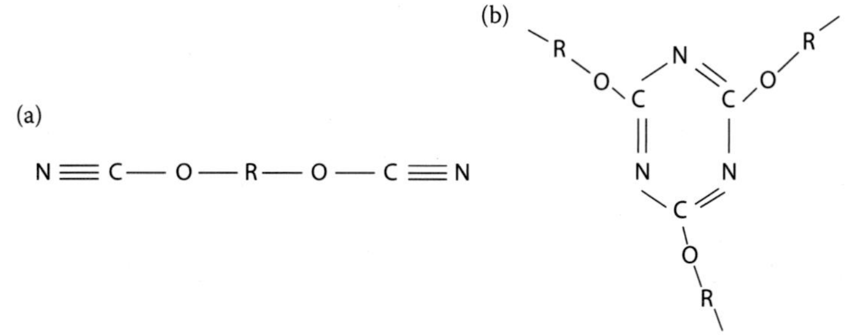

FIGURE 4.14
(a) Cyanate ester functional group and (b) triazine ring formed from the polymerization of these functional groups.

Most cyanate esters are formulated with either phenolic or epoxy resins to create significantly different final cured structures. Curing of reactants with phenolics leads to trimerization while trimerization is only one aspect of the cure reaction with epoxy resins (see Hamerton (1994) for further discussion).

Cure conditions are similar to BMIs and certain epoxies; with 1–2 h isothermal holds around 170°C. To bring the Tg above 200°C requires an elevated temperature post cure. The Tg of some cyanate ester resin systems exceed 275°C.

Cyanate ester adhesives see a significant amount of deployment in electronics manufacturing. The thermal stability and low moisture uptake makes cyanate esters compatible with electronics assembly processes such as solder reflow. Although cyanate esters are not particularly polar, adhesion with metallic substrates is generally strong with good adhesion maintained up to ~250°C. Cyanate ester adhesives are available in a number of forms including premixed, frozen resins, pastes, and films.

4.4.6 Silicones

The glass transition temperature of silicones tends to be well below zero. Correspondingly, these materials tend to be compliant over a wide range of temperatures and unlike the previously discussed materials, are not rigid after cure. Silicone adhesive systems are characterized by siloxane or the silicon and oxygen bond network. This is clearly observed in polydimethyl siloxane (Figure 4.15).

Silicone resins have long been known for their thermal stability in both adhesives and sealants. Owing to poor cohesive strength, silicone-based adhesive tend not to be used for structural applications. Non-structural or compressively loaded applications such as pressure sensitive masking or release tapes, sealants, and gaskets are highly suitable for silicone adhesives and sealants. The maximum continuous use temperature for silicone adhesive systems generally does not exceed 300°C although some temperature-stabilized varieties for continuous use at higher temperatures may be available from such manufacturers as DowCorning or NuSil. Properties such as peel strength tend to be consistent throughout the use temperatures.

Silicones are particularly substrate sensitive and generally require a primer to obtain a decent bond strength at room temperature. A number of cure systems are also used for silicone adhesives including tin and oxime catalyzed which require moisture and can be slow to cure, specifically when thick bond lines are used or when relative humidity is low. More expensive platinum catalyzed silicone adhesive systems are also available. These systems are thermally curable.

The following table summarizes a number of key data for the materials just discussed.

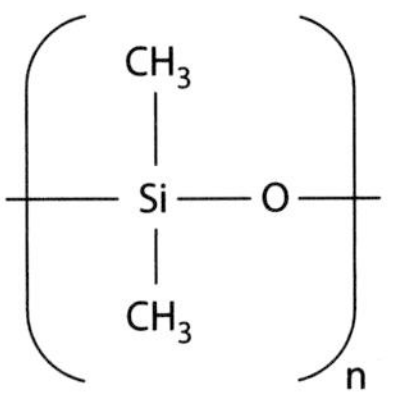

FIGURE 4.15
Silicone resin polymer—polydimethyl siloxane (PDMS).

Neat Resin	Typical Cure for T_{MaxUse}	Specific Gravity	Tg (°C)	CTE (ppm/°C)	Short Term $T_{continuous}$ (°C)	Long Term $T_{continuous}$ (°C)	RT Tensile Strength (MPa)	RT Elongation at break (%)	RT Tensile Modulus (GPa)	Fracture Toughness (J/m²)
Epoxy and epoxy phenol novolac	Varies with cure chemistry; High T epoxy requires pressure (0.14–0.7 MPa) and elevated temperature cure around continuous use temperature	1.11–1.4	60–230	45–70	120–290	175–230	48–90	1.5–8.0	2.6–3.8	70–210
BMI	0.49–0.70 MPa 175–190°C Post Cure 0.59–0.69 MPa 230–245°C	1.23–1.29	230–345 (dry)	30–65	315	230	35–90	1.5–3.0	3.4–4.1	70–105
PI	1.4–17 MPa 290–316°C Post Cure 1.4–17.2 MPa 316–400°C	1.19–1.45	315–420 (dry)	25–80	500–600	250–316	38–110	1.5–8.5	2.5–4.7	42.7–133.5
Cyanate ester	177–250°C	1.1–1.35	250–290	60–70	300	250	69–90	2.0–5.0	3.1–3.4	105–210
Silicone	RT	0.99–1.50	–125	80–300	300	250	1.03–10.3	100–1000		

Sources: ASM Handbook-Composites, Daniel Miracle and Steven Donaldson (eds.), ASM International Vol. 21. Materials Park, OH, 2001, Chapter 7; Engineered Materials Handbook: Volume 2, Engineered Plastics, Dostal, C.A. (ed.), ASM International, Metals Park, OH; Engineered Materials Handbook: Volume 3, Adhesives and Sealants, Dostal, C.A. (ed.), ASM International, Metals Park, OH.

4.5 High-Temperature Bonding Applications

Bonding for high-temperature applications is not significantly different than any bonding application where maximum joint strength and reliability is desired. The joint design is critical and will likely benefit from redesign where non-adhesive bonding joining methods were previously used. Joints which exhibit purely tensile shear will have the highest strength while peel is generally considered the weakest adhesive joint. Load-induced deflections may introduce cleavage stresses which, along with peel stresses, should be restricted to maximize the reliability of the adhesive joint (see Kinloch, 1987, for guidance on good and poor joint design).

If possible, the joint should be assembled in an environment where all aspects of adhesive bonding (adherend cleaning, surface preparation, priming, adhesive application, joint assembly and fixturing, and curing) can be controlled. When this is not possible, it is important to control as many of these elements as possible.

In addition to sufficiently roughening the bond surfaces of the adherends (to create a surface with the microscopic roughness necessary to maximize mechanical adhesion), these surfaces need to be sufficiently cleaned of rust, mill scale, dust, grease, oil, or other contaminants that adversely affect adhesion. Abrasive treatments such as wire brushing, sanding, and micro-bead blasting are effective in removing dirt and debris but generally need to be used in combination with chemical methods to remove oils or other contaminants that could leave weak boundary layers. Care should also be taken to ensure any abrasive medium used does not itself become a source of contamination that could impair bonding. Finally, the joint should be assembled in a timely manner to avoid atmospheric contamination of the freshly prepared adherend surfaces and the formation of weak boundary layers.

In addition to being able to survive the application temperature, the chosen adhesive should spontaneously wet the adherends. If available, a primer should be used with the adherend and adhesive to maximize wetting and provide covalent bonding. Primers or other adhesion promoters can be the key to performing in an extreme environment, including minimizing the deleterious effects of moisture. Silicones frequently require the use of primers for maximum bond strength and bond reliability. In many instances the manufacturer of an adhesive will recommend a specific primer for use with a particular substrate. The relationship between viscosity and cure of the adhesive must be such that running on vertical bonding areas does not occur, little-to-no absorption of the adhesive by porous adherends occur, and application and coverage is such that an optimum bond line is maintained. Stabilizing the joint during cure, through proper fixturing, is also crucial for maximizing joint strength.

For applications requiring performance over a broad temperature range, the use of two adhesives, one for low temperature and the other one for high temperatures can improve the joint strength. This is discussed at length in da Silva and Adams' (2007) paper.

Applications that require bonding at high temperatures ($T_{surface} > 100°C$) introduce additional complexity. The environment will likely be difficult to work in, so it should be expected that at least one of the adherends will not be optimally prepared for bonding. It may only be possible to wire brush and rag wipe a hot pipe to which a sensor may need to be attached. The large temperature difference between the elevated temperature adherend and the adhesive will result in a rapid temperature rise of the adhesive. Water, low molecular weight components, and other volatiles will vaporize, expanding the adhesive, express the adhesive from the joint, and leave pores in the adhesive. Improperly prepared two-part adhesives that are not properly degassed will exhibit rapid expansion of the entrapped air. Frozen adhesives should be allowed to thaw and warm to ambient before opening to avoid moisture absorption from condensation. Use of properly degassed, premixed, low outgassing (low volatility) high-temperature adhesives that exhibit less than 1 wt. % total mass loss (TML) will minimize the potential for these problems when used at high application temperatures. An adhesive consisting of 100% solids will generally meet this requirement. NASA maintains a website (http://outgassing.nasa.gov/) where the outgassing behavior of a number of adhesives is reported. The outgassing behavior may also be available from the manufacturer or reported on the technical data sheet.

4.6 Evaluating High-Temperature Adhesive Performance

After studying the various adhesive families and their known performance, reviewing adhesive manufacturer's websites and technical data sheets, consulting with sales engineers and colleagues, and considering all of the tribal or heritage knowledge from one's own organization or personal experience, a number of candidate materials will emerge as most promising for a particular application. *In situ* testing will always provide the most useful data for high-temperature adhesive performance and, ultimately, materials selection. Before committing to an expensive *in situ* test or qualification campaign, however, the candidate materials can be screened for cohesive (bulk) and adhesive performance.

4.6.1 Characterizing the Bulk Adhesive for Cohesive Performance

A manufacturer's technical data sheet is one of the first places to find information on the bulk properties of an adhesive to consider cohesive performance. The data sheet may even contain property and performance information at elevated temperatures, but this information is generally not provided with enough detail to properly screen the material for high temperature applications. For instance, it is unlikely a data sheet will ever provide enough information to determine the rate of degradation as a function of temperature.

A number of analytical tools are available to evaluate the high-temperature characteristics of the bulk adhesive in significantly useful detail. These include differential scanning calorimetry (DSC), thermogravimetric analysis (TGA), dynamic mechanical analysis (DMA), and thermomechanical analysis (TMA). For some of these techniques it may also be possible to control the environment (e.g., oxidizing or reducing atmospheres) to better simulate the application environment.

While some organizations may have the capability to perform these analyses in-house, they may also be easily and inexpensively performed at independent test facilities. In some cases an adhesive material manufacturer may also have performed these tests under relevant conditions and may make the data available upon request.

4.6.1.1 DSC

DSC measures the heat flow from a sample as a function of temperature. By measuring the heat released by a sample, exotherms, and the heat absorbed by a sample, endotherms, a number of physical properties can be determined. These properties include the glass transition (T_g), crystallization (T_c), and melting temperatures (T_m), heat of cure, heat of volatilization associated with the release of moisture or other low molecular weight volatiles (e.g., residual solvents), and thermal or oxidative degradation. The technique requires only a few milligrams of uncured adhesive. The material is typically weighed into a tared metal pan (generally aluminum) and covered by a pinhole-vented lid that is subsequently crimped in place. The sample pan is placed on a sensitive thermocouple in the differential scanning calorimeter, also referred to as the DSC. An empty reference pan, also with a crimp-sealed, pinhole-vented lid is placed on another thermocouple in the DSC. By measuring the difference in heater current required to maintain the sample and reference pans at the same temperature or measuring the temperature difference of the two pans as a function of temperature the heat flow to and from the sample is determined. Figure 4.16 illustrates the usefulness of the technique, particularly through the use of a multi-step temperature scan.

In this example, an uncured sample of a cyanate ester adhesive was heated from 0°C to 275°C at a heating rate of 10°C/min. This cycle supports elimination of any volatiles from the specimen and complete curing of the curing of the adhesive. The adhesive specimen tested is considered a low outgassing material, and this is supported by the lack of an appreciable endotherm below 100°C. The broad exotherm from ~110°C to 275°C is attributed to the sample's heat of cure. Cooling the sample back to 0°C from 275°C, using a refrigerated cooling system accessory, allows a second heating scan to be carried out using the

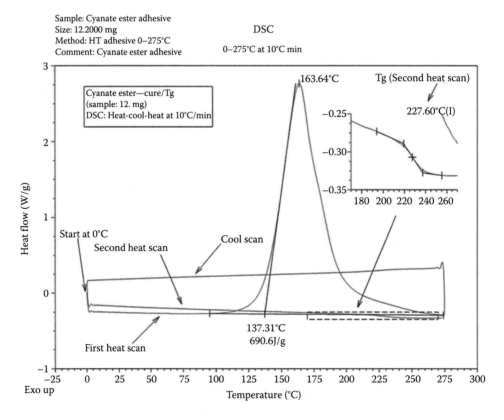

FIGURE 4.16
(**See color insert.**) DSC Heat-Cool-Heat results for a cyanate ester adhesive. (Courtesy of Bill Warner, JPL Analytical Chemistry Group.)

same conditions as the first heat cycle prior to any decomposition of the sample. The lack of an exotherm during the second scan confirms complete curing of the adhesive during the first scan while the endotherm at ~227°C is associated with the glass transition temperature (Tg) resulting from the cure cycle. While avoiding decomposition is an imperative for high-temperature performance, understanding the Tg is significant as certain effects, such as coefficient of thermal expansion mismatch, are intensified above the Tg.

It is important to note that the measured Tg of any polymeric adhesive is not only a function of the specific cure cycle but also of the measurement technique, as DMA and TMA generally give a different result. The measurement technique should be reported with any value of Tg.

4.6.1.2 TGA

TGA measures the mass of a sample as a function of temperature. This analytical technique is very complimentary to DSC. Endothermic and exothermic peaks associated with adsorption, desorption, volatilization, oxidation, and decomposition can be directly related to weight loss or gain observed by TGA. Other, non-mass related transformations such as Tg, Tc, and Tm will not be observed by TGA. The effluent from TGA can also be captured for chemical analysis to identify volatile components and decomposition products, but this may not be necessary for a screening study. Similar to DSC, the technique only requires a few milligrams of material. The uncured adhesive is generally loaded into a cleaned, tared, platinum metal hang-down pan. The pan is then loaded into the TGA furnace. The temperature of the sample is increased while continuously monitoring the mass. The rate of temperature change can be controlled and iso-thermal temperature holds can be included to simulate application conditions. This is illustrated in a three-stage TGA experiment (Figure 4.17) where the sample is rapidly heated from room temperature to 250°C at 100°C/min to approximate potential real-world cure conditions. A 1-hour iso-thermal hold follows to detect mass loss, which might be associated with volatiles. Heating the sample from 250°C to 1000°C at 20°C/min to observe weight loss associated with thermal degradation of the cured adhesive completes the experiment.

In this example, the experiment produces a small weight loss (~0.7%; Inset – Figure 4.17) during the initial heating stage of the experiment followed by very little sample weight loss, likely due to the loss of surface moisture, during the 250°C isothermal hold. Similar to DSC, this reflects the low moisture absorption and low volatility content of the adhesive. During the third phase of the experiment a small amount of mass loss (<4 wt. %) is observed from ~325 to 440°C. This can be attributed to incipient thermal degradation (most likely from non-uniform composition) and non-uniform heating during the experiment. While the onset of thermal degradation may be extrapolated to ~440°C, one might want to consider an upper use temperature, with respect to thermal degradation, to be in the vicinity of 325°C or repeat the experiment using a slower heating rate during the third phase. The sample completely decomposes to approximately 41% of the starting sample weight by ~973°C. This last data point is not particularly important in screening adhesives for high-temperature use; the onset of degradation is much more significant, but further illustrates the type of information which can be learned through TGA.

Lot to lot variation should also be considered in these thermal analyses. Without the appropriate variety of samples to characterize the variability of the material from lot to lot, appropriate safety factors or de-ratings should be considered when utilizing measured data for design.

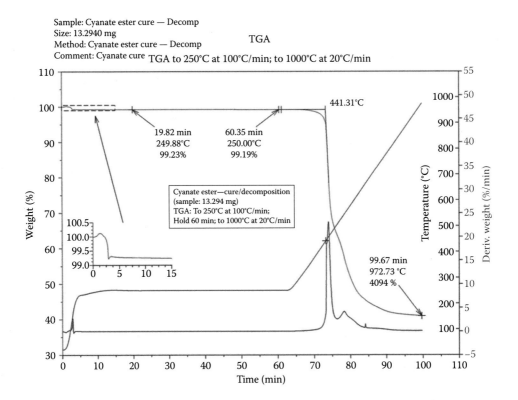

Sample: Cyanate ester cure — Decomp
Size: 13.2940 mg
Method: Cyanate ester cure — Decomp   TGA
Comment: Cyanate cure TGA to 250°C at 100°C/min; to 1000°C at 20°C/min

FIGURE 4.17
(**See color insert.**) TGA result for a cyanate ester adhesive. (Courtesy of Bill Warner, JPL Analytical Chemistry Group.)

4.6.1.3 DMA and TMA

DMA and TMA are two additional techniques that can be used to determine the glass transition temperature. In TMA a static load is applied to a cured sample and linear or volumetric dimensional changes are observed as a function of temperature. In DMA an oscillatory load is applied to a cured sample at a set frequency and the resultant deformation is observed as a function of time and temperature. This allows determination of viscoelastic effects such as damping in addition to stiffness (or modulus). Through TMA the glass transition temperature is determined from the observed change in coefficient in thermal expansion below and above the Tg. This is illustrated for a polyurethane block copolymer in Figure 4.18 with an observed Tg ~ −78°C.

Through DMA the glass transition temperature is observed as a large drop in the storage modulus. With a concurrent peak in the tan delta (tan δ), the ratio of the loss modulus to the storage modulus is also observed. This is illustrated for a semiconductor die attach epoxy adhesive in Figure 4.19 with an observed Tg ~ +143°C if the tan δ peak is used to determine the value.

While the storage modulus is not the same as the Young's modulus, it is a measure of stiffness and can be used to evaluate the change in stiffness as a function of temperature. The DMA can also reveal thermal degradation of mechanical properties, which may be useful in establishing upper-use limits for a particular material. DMA will generally

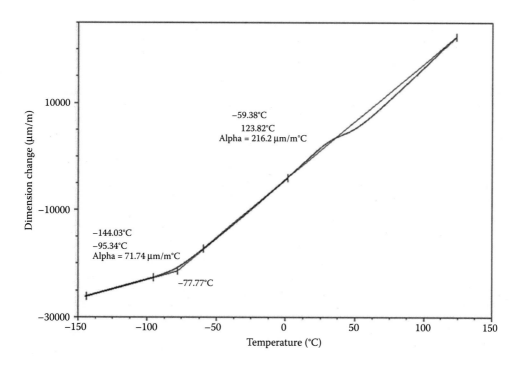

FIGURE 4.18
Example TMA thermograph for a polyurethane block copolymer. (Courtesy of Gary Plett, JPL Analytical Chemistry Group.)

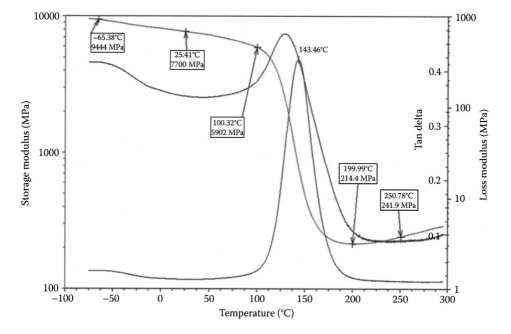

FIGURE 4.19
(**See color insert.**) Example DMA thermograph for a die attach epoxy adhesive. (Courtesy of Gary Plett, JPL Analytical Chemistry Group.)

provide more useful screening information than TMA, as the coefficients of thermal expansion above and below the Tg are often available with sufficient resolution from the manufacturer's technical data sheet.

Links to web-based resources for additional information on DSC, TGA, and DMA analysis are provided at the end of this chapter in the Internet references.

Adhesive strength and toughness are critical material characteristics for reliable joint design. Mechanical testing to determine material performance under relevant conditions, tensile, compressive, shear, or peel strength, or some combination of these, is important and non-trivial. Toughness, or amount of energy the adhesive can absorb through elastic and plastic deformation without breaking, is another vital material parameter for joint design. Toughness testing requires consistent sample preparation and good test procedures to generate useful data. A review of "Testing adhesive joints—Best practices" (da Silva 2012) is recommended prior to starting a mechanical test campaign.

4.6.2 Screening Tests for Adhesive Performance

The previous section discussed the evaluation of the high-temperature behavior of the bulk adhesive, which can be considered indicative of the cohesive performance. For high-temperature stability of an adhesive joint, the performance of the adhesive bond between the adherends and the adhesive also needs to be considered. The simplest method of evaluating the high-temperature performance of an adhesive bond is to compare lap shear test values of thermally aged specimens to a control specimen. The lap shear sample should be constructed using the same adherends as expected for the intended application. Lap shear specimens can also be used to evaluate adherend bond preparation including cleaning processes, surface preparations, and primers in thermally aged and un-aged conditions.

As mentioned previously, elevated temperature may not be the only environmental factor. In a humid environment, elevated temperature can increase water uptake into the adhesive. The adherend surface may see oxidation or reduction depending on the condition of the atmosphere. By thermally aging lap shear specimens in a relevant and controlled atmosphere to approximate the application environment, a number of candidate adhesive materials can be inexpensively screened.

T-peel tests are also widely used to evaluate adhesive performance of adhesive bonded joints for a variety of pre-treatments and adhesives. T-peel test specimens are constructed with thin flexible adherends and the test generally requires greater effort to perform than lap shear tests. Combining both lap shear and T-peel tests has been an effective method of screening automotive adhesives as a function of temperature (Grant et al. 2009; Banea and da Silva, 2010).

For a thermally variant environment thermal cycling provides another alternative test methodology, which can be used to evaluate the performance of a particular adhesive. Bond failure during thermal cycling is almost always due to the stress from coefficient of thermal expansion mismatch between the adherends. Strong adhesive bond strengths and compliance, or low elastic modulus, of the adhesive may improve the performance of these joints that experience significant thermal cycling.

For small specimens or microelectronic device applications, die shear testing is a reasonable alternative to lap shear testing. Unlike the lap shear specimen, only one adherend/adhesive interface is evaluated. An example of die shear test specimens used to screen a potential high temperature adhesive is shown in Figure 4.20. The intended application is

FIGURE 4.20
(**See color insert.**) Die shear test coupons bonded to an ill-prepared galvanized steel pipe under conditions representative of the intended application and installation environment.

the bonding of an ultrasonic transducer to a steam pipe in operation. The expected surface temperature of the steam pipe is between 200°C and 250°C with minimal surface preparation of the galvanized steel pipe allowed.

Additional examples of screening tests which consider the effect of temperature are noted in da Silva and Adams' (2005) work and Banea et al. 2010. An example of dual adhesive evaluation for broad temperature ranges use in aerospace applications is given in Marques et al. 2011, "Experimental study of silicone-epoxy dual adhesive joints for high temperature aerospace applications."

It is important to note that while there are standards and studies that relate to environmental exposure and test conditions, there is not a simple relationship between an accelerated aging test and service life. An in-situ test campaign or qualification study is ultimately necessary to validate the selected material for the end-use application.

Acknowledgments

The author thanks Linda Del Castillo, Jet Propulsion Lab (JPL), Pasadena, California; Erol Sancaktar, University of Akron, Ohio; Lucas F.M. da Silva, Editor in Chief of *The Journal of Adhesion*, University of Porto (FEUP), Portugal; and John Bishopp, Star Adhesion Limited, Cambridge, UK, for reviewing this chapter and providing valuable technical comments and suggestions.

Some of the research reported in this chapter was conducted at the Jet Propulsion Laboratory (JPL), California Institute of Technology, under a contract with the National Aeronautics and Space Administration (NASA). The author expresses his appreciation of the many scientists and engineers who contributed to the reported technologies.

References

Adams R.D., Coppendale J, Mallick V, Al-Hamdam H., *International Journal of Adhesion and Adhesives* 1992;12:185–190.

Arkles B., Hydrophobicity, Hydrophilicity and Silane Surface Modification, last accessed online 5/1/2013 www.gelest.com/goods/pdf/Library/advances/Hydrophobicity Hydrophilicity andSilanes.pdf

ASM Handbook–Composites, Daniel Miracle and Steven Donaldson (eds.), ASM International Vol. 21. Materials Park, OH, 2001, Chapter 7.

Banea M.D., Lucas F. M. da Silva, R. D. S. G. Campilho, Temperature dependence of the fracture toughness of adhesively bonded joints, *Journal of Adhesion Science and Technology* 2011–2026, 24:2010.

Banea M.D., Lucas F M da Silva, The effect of temperature on the mechanical properties of adhesives for the automotive industry, *Proceedings of the Institution of Mechanical Engineers, Part L, Journal of Materials: Design and Applications*, 2010;224: 51–62.

Bauer J., Hoper L., and Bauer M., Cyclotrimerization reactivities of mono-and difunctional cyanates, *Macromol Chem Physics*, 1998; 199: 2417–2423.

Bowditch M.R., The durability of adhesive joints in the presence of water, *Internaltional Journal of Adhesion and Adhesives*, 1996; 16: 73–79.

Cognard P. (ed.), *Adhesives and Sealants, Volume 1: Adhesives and Sealants: Basic Concepts and High Tech Bonding*, Elsevier, Amsterdam, Netherlands, 2005, Chapter 4.

Comyn J., *Adhesion Science*, Royal Society of Chemistry Paperbacks, Cambridge, UK, 1997.

Engineered Materials Handbook: Volume 2, Engineered Plastics, Dostal, C.A. (ed.), ASM International, Metals Park, OH.

Engineered Materials Handbook: Volume 3, Adhesives and Sealants, Dostal, C.A. (ed.), ASM International, Metals Park, OH.

Grant L.D.R., R.D. Adams, Lucas F.M. da Silva, Effect of the temperature on the strength of adhesively-bonded single lap and T joints for the automotive industry, *International Journal of Adhesion and Adhesives*, 2009;29: 535–542.

Hamerton I. (ed.), *Chemistry and Technology of Cyanate Ester Resins*, Chapman & Hall, Glasgow, UK, 1994.

Kinloch A.J., *Adhesion and Adhesives: Science and Technology*, Chapman & Hall, London, 1987.

Licari J.J. and Swanson D.W., *Adhesives Technology for Electronic Applications: Materials, Processing, Reliability* 2nd Edition, Elsevier, Waltham, MA, ISBN 9781437778892, 2011, pp. 1–403.

Lucas F.M. da Silva and R.D. Adams, Adhesive joints at high and low temperatures using similar and dissimilar adherends and dual adhesives, *International Journal of Adhesion and Adhesives* 2007, 27:216–226.

Lucas F.M. da Silva, Andreas Öchsner, R.D. Adams (eds.), *Handbook of Adhesion Technology*, Springer, Heidelberg, 2011.

Lucas F.M. da Silva, R.D. Adams, Measurement of the mechanical properties of structural adhesives in tension and shear over a wide range of temperatures, *Journal of Adhesion Science and Technology* 19(2): 109–142, 2005.

Lucas F.M. da Silva, D.A. Dillard, B.B. Blackman, and R.D. Adams (Eds), *Testing Adhesive Joints: Best Practices*, Wiley, Weinheim, Germany, 2012.

Marques E.A.S, D.N.M. Magalhães, Lucas F.M. da Silva, Experimental study of silicone-epoxy dual adhesive joints for high temperature aerospace applications, *Materialwissenschaft und Werkstofftechnik* 2001; 42:471–477.

Packham D.E., (ed.), *Handbook of Adhesion*, 2nd Edition, John Wiley & Sons, Ltd., West Sussex, England, ISBN 0471808741, 2005, pp. 1–638.

Petrie E.M., (ed.), *Handbook of Adhesives and Sealants*. 2nd Edition, McGraw Hill, New York, NY, ISBN 0071479163, 2007, pp. 1–1048.

Schneberger G.L., (ed.), *Adhesives in Manufacturing*, Marcel Dekker, New York, NY, ISBN 0824718941, 1983, pp. 1–682.

Stuart T.P, Crouch I.G., The design, testing and evaluation of adhesively bonded, interlocking, tapered joints between thick aluminium alloy plates, *International Journal of Adhesion and Adhesives* 1992, 12: 3.

Internet Resources

AI Technology (Epoxy, Cyanate Ester)—www.aitechnology.com/

DMA—http://www.perkinelmer.com/CMSResources/Images/44-74546GDE_IntroductionToDMA.pdf

Dow Corning (Cyanate Ester, Silicone)—www.dowcorning.com

DSC—http://www.perkinelmer.com/CMSResources/Images/44-74542GDE_DSC BeginnersGuide.pdf

Dupont (PI)—www2.dupont.com

Epoxy Technology (Epoxy, PI)—www.epotek.com

GE Sealants and Adhesives (Silicone)—www.gesealants.com

Gelest (Silanes, Surface Pretreatments)—www.gelest.com

Henkel (Epoxy)—www.henkel.com

Hexcel (Epoxy, BMI)—www.hexcel.com/

Huntsman (Araldite, Tactix Specialty Epoxy)—www.huntsman.com/advanced_materials/a/Home

Maverick (PI Resins)—www.maverickmolding.com

NuSil (Controlled Volatility Silicone)—www.nusil.com

Outgassing Data for Selecting Spacecraft Materials Online—outgassing.nasa.gov

Renegade Materials (Epoxy, BMI, PI adhesives)—www.renegadematerials.com

TGA—www.perkinelmer.com/CMSResources/Images/44-74556GDE_TGABeginners Guide.pdf

5

Oxidation of High-Temperature Aerospace Materials

James L. Smialek and Nathan S. Jacobson

CONTENTS

5.1 Introduction

High-temperature processes and applications require their associated components to reach elevated temperatures that produce consequences on material properties. High-temperature oxidation is one such consequence and results from chemical reactions with the surrounding environment. Depending on severity, material performance can be degraded by surface recession or internal embrittlement, placing limits on use temperatures and times. Coatings, cooling schemes, or gas modifications are often employed to counteract these effects and provide a safer operational envelope. The purpose of this chapter is to provide a general description of the physical process, with specific examples chosen to highlight the prominent classes of high-temperature materials. It is hoped it will serve as a brief introduction and overview. For more detailed treatment of fundamentals and a broader list of topics, the reader is referred to classic texts (Kofstad 1988; Lai 1990; Birks, Meier et al. 2006; Young 2008). More detailed reviews can also be found in the latest edition of Shreir's Corrosion, 2009. Here, dedicated chapters are found for defects and diffusion in oxide scales, stress effects, sulfidation, steam, and molten salt corrosion, steels, alumina and chromia scale forming alloys, coatings, intermetallics, and ceramics (Cottis et al. 2009).

Oxidation and corrosion of high-temperature materials is important in a wide range of applications. Many industrial processes such as rolling, forging, heat treating, chemical refining, and so on rely on high temperatures of the product material as well as the process apparatus. Heat engines, heaters, heat exchangers, exhaust catalysts, boilers, and fuel cells are some specific examples. In the aeronautics and space technology fields, these applications include the hot stages of aircraft engines, rocket engines, and re-entry surfaces on spacecraft. Turbine engines have continually advanced over the decades, extracting optimum performance from all components and materials. Issues associated with aircraft turbines have driven many studies of high-temperature corrosion and provided intense study and deep insights, as highlighted in this chapter.

An understanding of the oxidative and corrosive processes in an application involves first defining the environment (gas composition, temperatures, flow rates) to which the component is subjected. Aircraft turbine engines typically burn a high-purity hydrocarbon fuel. A plot of equilibrium combustion products versus fuel-to-air ratio is given in Figure 5.1 at typical adiabatic flame temperatures of ~1000–2400 K. These were calculated with the NASA CEA code (Gordon and McBride 1971) and a total pressure of 1 bar. Actual total pressures in a gas turbine may be 10 bar or more.

Note that the major reaction products—CO_2 and H_2O—are constant at ~10%. Fuel-rich combustion leads to excess $H_2(g)$ and $CO(g)$, whereas fuel-lean combustion leads to excess $O_2(g)$. There are also numerous minor species such as sulfur oxides and nitrogen oxides. Under some conditions, condensed phase salts may deposit. Figure 5.2 illustrates a gas turbine engine with the major high-temperature corrosion/degradation issues listed for each stage.

Hot-stage oxidation of rocket engines is more complex and depends on the particular type of rocket fuel used. There are a variety of rocket fuels, including RP-1 (kerosene), liquid oxygen (LOX)/liquid hydrogen, and various solid fuels. The most common of these is liquid oxygen/liquid hydrogen (LOX/H_2), which has been used on the U.S. Space Shuttle, Atlas Centaur, and Saturn Rockets. For rocket engines, the combustion conditions are typically given as oxidizer-to-fuel (O/F) ratio. Figure 5.3 shows a diagram of equilibrium products for the hydrogen/oxygen combustion as a function of O/F ratio, at typical adiabatic flame temperatures of ~1000–3400°C (Gordon and McBride 1971; Opila 2008).

In contrast to aircraft turbine engine hot stages and rocket engines, re-entry surfaces can reach extremely high temperatures (>2000°C, depending on re-entry trajectory) with significantly lower total gas pressures and very short exposure times (approximately minutes). Figure 5.4 shows some trajectories and the resultant atmospheres of dissociated oxygen and nitrogen. From a materials stability point of view, this is quite important as atomic oxygen and nitrogen are very reactive.

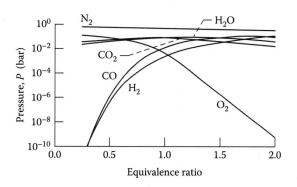

FIGURE 5.1
Calculated equilibrium pressures for primary combustion gas products as a function of equivalence ratio with the total pressure fixed at 1 bar. An equivalence ratio, φ, of 1.0 is defined as the stoichiometric fuel-to-air ratio with φ < 1 fuel lean and φ > 1 fuel rich. (Jacobson, N. S.: Corrosion of silicon-based ceramics in combustion environments. *J. Am. Ceram. Soc.* 1993. 76(1). 3–28. Copyright Wiley-VCH Verlag GmbH & Co. KGaA. Reproduced with permission.)

The shape of the re-entry surface is important in determining temperature. Large radii edges, such as those found on the U.S. Space Shuttle, do not reach temperatures as high as sharp edges. Currently, there is interest in developing sharp-edged re-entry vehicles for their higher maneuverability. These vehicles will necessarily require the development of higher-temperature materials. These hypersonic vehicle applications are illustrated in Figure 5.5.

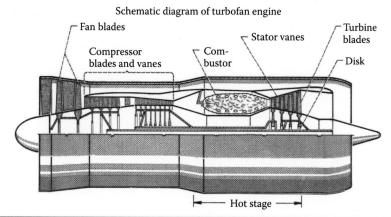

Component	Typical operating conditions			Critical problems
	Temperature (°C)	Stress (MPa)	Life (hr)	
Blades	900 – 1050	140 – 210	5000	Creep strength, stability, oxidation, hot corrosion, and thermal fatigue
Vanes	950 – 1100	35 – 70	5000	Thermal fatigue, oxidation, and hotcorrosion
Disks	400 – 650	420 – 1050	15,000	Low cycle fatigue hot corrosion
Combustors	850 – 1100	20 – 35	4000	Thermal fatigue oxidation

FIGURE 5.2
Diagram of gas turbine engine showing hot stage components and typical operating conditions with critical problems. Future engines continue to push the operating temperature, stresses, and life expectancies to high levels.

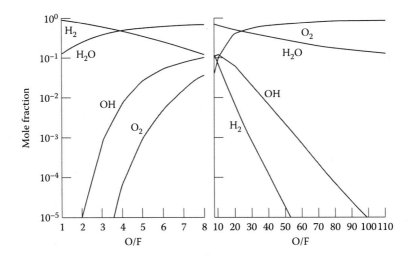

FIGURE 5.3

Rocket combustion gas composition as a function of oxidizer-to-fuel ratio. Typically, an LOX/H$_2$ runs in a regime with excess H$_2$ and water vapor. (From Opila, E. J. 2008. *SiC Recession in a Hydrogen/Oxygen Environment Combustion Environment*. 31st Annual Conference on Composites, Materials, and Structures, Daytona Beach, FL, American Ceramic Society.)

Regions with Chemical and Thermal Nonequilibrium		Chemical Species in High-Temperature AIR		
Region	Aerothermal Phenomenon	Region	Air Chemical Model	Species Present
Ⓐ	Chemical and thermal equilibrium	Ⓘ	Two species	O$_2$, N$_2$
Ⓑ	Chemical nonequilibrium with thermal equilibrium	ⒾⒾ	Five species	O$_2$, N$_2$, O, N, and NO
Ⓒ	Chemical and thermal nonequilibrium	ⒾⒾⒾ	Seven species	O$_2$, N$_2$, O, N, NO, NO$^+$, and e$^-$
		Ⓘ$_\mathrm{V}$	Eleven species	O$_2$, N$_2$, O, N, NO, O$_2^+$, N$_2^+$, O$^+$, N$^+$, NO$^+$, and e$^-$

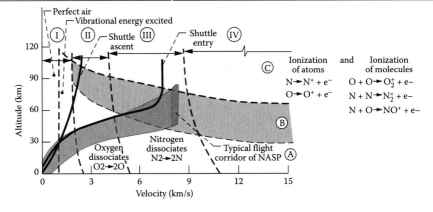

FIGURE 5.4

Flight trajectories for several hypersonic vehicles showing high velocities and dissociated air. (From Gupta, R. N. et al. 1990. A Review of the Reaction Rates and Thermodynamic and Transport Properties for an 11-Species Air Model for Chemical and Thermal Nonequilibrium Calculations to 30 000 K. NASA Reference Publication 1232.)

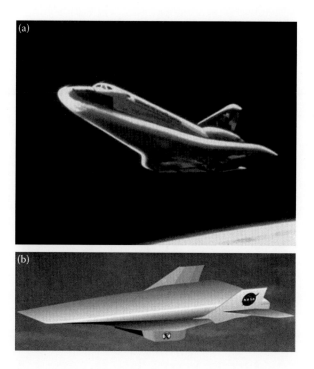

FIGURE 5.5
(**See color insert.**) Hypersonic vehicles. (a) Artist's rendition of the Space Shuttle Orbiter on re-entry with its large radii leading edges and blunt nose, which leads to lower temperature, but less maneuverability. (b) Artist's rendition of the proposed Hyper X vehicle with sharp edges and more maneuverability. (From Sozer, E. 2012. Non Equilibrium Gas and Plasma Dynamics Laboratory, http://ngpdlab.engin.umich.edu/static-pages/img/hypersonic-interaction-flows_hyper-x.png. Accessed December 14, 2012.)

Beyond these aeronautic and space applications, high-temperature oxidation and corrosion is important in a variety of other applications. Utility turbines and marine turbines involve many of the same considerations as aero turbines, but there are important differences due to generally lower-purity fuels and larger components. Heat exchangers may also involve combustion atmospheres. Chemical process applications are system specific and range from highly oxidizing to highly reducing/carburizing (Butt et al. 1992; Steinmetz and Rapin 1997).

5.2 Experimental Considerations

Clearly, materials cannot be tested in the actual application and it is the task of the corrosion scientist to devise testing methods that closely simulate the actual environment with minimal expense and small sample size. For aircraft gas turbines, materials are often tested in jet fuel burners, as illustrated in Figure 5.6. Such systems are generally heavily instrumented for accurate control and monitoring of system parameters (e.g., temperature, flow rates of air and fuel, video monitoring of sample). They can be modified to operate with a variety of fuels of different purity levels, including injected salt impurities and even test specimens under tension or compression (Fox et al. 2011).

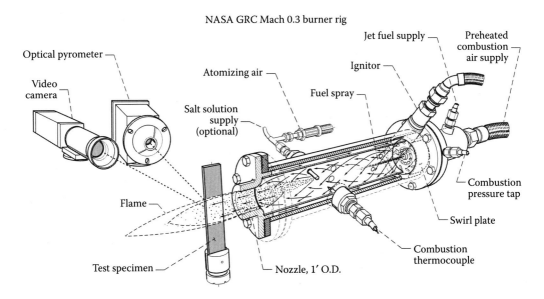

FIGURE 5.6
Schematic of NASA GRC Mach 0.3 Burner rig. (From Fox, D. S. et al. 2011. Mach 0.3 burner rig facility at the NASA Glenn Materials Research Laboratory. NASA/TM-2011-216986.)

Analogous test rigs for rocket engine materials are illustrated in Figure 5.7. As noted, there is a range of rocket engine configurations and, hence, a range of rocket engine material test rigs. The schematics in Figure 5.7 are for high-velocity and low-pressure or low-velocity and high-pressure H_2/O_2 tests. These are typically for short-duration pulses.

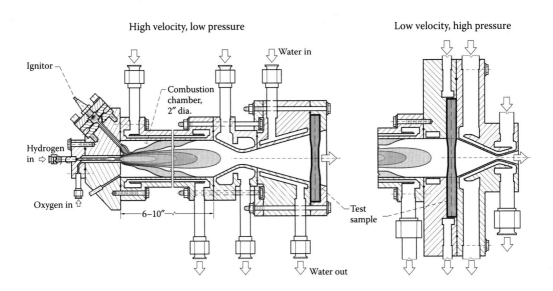

FIGURE 5.7
Schematic of two types of rocket test rigs with either high velocity and low H_2/O_2 pressure or low velocity and high H_2/O_2 pressure.

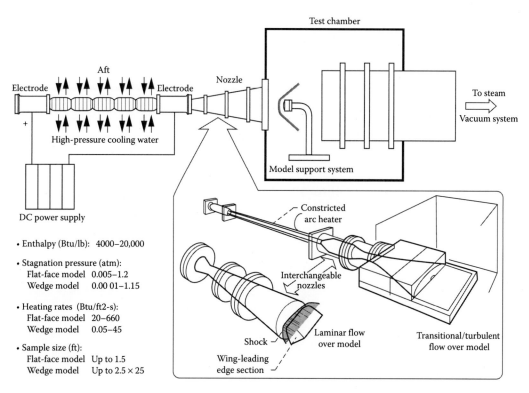

FIGURE 5.8
Schematic of arc-jet. (From www.nasa.gov/centers/ames. Accessed January 2013.)

Re-entry environments are best simulated with an arc-jet, which provides a high velocity of atoms, from dissociated O_2 and N_2, impinging on a sample. Dissociation occurs in a series of electrodes and the sample is subjected to a high temperature and dissociated air, which simulates re-entry environments. A schematic of an arc-jet is shown in Figure 5.8.

The test rigs described thus far are essential to provide accurate near-application reliability data. However, for understanding fundamental oxidation or corrosion mechanisms, laboratory furnaces are generally used. These can be box furnaces for static laboratory air exposures at high temperatures or tube furnaces if atmospheres and flow rates need to be controlled. In some cases, optical methods can be used to follow the growth of an oxide scale (Ogbuji and Opila 1995). In order to follow a corrosion process in real time, many groups use a thermogravimetric apparatus, which is a sensitive balance coupled to a vertical tube furnace. Here, mass (weight) change is typically associated with the amount of oxygen incorporated in a growing oxide scale. As a common example, the amount of oxygen gain experienced for alumina scales in hundreds of hours at 1100–1200°C is on the order of 1 mg/cm² or less. There are many designs of these systems: a typical one is illustrated in Figure 5.9. A variety of gas atmospheres can be used in this type of system, such as air, oxygen, CO_2, H_2, SO_2, and H_2O. For corrosive gases, a counter stream of an inert gas is run through the balance. For gases that condense, such as $H_2O(g)$, all plumbing outside of the furnace must necessarily be heated.

In actual application, components are often cycled through heating and cooling cycles. In the laboratory, this can be done manually by simply removing the sample from the furnace to simulate cooling and then reinserting the sample for heating. However, to

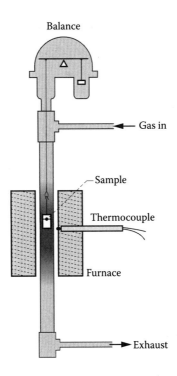

FIGURE 5.9
(See color insert.) Thermogravimetric apparatus for studying isothermal oxidation/corrosion under controlled conditions with in-situ monitor of mass changes. These mass changes as a function of time give the kinetics of the process.

realistically test for many hundreds or thousands of cycles, an automated test rig such as the one shown in Figure 5.10 is necessary. Samples are removed intermittently for weight change measurements.

After an oxidation/corrosion exposure, the resultant oxidation products are analyzed with a variety of microstructural analysis techniques (Bennett 1995; EFC 1995; Quadakkers and Viefhaus 1995; Rahmel and Kolarik 1995). These include x-ray diffraction (XRD), optical microscopy, scanning electron microscopy (SEM), and transmission electron microscopy (TEM).

Microscopy is essential in the examination of the surface or cross section of the oxidation/corrosion products growing on an alloy or ceramic substrate. For a thick product scale, a standard metallographic polished cross section is quite useful. If portions of the oxide are water-soluble, nonaqueous polishing compounds must be used. For a thin oxide scale, TEM with dimpling, focused ion beam thinning (FIB), or taper sections may be useful.

5.3 Major Features of High-Temperature Oxidation and Corrosion

5.3.1 Overview

In the strictest chemical sense, "oxidation" refers to the loss of an electron to another element. The most familiar cases would be those of transition element base metals (e.g., Fe,

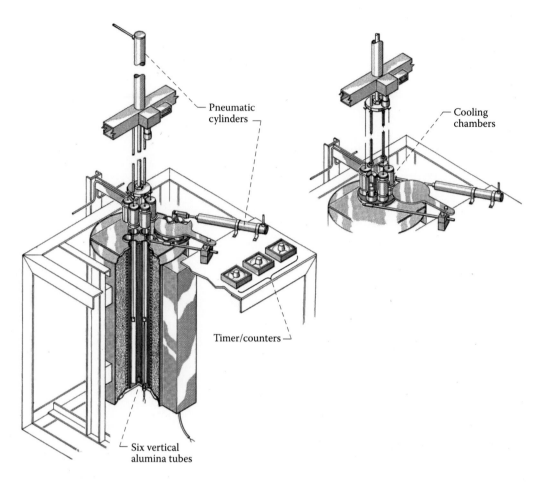

FIGURE 5.10
Automated cyclic test rig.

Co, Ni, Cu) reacting with oxygen present in the atmosphere to form ionic oxide scales such as FeO, Fe_2O_3, Fe_3O_4, NiO, and so on. However, depending on the gaseous environment, a wider breadth of reactions (sulfidation, nitridation, carburization) fall under this definition. In general, it can be stated that the thermodynamic stability of the product, defined by the free energy of formation, ΔG_f°, follows this trend: oxides > chlorides > sulfides > nitrides > carbides (Searcy et al. 1970). This chapter will deal primarily with oxidation and oxide scales as the most common reaction product, but complex industrial applications must deal with the entire spectrum of "oxidation" or reaction products.

The reaction can be described in most general terms by the series of equations below:

$$xM = xM^{+y} + xye^- \tag{5.1}$$

$$xM^{+y} + xye^- + \frac{1}{2}yO_2 = M_xO_y \tag{5.2}$$

$$\Delta G_f^0 = -RT\ln K_{eq} \tag{5.3}$$

$$K_{eq} = \frac{a_{M_xO_y}}{a_M^x (p_{O_2})_{eq}^{\frac{1}{2}y}} \tag{5.4}$$

The first equation refers to the loss of y electrons each from x atoms of metal M. The second equation shows the chemical reaction between the metal ion and molecular oxygen to form a stoichiometric oxide. The third equation defines the equilibrium constant, K_{eq}, in terms of the free energy of the reaction, and the last equation defines the equilibrium constant according to the activity ratios of the products and reactants.

These relations are of paramount importance in predicting which reactions will take place given the activity or partial pressure of reactants and established thermodynamic quantities for oxide phases as a function of temperature. In that regard, it is instructive to view the stability of a collection of oxides versus temperature and oxygen pressure using the simplified Ellingham/Richardson diagram in Figure 5.11 (Swalin 1962). Here, the thermodynamic stability, given as $-\Delta G_f^\circ$, is plotted as a function of temperature for a number of common oxides. The most stable oxide shown is MgO, while the least stable is Cu_2O. The nomograph form also allows one to quickly estimate the equilibrium $p(O_2)$ at which the oxide is stable. For the case of Al_2O_3 at 1000°C shown here, $p(O_2) \approx 10^{-34}$ atm. for equilibrium between oxide and pure metal. Thus, it is difficult to prevent oxidation of aluminum, even when present in very small amounts (activity) in an alloy. Similar exercises can be performed using CO/CO_2 or H_2/H_2O as oxidant mixtures.

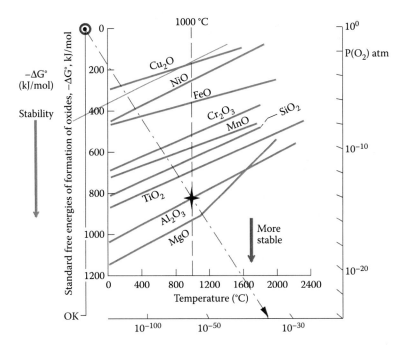

FIGURE 5.11
(See color insert.) Modified Ellingham/Richardson diagram showing the relative thermodynamic stability (free energy of formation) of various oxides. The corresponding equilibrium gas composition can be found using the nomograph by drawing a line from the focus point to the oxide curve at the temperature of interest and extending to the side axes for O_2, H_2O/O_2, and CO/CO_2 equilibrium pressures. (Swalin, R. A.: *Thermodynamics of Solids*. 1962. Copyright Wiley-VCH Verlag GmbH & Co. KGaA. Reproduced with permission.)

While thermodynamics dictate which oxides (or oxidation products) are most stable, the corresponding kinetic rates will also influence which phases prevail for a specific condition, especially regarding the phenomenon of transient oxidation. Kinetic considerations also determine the rate of material consumption once the stable phase forms as a continuous, rate-controlling layer. Once this layer is developed, rates of ionic diffusion (molecular for O_2 and H_2O in SiO_2 scales) generally dictate a parabolic rate of scale growth, k_p, according to the Wagner relation (Kofstad 1988), where x is scale thickness (or oxygen mass gain), z_c and z_a are the respective cation and anion valencies, and D_c and D_a are the respective cation and anion effective diffusion constants in the scale:

$$x^2 = k_p t; \quad k_{p,i} = 2x \frac{dx}{dt}$$

$$k_p = \int_{P_{O_2},\,int}^{P_{O_2},\,gas} \left\{ \frac{z_c}{-z_a} D_c + D_a \right\} d\ln P_{O_2} \tag{5.5}$$

Considerable effort has gone into rationalizing k_p in terms of independently determined diffusivities and vice versa, with partial success. Some ramifications may need to be considered regarding short circuit diffusion paths, such as grain boundaries. Other concerns are the effects of small amounts of impurities in the scales, which may serve as aliovalent dopants and affect the defect concentrations and diffusivities in a scale compared to pure bulk material. A final concern is that diffusivities are often determined from bulk material studies performed under atmospheric conditions. In contrast, the precise Wagner solution requires knowledge of D as a function of the pO_2 gradient in the scale, as indicated in Equation 5.5.

Despite these caveats, it is very instructive to compare materials based on the measured parabolic rate constants, as shown in Figure 5.12. Here, a wide experimental regime can

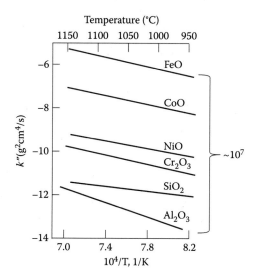

FIGURE 5.12

Arrhenius plot of growth rates for common oxide scales showing superiority of slower Al_2O_3, SiO_2, and Cr_2O_3 compared to base alloy oxides of Ni, Co, and Fe. (Smialek, J. L. and G. H. Meier: High temperature oxidation. In *Superalloys II*. ed. C. T. Sims, N. S. Stoloff and W. C. Hagel. 293–326. 1987. Copyright Wiley-VCH Verlag GmbH & Co. KGaA. Reproduced with permission.)

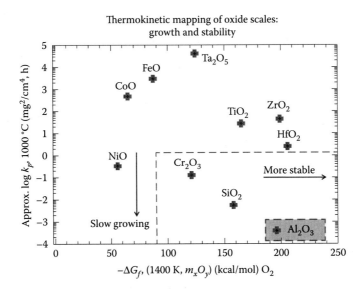

FIGURE 5.13
Thermokinetic map of various oxide scales showing parabolic rate versus free energy of formation. Optimum performance for stable slow-growing scales in lower right.

be easily displayed by means of Arrhenius plots of log k_p versus $1/T(K)$, where the slope gives the activation energy for oxidation and is related in part to diffusivity. It is apparent that Al_2O_3 scales grow up to 10^7 times slower than FeO scales, explaining why highly oxidation-resistant FeCrAlY heater alloys are preferred to simple Fe alloys or Fe–C steels. Furthermore, SiO_2 and Cr_2O_3 scales are also advantageous from the standpoint of slow growth rate.

For additional perspective, we present some scale properties on a thermokinetic map. This categorizes scale growth rates versus the thermodynamic stability for some typical oxides (Figure 5.13). The presumption is that the lower-growth and higher-stability oxides provide the most potential for protective behavior. Again we see Al_2O_3, SiO_2, and Cr_2O_3 in the optimum corner of the diagram and one or both properties lacking for the other oxides.

As these three oxides form the basis for most protective scales, a summary of their key properties is given in Table 5.1 (Ramanarayanan 1988; Wada et al. 2011; Smialek et al. 2013; Ogbuji and Opila, 1995), with respect to oxidation rates, transport, adhesion, water vapor, and salt corrosion. These features will be referred to in the subsequent discussions of specific material systems where we will review a number of different classes of common engineering materials, including some advanced concepts that remain under development. The intent is to provide a broad perspective overall, with detailed understanding for the more intensely studied areas.

5.3.3 Metals, Alloys, and Intermetallics

The common feature of the following sections is oxidation under ambient oxygen pressures typical of an earth-based environment, that is, 20% O_2 (0.2 atm, 0.2 bar, or 20 kPa). Under these conditions, virtually every metal in the periodic table will form a stable condensed oxide scale except for some of the precious metals such as Au and Pt. This means that any alloying elements to the base alloy are also subject to oxidation and may take part

TABLE 5.1

Kinetic and Chemical Sensitivities of Protective Al_2O_3, Cr_2O_3, and SiO_2 Scales

	Al_2O_3	Cr_2O_3	SiO_2
Rate controlling	Oxygen grain boundary dominates	Duplex, grain boundary Cr and O	Neutral oxygen permeation
$P(O_2)$ rate dependence	None observed[a]	None observed[a]	$[P(O_2)]^1$
Activation energy	380 kJ/mole for grain boundary oxygen	244 kJ/mole for undoped 263 kJ/mole for Y-doped	Low, 119 kJ/mol
Dopant/impurity effects	Growth reduced by 3×; high for adhesion	Growth reduced 10× by Y; high for adhesion	Very sensitive
Water vapor effects	Minor, transition aluminas transform faster, adhesion reduced in special cases	Major, volatility	Moderate, growth and volatility
Salt corrosion	Acidic fluxing to Al^{3+} Basic fluxing to $Al_2O_4^{2-}$	Acidic fluxing to Cr^{3+} Basic fluxing to $Cr_2O_4^{2-}$	Basic fluxing to SiO_3^{2-}

[a] Predicted diffusivity dependence: $P(O_2)^{-1/6}$ at metal interface (oxygen) $P(O_2)^{3/16}$ at gas surface (aluminum).

in both the transient and steady-state regime of scale growth. The sections are organized into classes of alloys for simplicity and application temperatures to some degree.

However, there are some generalities that can be summarized first. To varying degrees, base alloy elements used to formulate structural alloys, such as Ni, Co, Fe, Ti, Cu, and so on, are alloyed with multiple elements to provide a delicate balance of good mechanical properties and oxidation resistance. The oxidation behavior the base metals is characterized by simple, but fast-growing scales, such as NiO, CoO, Fe_2O_3, Cu_2O, and TiO_2. As discussed above, Al, Cr, and Si are the most useful in providing oxidation resistance. As these various elements are added at low levels, the oxidation behavior changes to internal oxidation of the more stable oxide. This amount increases with additive, as does the potential for mixed external oxides. For example, the spinel structure, for example, $NiAl_2O_4$, accommodates multiple cations (Co, Fe, Cr) over wide compositional ranges. Finally, at yet higher concentrations, a continuous external protective scale of the most stable oxide is developed with no internal oxides. Unfortunately, this often occurs at levels where substantial alloy phase changes result in the base metal and mechanical properties are seriously affected. For example, Ni–25Al, Fe–40Al, Ti–75Al, Ni–25Cr, Fe–25Cr, Ni–5Si, Fe–5Si, and Ti–67Si (atomic%) give the approximate high levels where steady-state scales of Al_2O_3, Cr_2O_3, and SiO_2 may be formed, under conditions of a reasonable temperature (above ~800–1000°C) and high oxygen pressure (above ~1 mbar O_2). For example, at lower temperatures, the establishment of a continuous protective layer may be seriously hampered by coexistent base metal oxides leading to an extensive period of transient oxidation. While most protective oxides would be stable well below ~10^{-15} bar O_2 for 1 bar total pressure (i.e., O_2 mixtures with inert gas), the base metal can evaporate at rates faster than oxidation at total pressures of ~10^{-3} bar.

To avoid mechanical property effects, a secondary oxygen getter is employed, such as Cr or Si to Ni–Al alloys to assist in exclusive Al_2O_3 scale growth (Smialek et al. 1997; Brady et al. 2000). Here, two effects on the sequence of transient oxidation ensue. One is the assistance of a second stable oxide in initial surface coverage of the alloy. This tends to "choke out" further growth of the fast-growing nonprotective NiO base metal oxide. The second effect is to lower the interfacial pO_2 to the equilibrium value for the secondary getter scale, given by Equation 5.4 and its metal activity in the alloy. Empirically, this secondary alloying element

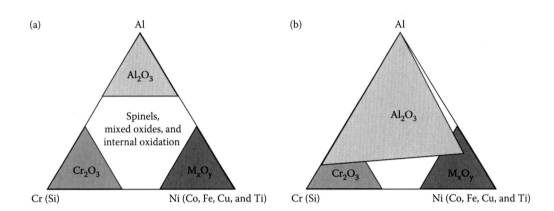

FIGURE 5.14
(**See color insert.**) Notional oxide phase schematic of how alloy compositions affect scale make-up. (a) symmetric, idealized: single oxides near corresponding single element corners; mixed oxides near mixed alloy compositions. (b) asymmetric, real: regions can be highly skewed, allowing for maximum effect with minimum alloy content. Here Cr (or Si) are shown to increase the Al_2O_3 field for a base alloy of pure Ni (Co, Fe, Cu, or Ti).

can be very powerful in reducing the total alloying content needed for a protective scale. This effect is shown schematically in the notional ternary diagrams of Figure 5.14a and b.

Ideally, if all components behaved similarly, one might expect symmetrical transitions to alternate external scales to look like Case A in a ternary map of scale formation. However, the synergistic role of the secondary getter produces a broadened protective zone (Case B) of the most powerful additive, as surmised from its position on a thermokinetic diagram. These concepts factor into the phenomenon known as transient oxidation. This is a period of transition during which oxides of the other elements are manifested, preceding steady-state growth of only the stable, healing, protective layer. Higher oxidation rates are associated with this period. While this phenomenon may proceed for a significant duration, the sooner it disappears, the more likely the alloy is useful as an oxidation-resistant material. Temperature may play a role in this regard, as lower temperatures allow transient effects to persist longer.

5.3.3.1 Copper Alloys

Copper, with a melting point of 1083°C, is not normally considered a high-temperature alloy, but deserves special consideration because the copper alloy NARloy–Z (Cu–3Ag–0.1Zr, wt.%) was used in the Space Shuttle main engine exhaust nozzle. This application of Cu materials derives from the intense thermal load of the hydrogen–oxygen rocket exhaust, exceeding 3000°C, that would destroy most materials. But backside cooling with –250°C liquid H_2 and the high thermal conductivity of Cu allows this alloy to survive. While classic oxidation with an intact surface scale is not observed, the material degrades by a phenomenon known as "blanching." Here, transients in the exhaust gas, perhaps from start-up and shut-down, produce a cyclic oxidation–reduction process with material consumption and weakening of thin cross sections. This process is allowed by the relative instability of CuO and Cu_2O scales (Figure 5.11) that can be easily reduced. Nonetheless, there are many oxidation studies on Cu alloys, including various Cu–Cr–Al combinations (Ogbuji and Humphrey 2000; Ellis 2005; Raj 2008a, b). In one example, it was seen that 17% Cr (wt.%) reduced growth rates, but was insufficient to form a healing layer of Cr_2O_3. Cr was located primarily in α-Cr second phase precipitates because of limited solubility of Cr in Cu, and could not participate in an overall healing layer. However, the addition of 5% Al did result in dramatic growth rate reduction by forming healing layers of aluminum oxides that prevented Cu oxides.

5.3.3.2 Iron Alloys (Steels)

Steels represent the most widespread use of metals on a weight basis, with significant applications at (somewhat) elevated temperatures (500°C), typically using Cr–V–Mo steels. Applications include boiler vessels and steam tubes, gas scrubbers, exhaust pipes, turbine shafts, and early turbine compressor blades. Though not generally considered a high-temperature material, sheer volume of use warrants some discussion here. Inexpensive, low alloy carbon steels can be used below ~400°C where oxidation is less of a problem. For higher temperatures up to ~650°C, higher Ni and Cr contents are required for strength and oxidation resistance, leading to more costly stainless steels. At one extreme, the oxidation behavior of pure iron is characterized by fast growing layers of FeO wustite, Fe_3O_4 magnetite, and Fe_2O_3 hematite. At the other extreme, Fe–25Cr can form continuous healing layers of slow-growing Cr_2O_3. Commercial alloys develop complex arrangements of $(Cr,Mn,Fe)_2O_3$, sesquioxides, and $Fe(Fe,Mn,Cr)_2O_4$ spinels, often disrupted by external fast-growing Fe-rich nodules. For the most part, load-bearing low-carbon steels have been relegated to relatively low-temperature regimes (~370°C), due in part to poor oxidation resistance. The high Cr, Ni austenitic stainless-steel alloys show better oxidation resistance and have useful strength at 650°C. Furthermore, various types of stainless steels can exhibit a wide variation in oxidation behavior (Figure 5.15) (Barrett 2003). To some extent, this can be associated with the Ni and Cr content, as well as cost, of the alloy.

However, considerable progress has been made recently in developing an alumina-forming austenitic (AFA) alloy with good high-temperature strength and oxidation resistance (Yamamoto et al. 2007; Brady et al. 2008). Here, a delicate balance between alloying for strength (Nb, C, B) and alloying for oxidation resistance (Al, Cr, Si, Hf, Y) has been

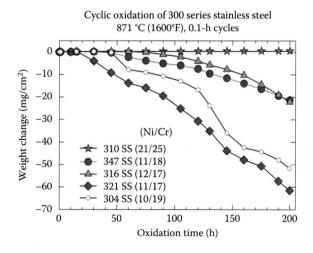

FIGURE 5.15
Trends in cyclic oxidation behavior of various common stainless steels and (Ni/Cr) content (871°C, 0.1 h cycles). (From Barrett, C. A. 2003. A High Temperature Cyclic Oxidation Data Base for Selected Materials Test at NASA Glenn Research Center. NASA/TM—2003-212546.)

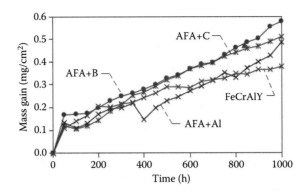

FIGURE 5.16
Cyclic oxidation of Al_2O_3-forming austenitic Fe–Ni–Cr–Al stainless steel in 950°C Air + 10% H_2O (10-h cycles). (With kind permission from Springer Science+Business Media: *Oxid. Met.* Increasing the upper temperature oxidation limit of alumina forming austenitic stainless steels in air with water vapor. 75(5–6), 2011, 337–357. Brady, M. P., K. A. Unocic et al.)

achieved, with exceptional improvements in oxidation behavior, as shown in Figure 5.16. Equally important, the AFA alloy costs about 2 times that of 347, but significantly less than commercial high-strength chromia formers, such as Haynes Alloy 230 or Incoloy 625.

To continue in this progression, perhaps some of the most oxidation-resistant alloys can also be considered in this Fe-based category, the FeCrAlY heater alloys. These generally have ~20–25% Cr, 5% Al, and 0.1 Y. Variations on this theme may employ Si, Ti, Zr, or Y_2O_3 at levels generally less than 1%, with trade names such as Kanthal, Hoskins 875, and Imphy FeCrAl. Other than a dispersion-strengthened MA956 alloy, FeCrAl is relatively weak and is not considered a structural alloy. However, it has superior qualifications for high temperature ($\geq$1100°C) applications, such as resistance heater elements, catalytic converter substrates, and heat exchangers. Their oxidation resistance is derived from the formation of a slow-growing adherent alumina scale. These scales will be discussed further in the following section on high-temperature nickel-based turbine alloys, where alumina scales and their fundamentals have been a primary focus of protective coatings and advanced single-crystal superalloys.

5.3.3.3 High-Temperature Nickel-Based Alloys

High-temperature Ni-based alloys can trace their origins to early nichrome alloys, where Cr additions produced Cr_2O_3 scales and provided some oxidation resistance at intermediate temperatures. More complex Fe–Ni–Cr Inconels offer cost-effective wrought materials, evolving with refractory element (Nb, Ta, Mo, W) additions for strengthening. Cast alloys often have aluminum additions around 5% (by weight). This results in a high volume fraction of a submicron, coherent γ'-Ni_3Al strengthening phase as well as increased oxidation resistance from Al_2O_3 scales.

Historically, a large body of oxidation research and alloy development was derived from the use of cast Ni-based superalloys in turbine engines. Key components include the combustor liner, turbine blade and vane airfoils, and the disks to which the airfoils are attached. The driving force for advanced material development is greater turbine efficiency and reduced emissions produced by higher operating temperatures. For the sake of example and continuity, the focus will be on the first-stage blades, that is, airfoils that see some of the highest loads, highest temperatures (~1100°C), and are the most expensive components. Typical single-crystal compositions revolve around 5–10 Co, 5–10 Cr, 5–6 Al, 6–12 Ta, 0–2 Ti, 0–2 Nb, 0–2 Mo, and 4–6 W wt.%, balance Ni. These have evolved from conventional cast or wrought alloys that often had higher Cr and lower Al, and occasionally higher W, Mo, Nb, Ti, or V. Complex phase chemistry and oxidation effects make simple generalizations difficult, but it can be said that Cr and Al provide oxidation resistance, while Ti, Nb, and Ta segregate to and strengthen the γ'-Ni_3Al cuboidal high volume percent fine precipitate strengthening phase. High-melting and slow-diffusing large Mo, W, and Re additions provide γ-Ni solid solution strengthening and creep resistance. Early single crystals contained 10% Cr and 2% Ti, but now are trending toward 2.5% Cr, no Ti or Nb, with 3–6% Re.

The interplay between Cr and Al follows much of what has been learned about transient oxidation and ternary oxide maps. Specifically, Cr assists Al in sealing an exposed metal surface by the simultaneous formation of Cr_2O_3 and Al_2O_3, both of which are thermodynamically preferred to fast-growing NiO, as seen in Figures 5.11 and 5.13 (Giggins and Pettit 1971). Eventually, a complete layer of Al_2O_3 is grown, and the oxidation rate is stabilized at a low level. Other oxidation features are the coexistence of $NiAl_2O_4$ spinel layers and $NiTa_2O_6$ tri-rutile as dispersed particles. These transient oxides appear to be more tolerable than NiO, $NiCr_2O_4$ spinel, or TiO_2 rutile phases by not increasing growth rates substantially or triggering massive spallation. These mechanisms are reflected in oxide maps that show a dramatically reduced Al content (5%) required to form exclusive Al_2O_3 scales by adding ~5% Cr, contrasted to ~20% Al needed without Cr (Figure 5.17). It is probably no coincidence that one of the most oxidation-resistant single-crystal superalloys, Rene'N5 + Y, has its Ni/Cr/Al content located at the Ni-rich tip of the Al_2O_3 field (region III) of the Ni–Cr–Al oxide map (Figure 5.17) (Giggins and Pettit 1971; Smialek and Meier 1987; Gleeson et al. 2009).

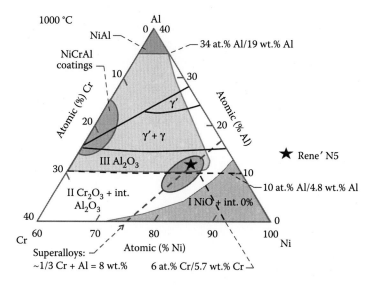

FIGURE 5.17
(See color insert.) Actual Ni-rich corner of the Ni–Cr–Al oxide map at 1000°C showing primary regions of Al_2O_3, Cr_2O_3, and NiO. Al_2O_3 region is extended to higher Ni contents when Cr is added. NiAl and NiCrAl coating compositions are superimposed as green and blue area; superalloys follow the dashed red line; single-crystal superalloy Rene' N5 is shown as star point. (With kind permission from Springer Science+Business Media: *J. Mat. Sci.* Compositional factors affecting the establishment and maintenance of Al_2O_3 scales on Ni-Al-Pt systems. 44(7), 2009, 1704–1710. Gleeson, B. et al.)

Another informative compositional feature is the average Cr and Al content of a wide range of superalloys, most predating single-crystal development. This is shown as the red dashed "superalloy" line and is given by $1/3\ Cr + Al = 8$ (in wt.%). That is to say, on average, the Cr and Al contents of past superalloys were not independently varied but often possess an inverse relation (Smialek et al. 1997). This had the consequence of masking independent effects on oxidation, often suggesting that low-Cr alloys were superior. In reality, this simply meant that high Al alloys were superior to low-Al alloys, with little information on independent Cr effects. Other compositional trends were analyzed statistically and showed that the relative degree of cyclic oxidation "attack" could be ordered according to a list of coefficients, a_i, where negative values refer to increased oxidation resistance and positive values indicate increased damage (Table 5.2) (Smialek et al. 1997; Barrett 2003).

The correlations in Table 5.2 apply only to the family of Ni-based superalloys in the typical Cr, Al, Mo, Nb, Ta, and W composition space. Outliers are easily produced when this space is exceeded by, say, the addition of one refractory element at exceptionally high values (such as 12 W in Mar M 200, 18 W in WAZ-20, or 18 Mo in NX188). Furthermore, a number of second-order interactions and cross-terms produce better correlations with the data, though it is then impossible to glean simple compositional trends at a glance. But here we see at least that, in general, Al is most beneficial, Ta is good, and Cr is marginal but not easily separable from Al. The addition of Mo, Nb, and Ti appear to be increasingly detrimental. These trends are consistent with highly beneficial Al_2O_3 scales, innocuous $NiTa_2O_6$, and detrimental $CrMoO_4$, $CrNbO_4$, $NiTiO_3$, and TiO_2 scales, as associated with respective cyclic oxidation regimes (Smialek and Meier 1987; Smialek et al. 1997).

An example of compositional trends in cyclic oxidation behavior is shown in Figure 5.18. Here the 1100°C, 200 h cyclic oxidation behavior of a number of alloys is presented from the vast 3000 test Barrett database (Barrett 2003), as searched, selected, and plotted from the Auping database user interface program (Smialek and Auping 2012). It is observed that better behavior (TAZ-8A, Mar-M 247, NASA-TRW-VIA) is associated with high 5–6% Al and low 0–1% Ti content, whereas the most severe weight loss (IN-792, IN-738) is associated with low 3–4% Al combined with high 3–5% Ti. Other detrimental effects can be associated with high 12% or 18% W (Mar-M 200, WAZ 20).

TABLE 5.2

Statistical Analysis of Average Effects of Alloying Elements on 1100°C Cyclic Oxidation of Some Commercial Cast Superalloys

Element	a_i
Al	−0.34
Ta	−0.16
Cr	−0.08
Mo	0.04
Nb	0.24
Ti	0.26

Source: Smialek, J. L. et al. 1997. *Design for Properties, ASM Handbook.* 20: 589–602. Materials Park, OH: ASM; Barrett, C. A. 2003. A High Temperature Cyclic Oxidation Data Base for Selected Materials Test at NASA Glenn Research Center. NASA/TM—2003-212546.

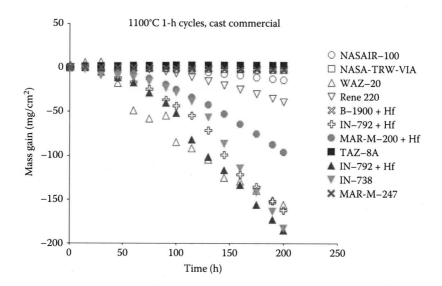

FIGURE 5.18
Cyclic oxidation of commercial Ni-based superalloys at 1100°C (1-h cycles). (Data from Barrett, C. A. 2003. A High Temperature Cyclic Oxidation Data Base for Selected Materials Test at NASA Glenn Research Center; Smialek, J. L. and J. V. Auping 2012. Cyclic Oxidation Database. NASA Glenn Research Center. NASA/TM—2003-212546.)

5.3.3.3.1 Coatings for Superalloys

At this point it can be seen that coatings are needed to provide some immunity and insurance from these potentially detrimental oxidation effects. Basically, there are two approaches, NiAl-based diffusion aluminide coatings and NiCrAl-based overlay coatings, as indicated by the green triangle and blue crescent regions, respectively, on the Ni–Cr–Al ternary map in Figure 5.17. The aluminides are based on the excellent oxidation behavior of the β-NiAl phase, having a large 50 at.% Al reservoir and able to form scales exclusively of α-Al$_2$O$_3$. These coatings have been in use for 50 years and are relatively straightforward to produce via halide salt-activated Al packs or AlCl$_3$ vapor CVD-type processing (Goward 1984; Das 2013). Some drawbacks of aluminides are the inherent brittleness of the NiAl intermetallic phase and the difficulty of tailored compositional control regarding dopants (Pt, Hf, Zr), impurities (S, C), and incorporation of substrate elements (Ti, Ta). Al depletion by interdiffusion with the substrate is well recognized as a trigger to oxidation failure. Scale spallation from thermal cycling is another contributing degradation mechanism.

The other coating strategy is a typical ~Ni–10Co–18Cr–14Al–0.1Y(+Si, Hf) at.% overlay, or variation, preferably produced by vacuum plasma spraying (VPS) or physical vapor deposition (PVD). These overlays generally consist of a β-γ/γ′ mixture, where the γ′ appears primarily during cooling as a fine precipitate, but dissolves at typical upper use temperatures. The benefits of this more expensive coating are hot corrosion resistance, greater ductility and compatibility with superalloy substrates, flexibility in compositional adjustments, and preferential coverage by masking. Drawbacks are line-of-sight deposition and the generally thicker coatings required for plasma spray processes. More recent formulations rely on strong, ductile γ/γ′ compositions, engineered for diffusional and thermodynamic stability with the substrate. These have Ta, Hf, W strengthening elements obtained by interdiffusion with the substrate, and enough Cr and Al (+ Pt, Si) for oxidation resistance (Kawagashi, Harada et al. 2008; Gleeson et al. 2009; Zhao et al. 2011).

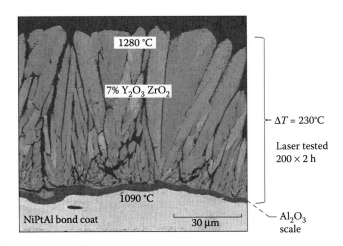

FIGURE 5.19
Microstructure of PVD YSZ coating after cyclic high heat flux laser testing. (200 1-h cycles; Ni(Pt)Al bond coat, N5 substrate; 1280°C surface, 1050°C metal temperature. (From Zhu, D. M. et al. 2001. *Surf. Coat. Tech.* 138(1): 1–8.)

Both diffusion and overlay coatings are well suited as alumina formers and have been employed as bond coats for thermal barrier coatings (TBCs). TBCs have gained widespread use in recent decades as a means to lower the metal temperature (~150°C) of air-cooled components. Yttria-stabilized zirconia (YSZ) has been favored over other ceramics because of superior insulative properties, high fracture resistance, and reasonable coefficient of thermal expansion (CTE) match to alloys (~10 vs. 16 ppm/°C). An example is shown in Figure 5.19. Here the YSZ was deposited by PVD, yielding the characteristic, strain-tolerant "feather-grained" microstructure on a Ni(Pt)Al CVD bond coat. By exposure in a high heat flux laser test, a thin adherent α-Al_2O_3 scale has formed between the bond coat and the TBC, which is necessary to maintain TBC integrity.

This coating system relies upon the uniformity of the CVD Ni(Pt)Al bond coat for a flat interface and minimized stress concentrations that might arise from any undo surface roughness, such as grain boundary ridges or rumpling. Conversely, plasma-sprayed overlay coatings are intrinsically rough in order to provide a good bond interface with plasma-sprayed YSZ top coats TBCs by mechanical bonding with the molten splats. Each coating system has distinctive failure modes due to these interfaces and their respective TBC microstructures, as the feathered structures fail by buckling and splat structures by lateral microcracks.

5.3.3.3.2 Alumina Scale Adhesion

Given the importance of alumina scales on superalloy, bond coat, and TBC performance, it is clear that alumina scale adhesion is critical to overall performance in any cyclic exposure. Thermal cycling and CTE mismatch can produce biaxial compressive stresses in the scale on the order of 4 GPa, which is more than sufficient to produce scale buckling and decohesion for poorly bonded scales (Christensen et al. 1997; Tolpygo et al. 1998). Scale adhesion had long been optimized historically by reactive element doping, for example, small ~0.1% additions of oxygen-active Y, La, Ce, Zr, Hf, and other elements. While numerous microstructural and diffusional effects are in play (Tien and Pettit 1972), the one phenomenon that remains critical for successful adhesion is the sulfur effect.

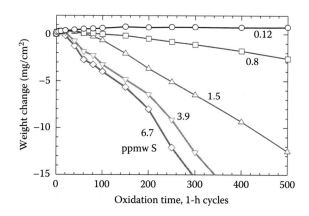

FIGURE 5.20
Effect of bulk sulfur content on alumina scale adhesion and 1100°C cyclic oxidation resistance of alloy PWA1480. (Smialek, J. L. Advances in the oxidation resistance of high-temperature turbine materials. *Surf. Int. Anal.* 2001. 31(7). 582–592. Copyright Wiley-VCH Verlag GmbH & Co. KGaA. Reproduced with permission.)

The oxygen-active elements are also sulfur-active and serve to prevent powerful sulfur segregation tendencies and interfacial weakening (Smeggil et al. 1986; Hong et al. 1990; Meier et al. 1995). Furthermore, sulfur removal, such as by hydrogen annealing, produces a direct correlation with improved adhesion. This was illustrated for a first-generation single-crystal alloy PWA 1480 (Smialek 2000). In this study, hydrogen annealing was performed at 1000–1300°C for various times and a range of sample thicknesses, resulting in a range of sulfur contents (the as-received S level was ~7 ppmw). In subsequent cyclic oxidation tests at 1100°C, the performance was seen to vary closely with residual sulfur content, as shown in Figure 5.20.

Once the sulfur content was reduced to about the 0.1–0.2 ppmw level, segregation was precluded and virtually all interfacial spallation was prevented. Given that sulfur removal is the only treatment that ever produced adhesion *without* reactive elements, the sulfur-gettering mechanism of reactive elements is strongly indicated. In addition, the total amount of sulfur in each sample was determined and correlated with the final weight loss for each sample. The criterion for excellent adhesion (no weight loss) corresponded to a quantity of sulfur equivalent to about 1 monolayer of total segregation, which is consistent with saturation levels that are believed to be on the order of ~0.5 monolayer (Smialek et al. 1994; Hou 2008). Currently, single-crystal superalloys are therefore processed with special attention to reduced sulfur content—including pure charge materials, melt crucibles, casting molds, and so on. Nominal S levels of 10 ppmw typical in the past are now closer to 1–2 ppmw. Special Ca-based flux desulfurization processing can result in 0.1–0.5 ppmw and cyclic oxidation behavior approaching that of hydrogen annealed material.

The sulfur content of the substrate may also affect the behavior of a coated system through interdiffusion, but is buffered by the protective factor of the coatings (Zhang et al. 2001). For example, Pt alloying in Ni(Pt)Al coatings or reactive element doping are both helpful in improving scale adhesion, even if the substrates contain moderate sulfur levels. Conversely, a clean substrate can be disabled by coatings with a high sulfur content, as may have happened in some traditional CVD aluminizing. Accordingly, process improvements for low sulfur aluminizing provided commensurate improvements in scale adhesion (Lee et al. 1998). The optimum performance can be demonstrated by applying low-sulfur coating processes to low-sulfur substrates. As would be expected, low-sulfur systems also result in improved TBC cyclic life, since alumina scale detachment would also result in spallation of the overlying TBC (Haynes et al. 2001; Smialek 2004).

Somewhat coincident with the detrimental sulfur effect on alumina scale adhesion, it was also observed that ambient moisture could adversely affect adhesion—in special cases. The phenomenon was most apparent at intermediate values of interfacial strength (sulfur content) and residual compressive stress on the scale (severity of oxidation). It was also proposed that some damage to the scale have occurred (cycling). Under these circumstances, alumina scales have been observed to spontaneously exfoliate when exposed to moisture or immersed in water. Much of the history and current hypothesis have been summarized in detail (Smialek 2006, 2010, 2011a). Here, a mechanism was proposed in which ambient moisture reacts with Al in the substrate to form $Al(OH)_3$, releasing H to the metal. Rapid hydrogen diffusion along the highly stressed interface then results in bond weakening and spontaneous spallation. Many aspects are shown to be analogous to electrochemical hydrogen embrittlement, environmental embrittlement of aluminides, and hydrogen-induced delamination of anodic films on Al. Moisture-induced failure of TBCs has also been demonstrated (Smialek 2011b).

5.3.3.3.3 Transition Alumina Scales

While *transient* oxidation refers to other scales formed at borderline alloy contents, another metastable scaling phenomenon deals with *transition* aluminas. Here, the initial scale is chemically alumina but has a cubic "defective spinel" structure. It nucleates preferentially, possibly due to strong epitaxial relations with the metal substrate (Doychak et al. 1989). Above ~900°C, this transition θ-alumina transforms after a period of time to the stable α-alumina rhombohedral corundum phase, with a notable reduction in growth kinetics (Rybicki and Smialek 1989; Brumm and Grabke 1992). This is often accompanied by a dramatic change in scale structure, as shown in Figure 5.21. Here, the θ morphology is

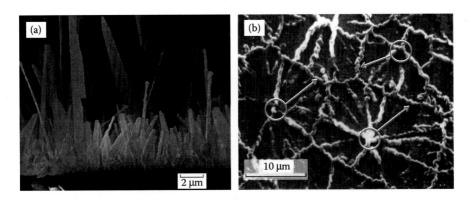

FIGURE 5.21
Distinctive Al_2O_3 scale microstructures formed on NiAl alloys. (a) Fracture cross section showing extensive outward growing plate-like transition θ-alumina from 16 h oxidation at 950°C. (With kind permission from Springer Science+Business Media: *Corr. Sci.* Effect of water vapor on the phase transformation of alumina grown on NiAl at 950°C. 53(9), 2011, 2943–2947. Zhou, Z. H. et al.) (b) Plan view of web structure formed after transformation to α-Al_2O_3, 1100°C for 100 h. (With kind permission from Springer Science+Business Media: *Oxid. Met.* Effect of the theta-alpha-Al_2O_3 transformation on the oxidation behavior of β-NiAl + Zr. 31(3–4), 1989, 275–304. Rybicki, G. C. and J. L. Smialek.) Overlayed circles suggest origins of spherulitic transformation; arrows indicate impingement boundaries after transformation.

characterized by outward-growing platelets, whereas the transformed α scale morphology is generally flatter, inward grooving, and displays a pinwheel or web-like structure of ridges. Doychak et al. described the evolution of such structures as a spherulitic-type nucleation (cf. crystoballite) and lateral growth accompanied by radial cracking from tensile transformation stresses (Doychak et al. 1989). As each transformation disk impinges on the adjoining transformed area, high-angle boundaries are formed with some vestiges of outward Al growth ridges. The ridge features are also prominent above healed radial cracks, now low-angle boundaries, which often gives a wheel-and-spokes appearance. Transition aluminas are more easily observed on binary NiAl than on ternary NiCrAl or FeCrAl alloys, though certain conditions may promote transition scales on these alloys as well. It is believed that transient α-Cr_2O_3 scales serve as a template for α-Al_2O_3 nucleation very early in the oxidation process. It is further noted that reactive element doping tends to segregate at grain boundaries and retard the transformation, while water vapor tends to increase surface diffusion and promote the transformation (Zhou et al. 2011).

5.3.3.4 Intermetallic Aluminide and Silicide Compounds

Intermetallic compounds are generally put into a class by themselves because they have less ductility and fracture toughness than most metals. These detriments originate from the stoichiometric bonding and highly ordered structures that make dislocation movement energetically unfavorable relative to brittle cleavage, especially at lower temperatures. Nonetheless, some have attractive properties that enable niche applications, such as coatings. Broad yet detailed overviews of the subject, including aluminides, silicides, and beryllides, have been provided (Grobstein and Doychak 1988; Meier 1988; Smialek et al. 1988; Doychak 1995; Welsch et al. 1996).

The most widely studied of the intermetallics would be the B2 β-NiAl compound, heralded for its high melting temperature and low density compared to superalloys. This compound can be somewhat ductilized by microalloying (<1%), with elements such as Fe, but are then weak in high-temperature creep. Alternatively, they can be strengthened for high-temperature behavior by microalloying with Hf or Zr, but are then too brittle at room temperature. No material has been developed that optimized both attributes. The low fracture toughness remains an unacceptable risk for turbine hardware, and most efforts for a stand-alone structural alloy have been abandoned. NiAl remains among the leading coating materials for turbine alloys (along with NiCoCrAlY γ/β two-phase compositions). About 5 at.% Pt is often incorporated into the coating (by a 7-μm electroplate prior to aluminizing), allowing Al_2O_3 formation over a wider compositional range, greater diffusional stability of high-Al phases, and increased scale adhesion (Gleeson et al. 2004). Recently, Hf has been incorporated as a beneficial dopant for reducing scale growth rate due by decreasing diffusion rates and for reducing rumpling by increasing creep strength (Mu et al. 2008). Other reactive elements, such as Y, Zr, Gd, Dy, Yb, and so on, also produced beneficial adhesion effects when added at low levels (≤0.1 at.%), with the most data available for the optimization of Zr at about 0.05 at.%. Si doping has also been found to be beneficial for improving the long-term ability to form protective alumina scales at reduced aluminum contents. 5 at.% Cr may be useful for high-temperature Type I hot corrosion resistance, but is less effective for Type II low-temperature hot corrosion, as discussed later.

γ'-Ni_3Al has been less utilized for oxidation resistance, though it is critical as the strengthening phase in Ni-based superalloys and undoubtedly plays a role in the overall oxidation resistance of these γ/γ' alloys. As a stand-alone alloy, it is not optimal from a mechanical properties standpoint, although Ni_3Al is more attractive than β-NiAl due to its higher ductility. While Ni₃Al is able to form some α-Al₂O₃ isothermally above ~1100°C, both Ni₃Al and Al-poor NiAl (≤40 at.% Al) suffer in cyclic oxidation where even a small amount of spallation triggers Al-depleted surfaces and the formation of rapid-growing NiAl₂O₄ spinel oxides.

Fe–Al compounds have been studied because of low cost. They are less prone to brittle failure than NiAl, but inferior to superalloys for high-temperature structural use. Their oxidation behavior is generally inferior to Ni–Al formulations, but their initial kinetics are still controlled by α-Al₂O₃ scales. Biaxial growth stresses produce substantial wrinkling of the scale. Combined with excessive HfO₂ internal oxide fingers from overdoping of Fe40Al–Zr,Hf,B alloys, premature mechanical disruption and spallation of the scales result (Smialek et al. 1990).

Ti–Al alloys and compounds have received a great deal of attention due to their low density and potential high-temperature strength, albeit at temperatures <1000°C. The schematic of scale thickness in Figure 5.22 presents an overview of relative scale make-up and architecture for the major binary phases. TiAl₃ is needed for the formation of exclusively alumina scales on binary compounds. One dominating common detrimental attribute for most of the Ti–Al system is the high propensity for oxygen solution and embrittlement or internal oxidation (Brindley et al. 1994; Brady et al. 1996).

α-2 Ti₃Al–Nb,Cr alloys have impressive strength/density ratios, but suffer from relatively rapid oxidation rates due to mixed TiO₂–Al₂O₃ scales. Although some oxidation benefits have accrued from various alloying schemes, none have prevented environmental embrittlement due to oxygen diffusion into the alloy. Orthorhombic Ti–25Al–25Nb alloys also have a very high strength-to-density ratio, but again suffer from oxygen embrittlement. γ-TiAl–2Cr–2Nb-type alloys are still of current interest because of reasonable mechanical properties and somewhat reduced sensitivity to embrittlement. The oxidation resistance of γ-TiAl alloys has been improved by the halide effect. Here, treatment with

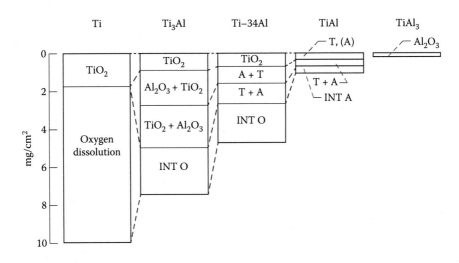

FIGURE 5.22
Schematic cross sections of scales formed over the binary Ti–Al system at 850°C for 100 h. Relative amounts of TiO₂ (T) and Al₂O₃ (A) scales are shown, with internal oxygen (Int. O). Scale thicknesses decrease with Al content, but mixed scales predominate for all alloys except TiAl₃. (From Smialek, J. L. et al. 1995. Service limitations for oxidation resistant intermetallic compounds. *High Temperature Ordered Intermetallics*. Pittsburgh: Materials Research Society.)

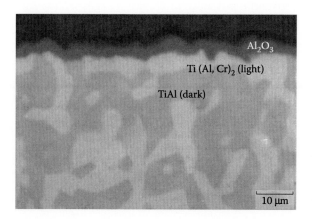

FIGURE 5.23
SEM/BSE cross-section micrograph of Ti–52Cr–12Al oxidized at 1000°C for 100 h. Al_2O_3 scale formed on $Ti(Cr,Al)_2$ Laves depletion zone layer. (Brady, M. P. et al.: High-temperature oxidation and corrosion of intermetallics. In *Corrosion and Environmental Degradation, Volume II.* ed. M. Shutze. 2000. 239–325. Copyright Wiley-VCH Verlag GmbH & Co. KGaA. Reproduced with permission.)

Cl or F compounds enhances Al (vapor) transport beneath the scale and favors protective α-Al_2O_3 formation (Schutze et al. 2002; Donchev et al. 2006; Leyens et al. 2006). Other ~Ti–50Al–15Cr alloying schemes employ the γ-Laves two-phase mixture that provides an Al-rich subscale phase that ensures alumina formation and produces a dramatic improvement in oxidation behavior, as shown in Figures 5.23 and 5.24. Such alloys have received some interest as oxidation-resistant coatings on structural alloys and do not appear to suffer from embrittlement.

As previously indicated, $TiAl_3$ alloys show exclusive slow-growing alumina scales. However, these compounds are extremely brittle, having both low toughness and low strength. $TiAl_3$ coatings on Ti–Al structural alloys generally improve oxidation resistance.

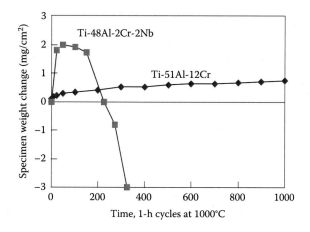

FIGURE 5.24
Cyclic oxidation data showing excellent long-term oxidation behavior of γ + Laves Ti–Al–Cr alloys such as Ti–51Al–12Cr at.%. (With kind permission from Springer Science+Business Media: *Corrosion and Environmental Degradation, Volume II.* High-temperature oxidation and corrosion of intermetallics. 1996, 239–325, Brady, M. P. et al.)

However, they are so brittle that they initiate cracks from thermal expansion mismatch stresses. One remarkable alloying route around this problem was the development of Ti–67Al–8Cr τ-phase alloys with up to a percent or so of ductility and remarkably good oxidation behavior (Parfitt et al. 1991; Smialek 1993). These compositions have also been studied as coatings for other Ti–Al phases and alloys. The τ-phase and the TiAlCr two-phase γ-Laves alloys coatings are intriguing and have received attention as protective coatings up to ~950°C. In addition, YSZ TBC coatings are also attractive because of their close CTE match with γ-TiAl substrates. However, widespread use of Ti-aluminides as working components had been limited to high-performance automotive valves and superchargers (Leyens et al. 2006). One γ-Ti47Al–2Cr–2Nb alloy has recently been incorporated as low-temperature, low-pressure stage 6 and 7 turbine airfoils in the new GEnx engine.

The previous discussions highlighted a number of oxidation-resistant aluminides, mostly attractive because of reduced density and cost compared to Ni-based superalloys. During the 1980s, a number of other efforts were dedicated toward developing oxidation-resistant aluminides with the potential of significantly higher melting points. Nb and Ta aluminides would fall into this category, as shown in Table 5.3.

However, simple binary refractory aluminides compounds cannot form healing layers of alumina and require various alloying schemes to produce second phases that serve as protective depletion zones during oxidation. At least one $NbAl_3$–Ti,Cr,Si,Y alloy demonstrated reasonable alumina scale growth at 1200°C (Hebsur et al. 1989). Nevertheless, the resulting compounds were always extremely brittle and have not reached sufficient advantages to remain competitive with other material options.

A number of metal silicides also have exceptionally high-melting temperatures, as shown in Table 5.3. Unlike the aluminides, these have demonstrated extremely low oxidation rates at relatively high temperatures (≥1200°C) due to the formation of healing layers of silica. However, they are also extremely brittle and remain a challenge for structural use. The most successful commercial silicide development is the alloying of $MoSi_2$ with boron and proprietary processing to form highly oxidation-resistant heating elements rated for >1600°C operation (e.g., trade name "Super Kanthal"). While still brittle, these materials are able to withstand repeated cycling at extremely high temperatures. Failure modes often result at cold ends of the elements due to thermal stresses. Some potential may exist for these silicides as coatings for ultra-high-temperature ZrB_2 and carbon/carbon composites, simply because there are few other materials able to survive temperatures above 1500°C.

While the refractory aluminides and silicides represent an intriguing class of high-temperature intermetallics, they are potentially vulnerable to an insidious low-temperature (~500–700°C) oxidation problem known as "pesting." This process is manifested as a rather rapid transformation of the compound into a dust made up of oxide mixtures and residual

TABLE 5.3

Melting Points of Selected Aluminides and Silicides

Aluminides		Silicides	
Nb_3Al	2060°C	$CrSi_2$	1570°C
$NbAl_3$	1680°C	$TiSi_2$	1760°C
$TaAl_3$	1608°C	$TaSi_2$	2040°C
		$MoSi_2$	2030°C

Source: Massalski, T. B. et al. 1990. *Binary Phase Diagrams.* Materials Park, OH: ASM International.

metal particles (Doychak and Grobstein 1989; Grabke and Meier 1995). The phenomenon arises from the *rapid* oxygen diffusion in refractory elements relative to the *slow* growth of protective alumina or silica scales at low temperatures. As the protective Al or Si element is depleted during oxidation, minute areas of refractory alloy are formed, having a high solubility for oxygen. This allows local ingress of oxygen, often along a distinctive $SiO_2 + RMO_x$ columnar morphology reminiscent of cellular precipitation. The refractory metal oxide RMO_x channels allow rapid inward diffusion of oxygen. This consumes the protective Al or Si element, but dispersed internally giving no external layer protective effect. Stresses build up from the volume expansion and cause swelling and possibly cracking. New surface area is created via cracking, and the process is repeated autonomously, with catastrophic results. Ironically, this pesting problem at 500°C has often been more problematic than surface scale oxidation above 1200°C. In a defect- and stress-free material, pesting may be avoided in the binary compound or at least delayed. However, it can be prevented entirely by alloying, and it has been addressed in $MoSi_2$ by alloying with low-temperature glass formers, like B and Al, to help retard dissolved oxygen (Hebsur 1994).

5.3.3.5 Refractory Metals and Silicide Coatings

The refractory metals (Nb, Ta, Mo, W, and Re) are natural candidates for high-temperature applications because of their extremely high melting points (2250–3400°C). These materials have among the highest creep strength at extreme temperatures of any other materials. Unfortunately, they dissolve high levels of oxygen and their oxides have fast growth rates. Nb_2O_5, Ta_2O_5, MoO_2, MoO_3, and WO_3 scales have such high growth rates that refractory alloys are rarely used in oxygen-containing environments without coatings. The scales typically form voluminous mounds of loose powder, which provide little protective ability and are therefore not useful from an engineering standpoint. MoO_3 and WO_3 are highly volatile and produce substantial metal recession simply by vaporization. Accordingly, most high-temperature applications for refractory elements are for inert gas or vacuum, where oxidation is not an issue. For example, W alloy heating elements power the highest-rated vacuum furnaces. However, a few applications are well served by their extremely high strength at high temperatures, where oxidation-resistant coatings become the enabling factor. Even with coatings, the potential for catastrophic failure generally limits the use of refractory elements to less critical components or short missions (e.g., rocket nozzles or exhaust nozzles of military turbines). History has shown that refractory metal silicides provide acceptable and reliable service, though not for the long term.

Coatings on refractory alloys are based on disilicides that form protective SiO_2 scales and use additives that enable lower-melting silicates to flow and seal through-cracks. The coatings are generally applied by slurry spraying followed by vacuum sintering, where other additives are used to lower the processing fusion temperature of the coating. Typical components include Mo, Nb, Ti, Fe, Co, Cr, and Si (Packer 1988). One coating, R512E, seems to have emerged as an industry standard and serves as a suitable example. It is a Si–20Cr–20Fe (wt.%) fused slurry coating that reacts somewhat with the base metal to form coatings with a higher remelt temperature. The weakness of this coating is the formation of through-cracks due to the CTE mismatch with substrates and the brittleness of the silicide phases. The Cr and Fe additions produce more fluid silicate glasses that seal the coating cracks at temperature. However, there are limits to this process, and continued cycling eventually grows the cracks wide and deep enough to allow substrate attack and subsequent failure. Multiple cracks can be seen in Figure 5.25a and b for both R512E and a pack silicided $MoSi(Ge)_2$ coating on Mo–Re (Cockeram and Rapp 1995).

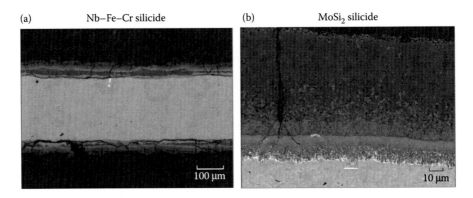

FIGURE 5.25
Cracks in silicide coatings produced on Mo–41Re refractory element alloy. (a) R512E Si–20Cr–20Fe oxidized for 0.5 h at 2300°F and (b) MoSi(Ge)$_2$ oxidized for 200 h at 2000°F. (From Glass, D. E. et al. 2001. Effectiveness of diffusion barrier coatings for Mo-Re Embedded in C/SiC and C/C. NASA/TM-2001–211264.)

While these coatings can survive hundreds of hours at 1500°C isothermally, cyclic exposures truncate lives considerably. For example, as previously discussed in Smialek et al. (1997), bars of FS85 (Nb–28Ta–10 W–1Zr by wt.) were slurry fusion-coated with a commercial R512E coating and tested in a half-hour cycle Mach 0.3 burner rig (Smialek et al. 1992). Approximately, 5 mg/cm^2 weight losses were incurred at 170, >200, 120, and 110 cycles for 982, 1093, 1260, and 1371°C exposures, respectively. Except for the 1093°C exposure, where crack healing survived, catastrophic oxidation pits or massive substrate attack occurred at the specimen edges, forming friable mixtures of Nb$_2$O$_5$, FeNbO$_4$, SiO$_2$, and amorphous silicates.

At this juncture, it is appropriate to introduce a hybrid material that co-optimizes strength and oxidation resistance, Mo(Zr)–3Si–1B wt.%, (Mo–12Si–8.5B–xZr atomic%) that was developed and studied intensively by (Park et al. 2002; Schneibel et al. 2005). It is a microcomposite phase mixture of Mo–Mo$_5$SiB$_2$–Mo$_3$Si with attractive high-temperature mechanical properties, respectable fracture toughness above 10 MPa · m$^{1/2}$, and melting above 2000°C. While protective silica scales are the objective, protection suffers with increasing B content from excessive B$_2$O$_3$ and MoO$_3$ volatility at high temperatures. Conversely, with decreasing B, too little scale and protective sealing occurs at low temperatures, allowing attack and mass loss of Mo. For example, improved behavior of Mo–Si–B was reported as a loss of 5 mg/cm^2 at 1300°C in 24 h, becoming worse at lower temperatures where sealing is delayed. This would be compared to a gain of only ~0.15 mg/cm^2 in 100 h for traditional SiC and Si$_3$N$_4$ silica formers that will be described in subsequent sections. Coatings are therefore still required for multiphase refractory silicides in long-term use.

It should be mentioned that similar approaches were undertaken for Nb-based alloys, using multiphase Nb(Ti,Cr)ss–Nb(Ti,Cr)$_5$Si$_3$–Nb(Ti)Cr$_2$ mixtures, which also achieved room-temperature toughness on the order of 10 MPa · m$^{1/2}$. Although many were unacceptable from an oxidation standpoint, the following alloy, 36Nb–22Ti–17Si–16Cr–5Ge–4.0Hf atom percent, showed promise. It exhibited weight change at least on the same magnitude as a number of single-crystal Ni-based superalloys in 1100 and 1177°C cyclic oxidation for 150 h (Chan 2004). Scale phases, however, were a long way from pure healing layers of silica, as they were composed of CrNbO$_4$ with small amounts of Nb$_2$O$_5$ and TiNb$_2$O$_7$, none of which are especially known for protective behavior. Intermittent spallation and low-temperature embrittlement/oxidation of Nb remain as concerns for Nb-silicides, and, therefore, coatings would still be needed.

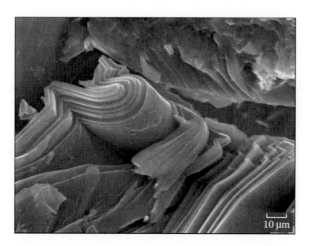

FIGURE 5.26
Deformation kinking of intercollated crystallographic planes in nano-laminate Cr$_2$AlC MAX-phase. (Fracture surface of three-point bend flexural test.) (Reprinted from *Mat. Sci. Eng.* A. 527(21–22), Yu, W. B. et al., Microstructure and mechanical properties of a Cr$_2$Al(Si)C solid solution. 5997–6001. Copyright 2010, with permission from Elsevier.)

5.3.3.6 MAX Compounds

In recent years, a new class of ceramic compounds has risen in popularity termed as "MAX" compounds, where typically M = various transition metals, A = Al or Si, and X = C or N. The most popular formulations are "312" Ti$_3$SiC$_2$ or Ti$_3$AlC$_2$ and "211" Ti$_2$AlC or Cr$_2$AlC (Barsoum and El-Raghy 2001). Their distinctive properties are high strain compliance and deformability under stress concentrations, but they also have numerous other attractive attributes. The mechanical, machining, and tribological properties are derived from the ability of intercollated planes in the ordered lattice to slip on an atomic scale or kink on a microscopic scale, thus absorbing enormous amounts of energy without crack propagation. An example of kinking in Cr$_2$AlC in flexure is shown in Figure 5.26 (Yu et al. 2010). Thus, these phases represent an interesting transition between metallic properties and those typical of covalent ceramic (SiC or Si$_3$N$_4$) carbides and nitrides.

Furthermore, they are stable to quite high temperatures >1500°C. The Ti$_3$SiC$_2$, Ti$_2$AlC, and Cr$_2$AlC have the ability to form protective SiO$_2$ or Al$_2$O$_3$ scales. The first two are available commercially, known as "MAXthal™" materials. In that regard, kinetics reminiscent of good silica or alumina forming oxidation-resistant materials have been documented, as shown in Figure 5.27 (Lin et al. 2007; Wang et al. 2012). Si or Al depletion ultimately leads to TiC, Ti$_5$Si$_3$, or Cr$_7$C$_3$ depletion zones, which may indicate when oxidative life has been reached.

5.4 Oxidation of SiO$_2$-Forming Ceramic Materials

5.4.1 Passive Oxidation

Silicon-based ceramics and composites include materials such as SiC, Si$_3$N$_4$, and SiC-fiber-reinforced SiC matrices. These materials show great promise for hot stage components in turbine engines and chemical process plants as well as re-entry shields due to their

(a)   (b)

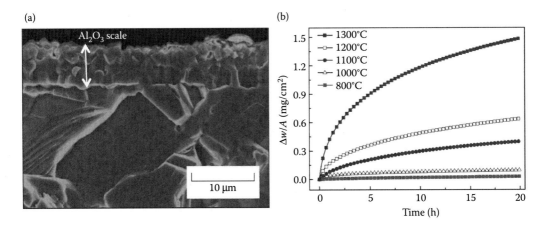

FIGURE 5.27
(See color insert.) Protective α-Al₂O₃ scales on "312" and "211" MAX phases. (a) Fracture section showing alumina surface scale formed on 312 Ti₃AlC₂ oxidized at 1250°C for 20 h. (Reprinted from *Corr. Sci.* 58, Wang, X. H. et al., Insights into high temperature oxidation of Al₂O₃-forming Ti₃AlC₂. 95–103. Copyright 2012, with permission from Elsevier.) (b) Isothermal oxidation kinetics of 211 Cr₂AlC oxidized at 1000–1250°C for 20 h. (Reprinted from *Acta Mat.* 55(18), Lin, Z. J. et al., High-temperature oxidation and hot corrosion of Cr₂AlC. 6182–6191. Copyright 2007, with permission from Elsevier.)

corrosion resistance and ability to retain strength even at high temperatures. A protective SiO₂ scale passively forms on the surface at high temperatures.

The growth of silica exhibits behavior significantly different than that of chromia and alumina. Much of the understanding of silica growth comes from the pioneering studies of Deal and Grove (1965), who derived the basic linear-parabolic model for silica growth on silicon:

$$x_o^2 + A\,x_o = B(t + \tau) \qquad (5.6)$$

Here, x_o is the oxide thickness, B/A is the linear rate constant, B is the parabolic rate constant, t is the time, and τ is the shift in time that corresponds to the initial oxide thickness. Unlike the growth of alumina and chromia, silica growth is characterized by permeation of neutral oxygen through the growing silica scale. The silica scale is usually amorphous in structure, made up of stochastically situated SiO₄ tetrahedra with an open, disordered structure between units. The "channels" through the glassy scale provide pathways for the neutral oxygen to move. Parabolic growth dominates after longer times and at higher temperatures and the parabolic constant is given by

$$B \equiv \frac{2D_{eff}C^*}{N_1} \qquad (5.7)$$

Here, D_{eff} is the effective permeation coefficient, which has been measured independently by Norton (Norton 1961), C^* is the equilibrium concentration of oxygen in the oxide, and N_1 is the number of oxidant molecules in a unit volume of oxide. Thus, the rate of growth of a silica scale is proportional to the pressure of oxygen. This contrasts with chromia and alumina growth, whose growth rates are generally independent of the oxygen pressure. The temperature dependence of the parabolic rate constant results in a low activation energy

(119 kJ/mol) as bond breaking is not involved in the diffusion process. This too contrasts with chromia and alumina growth (see summary in Table 5.1).

Deal and Grove (1965) also derived the effects of water vapor from the above equation on silica oxidation. While the effective permeation coefficient is smaller for water than oxygen, the solubility of water in silica is much greater and hence the parabolic rate constant is greater for water than oxygen. It should also be noted that silica scales are quite susceptible to impurity effects. Many metal ions act like network modifiers, opening up the network for additional transport paths and increasing oxidation rates. For this reason, it is important to study silica growth in a very clean environment (Opila 1995).

Silica growth on SiC and Si_3N_4 has been extensively studied due to the wide number of high-temperature applications. Figure 5.28 is an Arrhenius-type plot of parabolic rate constants for Si, SiC, and Si_3N_4.

The similarity between Si oxidation and SiC is clear and has been observed by other investigators (Motzfeldt 1964; Ogbuji and Opila 1995). The small rate differences are attributed to the additional oxygen needed to oxidize the carbon in SiC.

The silica scales discussed up to this point are amorphous. However, a variety of factors lead to crystallization of SiO_2, including temperature, pressure, and mechanical effects. An example of spherulitic (i.e., disk-shaped) crystallite formation in a growing SiO_2 film is shown in Figure 5.29. The exact cause and effect of this crystallization is still controversial. Some investigators correlate this crystallization to a change in reaction rate (Costello and Tressler 1981; Narushima et al. 1989). These investigators observed a change in slope of the Arrhenius plot at higher temperatures, which they attributed to crystallization. However, other work indicates that a continuous film of amorphous silica exists to higher temperatures and crystallization may not be extensive enough to account for a slowing of reaction rates (Ogbuji and Opila 1995; Ogbuji 1997). These studies do not measure a change in slope in high-purity environments and, therefore, the change in slope observed by others may be attributed to impurity effects.

The oxidation rate of Si_3N_4 shown in Figure 5.28 is slower than SiC, which has been an area of some controversy. Microstructurally, the distinguishing characteristic of the oxide scale on Si_3N_4 is an intermediate layer of silicon oxynitride. This has been described as

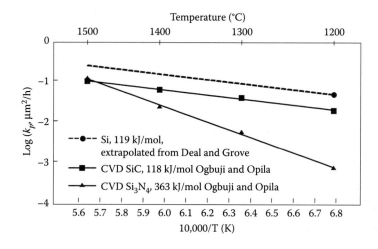

FIGURE 5.28
Plot of oxidation rates for Si, SiC, and Si_3N_4. (Adapted from Ogbuji, L. U. J. T. and E. J. Opila 1995. *J. Electrochem. Soc.* 142(3): 925–930.)

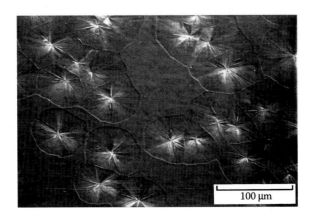

FIGURE 5.29
CVD SiC oxidized for 24 h at 1400°C showing formation of spherulitic silica scales that are in compression on cooling, which leads to buckling and spallation. (Jacobson, N. S.: Corrosion of silicon-based ceramics in combustion environments. *J. Am. Ceram. Soc.* 1993. 76(1). 3–28. Copyright Wiley-VCH Verlag GmbH & Co. KGaA. Reproduced with permission.)

either a discrete phase of composition Si_2ON_2 or a solid solution of SiO_2 and Si_3N_4. It is unclear why net rates are so much lower than Si and SiC if oxidation is controlled by oxygen permeation through the growing SiO_2 scale in all cases. Some investigators have suggested that Si_2ON_2 acts as a diffusion barrier, but an analysis of a diffusion barrier process leads to very high pressures at the interface SiO_2/Si_2ON_2 interface (Luthra 1991). Evidence suggests that the scale consists of a solid solution with a gradual transition from Si_3N_4 to SiO_2 and that oxidation occurs with a substitution of O for N across this solution (Ogbuji and Jayne 1993). This type of oxidation mechanism, as compared to permeation in Si and SiC, can account for the differing rates.

Another important difference between silica scales and chromia/alumina scales is their behavior under thermal cycling. Figure 5.30 compares the thermal expansion of

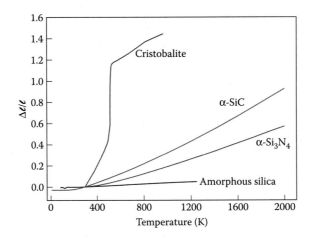

FIGURE 5.30
Thermal expansion curves for cristobalite, α-SiC, α-Si_3N_4, and quartz (amorphous silica). (Jacobson, N. S.: Corrosion of silicon-based ceramics in combustion environments. *J. Am. Ceram. Soc.* 1993. 76(1). 3–28. Copyright Wiley-VCH Verlag GmbH & Co. KGaA. Reproduced with permission.)

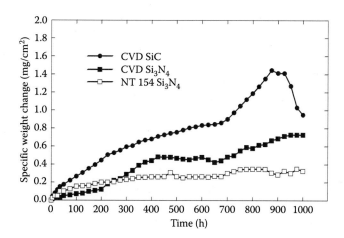

FIGURE 5.31
Cyclic oxidation of CVD SiC, CVD Si_3N_4, and NT-154 (MgO)–Si_3N_4 at 1300°C in 5 h cycles. (From Opila, E. J. et al. 1993. *Cer. Eng. Sci. Proc.* 49(7–8): 2363–2369; Opila, E. and N. Jacobson: Corrosion of ceramic materials. In *Corrosion and Environmental Degradation*, ed. M. Shutze. 328–387. 2000. Copyright Wiley-VCH Verlag GmbH & Co. KGaA. Reproduced with permission.)

cristobalite, α-SiC, α-Si_3N_4, and quartz. In application, SiC and Si_3N_4 ceramics would have some cristobalite in the scale due to impurities. The curves indicate the cristobalite would be in tension during cooling which would form cracks. However, upon reheating, the cracks would heal, which results in good thermal cycling behavior as shown in Figure 5.31 (Opila et al. 1993). This is in contrast to superalloy scales that are in compression on cooling, which leads to buckling and spallation.

The discussion regarding the oxidation of silicon-based ceramics thus far has centered on high-purity SiC and Si_3N_4. However, many commercial ceramics contain additives (e.g., ~0.5% B and 3% C to SiC by wt.) to promote sintering. These have been found to have relatively minor effects on oxidation as compared to the rate for CVD SiC; however, they do lead to more bubbling at the SiO_2/SiC interface. Refractory oxides, such as MgO and Y_2O_3, are often added to Si_3N_4 to promote sintering. These have been shown to diffuse into the growing oxide scale (Cubicciotti et al. 1977) and the diffusion rate of the cation into the growing scale may actually control the rate of oxidation. Figure 5.32 shows the surface of an oxidized Si_3N_4 coupon with Y_2O_3 additives.

5.4.1.1 Oxidation in H_2O/O_2

The effects of water vapor on the oxidation of Si-based ceramics and composites are important for applications in combustion environments. As discussed in Section 5.1, these environments contain ~10% water vapor. Opila (Opila 1994; Opila and Hann 1997) has shown that the effects of water vapor on the oxidation of Si-based ceramics are threefold: (1) enhanced transport of container/furnace impurities to the sample and increased oxidation rates, (2) enhanced silica scale growth due to the higher solubility of H_2O in the scale, as discussed before, and (3) formation of $Si(OH)_4(g)$ and resultant consumption of the scale. In the third case, the scale grows according to a standard parabolic law following the reaction

$$SiC(s) + 3/2\ O_2(g) = SiO_2(s) + CO(g) \tag{5.8}$$

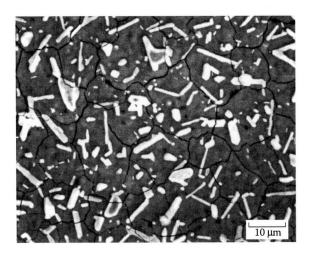

FIGURE 5.32

Si_3N_4 containing Y_2O_3 additives oxidized for 97 h at 1300°C in dry oxygen. The bright crystallites are yttrium silicate precipitates that form as Y^{+3} diffuses into the growing silica scale. (Jacobson, N. S.: Corrosion of silicon-based ceramics in combustion environments. *J. Am. Ceram. Soc.* 1993. 76(1). 3–28. Copyright Wiley-VCH Verlag GmbH & Co. KGaA. Reproduced with permission.)

As the scale forms, it is consumed via the formation of $Si(OH)_4(g)$, which follows a linear rate law:

$$SiO_2(s) + 2\,H_2O(g) = Si(OH)_4(g) \tag{5.9}$$

This leads to "paralinear" kinetics in the laboratory (Opila and Hann 1997). Examples of this behavior are shown for SiC, Si_3N_4, and SiO_2 alone for the linear component in Figure 5.33.

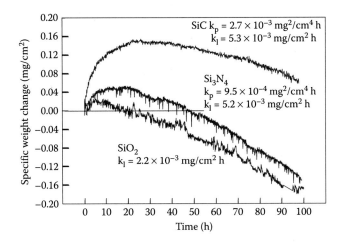

FIGURE 5.33

Paralinear behavior of SiC and Si_3N_4 and linear behavior of SiO_2 at 1200°C, 50% H_2O/50% O_2 at 4.4 cm/s flow rate. (Opila, E. and N. Jacobson: Corrosion of ceramic materials. In *Corrosion and Environmental Degradation*, ed. M. Shutze. 328–387. 2000. Copyright Wiley-VCH Verlag GmbH & Co. KGaA. Reproduced with permission.)

In combustion applications, this process can lead to recession of the ceramic over long times (Opila et al. 1999; Robinson and Smialek 1999). In such a situation, the linear portion of the curve dominates and Si(OH)$_4$(g) transport through the static boundary layer is rate limiting. Thus, the linear weight loss has the following functional form (Geiger and Poirier 1973) for laminar flow:

$$k_l = 0.664\,(Re)^{0.5}\,(Sc)^{0.33}\,\frac{D\rho}{L} \tag{5.10}$$

Here, Re is the Reynolds number, Sc is the Schmidt number, D is the gas phase diffusivity of Si(OH)$_4$ in the boundary layer, ρ is the density of Si(OH)$_4$ near the surface, and L is a characteristic dimension. The rate constant dependence on velocity (v), partial pressure of H$_2$O, and total pressure (P_{total}) can be extracted based on the above relationship and the stoichiometry of Si(OH)$_4$ formation (assuming p(H$_2$O) varies as total pressure):

$$k_l \propto \mathrm{v}^{0.5}\,P_{total}^{1.5} \tag{5.11}$$

Robinson and Smialek (1999) examined CVD SiC weight loss rates in a pressurized burner over a range of conditions. Their data are shown in Figure 5.34.

From this analysis, they derived the following semiempirical expression:

$$k_l[\mu\,\mathrm{m\,h^{-1}}] = 6.32 \cdot \exp(-108000\,[\mathrm{J/mole}]/RT) \cdot P^{1.5} \cdot \mathrm{v}^{0.5} \tag{5.12}$$

Yuri and Hisamatsu (2003) conducted a similar study on SiC and other ceramics in a combustion environment in turbulent flow, obtaining similar results due to the formation of volatile hydroxides.

A viable solution to the water vapor attack problem is the use of refractory coatings. A refractory-coated silicon-based ceramic combines the chemical resistance of a refractory oxide and the desirable mechanical properties of a silicon-based ceramic or composite.

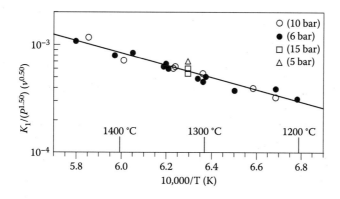

FIGURE 5.34
Linear rate constant normalized to P$^{1.5}$ and v$^{0.5}$ as a function of inverse temperature. (Robinson, R. C. and J. L. Smialek: SiC recession caused by SiO$_2$ scale volatility under combustion conditions: I. Experimental results and empirical model. *J. Am. Ceram. Soc.* 1999. 82(7). 1817–1825. Copyright Wiley-VCH Verlag GmbH & Co. KGaA. Reproduced with permission.)

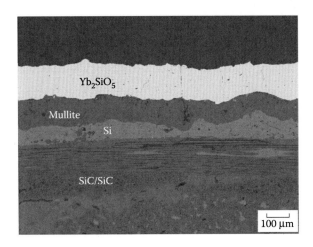

FIGURE 5.35
EBC coating on a SiC composite. This is a multilayer coating, with a Si bond coat to promote adherence, a mullite reaction barrier, and an outer rare earth silicate layer with a reduced silica thermodynamic activity. (From Jacobson, N. S. et al. 2005. *ASM Handbook. Volume 13B Corrosion: Materials.* ed. S. D. Cramer and J. Covino, B. S. Materials Park, OH: ASM International. 565–578).

These coatings are aptly called "environmental barrier coatings" or "EBCs." A number of candidate coatings have been proposed (Lee et al. 1995; Lee and Miller 1996; Lee 2000). Often these are multilayer coatings using various compounds and layers to provide adherence and match thermal expansion. The outer topcoat must either eliminate the silica component or be a compound with a significantly lower silica thermodynamic activity. An example of a proposed EBC is shown in Figure 5.35.

5.4.2 Active Oxidation

The oxidation of silica formers can be categorized as "passive" or "active," drawing from the lexicon of aqueous corrosion of metals. In the broad sense, passive refers to condensed external layers with some protective barrier attributes. Active refers to an uncondensed vapor or solute that is dispersed away from the reaction surface and has no solid-state barrier layer effect. The SiO_2 scales discussed up to this point are passive layers. However, active oxidation is another distinct oxidation mode possible for Si-based ceramics and composites. In addition to condensed-phase SiO_2, the silicon–oxygen system forms a very stable gaseous suboxide, SiO(g). At high temperatures and reduced oxidant pressures, Si-based ceramics and composites no longer form the protective SiO_2 scale (passive oxidation), but rather a volatile SiO(g) product (active oxidation), which leads to rapid component degradation. This phenomenon is a concern in low oxygen potential environments such as re-entry or heat-treating environments. The major needs in this field are defining the transition points from passive-to-active oxidation and determining the rates of active oxidation.

It is again useful to develop an understanding of active oxidation for SiC and Si_3N_4 based on the fundamental work for pure silicon (Wagner 1958; Turkdogan et al. 1963; Hinze and Graham 1976; Singhal 1976). There are two models for the transition from active-to-passive oxidation, which are shown schematically in Figure 5.36.

The Wagner model (Wagner 1958) begins with a bare Si(s) surface oxidizing at a low oxygen potential to form SiO(g). As the oxygen potential is increased, the amount of SiO(g) eventually will equal that required for thermodynamic equilibrium between Si and SiO_2:

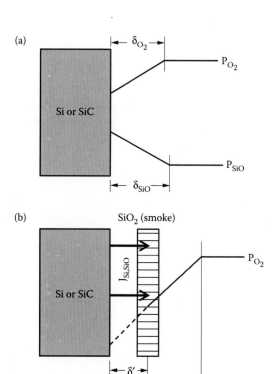

FIGURE 5.36
Schematic of two commonly used models for active-to-passive transitions. (a) Wagner model based on thermo-dynamic equilibrium condition for SiC/SiO$_2$ stability. (b) Turkogan model based on formation of SiO$_2$ smoke. (From Jacobson, N. S. et al. 2013. *J. Am. Ceram. Soc.* 96(3): 838–844.)

$$Si(s) + SiO_2(s) = 2\ SiO(g) \tag{5.13}$$

At this point, SiO$_2$ can form and the transition to passive oxidation occurs. Equations based on gas transport through the static boundary layer gives

$$P_{O_2}^{active-to-passive} = \frac{1}{2}\left(\frac{D_{SiO}}{D_{O_2}}\right)^{1/2} P_{SiO}^{eq} \tag{5.14}$$

Here, D_{SiO} and D_{O_2} are the gas-phase diffusivities of SiO(g) and O$_2$(g), respectively, and P_{SiO}^{eq} is the equilibrium pressure of SiO(g) calculated from the reaction. The Turkdogan model (Turkdogan et al. 1963; Hinze and Graham 1976) is based on the formation of Si(g) at the SiC surface and diffusion outward. This reacts with O$_2$ diffusing inward and forms SiO$_2$(smoke) a short distance away from the surface. The Si(g) flux outward is described by the Langmuir equation and the O$_2$ flow inward is described by a boundary layer-limited flux. These are equated and the active-to-passive transition is given by

$$P_{O_2}^{active-to-passive} = \frac{P_{Si}}{h_{O_2}}\left(\frac{RT}{2\pi M_{Si}}\right)^{1/2} \tag{5.15}$$

Here, P_{Si} is the equilibrium pressure of silicon at the SiC surface, R is the gas constant, T is the absolute temperature, M_{Si} is the molecular weight of Si(g), and h_{O_2} is the mass transfer coefficient for O_2, which is calculated from the equation

$$h_{O_2} = 0.664 \left(\frac{D_{O_2}^4 \rho}{\eta} \right)^{1/6} \left(\frac{v}{L} \right)^{1/2} \tag{5.16}$$

Here, D_{O_2} is the diffusion coefficient of oxygen, ρ is the gas density, η is the viscosity, v is the linear gas velocity past the sample, and L is a characteristic dimension of the sample. Equation 5.15 applies directly to pure Si, for SiC, the vapor pressure of Si is too low (Hinze and Graham 1976).

Wagner's theory treats the active-to-passive transition as entirely different phenomenon than the passive-to-active transition. The active-to-passive transition is based on the condition for equilibrium discussed above, while the passive-to-active transition is based on the decomposition of the SiO_2 scale:

$$SiO_2(s) = SiO(g) + \frac{1}{2}O_2(g) \tag{5.17}$$

These treatments explain the hysteresis between the active-to-passive and passive-to-active transitions for pure silicon and point out that these transitions differ by several orders of magnitude.

These basic concepts have been extended to SiC and Si_3N_4 by a number of investigators (Gulbransen and Jansson 1972; Hinze and Graham 1976; Singhal 1976; Keys 1977; Vaughn and Maahs 1990; Balat 1996; Jacobson and Myers 2010). For the analog to the Wagner condition for Si/SiO_2 equilibrium, three different equilibria are possible for the SiC/SiO_2:

$$SiC(s) + 2SiO_2(s) = 3SiO(g) + CO(g) \tag{5.18a}$$

$$SiC(s) + SiO_2(s) = 2SiO(g) + C(s) \tag{5.18b}$$

$$2SiC(s) + SiO_2(s) = 3Si(l,s) + 2CO(g) \tag{5.18c}$$

Different transition equations can be written based on each of these. Figure 5.37 compares these calculated boundaries to several literature measurements. The solid line corresponds to the decomposition of SiO_2. Although the Wagner expressions approximate the transition data, they do not completely describe the observed phenomena and microstructures (Hinze and Graham 1976; Jacobson and Myers 2010). More recent work explains some of these observations (Balat 1996; Charpentier et al. 2010a, b; Harder et al. 2012; Jacobson et al. 2012).

Recently, Jacobson and colleagues (Harder et al. 2012; Jacobson et al. 2012) have examined the active-to-passive and passive-to-active transitions for SiC in detail. In contrast to pure Si, these transitions occur at a similar $P(O_2)$, but must necessarily be due to different phenomena. The active-to-passive transition is similar to that discussed by Hinze and Graham for Si. It appears that the active-to-passive transition is initiated by the formation of SiO(g) oxidizing away from the surface to form SiO_2 rods. These continue to grow until a passive SiO_2 scale is formed, which is shown schematically in Figure 5.38. The passive-to-active transition appears to be initiated by the breakdown of the passive scale as gases are generated by reaction between the SiO_2 scale and SiC substrate, which is shown schematically in Figure 5.39. A similar mode of SiO_2 rod formation has been

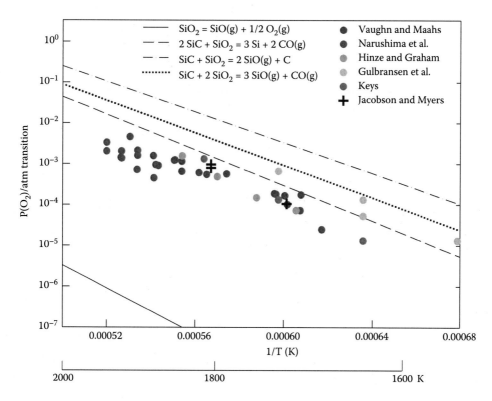

FIGURE 5.37
(**See color insert.**) Plot of literature active/passive transitions a function of inverse temperature and compared to boundaries derived from the SiC/SiO_2 equilibrium and also SiO_2 decomposition.

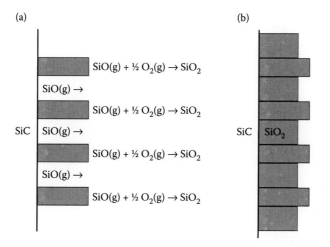

FIGURE 5.38
Schematic of rod formation mechanism during the transition from active-to-passive oxidation. (a) Initial rod formation (b) Space between rods fills in and oxide becomes protective. (From Jacobson, N. S. et al. 2013. *J. Am. Ceram. Soc.* 96(3): 838–844.)

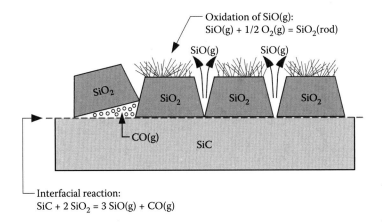

FIGURE 5.39
Proposed mechanism of passive-to-active transition for SiC. (From Harder, B. J. et al. 2013. *J. Am. Ceram. Soc.* 96(2): 606–612.)

reported in the oxidation of Ni–Si alloys at a relatively low partial pressure of oxygen (Carter et al. 2001).

Once the transition points have been established with active oxidation, the second critical issue is the rate of active oxidation. The studies described thus far utilize O_2/inert gas mixtures to attain low oxygen potentials. Rates can be described by boundary layer-limited transport equations (Jacobson and Myers 2010). Active oxidation can also be initiated by lowering the total pressure. In a reduced pressure environment, active oxidation rates are significantly higher.

In applications, it is best to work within regions where active oxidation does not occur, as it can be very destructive. There may also be coatings that limit active oxidation.

5.4.3 SiC-Based Composites

The major problem that limits the use of monolithic ceramics in structural applications is the low fracture toughness of the monolithics. A good deal of research has been devoted to the development of ceramic matrix composites (CMCs) to improve their fracture toughness (Evans 1990). Most research has focused on continuous fiber-reinforced SiC matrices to achieve high toughness (van Roode et al. 2003). The fiber is either carbon or SiC with a C or BN coating. Degradation issues such as silica growth, water vapor attack, salt corrosion, and active oxidation, are all still a concern since the outer surface of the CMC is SiC protected by a SiO_2 layer. However, these composites introduce a new factor with an easily oxidizable (C or BN) second phase.

In the case of the carbon fiber or carbon fiber coatings on SiC fiber, oxygen must be prevented from accessing the interior of the composite, which can occur via matrix cracking. Below ~400°C, carbon oxidizes very slowly. At temperatures up to ~1100°C, the cracks due to processing remain open and degradation of the carbon fibers or carbon fiber coatings may occur. Above ~1100°C, a film of silica will grow on the SiC and effectively seal the cracks. Filipuzzi et al. (Filipuzzi et al. 1994; Filipuzzi and Naslain 1994) have developed an analytical model for this process, which encompasses diffusion into the pore between the fiber and the matrix and simultaneous growth of silica on the pore wall. This is shown schematically in Figure 5.40.

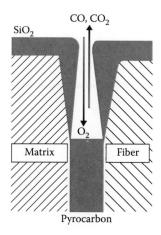

FIGURE 5.40

Model of carbon consumption and oxidation of silica formation in a composite composed of SiC matrix, SiC fibers, and carbon coatings. (Adapted from Filipuzzi, L. et al. 1994. *J. Am. Ceram. Soc.* 77(2): 459–466; Filipuzzi, L. and R. Naslain: Oxidation mechanisms and kinetics of 1D-SiC/C/SiC composite—Materials. II. modeling. *J. Am. Ceram. Soc.* 1994: 77(2). 467–480. Copyright Wiley-VCH Verlag GmbH & Co. KGaA. Reproduced with permission.)

Boron nitride (BN) fiber coatings should offer an advantage over carbon fiber coatings since the former oxidizes to a condensed phase and not a gaseous phase:

$$2BN(s) + \frac{3}{2} O_2(g) = B_2O_3(l) + N_2(g)$$

$$(5.19)$$

However, the liquid B_2O_3 is still not a desirable product, as it reacts with SiO_2 to form a low-melting borosilicate as well as a very stable boron hydroxide in the presence of water vapor (Jacobson et al. 1999a,b). Figure 5.41 illustrates these two cases. Recent studies have shown that on actual composites, BN oxidation effects may extend far into the composite (Opila, 2011).

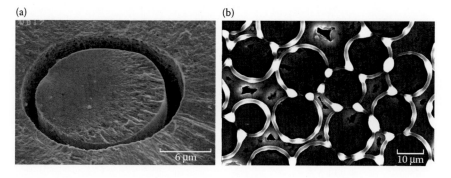

FIGURE 5.41

Micrographs illustrating high-temperature reaction effects on BN fiber coatings (a). Volatilization and removal of coating after exposure to humid laboratory air at 500°C. (b) Borosilicate glass formation after 100 h in oxygen at 816°C. (Jacobson, N. S. et al.: High-temperature oxidation of boron nitride: II, boron nitride layers in composites. *J. Am. Ceram. Soc.* 1999. 82(6). 1473–1482. Copyright Wiley-VCH Verlag GmbH & Co. KGaA. Reproduced with permission.)

5.5 Corrosion in Complex Gas/Deposit Environments

5.5.1 Corrosion in Mixed Gases

Often alloys are exposed to complex atmospheres containing more than one reactive species. Examples are SO_2–O_2, Cl_2–O_2, CO_2–CO, O_2–N_2, and various industrial environments with a complex mixture of gases. Depending on the metal or alloy, the products may be a complex mixture of phases. Corrosion in oxidizing–sulfidizing environments has been critically reviewed by Gesmundo et al. (1989). The situation is best described thermodynamically with a predominance diagram (Gulbransen and Jansson 1973), as illustrated with the Co–O–S diagram in Figure 5.42 at 650°C (Jacobson and Worrell 1984). This diagram is very useful as it shows that the oxide is much more stable than the sulfide, since the oxide forms at ten orders of magnitude lower $P(O_2)$ as compared to the $P(S_2)$ needed to form the sulfide. Thus, the sulfide is only expected at low oxygen potentials—either set in the gas or in a region near the scale/metal interface. Figure 5.43 is a cross section of a scale formed on Co in SO_2. Note the inner layer of sulfide and outer layer of a mixed sulfide oxide. Similar behavior was first observed for Ni (Luthra and Worrell 1978), and in both cases the sulfides are interconnected, creating rapid diffusion rates and hence rapid corrosion rates. One important question is the formation or lack of formation of sulfate at the gas/scale interface, which is predicted from thermodynamics in many cases. This is controversial (Luthra and Worrell 1978, 1979; Lillerud et al. 1984) as there may be kinetic barriers, but the issue is important for the formation of metal sulfates in Type II hot corrosion.

Other mixed gases behave somewhat similarly. Oxides are generally more stable than nonoxide phases and hence the condensed second phase (e.g., metal chloride, metal carbide) forms in the low oxygen potential portion of the scale. In the case of Co, the reaction with Cl_2–O_2 leads to an oxide scale with a condensed phase of cobalt chloride below the

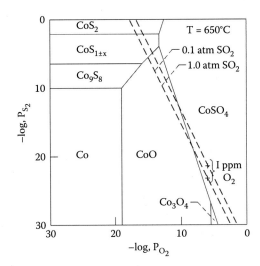

FIGURE 5.42
Predominance diagram for Co–O–S at 650°C. (From Jacobson, N. S. and W. L. Worrell 1984. *J. Electrochem. Soc.* 131(5): 1182–1188.)

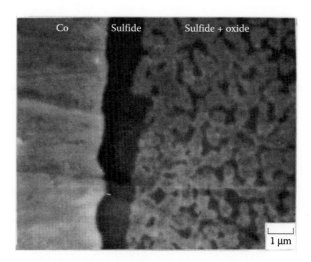

FIGURE 5.43
Cross section of Co + SO$_2$ after 2 h at 750°C. (From Jacobson, N. S. and W. L. Worrell 1984. *J. Electrochem. Soc.* 131(5): 1182–1188.)

oxide (Jacobson et al. 1986). A number of metals and alloys form both oxides and nitrides in mixed O$_2$–N$_2$ environments. For these mixed gases, the predominance diagram is again a useful map to understand the behavior of both metals and alloys.

5.5.2 Hot Corrosion of Alloys

5.5.2.1 Type I Hot Corrosion

In a combustion situation, ingested ambient NaCl reacts with sulfur in the fuel to form the highly stable Na$_2$SO$_4$ species (Kohl et al. 1975):

$$2\,NaCl(g) + SO_2(g) + \frac{1}{2}O_2(g) + H_2O(g) = Na_2SO_4(g) + 2\,HCl(g) \tag{5.20}$$

The NaCl is typically from a marine environment, but may also be from Na impurities in the fuel. The Na$_2$SO$_4$ may then deposit on engine parts and lead to extensive corrosion. It should also be noted that sulfates such as Na$_2$SO$_4$ are also contained in fine atmospheric particulates that, in some parts of the world, are found at very high levels. Type I hot corrosion occurs, by definition, between the melting point of Na$_2$SO$_4$ and the highest temperature of deposition, which is a function of pressure, sulfur, and sodium concentrations (Jacobson 1989). The rate of deposition is important and has been the topic of numerous theoretical and experimental studies (Stearns et al. 1981). In addition, other salts or oxides can form deposits, such as vanadates from low-purity fuels. The general considerations discussed in this brief summary can be extended to other deposits beyond Na$_2$SO$_4$.

The mechanism of corrosion by Na$_2$SO$_4$ is complex and has been extensively studied by numerous investigators (Stringer 1977; Rapp 1986; Pettit and Giggins 1987; Pettit 2011). Generally, hot corrosion is described by an initiation and propagation stage. The initiation stage involves the deposition of the salt on the surface of the alloy and initial breakdown of the oxide. The major attack occurs during the propagation stage. This is described by two major mechanisms: fluxing and sulfidation. The fluxing mechanism involves the dissolution of the protective oxide in the molten salt and has been examined in detail by Rapp and colleagues (Rapp 1986, 2002). This concept relates to acid/base character of oxides (Flood and Förland 1947). Figure 5.44 is a solubility diagram that shows the solubility of various oxides in a Na_2SO_4 melt, as a function of activity of Na_2O, which is the basic constituent of Na_2SO_4. In a melt of Na_2SO_4, the basicity is set by the overpressure of $P(SO_3)$:

$$Na_2SO_4(l) = Na_2O(s) + SO_3(g) \quad a(Na_2O) = \frac{K}{P(SO_3)} \tag{5.21}$$

Here, K is the tabulated equilibrium constant for this reaction. In Figure 5.44, the increase in solubility on the right-hand side of each curve is due to the lower basicity and hence is from acidic dissolution:

$$Al_2O_3 = 2\,Al^{+3} + 3\,O^{2-} \tag{5.22}$$

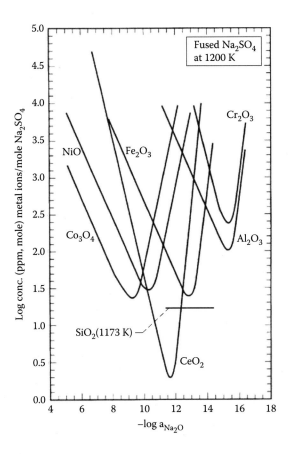

FIGURE 5.44
Solubility curves for common oxides in Na_2SO_4. (Reprinted from *Corr. Sci.* 44(2), Rapp, R. A. Hot corrosion of materials: a fluxing mechanism? 209–221, Copyright 2002, with permission from Elsevier.)

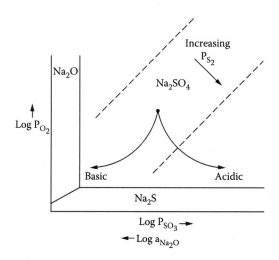

FIGURE 5.45
Section of a predominance diagram showing how conditions at the melt/alloy interface can shift the basicity of Na_2SO_4. (Pettit, F. S. and C. S. Giggins: Hot corrosion. *Superalloys II*, ed. C. T. Sims, N. S. Stoloff and W. C. Hagel. 327–358. 1987. Copyright Wiley-VCH Verlag GmbH & Co. KGaA. Reproduced with permission.)

The increase in solubility of the left-hand side of the curve is due to basic dissolution:

$$Al_2O_3 + O^{2-} = 2\,AlO_2^- \tag{5.23}$$

Rapp further extends this dissolution to show that extensive corrosion occurs when the gradient of dissolved oxide through the melt is negative, such that dissolution occurs at the melt/alloy interface and reprecipitation of the oxide occurs at the gas/melt interface. This leads to rapid degradation of the alloy.

The actual hot corrosion process on a superalloy is more complex, with numerous side reactions occurring. The actual basicity of the Na_2SO_4 melt may change with conditions at the melt interface, as shown in Figure 5.45. Superimposing predominance diagrams for Ni or Co over the Na_2SO_4 predominance diagram can be correlated with the scale morphology and composition and this can be used to derive the reaction mechanism (Pettit and Giggins 1987). Also, nickel-based alloys often show internal sulfidation from Type I hot corrosion. Thus, the three mechanisms to consider for hot corrosion are acidic fluxing, basic fluxing, and sulfidation (Pettit and Giggins 1987).

The solubility behavior for silica is also shown in Figure 5.44. This is a strongly acidic oxide and shows very limited acidic solubility. However, it is very soluble in a basic molten salt. This will be discussed in more detail later in the section on hot corrosion of Si-based ceramics and composites.

There is no simple solution to the problem of hot corrosion. Coatings such as CoCrAlY have been effective in reducing Type I hot corrosion although they show problems with Type II hot corrosion, which will be discussed in the next section. In general, higher-Cr alloys show better Type I and II hot corrosion resistance (Stringer 1977; Luthra and Wood 1984).

5.5.2.2 Type II Hot Corrosion

Type II hot corrosion is a low-temperature form of hot corrosion (LTHC) generally appearing in the range of 600–800°C. The deposition of sulfate salts is still a requisite feature;

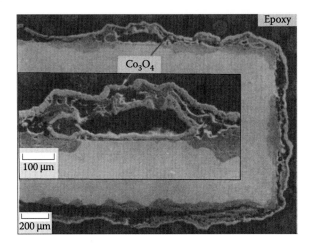

FIGURE 5.46
Massive low-temperature hot corrosion product on Co–22Cr–12Al wt.% alloy after 6.3 h at 750°C; Na_2SO_4 film in 0.15% SO_2/O_2. (From Luthra, K. L. 1982. *Metall. Trans. A.* 13(10): 1843–1852.)

however, pure Na_2SO_4 is initially solid below 884°C. Other components from ingested sea salt (Mg) may serve to form a low-melting $MgSO_4$–Na_2SO_4 eutectic at 660°C, thereby reducing the temperature range at which molten sulfates and accelerated attack may occur. Additional routes to form low-melting deposits are sulfation of Ni and Co oxides that may form during application to form other $NiSO_4$–Na_2SO_4 (660°C) and $CoSO_4$–Na_2SO_4 (585°C) eutectics. The latter has emerged as an insidious attack mode in service for Co-containing superalloys and CoNiCrAlY coatings exposed to high salt ingestion and sulfur (SO_2) content in the fuels. An example of this type of attack is shown in Figure 5.46. Here, a model CoCrAl coating composition, which might be used to prevent high-temperature corrosion and oxidation, is shown to be severely attacked with degraded regions easily on the order of 100 μm in thickness. Furthermore, it is reported that there is no depletion zone under the attack, as might be expected for high-temperature corrosion where diffusion comes into play. The test conditions were 6.3 h at 750°C, in 0.15% SO_2/O_2 with a Na_2SO_4 deposit (Luthra 1982a).

The chemical mechanisms that come into play have been set forth by Luthra (1982b) and others (Jones and Gadomski 1982; Chiang et al. 1983; Misra and Stearns 1984; Misra and Whittle 1984). The controlling reaction is the sulfation of CoO or Co_3O_4 to form a liquid $CoSO_4$–Na_2SO_4:

$$CoO + SO_3 = CoSO_4 \quad p(SO_3)_{eq} = 1.5 \times 10^{-5} \text{ atm (750°C)} \tag{5.24a}$$

or

$$\frac{1}{3} Co_3O_4 + SO_3 = CoSO_4 + \frac{1}{6}O_2 \quad p(SO_3)_{eq} = 5 \times 10^{-5} \text{ atm (750°C)} \tag{5.24b}$$

The thermodynamics of the reaction dictate the equilibrium constant, from which the equilibrium $p(SO_3)$ can be determined. According to the above reactions, the Co-sulfate should not form, the binary eutectic will not occur, and no LTHC results for values of $p(SO_3)$ below (1.5 or 5) $\times 10^{-5}$ atm. The same critical value for NiO sulfation is an order of magnitude greater, at 4×10^{-4} atm. This and the lower Co–Na–sulfate eutectic temperature explain why Co alloying elements had first been associated with more prevalent LTHC events than Co-free alloys. Furthermore, the equilibrium liquid fraction of the salt increases with $p(SO_3)$ and the degree of corrosion is dependent on the quantity of liquid sulfate. Thus, the $p(SO_2)$ of the environment, which fixes the $p(SO_3)$, is a critical factor influencing Type II LTHC. It has been estimated that the $p(SO_2)$ existing in a 10 atm. turbine engine using fuel having 1 ppm S impurity is equivalent to a laboratory $p(SO_2)$ of about 0.15% at 1 atm (Luthra and Wood 1984). At 750°C, this produces an equilibrium $p(SO_3)$ of about 3×10^{-4} atm (0.03%), which is enough to cause LTHC of Co.

The temperature effects on equilibrium phases can be significant and are worth considering. One direct effect of elevated temperature is to exceed the melting point of the eutectic salt and increase the liquid fraction according to the binary phase diagram of the salt system in play. An indirect effect is the reduction in the amount of SO_3 that results from the SO_2–O_2 equilibrium. Alternatively, the equilibrium $p(SO_3)$ for the sulfation reactions above is seen to increase with temperature. The result of the latter two considerations is a type of $p(SO_3)$ "squeeze play" in which the amount needed rises and the amount available decreases with increase in temperature. This has been quantified for the Co–O–S system by Luthra and is portrayed in the $p(SO_3)$ versus T reaction map of Figure 5.47. Here, the $p(SO_3)$ produced by 0.15% SO_2 is predicted to fall by about an order of magnitude from 650°C to 900°C, Curve A. The amount needed to just begin to form a liquid increases to a maximum at about 850°C for the curve labeled CoO and Co_3O_4. Thus, the liquid fractions

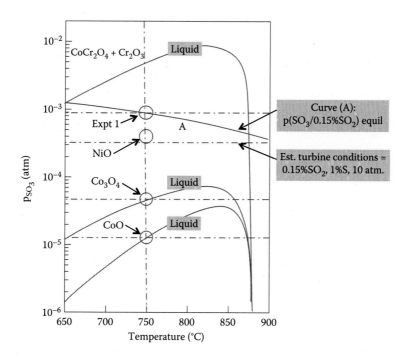

FIGURE 5.47
(**See color insert.**) SO_3—Temperature reaction map for LTHC of Co. Curve A represents the equilibrium $P(SO_3)$ for an 0.15% SO_2/O_2 environment at 1 atm. It is above that needed to form liquid $CoSO_4$–Na_2SO_4 salt solutions from CoO and Co_3O_4. Engine conditions also exceed these requirements for Co sulfation, but not Ni sulfation. (Reprinted from *Thin Sol. Films*, 119(3), Luthra, K. L. and J. H. Wood, High chromium cobalt-base coatings for low-temperature hot corrosion, 271–280, Copyright 1984, with permission from Elsevier.)

at points between Curve A and these two limits would decrease, and corrosive attack may be reduced as the temperature is raised to about 850°C. Above 884°C, pure Na_2SO_4 is 100% liquid without the need for CoO or NiO sulfation and eutectic salt formation.

Thus, we see that Ni and Co incorporate into the Na_2SO_4 film and trigger melt formation at various $p(SO_3)$. Misra has also analyzed Al_2O_3 sulfation in SO_3 and determined the dissolved mole fraction of $Al_2(SO_4)_3$ to be 0.133 in liquid Na_2SO_4–$NiSO_4$ at 750°C, but at an unrealistically high $p(SO_3)$ of 3.8×10^{-2} atm (Misra 1982). Furthermore, he studied alumina solubility in a Na_2SO_4–$NiSO_4$ melt and found it to be only 8.79×10^{-4} (0.09%) moles of Al per mole of salt at 750°C and $p(SO_3) = 3.64 \times 10^{-3}$ atm, compared to a calculated value of 4.03×10^{-3} (0.40%). These small amounts at high $p(SO_3)$ indicate that well-formed alumina scales are not especially susceptible to sulfation or dissolution in molten salts. Finally, he states that no molten Cr–Na-sulfates could form even at 10% SO_2. And the solubility of Cr_2O_3 in a Na_2SO_4–$NiSO_4$ melt was only 3 ppm at 750°C using an extremely high $p(SO_3)$ of 3.8×10^{-2} atm.

Alloying effects for Type II LTHC resistance are somewhat enigmatic and counterintuitive. At first, the beneficial effects of increasing Cr content are as expected, as seen in Figure 5.48. The scales formed on Co–40Cr were composed of Cr_2O_3 that will not form Cr–Na-sulfate eutectics at comparable $p(SO_3)$. The same is true for Co–15Al alloys for the same reasons as shown in Figure 5.49. However, when Al is added to Co–40Cr or Cr is added to Co–15Al, (cf. Figures 5.48b and 5.49b), LTHC ensues. Since the ternary additions are well known to encourage healing layers compared to just the binary, one conclusion is that stopping transient oxidation actually encourages LTHC. Thus, some faster-growing transient $CoCr_2O_4$ on the Co–Cr binary or $CoAl_2O_4$ on the Co–Al binary may be beneficial in separating the salts from the alloy. In contrast, the growth of pure Cr_2O_3 or Al_2O_3 scales

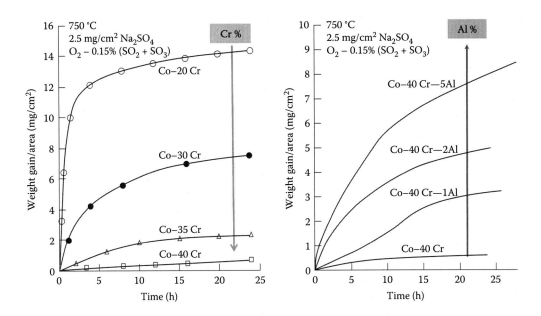

FIGURE 5.48

Effect of composition on 750°C Na_2SO_4 corrosion of Co–Cr alloys (0.15% SO_2/O_2). (Reprinted from *J. Electrochem. Soc.* 132(6), Luthra, K. L., Kinetics of the low-temperature hot corrosion of Co-Cr-Al Alloys, 1293–1298, Copyright 1985, with permission from Elsevier.) (a) Corrosion rates dramatically reduced by 40 wt.% Cr additions to Co. (b) Corrosion rates dramatically increased by 5 wt.% Al additions to Co–40Cr.

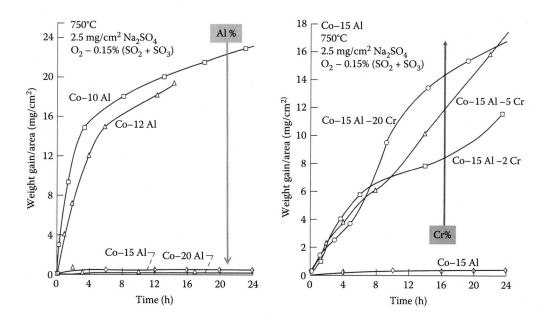

FIGURE 5.49
Effect of composition on 750°C Na$_2$SO$_4$ corrosion of Co–Al alloys (0.15% SO$_2$/O$_2$). (From Luthra, K. L. 1985. *J. Electrochem. Soc.* 132(6): 1293–1298.) (a) Corrosion rates dramatically reduced by 15 wt.% Al additions to Co. (b) Corrosion rates dramatically increased by 5 wt.% Cr additions to Co–15Al.

may be too slow at these temperatures to prevent attack. In summary, the usual strategy employed for oxidation and Type I hot corrosion resistance does not seem to apply for Type II LTHC. Other than high-Cr alloys, there does not appear to be a clear consensus on future directions or fundamental solutions. Furthermore, these alloying trends do not consider changes in alloy phase constitution with alloying addition. This may become significant as multiphase structures necessarily cause element distributions—particularly Cr and Al—to be nonuniform and partitioned below limits required for protective behavior.

5.5.2.3 Hot Corrosion of SiC and Si$_3$N$_4$ Ceramics

The effects of molten salt corrosion on silica scales must also be considered. Unlike alumina and chromia, silica is a strongly acidic oxide. This has important implications for molten salt corrosion. As discussed earlier, salt corrosion occurs by a fluxing mechanism where the aggressive salt dissolves the protective oxide. Since silica is acidic, it is generally inert to acidic molten salts, but quite reactive with basic molten salts (Jacobson 1993). In the case of Na$_2$SO$_4$, the thermodynamic activity of Na$_2$O is an index of basicity. Thus a high activity of Na$_2$O will react with SiO$_2$:

$$Na_2O + SiO_2 = Na_2SiO_3 \tag{5.25}$$

At 900°C, the threshold activity of Na$_2$O for this reaction is calculated to be 8.71 × 10^{-11} (Bale et al. 2002). If a(Na$_2$O) is greater than this, a reaction will occur; if a(Na$_2$O) is less than this, no reaction will occur. When a reaction does occur, a liquid film forms in place of the protective SiO$_2$ film. Diffusion rates through the liquid are, of course, significantly greater than through the solid SiO$_2$ and hence oxidation occurs much more readily. This coupled

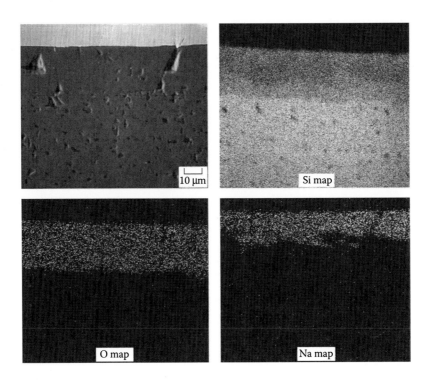

FIGURE 5.50
Polished cross section of the corrosion scale on sintered α-SiC with carbon and boron additives. This sample was corroded with Na_2CO_3, which gives a high activity of Na_2O, for 48 h at 1000°C. Note the thick outer layer of sodium silicate and the thick inner layer of silica. (Jacobson, N. S.: Corrosion of silicon-based ceramics in combustion environments. *J. Am. Ceram. Soc.* 1993. 76(1). 3–28. Copyright Wiley-VCH Verlag GmbH & Co. KGaA. Reproduced with permission.)

oxidation/dissolution can continue until the available Na_2O is consumed or the ceramic is consumed (Mayer and Riley 1978). Figure 5.50 shows a thick silicate scale on a coupon of SiC. Note that after the Na_2O is consumed, the sample continues to oxidize. The presence of Na^+ likely enhances the growth rate of the SiO_2, leading to an unusually thick SiO_2 scale below the silicate.

The rapid consumption of the ceramic does not occur with a smooth planar interface. Rather, it is characterized by pitting and grain boundary attack, as shown in Figure 5.51 (Jacobson and Smialek 1986). In a monolithic ceramic, these pits may be fracture origins and can lead to significant strength degradation (Smialek and Jacobson 1986).

EBCs may also be a solution to the salt corrosion problem (Lee et al. 1994). However, providing a barrier to Na^+ is very challenging given the high mobility of this cation at high temperatures.

5.5.2.4 CMAS and Volcanic Ash Considerations

A different form of deposit attack has emerged over the last two decades regarding turbine engines and is gaining widespread attention. The ingestion of fine particulate airborne materials, especially in geographic regions with desert sand conditions, results in deposition on turbine components. There can be compositional variations, but the generic material has been termed "CMAS," referring to calcium-magnesium-aluminosilicate. Actual

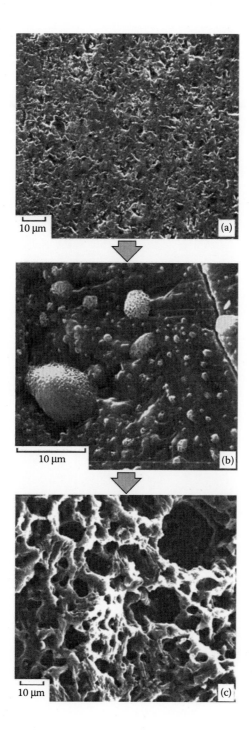

FIGURE 5.51
Sequence of sintered α-SiC with carbon and boron additives (a). As received surface. (b) Thick glassy layer formed after corrosion with $Na_2SO_4/0.01SO_3-O_2$ for 38 h at 1000°C. (c) After corrosion product has been cleanly removed, revealing heavily etched and pitted SiC. (Jacobson, N. S.: Corrosion of silicon-based ceramics in combustion environments. *J. Am. Ceram. Soc.* 1993. 76(1). 3–28. Copyright Wiley-VCH Verlag GmbH & Co. KGaA. Reproduced with permission.)

deposits contained NiO, Fe_2O_3, (Ca, Mg)CO_3 dolomite, and (Ca,Mg)SO_4 in various proportions (Smialek 1991; Smialek et al. 1992; Stott et al. 1992; Levi et al. 2012). Average model materials are now targeted at 35CaO–10MgO–7Al_2O_3–48SiO_2 mole percent and are used as simulants in mechanistic deposit studies. While the deposits often indicate a glassy nature (molten flow with bubbles), crystalline quartz, CaO · MgO · 2SiO_2 (diopside), and $CaSO_4$ phases have been identified by XRD. Reheating deposits found substantial softening/melting to occur as low as 1130°C or as high as ~1200°C (Smialek et al. 1992).

The corrosive properties of CMAS and volcanic ash on alloys has not been fully explored. There is some indication that fast-growing, less protective alumina scales have formed on NiAl coatings underneath thick CMAS deposits, with $CaSO_4$ suggested as the corrosive compound (Smialek et al. 1992, 1994). However, $CaSO_4$ melts at 1450°C, so a molten sulfate attack could not occur and some type of Ca-doped scale mechanism must apply. Regardless of any chemical attack, massive deposits may physically block airfoil-cooling holes and lead to degradation by overheating.

More recently, much attention has been given to CMAS effects on TBCs, as these coatings have become more prevalent and flights in high particulate areas more frequent. Here, the TBC surface temperature can easily exceed 1200–1250°C, which can cause the glassy silicate to become fluid. CMAS can then penetrate the grain or splat boundary porosity commonly associated with strain-tolerant EB-PVD or plasma-sprayed YSZ coatings. During cooling, the compressive stress in the coating can no longer be accommodated as the infiltrated glass becomes solid. Buckling, cracking, and coating failure result.

Furthermore, for sustained high-temperature exposure in laboratory tests with synthetic CMAS, it has been demonstrated by Krämer and Yang that

1. "CMAS rapidly and extensively infiltrates the TBC structure as soon as melting occurs,

2. CMAS severely attacks the TBC at lower temperatures (1240°C) and in times as short as 4 h,

3. CMAS obliterates the columnar morphology and converts the original t'-YSZ [feathered grains] into monoclinic globular particles in the upper region of the coating,

4. the attack is largely suppressed in the bulk of the TBC and

5. reactivates at the bottom of the coating, where CMAS dissolves the alumina substrate and converts the original t' [columnar grains] into larger cubic YSZ globules" (Kramer, Yang et al. 2006). See, for example, the cross section in Figure 5.52.

These structural and chemical attack mechanisms all serve to decrease TBC durability.

Further complexities were found from the detailed FIB-XTEM examination of a TBC-coated airfoil pulled from service (Braue 2009). Here, in addition to CMAS, high levels of $CaSO_4$ were documented at the gas surface and infiltrated to the base of the TBC in Figure 5.53. It was speculated that a Na–Mg–Ca sulfate eutectic at 738°C might precede the high-melting Ca sulfate, fully penetrate the TBC, then distill out the high-vapor-pressure (Na, Mg) components. Additional features were the formation of $CaZrO_3$, Ca-stabilized ZrO_2, $CaSiO_3$ wollastonite, Ca–Al–Fe-silicate, and $Ca_2ZrSi_4O_{12}$.

With the eruption of the Eyjafjallajokull volcano in Iceland, heightened interest in airborne volcanic ash deposits led to more studies of deposits on TBCs (Mechnich et al. 2011). The ash was similar to CMAS, but with notably lower CaO content and higher Na alkali content. (However, comparison of alkali contents may be problematic, as they may be volatilized from actual airfoil deposits). Simulated ash, shown in Table 5.4, was constructed

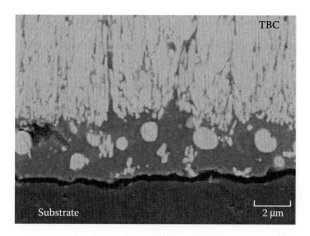

FIGURE 5.52
Spheroidized YSZ precipitates in glassy reaction zone under a TBC. CMAS degradation of YSZ; 1240°C for 4 h. (Kramer, S., J. Yang et al.: Thermochemical interaction of thermal barrier coatings with molten CaO-MgO-Al$_2$O$_3$-SiO$_2$ (CMAS) deposits. *J. Am. Ceram. Soc.* 2006. 89(10). 3167–3175. Copyright Wiley-VCH Verlag GmbH & Co. KGaA. Reproduced with permission.)

from analyzed compositions and deposited on coatings. The deposits softened at 930°C and wetted and flowed into the coating at 1100°C with a penetration depth proportional to temperature. In 1200°C exposures for 1 h, the ash–YSZ interface reacted to form albite NaAlSi$_3$O$_8$, anorthite CaAl$_2$Si$_2$O$_8$, zircon ZrSiO$_4$, and spinel ~(Fe,Mg)(Fe,Al)$_2$O$_4$. The primary degradation mode appeared to be infiltration and leaching out of the stabilizing element (Y) in YSZ. In these short treatments, the distinctive feathered columnar YSZ grain morphology had not been seriously affected, but the reaction products may take their toll at higher temperatures and longer times.

Advanced TBC formulations with greater high-temperature stability are based on the sinter-resistant pyrochlore structure such as Y$_2$Zr$_2$O$_7$ or Gd$_2$Zr$_2$O$_7$. These materials also appear

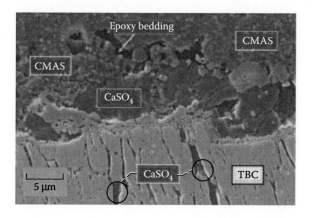

FIGURE 5.53
In-service HPT airfoil showing CMAS/CaSO$_4$ attack of YSZ TBC surface forming CaZrO$_3$ and Ca-fully stabilized ZrO$_2$ reaction products. EB-PVD YSZ, LPPS NiCoCrAlY(Hf,Si) bond coat, PWA 1484 airfoil; PW4000 turbine; 17,000 h. (With kind permission from Springer Science+Business Media: *J. Mat. Sci.*, Environmental stability of the YSZ layer and the YSZ/TGO interface of an in-service EB-PVD coated high-pressure turbine blade, 44(7), 2009, 1664–1675, Braue, W.)

TABLE 5.4

Compositions of Model and Actual CMAS and NTF-CMAS Fly Ash (with Na, Ti, Fe-Oxides)

Values in Wt% (Mol%)[a]	Lab-Scale CMAS	Example of an In-Service Deposition an HPT Airfoil	Eyjafjallajokull Volcanic Ash (April 15, 2010)	Artificial Volcanic Ash
SiO_2	48.50 (48.15)	22.40	57.69	57.90 (63.05)
Al_2O_3	11.80 (6.90)	10.10	15.48	15.50 (9.95)
CaO	33.20 (35.32)	33.60	5.15	5.20 (6.07)
MgO	6.50 (9.62)	9.90	2.15	2,15 (3.49)
FeO	—	15.40	9.61	9.65 (8.79)
TiO_2	—	3.0	1.59	1.60 (1.31)
Na_2O	—	—	5.25	5.25 (5.54)
K_2O	—	—	1.72	1.75 (1.22)
MnO	—	—	0.27	0.27 (0.25)
P_2O_5	—	0.7	0.73	0.73 (0.33)
Traces (S, Ni, Zr, etc.)	—	4.9	0.36	—

Source: Mechnich, P., W. Braue et al.: High-temperature corrosion of EB-PVD yttria partially stabilized zirconia thermal barrier coatings with an artificial volcanic ash overlay. *J. Am. Ceram. Soc.* 2011. 94(3). 925–931. Copyright Wiley-VCH Verlag GmbH & Co. KGaA. Reproduced with permission.

[a] Values in mol% are only given for synthetic materials without unspecified trace elements.

to be resistant to CMAS and ash deposits by rapidly forming refractory $Ca_2Gd_8(SiO_4)_6O_2$ (apatite) and $CaAl_2Si_2O_8$ (anorthite) phases that halt liquid penetration of the glassy deposits (Gledhill et al. 2011; Drexler et al. 2012).

Finally, it is noted that proposed EBC on SiC- or Si_3N_4-based materials discussed below may also have a problem with CMAS. The primary constituent of the first-generation EBC is the hexacelsian Ba,Sr-aluminosilicate (BSAS). CMAS exposure has been noted to penetrate grain boundaries and dissolve outer BSAS layers, and reprecipitate a Ca-rich celsian phase and Ba-rich anorthite in the CMAS glass (Grant et al. 2007; Harder et al. 2011). The penetration of porosity associated with any EB-PVD or plasma-sprayed structure is also likely to be a problem similar to those mechanisms discussed for TBC structures.

5.6 Re-Entry Materials

A spacecraft travels through space at tremendous speeds and experiences significant deceleration upon re-entering the earth's (or other body's) atmosphere. A shock wave develops around the craft that results in a highly complex atmosphere. This involves significant heating and an atmosphere containing molecules partially or fully dissociated by the shock into atoms and charged species. A good deal of research has gone into understanding this re-entry environment for the earth's atmosphere as well as that of neighboring planets (e.g., Mars) (Favaloro 2000; Grabow 2006).

Protection of the spacecraft from the heat of re-entry is a complex task and involves (1) the re-entry trajectory, (2) the actual spacecraft design, and (3) the selection of heat shield material. A shallow re-entry trajectory, such as that of the Apollo vehicles or the Space Shuttle Orbiters, tends to have a lower heating rate and longer heating times while a sharp re-entry trajectory, such as that of a ballistic missile, tends to have a high heating rate and a short heating time. Spacecraft design with sharp edges experience high local heating, but exhibit better maneuverability. Blunt nose vehicles, such as the Apollo capsules and the Space Shuttle Orbiters, tend to have less maneuverability but lower heating rates. The actual selection of heat shield materials is a unique field of study, although there are numerous considerations that are similar to the high-temperature oxidation/corrosion issues discussed in other parts of this chapter.

In general, heat shield materials fall into two categories—ablatives and reusable. Beyond the ability to withstand a high heat flux, there are numerous other requirements. These include adherence of the heat shield to the vehicle, high shear forces, and erosion/chemical reactions due to atmospheric constituents such as rain, ice, and atomic and molecular species.

Ablative materials dissipate heat by losing or transforming their surface material. Key parameters are the heat of ablation, which is the heat absorbed per unit weight, and the thermal conductivity. Ablators are classified by how the material loss or transformation occurs, as illustrated in Figure 5.54.

Subliming ablators involve an endothermic heat of vaporization, where the heat of ablation vaporizes the material. In addition, the gases generated carry heat away and thicken the boundary layer around the vehicle surface. Examples of subliming ablators are Teflon, graphite, and carbon/carbon composites. When used alone, graphite tends to crack due to isentropic thermal expansion, but carbon fiber reinforced carbon composites can be designed to minimize this problem. Melting ablators are similar to subliming ablators, except that the heat of ablation goes into forming a liquid film on the surface. Examples of melting ablators are nylon and quartz.

Charing ablators are most commonly used for re-entry surfaces. Again, an endothermic reaction forms a surface char and gases at the surface, which leads to an increase in thermal conductivity. Examples are carbon-phenolic materials and various mixtures of epoxy, cork, and even wood. The Apollo capsules used a complex structure based on an epoxy resin with phenolic microballoons and silica fibers in a honeycomb structure of fiberglass.

Finally, intumescent ablators are similar to charing ablators, except that they undergo a volume expansion and swell to form an extra layer, which leads to a decrease in thermal

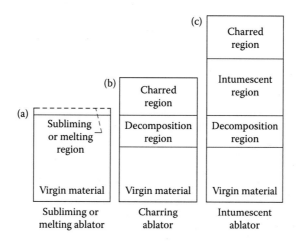

FIGURE 5.54
Schematic of different types of ablators. (From Favaloro, M. 2000. Ablative Materials Kirk-Othmer Encyclopedia of Chemical Technology.)

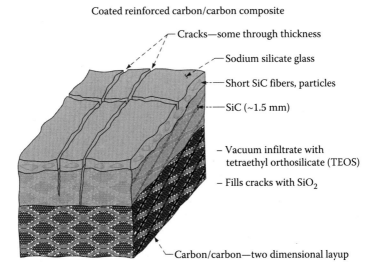

Coated reinforced carbon/carbon composite

— Cracks—some through thickness

— Sodium silicate glass

— Short SiC fibers, particles

— SiC (~1.5 mm)

– Vacuum infiltrate with
  tetraethyl orthosilicate (TEOS)

– Fills cracks with SiO_2

— Carbon/carbon—two dimensional layup

FIGURE 5.55
(See color insert.) Schematic of RCC.

conductivity. These are somewhat different than the materials discussed thus far and are primarily used for fire protection, rather than re-entry shields.

Reusable heat shields. In addition to ablatives, reusable heat shields have been used for protection from re-entry environments. The Space Shuttle Orbiter was the most well-known application of reusable reentry heat shields (Jenkins 1993). The outer skin of the Orbiter is composed of various insulation components, including thermal tiles, thermal blankets, and reinforced carbon/carbon (RCC) on the hottest parts—the wing leading edge and nose cap. The well-known tiles are highly porous and constructed of high-temperature fibers based on silica and aluminosilicate. Most of the heat of re-entry is taken by the RCC materials and shall be briefly discussed as an example of a highly successful reusable heat shield.

A schematic of the RCC material is shown in Figure 5.55. It is a two-dimensional weave of SiC fibers that is repeatedly infiltrated with a carbon precursor to fill open porosity.

Oxidation protection is provided with a conversion coating of SiC. The SiC necessarily cracks due to the thermal expansion mismatch with the carbon. These cracks are filled with a vacuum infiltration of tetra-ethyl orthosilicate (TEOS), which decomposes to silica upon heating. The surface is painted with a sodium silicate glass that flows on heating and fills the cracks. This is shown schematically in Figure 5.56.

In general, the infiltrated TEOS and surface sodium silicate is quite effective in reducing and nearly eliminating oxidation of the carbon/carbon substrate. Figure 5.57 shows weight loss curves in oxygen of RCC with and without the glass sealants.

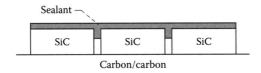

Sealant ⌐

| SiC | SiC | SiC |

Carbon/carbon

FIGURE 5.56
Schematic showing glass sealant filling cracks between SiC.

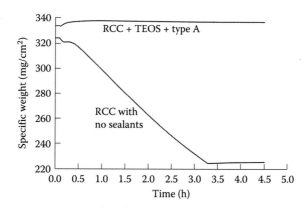

FIGURE 5.57
Effect of sealants on the laboratory oxidation of RCC in oxygen at 1200°C.

It is well known that oxidation of carbon fibers leads to a pointed morphology, as illustrated in Figure 5.58 (Glime and Cawley 1995; Jacobson and Curry 2006). This "signature" of carbon oxidation makes it easy to look for oxidation in actual mission-exposed RCC panels, as shown in Figure 5.59. Only a small cavity was found at the base of the crack. These panels are from the hottest part of the wing leading edge and had flown 19 missions. This is the point where the sealant needs to be reapplied.

As noted above, when the vehicle re-enters the atmosphere, there is a layer of molten sodium silicate on the surface of the RCC. After multiple re-entries, this sealant becomes depleted, (Williams et al. 1994) by both vaporization of the sodium silicate and shear forces pushing the sealant on the panels.

Owing to the consequences of oxidation of the RCC substrate, models for oxidation of a carbon substrate through a crack in a SiC coating (Medford 1975; Jacobson et al. 2008) have been developed and the key features of the most recent model will be summarized. At higher temperatures, gas-phase diffusion controls the oxidation of carbon (Walker Jr. et al.

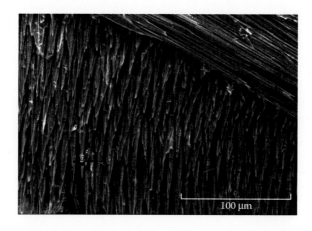

FIGURE 5.58
Characteristic "pointed" morphology caused by oxidation of carbon fibers at 1100°C for 0.5 h. (Reprinted from *Carbon*, 44(7), Jacobson, N. S. and D. M. Curry, Oxidation microstructure studies of reinforced carbon/carbon, 1142–1150, Copyright 2006, with permission from Elsevier.)

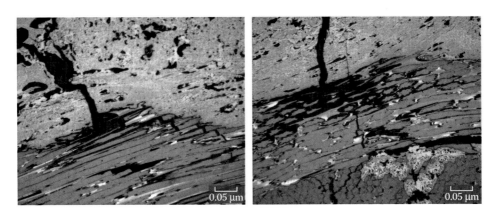

FIGURE 5.59
Oxidation of RCC (Panel 10 L) after 19 missions, showing a small cavity at the base of the crack in the SiC coating.

1959). The reaction of carbon must be described in two steps due to the thermodynamic incompatibility of C, CO, and CO_2. The reactions are as follows:

$$C(s) + CO_2(g) = 2\,CO(g) \tag{5.26a}$$

$$CO(g) + \frac{1}{2}O_2(g) = CO_2(g) \tag{5.26b}$$

Thus, $CO_2(g)$ is the oxidizing agent, as shown schematically in Figure 5.60.

Transport equations with appropriate boundary conditions were developed and must include both a diffusive term and a convective term:

$$J_i = D_i^{\text{eff}}\left(\frac{\partial c_i}{\partial x}\right) + v_i^{\text{ave}} c_i \tag{5.27}$$

Here, J_i is the flux of a species i, D_i^{eff} is the effective gas-phase diffusivity, c_i is the concentration of species I, and v_i^{ave} is the average molar velocity. Using the appropriate boundary

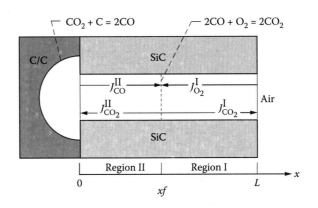

FIGURE 5.60
Model of carbon oxidation through a crack. (Reprinted from *Surf. Coat. Tech.*, 203(3–4), Jacobson, N. S. et al., Oxidation through coating cracks of SiC-protected carbon/carbon, 372–383, Copyright 2008, with permission from Elsevier.)

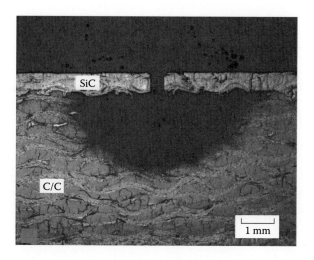

FIGURE 5.61

Optical micrograph showing oxidation void in RCC formed after 2.5 h at 1200°C in air with model crack intentionally cut in coating. (Reprinted from *Surf. Coat. Tech.*, 203(3–4), Jacobson, N. S. et al., Oxidation through coating cracks of SiC-protected carbon/carbon, 372–383, Copyright 2008, with permission from Elsevier.)

conditions, an analytical expression for the outward flux of CO_2 was derived, which described the consumption of carbon. This model accurately described the growth of a cavity below a machined slot, as shown in Figure 5.61. To extend the model further to an actual cracked RCC coating, a laboratory specimen was analyzed in regard to crack spacing and crack width following the methods of Yugartis et al. (1994). Clearly, the oxidation patterns are no longer as clean as those in Figure 5.61, appearing like Figure 5.59. Nonetheless, agreement with the model is reasonable.

The use of heat shields containing SiC at low pressures and high temperatures of re-entry also suggests the possibility of active oxidation, as discussed previously. Clearly, this is potentially a serious problem and can be ameliorated by an appropriate re-entry trajectory or, if necessary, additional coatings.

5.7 Closing Remarks

We have seen that past, current, and proposed high-temperature systems must take notice of oxidation and corrosion issues to maintain long-term durability. The gas turbine engine provides a major driving force for developing and understanding advanced oxidation-resistant materials. Protection from reaction with everpresent oxygen is based primarily on slow-growing Cr_2O_3, Al_2O_3, or SiO_2 scales—generally in decreasing order of kinetics and increasing use temperature. Various alloying and coating schemes have enabled Ni-based superalloy components to survive homologous surface temperatures of 0.9 T_m (~1200°C) due to stable Al_2O_3 scale formation and the use of ZrO_2-based ceramic TBCs. This has occurred steadily over 50 years of experience with stainless steels, inconels, early superalloys, and single-crystal superalloys.

Refractory metals have outstanding melting temperatures, but have been unable to be developed with anywhere near the oxidation resistance of the superalloys systems. The

Mo–Si–B alloys come the closest, but walk a thin line between excessive losses from MoO_3 and B_2O_3 vaporization at high temperature and incomplete protection from internal oxidation and embrittlement at low temperature. These alloys also lack the tensile strength and toughness of pure refractory metals. Silicide coatings provide some protection to refractory metals, but not long term because of CTE mismatch cracking that allows fatal substrate attack. The aluminide and silicide intermetallics present an attractive combination of low density and high melting point, but are plagued with low fracture toughness and catastrophic pest oxidation at low temperatures. $MoSi_2$-based materials, again with boron, represents one long-standing success, with at least enough fracture toughness to enable application as unstressed 1600°C furnace heating elements. γ-TiAlX alloys also show promise as 800°C structural materials for near term application in automotive superchargers and low-pressure turbine stages.

SiC and Si_3N_4 ceramics are well-characterized materials with excellent high-temperature (1400°C) oxidation resistance due to extremely slow-growing SiO_2 scales. Their future use in engines, under intense developmental efforts, is predicated on successful CMC fabrication technology and protection from fiber attack and SiO_2 scale volatility. The latter requires a fail-safe EBC for $T > 1200°C$, and, while BSAS and $Yb_2Si_2O_7$ types show promise as coatings, the demand for increased temperature continues to uncover ever more deficiencies.

While these considerations can bring perspective to nominal performance, more sporadic or uncontrolled attack from ingested sea salt (hot corrosion) or airborne dust (CMAS) must also be considered. These can be catastrophic because of the rapidity at which protective scales may be chemically undermined by hot corrosion, spanning a wide temperature regime, or the detriment to the strain tolerance of ceramic coatings by infiltrated molten glass deposits.

Acknowledgments

The authors thank Sebastien Dryepondt, Oak Ridge National Laboratory, Oak Ridge, Tennessee; Bryan Harder, NASA Glenn Research Center, Cleveland, Ohio; Paul Gannon, Montana State University, Bozeman, Montana; and Brian Gleeson, University of Pittsburgh, Pittsburgh, Pennsylvania, for reviewing this chapter and for providing valuable technical comments and suggestions. The authors also acknowledge current and past colleagues at NASA Glenn for key collaborations and for providing the historical foundation for many of the topics covered in this chapter. These include Beth Opila, Dennis Fox, Jim Nesbitt, Bob Miller, Carl Lowell, Chuck Barrett, and Ralph Garlick. The latter has been made possible through decades of support through the NASA Glenn (Lewis) Research and Technology Directorate and numerous turbine materials technology programs.

References

Balat, M. J. H. 1996. Determination of the active-to-passive transition in the oxidation of silicon carbide in standard and microwave-excited air. *J. Eur. Ceram. Soc.* 16(1): 55–62.

Bale, C. W., P. Chartrand et al. 2002. FactSage thermochemical software and databases. *Calphad* 26(2): 189–228.

Barrett, C. A. 2003. A High Temperature Cyclic Oxidation Data Base for Selected Materials Test at NASA Glenn Research Center. NASA/TM—2003-212546.

Barsoum, M. W. and T. El-Raghy 2001. The MAX phases: Unique new carbide and nitride materials—Ternary ceramics turn out to be surprisingly soft and machinable, yet also heat-tolerant, strong and lightweight. *Am. Scientist* 89(4): 334–343.

Bennett, M. J. 1995. Surface microsurgery preparation procedures for high temperature corrosion characterisation. In *Guidelines for Methods of Testing and Research in High Temperature Corrosion*. ed. H. J. Grabke and D. B. Meadowcroft. Leeds, UK: Maney Publishing. EFC-14: 158–176.

Birks, N., G. H. Meier et al. 2006. *Introduction to the High Temperature Oxidation of Metals*. Cambridge, UK: Cambridge University Press.

Brady, M. P., W. J. Brindley et al. 1996. The oxidation and protection of gamma titanium aluminides. *JOM* 48(11): 46–50.

Brady, M. P., B. Gleeson et al. 2000. Alloy design strategies for promoting protective oxide scale formation. *JOM* 52(1): 16–21.

Brady, M. P., B. A. Pint et al. 2000. High-temperature oxidation and corrosion of intermetallics. In *Corrosion and Environmental Degradation, Volume II*. ed. M. Shutze. Weinheim, Germany: Wiley-VCH. 239–325.

Brady, M. P., K. A. Unocic et al. 2011. Increasing the upper temperature oxidation limit of alumina forming austenitic stainless steels in air with water vapor. *Oxid. Met.* 75(5–6): 337–357.

Brady, M. P., Y. Yamamoto et al. 2008. The development of alumina-forming austenitic stainless steels for high-temperature structural use. *JOM* 60(7): 12–18.

Braue, W. 2009. Environmental stability of the YSZ layer and the YSZ/TGO interface of an in-service EB-PVD coated high-pressure turbine blade. *J. Mat. Sci.* 44(7): 1664–1675.

Brindley, W. J., J. L. Smialek et al. 1994. *Oxidation and Embrittlement of Orthorhombic-Titanium Alloys*. NASA HiTemp Review, CP 10146.

Brumm, M. W. and H. J. Grabke 1992. The oxidation behavior of NiAl.I. Phase-transformations in the alumina scale during oxidation of NiAl and NiAl-Cr Alloys. *Corr. Sci.* 33(11): 1677–1690.

Butt, D. P., R. E. Tressler et al. 1992. Corrosion of SiC materials in N2–H2–CO gaseous environments: I, Thermodynamics and kinetics of reactions. *J. Am. Ceram. Soc.* 75(12): 3257–3267.

Carter, P., B. Gleeson et al. 2001. Rapid growth of SiO_2 nanofibers on silicon-bearing alloys. *Oxid. Met.* 56(3–4): 375–394.

Chan, K. S. 2004. Cyclic oxidation response of multiphase niobium-based alloys. *Metall. Mat. Trans. A* 35(2): 589–597.

Charpentier, L., M. Balat-Pichelin et al. 2010a. High temperature oxidation of SiC under helium with low-pressure oxygen-Part 1: Sintered α-SiC. *J. Eur. Ceram. Soc.* 30(12): 2653–2660.

Charpentier, L., M. Balat-Pichelin et al. 2010b. High temperature oxidation of SiC under helium with low-pressure oxygen. Part 2: CVD β-SiC. *J. Eur. Ceram. Soc.* 30(12): 2661–2670.

Chiang, K., F. Pettit et al. 1983. Low temperature hot corrosion. *High Temperature Corrosion*. ed. R. Rapp, Houston, TX: NACE, 519–530.

Christensen, R. J., V. K. Tolpygo et al. 1997. The influence of the reactive element yttrium on the stress in alumina scales formed by oxidation. *Acta Mat.* 45(4): 1761–1766.

Cockeram, B. V. and R. A. Rapp 1995. Oxidation-resistant boron-doped and germanium-doped silicide coatings for refractory-metals at high-temperature. *Mat. Sci. Eng. A* 192: 980–986.

Costello, J. A. and R. E. Tressler 1981. Oxidation-kinetics of hot-pressed and sintered α-SiC. *J. Am. Ceram. Soc.* 64(6): 327–331.

Cottis, R., R. Lindsay et al., eds. 2009. *Shrier's Corrosion*. Amsterdam: Elsevier.

Cubicciotti, D., K. H. Lau et al. 1977. The rate controlling process in the oxidation of hot pressed silicon nitride. *J. Electrochem. Soc.* 124(12): 1955–1956.

Das, D. K. 2013. Microstructure and high temperature oxidation behavior of Pt-modified aluminide bond coats on Ni-base superalloys. *Prog. Mat. Sci.* 58: 151–182.

Deal, B. E. and A. S. Grove 1965. General relationship for the thermal oxidation of silica. *J. Appl. Phys.* 36(12): 3770–3778.

Donchev, A., E. Richter et al. 2006. Improvement of the oxidation behaviour of TiAl-alloys by treatment with halogens. *Intermetallics* 14(10–11): 1168–1174.

Doychak, J. 1995. High temperature oxidation. *Intermetallic Compounds, Principles and Practice, Vol. 1— Principles.* ed. J. H. Westbrook and F. L. Fleischer, New York: John Wiley & Sons Ltd.

Doychak, J. and T. Grobstein 1989. The oxidation of high-temperature intermetallics. *JOM* 41(10): 30–31.

Doychak, J., J. L. Smialek et al. 1989. Transient oxidation of single-crystal β-NiAl. *Metall. Trans. A* 20(3): 499–518.

Drexler, J. M., A. L. Ortiz et al. 2012. Composition effects of thermal barrier coating ceramics on their interaction with molten Ca-Mg-Al-silicate (CMAS) glass. *Acta Mat.* 60(15): 5437–5447.

EFC 1995. *Guidelines for Methods of Testing and Research in High Temperature Corrosion*, Leeds, UK: Maney Publishing.

Ellis, D. L. 2005. *GRCop-84: A High-Temperature Copper Alloy.* NASA/TM—2005-213566.

Evans, A. G. 1990. Perspective on the development of high-toughness ceramics. *J. Am. Ceram. Soc.* 73(2): 187–206.

Favaloro, M. 2000. Ablative Materials Kirk-Othmer Encyclopedia of Chemical Technology. http://www.scribd.com/doc/30129331/Ablative-Materials. (Accessed October 17, 2012).

Filipuzzi, L., G. Camus et al. 1994. Oxidation mechanisms and kinetics of 1D-SiC/C/SiC composite—Materials. I. An experimental approach. *J. Am. Ceram. Soc.* 77(2): 459–466.

Filipuzzi, L. and R. Naslain 1994. Oxidation mechanisms and kinetics of 1D-SiC/C/SiC composite—Materials. II. Modeling. *J. Am. Ceram. Soc.* 77(2): 467–480.

Flood, H. and T. Förland 1947. The acidic and basic properties of oxides. *Acta Chem. Scand* 1: 592–604.

Fox, D. S., R. A. Miller et al. 2011. Mach 0.3 burner rig facility at the NASA glenn materials research laboratory. NASA/TM-2011-216986.

Geiger, G. H. and D. R. Poirier 1973. *Transport Phenomena in Metallurgy.* Reading, Massachusetts: Addison-Wesley Publishing Company.

Gesmundo, F., D. J. Young et al. 1989. The high temperature corrosion of metals in sulfidizing-oxidizing environments: A critical review. *H. Temp. Mat. Proc.* 8: 149–190.

Giggins, C. S. and F. S. Pettit 1971. Oxidation of Ni-Cr-Al Alloys between 1000°C and 1200°C. *J. Electrochem. Soc.* 118(11): 1782–1790.

Glass, D. E., R. N. Shenoy et al. 2001. Effectiveness of diffusion barrier coatings for Mo-Re Embedded in C/SiC and C/C. NASA/TM-2001-211264.

Gledhill, A. D., K. M. Reddy et al. 2011. Mitigation of damage from molten fly ash to air-plasma-sprayed thermal barrier coatings. *Mat. Sci. Eng. A* 528(24): 7214–7221.

Gleeson, B., N. Mu et al. 2009. Compositional factors affecting the establishment and maintenance of Al_2O_3 scales on Ni-Al-Pt systems. *J. Mat. Sci.* 44(7): 1704–1710.

Gleeson, B., W. Wang et al. 2004. Effects of platinum on the interdiffusion and oxidation behavior of Ni-Al-based alloys. *Mat. Sci. Forum* 461–464: 213–222.

Glime, W. H. and J. D. Cawley 1995. Oxidation of carbon-fibers and films in ceramic-matrix composites—A weak-link process. *Carbon* 33(8): 1053–1060.

Gordon, S. and B. J. McBride 1971. Computer Program for Calculation of Complex Chemical Equilibrium Compositions, Rocket Performance, Incident and Reflected Shocks, and Chapman-Jouguet Detonations. NASA SP-273.

Goward, G. W. 1984. Protective Coatings for high temperature gas turbine alloys: A review of the state of technology. *Surface Engineering: Surface Modification of Materials.* ed. R. Kossowsky and S. C. Singhal, Martinus Nijhoff. 408.

Grabke, H. J. and G. H. Meier 1995. Accelerated oxidation, internal oxidation, intergranular oxidation, and pesting of intermetallic compounds. *Oxid. Met.* 44(1–2): 147–176.

Grabow, R. 2006. Ablative Heat Shielding for Spacecraft Re-Entry. http://courses.ucsd.edu/rherz/mae221a/reports/Grabow_221A_F06.pdf. (Accessed October 17, 2012).

Grant, K. M., S. Kramer et al. 2007. CMAS degradation of environmental barrier coatings. *Surf. Coat. Tech.* 202(4–7): 653–657.

Grobstein, T. and J. Doychak 1988. *Oxidation of High Temperature Intermetallics.* Warrendale, PA: The Minerals, Metals, and Materials Society.

Gulbransen, E. A. and S. A. Jansson 1972. The high-temperature oxidation, reduction, and volatilization reactions of silicon and silicon carbide. *Oxid. Met.* 4(3): 181–201.

Gulbransen, E. A. and S. A. Jansson 1973. Thermochemical considerations of high temperature gas–solid reactions. Metallurgical Society AIME, Boston, MA.

Gupta, R. N., Y. M. Yos et al. 1990. A Review of the Reaction Rates and Thermodynamic and Transport Properties for an 11-Species Air Model for Chemical and Thermal Nonequilibrium Calculations to 30 000 K. NASA Reference Publication 1232.

Harder, B. J., N. S. Jacobson et al. 2013. Oxidation transitions for SiC: Part II. Passive-to-active transitions. *J. Am. Ceram. Soc.* 96(2): 606–612.

Harder, B. J., J. Ramirez-Rico et al. 2011. Chemical and mechanical consequences of environmental barrier coating exposure to calcium-magnesium-aluminosilicate. *J. Am. Ceram. Soc.* 94: S178–S185.

Haynes, J. A., M. J. Lance et al. 2001. Characterization of commercial EB-PVD TBC systems with CVD (Ni,Pt)Al bond coatings. *Surf. Coat. Tech.* 146: 140–146.

Hebsur, M. 1994. *Pest Resistant and Low CTE MoSi$_2$-Matrix for High Temperature Structural Applications.* Intermetallic Matrix Composites III, MRS Proceedings 350: 177.

Hebsur, M. G., J. R. Stephens et al. 1989. Influence of alloying elements on the oxidation behavior of NbAl$_3$. In *Oxidation of High Temperature Intermetallics.* ed. T. Grobstein and J. Doychak. Warrendale, PA: The Minerals, Metals, and Materials Society.

Hinze, J. W. and H. C. Graham 1976. The active oxidation of Si and SiC in the viscous gas-flow regime. *J. Electrochem. Soc.* 123(7): 1066–1073.

Hong, S. Y., A. B. Anderson et al. 1990. Sulfur at nickel alumina interfaces—Molecular-orbital theory. *Surf. Sci.* 230(1–3): 175–183.

Hou, P. Y. 2008. Segregation phenomena at thermally grown Al$_2$O$_3$/Alloy interfaces. *Ann. Rev. Mat. Res.* 38: 275–298.

Jacobson, N. S. 1989. Sodium sulfate-deposition and dissolution of silica. *Oxid. Met.* 31(1–2): 91–103.

Jacobson, N. S. 1993. Corrosion of silicon-based ceramics in combustion environments. *J. Am. Ceram. Soc.* 76(1): 3–28.

Jacobson, N. S. and D. M. Curry 2006. Oxidation microstructure studies of reinforced carbon/carbon. *Carbon* 44(7): 1142–1150.

Jacobson, N., S. Farmer et al. 1999a. High-temperature oxidation of boron nitride: I, monolithic boron nitride. *J. Am. Ceram. Soc.* 82(2): 393–398.

Jacobson, N. S., D. S. Fox et al. 2005. Performance of ceramics in severe environments. In *ASM Handbook. Volume 13B Corrosion: Materials.* ed. S. D. Cramer and J. Covino, B. S. Materials Park, OH: ASM International. 565–578.

Jacobson, N. S., B. J. Harder et al. 2013. Oxidation transitions for SiC. Part I. Active-to-passive transitions. *J. Am. Ceram. Soc.* 96(3): 838–844.

Jacobson, N. S., M. J. McNallan et al. 1986. The formation of volatile corrosion products during the mixed oxidation-chlorination of cobalt at 650°C. *Metall. Trans. A.* 17(7): 1223–1228.

Jacobson, N. S., G. N. Morscher et al. 1999b. High-temperature oxidation of boron nitride: II, boron nitride layers in composites. *J. Am. Ceram. Soc.* 82(6): 1473–1482.

Jacobson, N. and D. Myers 2010. Active oxidation of SiC. *Oxid. Met.* 75: 1–25.

Jacobson, N. S., D. J. Roth et al. 2008. Oxidation through coating cracks of SiC-protected carbon/carbon. *Surf. Coat. Tech.* 203(3–4): 372–383.

Jacobson, N. S. and J. L. Smialek 1986. Corrosion PItting of SiC by Molten Salts. *J. Electrochem. Soc.* 133(12): 2615–2621.

Jacobson, N. S. and W. L. Worrell 1984. Reaction of cobalt in SO$_2$ atmospheres at elevated temperatures. *J. Electrochem. Soc.* 131(5): 1182–1188.

Jenkins, D. R. 1993. *The History of Developing the National Space Transportation System.* Marceline, MO: Walsworth Publishing.

Jones, R. L. and S. T. Gadomski 1982. Reactions of SO$_2$(SO$_3$) with NiO-Na$_2$SO$_4$ in nickel-sodium mixed sulfate formation and low temperature hot corrosion. *J. Electrochem. Soc.* 129(7): 1613–1618.

Kawagashi, K., H. Harada et al. 2008. EQ coating: A new concept for SRZ-free coating systems. In *Superalloys 2008*, Warrendale, PA: The Minerals, Metals, and Materials Society.

Keys, L. H. 1977. The oxidation of silicon carbide. In *Properties of High Temperature Alloys with an Emphasis on Environmental Effects*. ed. Z. A. Foroulis and F. S. Pettit, 681–696. Princeton, NJ: Electrochemical Society.

Kofstad, P. 1988. *High Temperature Corrosion*. London: Elsevier Applied Science.

Kohl, F. J., C. A. Stearns et al. 1975. Sodium sulfate: Vaporization thermodynamics and role in corrosive flames. In *Metal-Slag-Gas Reactions and Processes*. ed. Z. A. Foroulis and W. W. Smeltzer, 649–664. Princeton, NJ: Electrochemical Society.

Kramer, S., J. Yang et al. 2006. Thermochemical interaction of thermal barrier coatings with molten CaO-MgO-Al$_2$O$_3$-SiO$_2$ (CMAS) deposits. *J. Am. Ceram. Soc.* 89(10): 3167–3175.

Lai, G. Y. 1990. *High-Temperature Corrosion of Engineering Alloys*. Materials Park, OH: ASM.

Lee, K. N. 2000. Current status of environmental barrier coatings for Si-based ceramics. *Surf. Coat. Tech.* 133: 1–7.

Lee, K. N., N. S. Jacobson et al. 1994. Refractory Oxide Coatings on SiC Ceramics. *MRS Bull.* 19(10): 35–38.

Lee, K. N. and R. A. Miller 1996. Development and environmental durability of mullite and mullite/ YSZ dual layer coatings for SiC and Si$_3$N$_4$ ceramics. *Surf. Coat. Tech.* 86(1–3): 142–148.

Lee, K. N., R. A. Miller et al. 1995. New generation of plasma sprayed mullite coatings on silicon carbide. *J. Am. Ceram. Soc.* 78(3): 705–710.

Lee, W. Y., Y. Zhang et al. 1998. Effects of sulfur impurity on the scale adhesion behavior of a desulfurized Ni-based superalloy aluminized by chemical vapor deposition. *Metall. Mat. Trans. A.* 29(3): 833–841.

Levi, C. G., J. W. Hutchinson et al. 2012. Environmental degradation of thermal-barrier coatings by molten deposits. *MRS Bull.* 37(10): 932–941.

Leyens, C., R. Braun et al. 2006. Recent progress in the coating protection of gamma titanium aluminides. *JOM.* 58(1): 17–21.

Lillerud, K. P., B. Haflan et al. 1984. On the reaction-mechanism of nickel with SO$_2$ + O$_2$-SO$_3$. *Oxid. Met.* 21(3–4): 119–134.

Lin, Z. J., M. S. Li et al. 2007. High-temperature oxidation and hot corrosion of Cr$_2$AlC. *Acta Mat.* 55(18): 6182–6191.

Luthra, K. L. 1982a. Low-temperature hot corrosion of cobalt-base alloys part I. Morphology of the reaction-product. *Metall. Trans. A.* 13(10): 1843–1852.

Luthra, K. L. 1982b. Low-Temperature hot corrosion of cobalt-base alloys part II. Reaction-mechanism. *Metall. Trans. A.* 13(10): 1853–1864.

Luthra, K. L. 1985. Kinetics of the low-temperature hot corrosion of Co-Cr-Al Alloys. *J. Electrochem. Soc.* 132(6): 1293–1298.

Luthra, K. L. 1991. Some new perspectives on oxidation of silicon-carbide and silicon-nitride. *J. Am. Ceram. Soc.* 74(5): 1095–1103.

Luthra, K. L. and J. H. Wood 1984. High chromium cobalt-base coatings for low-temperature hot corrosion. *Thin Sol. Films* 119(3): 271–280.

Luthra, K. L. and W. L. Worrell 1978. Simultaneous sulfidation-oxidation of nickel at 603°C in argon-SO$_2$ atmospheres. *Metall. Trans. A.* 9(8): 1055–1061.

Luthra, K. L. and W. L. Worrell 1979. Simultaneous sulfidation-oxidation of nickel at 603°C in SO$_2$-O$_2$-SO$_3$ atmospheres. *Metall. Trans. A.* 10(5): 621–631.

Massalski, T. B., H. Okamoto et al. 1990. *Binary Phase Diagrams*. Materials Park, OH: ASM International.

Mayer, M. I. and F. L. Riley 1978. Sodium-assisted oxidation of reaction-bonded silicon nitride. *J. Mat. Sci.* 13(6): 1319–1328.

Mechnich, P., W. Braue et al. 2011. High-temperature corrosion of EB-PVD yttria partially stabilized zirconia thermal barrier coatings with an artificial volcanic ash overlay. *J. Am. Ceram. Soc.* 94(3): 925–931.

Medford, J. E. 1975. *Prediction of Oxidation Performance of Reinforced Carbon-Carbon Material for Space Shuttle Leading Edges*. Paper No. 75–730, 10th AIAA Thermophysics Conference, Denver, CO.

Meier, G. H. 1988. Fundamentals of the oxidation of high temperature intermetallics. In *Oxidation of High Temperature Intermetallics*. ed. T. Grobstein and J. Doychak, 1–16. Warrendale, PA: The Minerals, Metals, and Materials Society.

Meier, G. H., F. S. Pettit et al. 1995. The effects of reactive element additions and sulfur removal on the adherence of alumina to Ni- and Fe-base alloys. *Mat. Corr.* 46(4): 232–240.

Misra, A. K. 1982. Sodium sulfate induced corrosion of alloys at intermediate temperatures. Lawrence Berkeley Laboratory, LBL-14591.

Misra, A. K. and C. A. Stearns 1984. Mechanism of Na_2SO_4-induced catastrophic corrosion of Mo containing nickel-base superalloys at elevated-temperatures. *J. Electrochem. Soc.* 131(3): C81-C81.

Misra, A. K. and D. P. Whittle 1984. Effects of SO_2 and SO_3 on the Na_2SO_4 induced corrosion of nickel. *Oxid. Met.* 22(1–2): 1–33.

Motzfeldt, K. 1964. On the rates of oxidation of silicon and of silicon carbide in oxygen, and correlation with the permeability of silica glass. *Act. Chem. Scand.* 18: 1596–1606.

Mu, N., T. Izumi et al. 2008. Compositional factors affecting the oxidation behavior of Pt-Modified γ-Ni + γ'-Ni_3Al-based alloys and coatings. *Mat. Sci. Forum* 595–598: 239–247.

Narushima, T., T. Goto et al. 1989. High-temperature passive oxidation of chemically vapor-deposited silicon-carbide. *J. Am. Ceram. Soc.* 72(8): 1386–1390.

Norton, F. J. 1961. Permeation of gaseous oxygen through vitreous silica. *Nature* 171: 701.

Ogbuji, L. U. J. T. 1997. Effect of oxide devitrification on oxidation kinetics of SiC. *J. Am. Ceram. Soc.* 80(6): 1544–1550.

Ogbuji, L. U. J. T. and D. L. Humphrey 2000. *Oxidation Behavior of GRCop-84 (Cu-8Cr-4Nb) at Intermediate and High Temperatures*. NASA/CR—2000-210369.

Ogbuji, L. U. T. and D. T. Jayne 1993. Mechanism of incipient oxidation of bulk chemical vapor-deposited Si_3N_4. *J. Electrochem. Soc.* 140(3): 759–766.

Ogbuji, L. U. J. T. and E. J. Opila 1995. A Comparison of the oxidation kinetics of SiC and Si_3N_4. *J. Electrochem. Soc.* 142(3): 925–930.

Opila, E. J. 1994. Oxidation-kinetics of chemically vapor-deposited silicon-carbide in wet oxygen. *J. Am. Ceram. Soc.* 77(3): 730–736.

Opila, E. 1995. Influence of alumina reaction tube impurities on the oxidation of chemically-vapor-deposited silicon carbide. *J. Am. Ceram. Soc.* 78(4): 1107–1110.

Opila, E. J. 2008. *SiC Recession in a Hydrogen/Oxygen Environment Combustion Environment*. 31st Annual Conference on Composites, Materials, and Structures, Daytona Beach, FL, American Ceramic Society.

Opila, E. J., Boyd, M. K. 2011. Oxidation of SiC fiber-reinforced SiC matrix composites with a BN interphase. *Mat. Sci. Forum*. 696, ed. T. Maruyama, M. Yoshiba, K. Kurokawa, Y. Kawahara, N. Otsuka, Trans. Tech. Publications, Switzerland, pp. 342–347.

Opila, E. J., D. S. Fox et al. 1993. Cyclic Oxidation of Monolithic SiC and Si_3N_4 Materials. *Cer. Eng. Sci. Proc.* 49(7–8): 2363–2369.

Opila, E. J. and R. E. Hann 1997. Paralinear oxidation of CVD SiC in water vapor. *J. Am. Ceram. Soc.* 80(1): 197–205.

Opila, E. and N. Jacobson 2000. Corrosion of ceramic materials. In *Corrosion and Environmental Degradation*. ed. M. Shutze. Weiheim: Wiley-VCH. II: 328–387.

Opila, E. J., J. L. Smialek et al. 1999. SiC recession caused by SiO_2 scale volatility under combustion conditions: II. Thermodynamics and gaseous-diffusion model. *J. Am. Ceram. Soc.* 82(7): 1826–1834.

Packer, C. M. 1988. Overview of silicide coatings for refractory metals. *Oxidation of High Temperature Intermetallics*. ed. T. Grobstein and J. Doychak. 235–244. Warrendale, PA: The Minerals, Metals, and Materials Society.

Parfitt, L. J., J. L. Smialek et al. 1991. Oxidation behavior of cubic phases formed by alloying Al_3Ti with Cr and Mn. *Scrip. Met. et Mat.* 25(3): 727–731.

Park, J. S., R. Sakidja et al. 2002. Coating designs for oxidation control of Mo-Si-B alloys. *Scripta Mat.* 46(11): 765–770.

Pettit, F. 2011. Hot corrosion of metals and alloys. *Oxid. Met.* 76(1–2): 1–21.

Pettit, F. S. and C. S. Giggins 1987. Hot corrosion. *Superalloys II.* ed. C. T. Sims, N. S. Stoloff and W. C. Hagel, 327–358. New York: John Wiley & Sons.

Quadakkers, W. J. and H. Viefhaus 1995. The application of surface analysis techniques in high temperature corrosion research. In *Guidelines for Methods of Testing and Research in High Temperature Corrosion EFC 14.* ed. H. J. Grabke and D. B. Meadowcroft, 189–217. Leeds, UK: Maney Publishing.

Rahmel, A. and V. Kolarik 1995. Metallography, electron microprobe and x-ray structure analysis. In *Guidelines for Methods of Testing and Research in High Temperature Corrosion EFC 14.* ed. H. J. Grabke and D. B. Meadowcroft, 147–157. Leeds, UK: Maney Publishing.

Raj, S. V. 2008a. Comparison of the isothermal oxidation behavior of as-cast Cu-17%Cr and Cu-17%Cr-5%Al Part I: Oxidation kinetics. *Oxid. Met.* 70(1–2): 85–102.

Raj, S. V. 2008b. Comparison of the isothermal oxidation behavior of as-cast Cu-17%Cr and Cu-17%Cr-5%Al. Part II: Scale microstructures. *Oxid. Met.* 70(1–2): 103–119.

Ramanarayanan, T. A., Ayer, R. et al. 1988. The influence of yttrium on oxide scale growth and adherence. *Oxid. Met.* 29(5/6): 445–472.

Rapp, R. A. 1986. Chemistry and electrochemistry of the hot corrosion of metals. *Corrosion* 42(10): 568–577.

Rapp, R. A. 2002. Hot corrosion of materials: a fluxing mechanism? *Corr. Sci.* 44(2): 209–221.

Robinson, R. C. and J. L. Smialek 1999. SiC recession caused by SiO_2 scale volatility under combustion conditions: I. Experimental results and empirical model. *J. Am. Ceram. Soc.* 82(7): 1817–1825.

Rybicki, G. C. and J. L. Smialek 1989. Effect of the theta-alpha-Al_2O_3 Transformation on the oxidation behavior of β-NiAl + Zr. *Oxid. Met.* 31(3–4): 275–304.

Schneibel, J. H., R. O. Ritchie et al. 2005. Optimizaton of Mo-Si-B intermetallic alloys. *Metall. Mat. Trans. A.* 36(3): 525–531.

Schutze, M., G. Schumacher et al. 2002. The halogen effect in the oxidation of intermetallic titanium aluminides. *Corr. Sci.* 44(2): 303–318.

Searcy, A. W., D. V. Ragone et al. 1970. *Chemical and Mechanical Behavior of Inorganic Materials.* New York: Wiley-Interscience.

Singhal, S. C. 1976. Thermodynamic analysis of the high-temperature stability of silicon nitride and silicon carbide. *Ceramurg. Int.* 2(3): 123–130.

Smeggil, J. G., A. W. Funkenbusch et al. 1986. A Relationship between indigenous impurity elements and protective oxide scale adherence characteristics. *Metall. Mat. Trans. A.* 17(6): 923–932.

Smialek, J. L. 1991. *The Chemistry of Saudi Arabian Sand: A Deposition Problem on Helicopter Turbine Airfoils.* NASA TM/-2009–105234.

Smialek, J. L. 1993. Oxidation behavior of $TiAl_3$ coatings and alloys. *Corr. Sci.* 35(5–8): 1199–1208.

Smialek, J. L. 2000. Maintaining adhesion of protective Al_2O_3 scales. *JOM* 52(1): 22–25.

Smialek, J. L. 2001. Advances in the oxidation resistance of high-temperature turbine materials. *Surf. Int. Anal.* 31(7): 582–592.

Smialek, J. L. 2004. Improved oxidation life of segmented plasma sprayed 8YSZ thermal barrier coatings. *J. Therm. Spray Tech.* 13(1): 66–75.

Smialek, J. L. 2006. Moisture, induced delayed spallation and interfacial hydrogen embrittlement of alumina scales. *JOM* 58(1): 29–35.

Smialek, J. L. 2010. Moisture-Induced Alumina Scale Spallation: The Hydrogen Factor. NASA/TM-2010–216260.

Smialek, J. L. 2011a. Hydrogen and moisture-induced scale spallation: Cathodic descaling of a single crystal superalloy. *Electrochim. Act.* 56(4): 1823–1834.

Smialek, J. L. 2011b. Moisture-induced TBC spallation on turbine blade samples. *Surf. Coat. Tech.* 206(7): 1577–1585.

Smialek, J. L., F. A. Archer et al. 1992. *The Chemistry of Saudi Arabian Sand: A Deposition Problem on Helicopter Airfoils.* 3rd International SAMPE Metals and Metals Processing Conference, Toronto, Canada.

Smialek, J. L., F. A. Archer et al. 1994. Turbine airfoil degradation in the Persian-Gulf-War. *JOM* 46(12): 39–41.

Smialek, J. L. and J. V. Auping 2012. Cyclic Oxidation Database. NASA Glenn Research Center.

Smialek, J. L., C. A. Barrett et al. 1997. Design for oxidation. In *Design for Properties, ASM Handbook*. 20: 589–602. Materials Park, OH: ASM.

Smialek, J. L., M. Cuy et al. 1992. Burner Rig Oxidation of R512E Coated FS85 Nb Alloy, NASA Lewis Research Center.

Smialek, J. L., J. Doychak et al. 1988. Oxidation of Intermetallic Compounds. HiTemp Review, NASA CP-10025.

Smialek, J. L., J. Doychak et al. 1990. Oxidation behavior of FeAl + Hf, Zr, B. *Oxid. Met.* 34(3–4): 259–275.

Smialek, J. L. and N. S. Jacobson 1986. Mechanism of strength degradation for hot corrosion of α-SiC. *J. Am. Ceram. Soc.* 69(10): 741–752.

Smialek, J. L., N. S. Jacobson et al. 2013. Oxygen permeability and grain boundary diffusion applied to alumina scales. NASA/TM 2013-217855.

Smialek, J. L., D. T. Jayne et al. 1994. Effects of hydrogen annealing, sulfur segregation and diffusion on the cyclic oxidation resistance of superalloys—A review. *Thin Sol. Films* 253(1–2): 285–292.

Smialek, J. L. and G. H. Meier 1987. High temperature oxidation. In *Superalloys II*. ed. C. T. Sims, N. S. Stoloff and W. C. Hagel. 293–326. New York: John Wiley & Sons.

Smialek, J. L., J. A. Nesbitt et al. 1995. Service limitations for oxidation resistant intermetallic compounds. *High Temperature Ordered Intermetallics*. Pittsburgh: Materials Research Society.

Sozer, E. 2012. Non Equilibrium Gas and Plasma Dynamics Laboratory http://ngpdlab.engin.umich.edu/static-pages/img/hypersonic-interaction-flows_hyper-x.png. (Accessed December 14, 2012).

Stearns, C. A., F. J. Kohl et al. 1981. Combustion-system processes leading to corrosive deposits. *High Temperature Corrosion*. ed. R. A. Rapp. 441–450. Houston, TX: National Association of Corrosion Engineers.

Steinmetz, P. and C. Rapin 1997. Corrosion of metallic materials in waste incinerators. In *High Temperature Corrosion and Protection of Materials 4, Part 2*. R. Streiff, J. Stringer, R. C. Krutenat, M. Caillet and R. A. Rapp. 505–517. Switzerland: Trans Tech Publications.

Stott, F. H., D. J. de Wet et al. 1992. *The Effects of Molten Silicate Deposits on the Stability of Thermal Barrier Coatings for Tubine Applications at very High Temperature*. 3rd International SAMPE Metals and Metals Processing Conference, Toronto, Canada.

Stringer, J. 1977. Hot Corrosion of high-temperature alloys. In *Annual Review of Materials Science 7*. 477–509. Palo Alto, CA: Annual Reviews.

Swalin, R. A. 1962. *Thermodynamics of Solids*. New York: John Wiley & Sons.

Tien, J. K. and F. S. Pettit 1972. Mechanism of oxide adherence on Fe-25Cr-4Al (Y or Sc) Alloys. *Metall. Trans.* 3(6): 1587–1599.

Tolpygo, V. K., J. R. Dryden et al. 1998. Determination of the growth stress and strain in α-Al_2O_3 scales during the oxidation of Fe-22Cr-4.8Al-0.3Y alloy. *Act. Mat.* 46(3): 927–937.

Turkdogan, E. T., P. Grieveson et al. 1963. Enhancement of diffusion-limited rates of vaporization of metals. *J. Phys. Chem.* 67(8): 1647–1654.

van Roode, M., J. K. Ferber et al. 2003. *Ceramic Gas Turbine Component Development and Characterization*. New York: ASME Press.

Vaughn, W. L. and H. G. Maahs 1990. Active-to-passive transition in the oxidation of silicon carbide and silicon nitride in air. *J. Am. Ceram. Soc.* 73(6): 1540–1543.

Wada, M. T. Matsudaira et al. 2011. Mutual grain-boundary transport of aluminum and oxygen in polycrystalline Al_2O_3 under oxygen potential gradients at high temperatures. *J. Ceram. Soc. Jpn.* 119(11): 832–839.

Wagner, C. 1958. Passivity during the oxidation of silicon at elevated temperatures. *J. Appl. Phys.* 29: 1295–1297.

Walker Jr., P. L., F. Rusinko Jr. et al. 1959. Gas reactions of carbon. *Adv. Catal. 11*. ed. D. D. Eley, P. W. Selwood, and P. B. Weisz, 133–221. New York: Academic Press.

Wang, X. H., F. Z. Li et al. 2012. Insights into high temperature oxidation of Al_2O_3-forming Ti_3AlC_2. *Corr. Sci.* 58: 95–103.

Welsch, G., J. L. Smialek et al. 1996. High temperature oxidation and properties. In *Oxidation and Corrosion of Intermetallic Alloys*. ed. G. Welsch and P. D. Desai, 121–266. Lafayette, IN: NIAC/CINDAS.

Williams, S. D., D. M. Curry et al. 1994. *Ablation Analysis of the Shuttle Orbiter Oxidation Protected Reinforced Carbon-Carbon System*, AIAA 94–2084, 6th AIAA/ASME Joint Thermophysics and Heat Transfer Conference, Colorado Springs, CO.

Yamamoto, Y., M. P. Brady et al. 2007. Alumina-forming austenitic stainless steels strengthened by laves phase and MC carbide precipitates. *Metall. Mat. Trans. A.* 38(11): 2737–2746.

Young, D. J. 2008. *High Temperature Oxidation and Corrosion of Metals*. Oxford, UK: Elsevier.

Yu, W. B., S. B. Li et al. 2010. Microstructure and mechanical properties of a $Cr_2Al(Si)C$ solid solution. *Mat. Sci. Eng. A.* 527(21–22): 5997–6001.

Yurgartis, S. W., M. D. Bush et al. 1994. Morphological description of coating cracks in sic coated carbon-carbon composites. *Surf. Coat. Tech.* 70(1): 131–142.

Yuri, I. and T. Hisamatsu 2003. *Recession Rate Prediction for Ceramic Materials in Combustion Gas Flow GT2003-38886*. ASME Turbo Expo 2003, Atlanta, GA: ASME.

Zhang, Y., J. A. Haynes et al. 2001. Effects of Pt incorporation on the isothermal oxidation behavior of chemical vapor deposition aluminide coatings. *Metall. Mat. Trans. A.* 32(7): 1727–1741.

Zhao, X., J. Liu et al. 2011. Evolution of interfacial toughness of a thermal barrier system with a Pt-diffused γ/γ′ bond coat. *Act. Mat.* 59(16): 6401–6411.

Zhou, Z. H., H. B. Guo et al. 2011. Effect of water vapor on the phase transformation of alumina grown on NiAl at 950°C. *Corr. Sci.* 53(9): 2943–2947.

Zhu, D. M., R. A. Miller et al. 2001. Thermal conductivity of EB-PVD thermal barrier coatings evaluated by a steady-state laser heat flow technique. *Surf. Coat. Tech.* 138(1): 1–8.

6

High-Temperature Materials Processing

Olivia A. Graeve and James P. Kelly

CONTENTS

6.1 Introduction

Ultra-high-temperature ceramics (UHTCs) are a class of materials that have a variety of applications in extreme environments of temperature and pressure. Some researchers choose to define UHTCs by their melting temperature, citing temperatures ranging from 2500°C to 3000°C (Pulci et al. 2011; Squire and Marschall 2010; Savino et al. 2010). However, the lower end of this range includes oxide ceramics, which are not considered UHTC materials for most purposes. Other researchers consider that UHTCs must be able to function at operating temperatures of 1800–2500°C (Pulci et al. 2011; Ramírez-Rico et al. 2011; Jayaseelan et al. 2011; Yang et al. 2010; Alfano et al. 2010), but many of the materials classified as UHTCs cannot withstand prolonged exposure at these temperatures if the atmosphere is oxidizing. Another way of defining UHTCs is by application, such as materials for hypersonic flight, space reentry vehicles, or rocket propulsion (Pulci et al. 2011; Jayaseelan et al. 2011; Zou et al. 2011; Lawson et al. 2011; Eakins et al. 2011). However, this definition overlooks the fact that UHTCs have the potential to be replacement materials for applications that typically do not require UHTCs.

We will arbitrarily choose to define UHTCs as borides, carbides, nitrides, and silicides with melting temperatures above 2000°C. It is important to stress that this temperature is much lower than typically reported for UHTC materials, but is necessary in order to include typical materials included as additives in UHTC composites, to improve their behavior at high temperatures. For example, silicon carbide has a lower melting temperature than zirconium and hafnium diborides, but is used as an additive to improve the oxidation resistance of these materials at high temperatures (Fahrenholtz et al. 2004; Opeka

TABLE 6.1

Melting Temperatures of UHTCs

Compound	T_m (°C)	Compound	T_m (°C)	Compound	T_m (°C)
HfB_2	3370	HfC	3830	NbN	2630
LaB_6	2450	SiC	2550	Nb_2N	2630
MoB	2180	TaC	3830	TaN	2950
NbB_2	3000	TiC	2940	Ta_2N	2950
TaB_2	3270	WC	2790	TiN	3290
TiB_2	3230	AlN	2200	VN	2350
ZrB_2	3000	BN	2970	ZrN	2960
B_4C	2450	HfN	3330	$MoSi_2$	2020

Source: Adapted from Varma A., J.-P. Lebrat, *Chem. Eng. Sci.*, 47(9–11), 1992, 2179–2194.

et al. 1999). Aluminum nitride and disilicides having even lower melting temperatures (closer to 2000°C) can also provide improved behavior at high temperatures (Gasch and Johnson 2010; Zhang et al. 2009; Wu et al. 2009). A partial list of materials meeting our criteria and commonly considered UHTCs is provided in Table 6.1. As UHTC composites are designed to withstand the harsh conditions of multiple applications and ranges of conditions, it may become necessary to continue to modify the definition of UHTCs.

The most common UHTCs are IV–VI transition metal nonoxides. Carbide and nitride compounds, both of which fall in this category, can be considered interstitial compounds (Pierson 1996). The electronegativity difference between the two elements (i.e., the transition metal and the carbon or nitrogen) is large; thus, the sizes of the carbon and nitrogen atoms are small enough to sit in the interstitial sites of the metal lattice. The bonding in these materials is metallic, giving rise to high electrical and thermal conductivities, but also partly covalent and ionic leading to the refractory and brittle nature of the compounds. The bonding of group IV–VI diborides is similarly mixed (Aronsson et al. 1965; Vajeeston et al. 2001). The presence of boron–boron, metal–boron, and metal–metal bonds results in purely covalent bonding, a mixture of covalent and ionic (<8% ionic), and predominantly metallic types of bonding, respectively.

Other UHTC materials, such as silicon carbide and boron carbide, are characterized by low electronegativity, small elemental size differences, and bonding that is essentially covalent (Pierson 1996). This results in high melting temperature, but a lack of metallic-like properties compared to the interstitial compounds. In addition to the more common compounds listed in Table 6.1, there are about 240 binary UHTC materials with melting temperatures above 2000°C and at least 130 with melting temperature above 2500°C, if you also consider the oxide, sulfide, and phosphide compounds (Andrievski 1999). The use of oxides, sulfides, and phosphides by themselves is not as well developed for UHTC applications, but these materials can be added to UHTC composites to optimize high-temperature properties and behavior.

6.2 Powder Synthesis

6.2.1 Introduction

One particular issue with the manufacturing of UHTC composites is that their high melting temperatures and lack of stability under some conditions (i.e., oxygen environments) make them expensive to process, significantly hindering the ability to utilize them in a variety of applications. Producing very fine and well-dispersed powders has the potential to reduce sintering temperatures and times during consolidation, thus, reducing processing costs. From an economic standpoint, this is highly desirable. However, reducing powder size brings its own set of unique problems, including high-surface areas that can result in higher amounts of oxide contamination.

Synthesis can be accomplished by a variety of techniques and can be separated into three categories: top-down, bottom-up, and intermediate approaches. In the top-down approach, a large structure is broken into smaller components using physical or chemical methods. In the bottom-up approach, atomic or molecular species are assembled into structures such as particles, wires, or rods. Intermediate approaches may involve a combination of top-down and bottom-up syntheses. The entire breadth of powder synthesis cannot be covered easily and therefore only an introduction will be provided here, with a focus on the important considerations that should be made during synthesis. To obtain further details on the various synthesis methods, there are many resources available to the interested reader (Kelly and Graeve 2012; Sonber and Suri 2011; Selvaduray and Sheet 1993; Haussonne 1995; Cousin and Ross 1990; Segal 1986; Ring 1996; Warner 2011; Lu 2012; Lee and Pope 1994; Sugimoto 2000; Andrievski 1994). Listings of available techniques for top-down and bottom-up methodologies can be found in Tables 6.2 and 6.3, respectively.

An ideal synthesis method that will result in optimum consolidation should produce fine powder crystallites with low levels of agglomeration, high-phase purity, and high chemical purity (including clean surfaces). From the perspective of cost it should also be energy efficient, fast, scalable, and flexible for production of a diverse group of materials. All typical synthesis methods satisfy some of these criteria, but often fall short of a few key desired characteristics. Below we describe a few techniques used for the preparation of one particular UHTC, namely TaC, after which we expand our discussion to other material compositions.

6.2.2 Tantalum Carbide: Model System for UHTC Syntheses

Detailed studies of TaC-based composites has a long history extending back to the early twentieth century with a main focus in the early years on the characterization of microstructure and physical properties of this material (Bowman 1961; Toth et al. 1968; Nikol'skaya and Avarbé 1969; Petrov et al. 1969; Sheindlin et al. 1969, 1973; Rowcliffe and Warren 1970; Baxter 1971; Becher 1971; Rowcliffe and Hollox 1971a; 1971b; Jun and Shaffer 1971a; 1971b; Martin 1973; Samsonov 1973; Sirdeshmukh and Rao 1973; Bukatov et al. 1975; Pilyankevich et al. 1975; Blais and Bolbach 1977; Borisova and Borisov 1977; Veretennikov et al. 1977; Danil'chenko et al. 1977; Petrova and Chekhovskoi 1978; Chaikovskii et al. 1979; Pavlov et al. 1979; Allison et al. 1982; Markhasev et al. 1983; Modine et al. 1984; Gruzalski et al. 1985; Gruzalski and Zehner 1986; Hoffman and Williams 1986; Lipatnikov et al. 1987; Gusev et al. 1987; 1988; Danil'yants et al. 1988; Allison et al. 1988a,b; Ortman et al. 1988; Lipatnikov et al. 1986; Rempel et al. 1988, 1990; Garbe and Kirschner 1989; Gololobov et al. 1989; Lipatnikov et al. 1989; Gusev 1989, 1993; Khusainov et al. 1990; Gusev et al. 1991, 1996a,b, 2007; Mackie et al. 1991; Kudryavtsev et al. 1992; Price et al. 1993; Kim et al. 1994; Khyzhun et al. 1996; Desmaison-Brut et al. 1997; Khyzhun 1997; Rafaja et al. 1998; Jang et al. 1999; Rocher et al. 2002; Lipatnikov and Rempel 2005; López-de-la-Torre et al. 2005; Sahnoun et al. 2005; Nakamura and Yashima 2008; Hackett et al. 2009; Krechkivska and Morse 2010).

Some of the earliest efforts describe the synthesis of TaC by heating a mixture of tantalum pentoxide, tantalite, carbon and sodium bicarbonate. Since then, the reduction of

TABLE 6.2

Examples of Top-Down Methods

Technique	Specific Examples
Mechanical energy methods	Ball milling
	Rolling and beating
	Extrusion and drawing
	Mechanical machining/polishing/grinding
	Mechanical cutting
	Compaction and consolidation
	Atomization
Thermal fabrication methods	Annealing
	Chill-block melt spinning
	Electrohydrodynamic atomization
	Electrospinning
	Liquid dynamic compaction
	Gas atomization
	Evaporation
	Template extrusion
	Sublimation
	Polymer carbonization
High-energy and particle methods	Arc-discharge
	Laser ablation
	Solar energy vaporization
	RF sputtering
	Ion milling
	Electron-beam evaporation
	Reactive ion etching (RIE)
	Pyrolysis
	Combustion
	High-energy sonication
Lithographic methods	LIGA (lithography, electroplating, and molding) techniques
	Photolithography
	Immersion lithography
	Deep ultraviolet lithography
	Extreme ultraviolet lithography
	Electron-beam lithography
	Electron beam projection lithography
	Focused ion beam lithography
	Microcontact printing methods
	Nanoimprint lithography
	Nanosphere lithography
	Scanning atomic force microscopy nano-stencil
	Scanning probe nano-lithographies
Chemical methods	Chemical etching
	Chemical–mechanical polishing
	Electropolishing
	Anodizing

Source: Kelly, J.P., O.A. Graeve, Effect of powder characteristics on nanosintering, in: *Sintering*, Eds. R.H.R. Castro and K. Van Benthem, Springer, New York, ISBN 9783642310089, 2012, 57–92.

an oxide and subsequent carburization or the direct carburization of tantalum metal at elevated temperatures (even well above 1500°C) using solid carbon or hydrocarbon gases has been used by many researchers (Eick and Youngblood 2009; Li et al. 2007; Xiang et al. 2006; Fernandes et al. 2006; Kwon et al. 2004; Ciaravino et al. 2002; Chang et al. 2002; Rodríguez et al. 2001; Preiss et al. 1997; Hassine et al. 1995; Shvab and Kislyi 1974). Due to the

TABLE 6.3

Examples of Bottom-Up Methods

Technique	Specific Examples
Gas-phase methods	Atomic layer deposition
	Thermolysis–pyrolysis
	Organometallic vapor-phase epitaxy
	Molecular beam epitaxy
	Ion implantation
	Gas-phase condensation
Liquid-phase methods	Molecular self-assembly
	Supramolecular chemistry
	Nucleation and sol–gel processes
	Reduction of metal salts
	Single-crystal growth
	Electrodeposition/electroplating
	Molten salt solution electrolysis
	Template synthesis
	Combustion
Physical vapor deposition	Vacuum evaporation
	Sputtering
	Molecular beam epitaxy (MBE)
Chemical vapor deposition	Atmospheric pressure chemical vapor deposition (APCVD)
	Low-pressure chemical vapor deposition (LPCVD)
	Plasma-assisted (enhanced) chemical vapor deposition (PECVD)
	Photochemical vapor deposition (PCVD)
	Laser chemical vapor deposition (LCVD)
	Metal-organic chemical vapor deposition (MOCVD)
	Chemical beam epitaxy (CBE)
	Chemical vapor infiltration (CVI)

Source: Kelly, J.P., O.A. Graeve, Effect of powder characteristics on nanosintering, in: *Sintering*, Eds. R.H.R. Castro and K. Van Benthem, Springer, New York, ISBN 9783642310089, 2012, 57–92.

high processing temperatures and nature of the precursors, the particle sizes achieved by this method are relatively large, but the amount a surface oxide is typically small because of the low surface area of the powders. Reactions have been performed at temperatures as low as 700°C by using more reactive precursors to form nanoparticles, but reactions above 1500°C are favored for producing powders with low oxygen and low free carbon. When highly reactive precursors and low temperatures are used, oxygen content may be higher and purification may be necessary.

The reaction between elemental tantalum and carbon is exothermic (Varma and Lebrat 1992; Moore and Feng 1995a). This makes possible the synthesis of tantalum carbide by self-propagating high-temperature synthesis (SHS) (Graeve and Munir 2002; Aruna and Mukasyan 2008; Yeh and Liu 2006; Xue and Munir 1996; Moore and Feng 1995b; Shkiro et al. 1979). The underlying basis of the SHS process is the ability of exothermic reactions to sustain themselves in the form of a reaction (combustion) wave. Since the temperature of the combustion wave can be extremely high (as high as 3000 K) and the rate of wave propagation can be relatively rapid (as high as 25 cm/s), this process offers an opportunity to investigate reactions at extreme thermal gradients (Munir and Anselmi-Tamburini 1989). The process is shown schematically in Figure 6.1 for reactions in which all the reactants

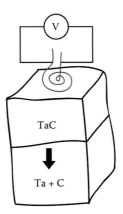

FIGURE 6.1
(**See color insert.**) Schematic representation of a Ta + C self-propagating high-temperature synthesis (SHS) reaction to form TaC. The conductive coil on top (usually made of tungsten) represents one of several possible sources of heat that could initiate the SHS reaction.

start as solid powders (e.g., Ta + C, Ti + C, Ti + 2B, Hf + C, Ta + 2B, etc.). The figure shows a combustion front that has progressed partially down the sample, leaving behind the product of combustion. Although such reactions must be highly exothermic to be self-sustaining, they do not self-initiate without the addition of external energy. This can be accomplished by heating one end of the sample by thermal radiation or through the use of laser energy. However, once initiated, these reactions will generally self-propagate, although some reactions require some level of preheating to increase the adiabatic temperature or the application of an additional heating mechanism, such as the passage of current (Graeve and Munir, 2002). Usually, the technique requires fine starting powders so that a good mixture of the precursor metal powders and the nonmetal (i.e., carbon, boron) powders can take place.

Simultaneous pressure application and/or electromagnetic field assistance have been applied to improve the combustion wave propagation, completion of reaction for enhanced phase purity, and higher densities for achieving simultaneous sintering. Secondary additives producing gases and liquids during the reaction have also been used and can affect the final product density and composition, but add complexity to the reactions. The primary disadvantage of this technique is the lack of process control after the reaction is initiated. Limited control of temperature has been achieved by diluting the reactants with the final product to reduce temperature or by performing the synthesis at elevated temperatures to increase the reaction temperature.

Another technique used for synthesizing TaC powders is via gas-phase reactions (Ishigaki et al. 2005; Chen et al 2002; Ahlén et al. 2000; Fukunaga et al. 1999; Johnsson et al. 1997; Johnsson and Nygren 1997; Tatarintseva et al. 1983; Alekseev et al. 1980a,b; Grossklauss and Bunshah 1975; Takahashi and Sugiyami 1974) as well as films (Naiki et al. 1972; Grossklauss and Bunshah 1975; Tatarintseva et al. 1983; Gesheva and Vlakhov 1987; Dua and George 1994) and single crystals (Savitsky and Burkhanov 1978). This is typically a low-temperature synthesis method that utilizes vapors of tantalum chloride and hydrocarbons in hydrogen/argon gas streams, although direct vaporization of tantalum metal has also been applied. Precursors are typically more expensive, but the technique is particularly useful for applying dense coatings and is one of the few techniques capable of providing unique fiber/whisker morphologies with transition metal catalysts. High purities

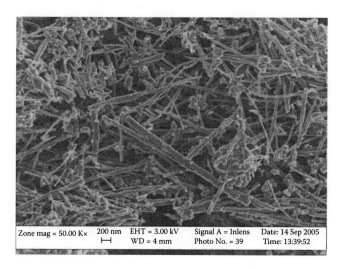

FIGURE 6.2
Scanning electron micrograph of TaC fibers obtained by reaction of tantalum with carbon nanotubes using a molten LiCl–KCl–KF salt system as a reaction medium. (Reprinted from *Carbon*, 47(1), Li, X. et al., A convenient, general synthesis of carbide nanofibres via templated reactions on carbon nanotubes in molten salt media, 201–208, Copyright 2009, with permission from Elsevier.)

can be difficult to achieve, however. Products often contain a mixture of the monocarbide phase, sub-carbide phase, tantalum metal, and/or carbon residuals.

A better example of the preparation of fibers is provided in Figure 6.2 (Li et al. 2009), where one can see that the product is mostly TaC fibers. This process involves the reaction of Ta and carbon nanotubes using a molten LiCl–KCl–KF salt system as reaction medium. The fibers are polycrystalline and there is still some low-aspect ratio powders, but the majority appears to be fibers of diameter between 40 and 90 nm. TaC coatings have also been prepared by deposition in molten salts (Stern and Gadomski 1983).

A sub-class of SHS reactions known as solid-state metathesis (SSM) reactions also makes use of molten salts (Nartowski et al. 1999; Gillan and Kaner 1996). The premise is to use anion exchange reactions to form the compound of interest and a co-produced salt. Besides anion exchange reactions, a secondary mechanism includes reductive recombination, where the ions are reduced to elemental form followed by reaction from the elements. The reaction is typically self-sustaining if the reaction temperature exceeds the melting temperature of the co-produced salt. The advantage over traditional SHS is that the co-produced salt from the anion exchange is an effective heat brake applied to the reaction limiting the maximum reaction temperature to that of the boiling point of the salt. The lower temperatures favor formation of small particles. However, the reactions are not as exothermic as traditional SHS reactions. For reaction temperatures below the melting point of the co-produced salt, extended times at moderately elevated temperatures may be necessary to complete the reaction. Ma et al. (2007) adds to the complexity of metathesis reactions by suggesting compositions that produce a vapor phase.

More recently, solvothermal synthesis methodologies have been reported for synthesizing TaC powders (Kelly and Graeve 2011; Kelly et al. 2010). This process relies on SHS reactions within a melt. Similar to traditional SHS synthesis, the reaction is fast, assisting scalability by allowing for large quantities of powders to be produced rapidly (Aruna and Mukasyan 2008; Nartowski et al. 1999; Moore and Feng 1995a). The metallic melt used during the reaction helps in prevention of grain growth, overcoming the difficulty of producing fine powders from traditional SHS.

Other laboratory-scale synthesis techniques have been presented in the literature (Shi et al. 2004; Li et al. 2008; Lei et al. 2008; Nelson and Wagner 2002). However, since these techniques use more hazardous chemicals and/or result in low yields, they are not covered in detail here. Table 6.4 summarizes the more common techniques and advantages as disadvantages of each.

TABLE 6.4

Advantages and Disadvantages of Various Tantalum Carbide Synthesis Techniques

Technique	Advantages	Disadvantages
High-temperature carburization	Low free carbon content Inexpensive raw materials Large-scale production has been demonstrated	Long times at high temperature are energy intensive Results in large particle sizes Post-synthesis communion required for obtaining small particle sizes (i.e., deaggregation
Low-temperature carburization	Similar to well-established high-temperature carburization technique Can produce small particle sizes	Requires preparation of highly reactive precursors Requires leaching and further chemical purification Extended times at intermediate temperatures Possible particle agglomeration
Self-propagating High-temperature synthesis	Requires only ignition Fast Simultaneous densification possible Self-purifying Large-scale production has been demonstrated Inexpensive raw materials	Simultaneous densification difficult for TaC A 2700 K adiabatic flame temperature results in large and agglomerated particles Phase purity difficult to achieve Lack of control after ignition Typically aggregated particles
Gas-phase synthesis	Can obtain fiber/whisker morphology Can be used to apply coatings Low processing temperatures Fine particle size achievable Industrial scale feasible	Typically uses more expensive precursors High-phase purity is difficult to achieve Potentially hazardous gases are produced Possible particle agglomeration/aggregation
Molten salt/metal synthesis	Can be used to apply coatings Nanopowders or single crystals can be obtained Fiber/whisker morphology possible Relatively low temperatures used	Requires leaching and further chemical purification Requires extended periods of time at intermediate temperatures possibly leading to aggregation
Metathesis reactions, sub-class of molten salt/metal synthesis and self-propagating high-temperature synthesis	Exothermic reactions provide energy Co-produced salt limits temperature and can produce small particle sizes Combines advantages of multiple techniques	Limited exothermicity may require moderate temperatures to be sustained for extended periods of time Requires leaching and further chemical purification Precursor limitations May lead to particle aggregation

6.2.3 Synthesis of Carbides and Borides

Table 6.4 exemplifies, by extension, the synthesis of a variety of carbides, not just the specific example of TaC. For example, the preparation of HfC by SHS is well understood (Kecskes et al. 1990) and generally results in the synthesis of powders, although the preparation of partially dense pellets is also feasible, as illustrated in Figure 6.3.

Similarly, boride powders such as LaB_6 can be obtained by explosively reacting metal nitrates and an organic fuel (Kanakala et al. 2010, 2011), with the type of fuel and the fuel-to-nitrate ratio representing the most important controlling parameters for achieving a self-sustaining process. During the synthesis of LaB_6, for example, lanthanum nitrate and boron powders are mixed with an organic fuel resulting in the following chemical reaction:

$$La(NO_3)_3 + 6 \times B + fuel \xrightarrow{\Delta} LaB_6 + x \cdot NO_2 + y \cdot H_2O + z \cdot CO_2 + other gases \quad (6.1)$$

A variety of rare-earth hexaboride powders can be prepared in a few seconds taking advantage of this type of combustion reaction using a very low amount of fuel and amorphous or rhombohedral boron powders. The addition of larger amounts of fuel results in the formation of a significant amount of undesirable oxide phases due to the higher reaction temperatures reached during synthesis. Thus, the reaction temperature must be kept very low in order to promote the formation of the boride. It should be mentioned that the technique has been utilized since the 1990s for preparing many types of oxides, as has been shown in countless studies (Shea et al. 1996; Lopez et al. 1997; Hirata et al. 2001; Graeve et al. 2006, 2010; Sinha et al. 2009). However, the synthesis of borides is more recent and requires the efficient mixing of boron powders into the nitrate and fuel mixture. Resulting LaB_6 powders have a cubic morphology that mirrors the cubic symmetry of the unit cell, illustrated in Figure 6.4.

Other special morphologies, such as nanorods of borides, including HfB_2 (Figure 6.5), can be obtained using a solvothermal process that involves $HfCl_4$ and B, as follows:

$$HfCl_4 + 2B + 2Mg \rightarrow HfB_2 + 2MgCl_2 \quad (6.2)$$

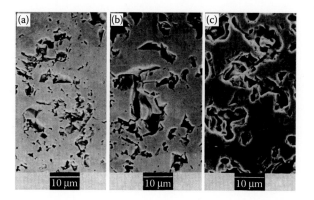

FIGURE 6.3
Polished surfaces of (a) HfC, (b) 3HfC: 1TiC, and (c) HfC: TiC prepared by self-propagating high-temperature synthesis. (Kecskes L.J., R.F. Benck, P.H. Netherwood. Dynamic compaction of combustion-synthesized hafnium carbide. *J. Am. Ceram. Soc.* 1990. 73(2). 383–387. Copyright Wiley-VCH Verlag GmbH & Co. KGaA. Reproduced with permission.)

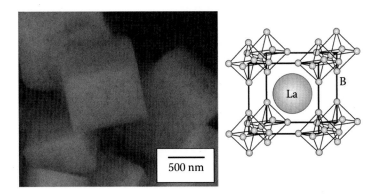

FIGURE 6.4
(**See color insert.**) Scanning electron micrograph of LaB_6 powders obtained by solution combustion synthesis and the LaB_6 simple cubic unit cell. (From Kanakala R., G. Rojas-George, O.A. Graeve, *J. Am. Ceram. Soc.*, 93(10), 2010, 3136–3141; Kanakala R. et al., *ACS Appl. Mater. Interfaces*, 3(4), 2011, 1093–1100.)

The reactant powders are first milled for 2 h and subsequently calcined for 1 h at 1373 K. After reaction, the powders consist of HfB_2 (65 wt%), HfO_2 (monoclinic, 11 wt%) and MgO (24 wt%), thus, the process does not result in pure material (Bégin-Colin et al. 2004). A variety of similar reactions can also result in mixtures of powder morphologies, such as illustrated in Figure 6.5 (Blum and Kleebe 2004), although the powders still contain impurities. It appears that pure powders can be obtained by reacting $HfCl_4$ and $NaBH_4$ in an autoclave at 600°C (Chen et al. 2004).

Obtaining HfB_2 and other borides by reducing the oxide is also feasible and the most common methodology used for obtaining these materials. The conventional carbothermal/borothermal reduction involves the following reaction:

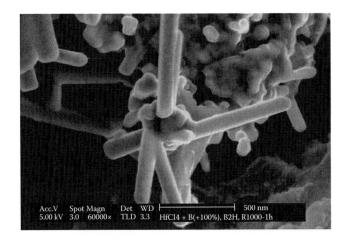

FIGURE 6.5
Scanning electron micrograph of powders and rods of $HfB_2/HfO_2/MgO$ obtained by mechanically mixing $HfCl_4$, B, and Mg. (With kind permission from Springer Science+Business Media: *J. Mater. Sci.*, Mechanically activated synthesis of ultrafine rods of HfB2 and milling induced phase transformation of monocrystalline anatase particles, 39(16–17), 2004, 5081–5089, Bégin-Colin S. et al.)

FIGURE 6.6
Scanning electron micrograph of HfB_2/HfO_2 powders obtained by reaction of Hf and B_2O_3. (With kind permission from Springer Science+Business Media: *J. Mater. Sci.*, Chemical reactivities of hafnium and its derived boride, carbide and nitride compounds at relatively mild temperature, 3919, 2004, 6023–6042, Blum Y.D., H.-J. Kleebe.)

$$HfO_2 + \frac{1}{2}B_4C + \frac{3}{2}C \rightarrow HfB_2 + 2CO \tag{6.3}$$

The problem with this process is that it usually results in impurities such as HfO_2 and HfC in the final products (Figure 6.6). A modification of this reaction can result in higher purity boride powders of a uniform crystallite size distribution, as described in Equation 6.4 and illustrated in Figure 6.7. However, while the crystallite size is very uniform, the level of agglomeration is difficult to determine from the micrographs. The powders might be lightly agglomerated or they might consist of hard agglomerates of a large particle size. Nonetheless, this is an excellent example of pure and uniform powders that have the potential for excellent sinterability.

$$(1 + x) \cdot HfO_2 + \frac{5}{7} \cdot B_4C + 5x \cdot C$$
$$\rightarrow (1 + x) \cdot HfB_2 + \left(\frac{3}{7} - x\right) \cdot B_2O_3 + \left(\frac{5}{7} + 5x\right) \cdot CO\left(0 \leq x \leq \frac{3}{7}\right) \tag{6.4}$$

Finally, polymeric precursors are sometimes useful for the preparation of both borides and carbides. For example, the use of PMCS (polymethylcarbosilane) or PND (poly(norbornenyldecaborane)), has been used for the syntheses of SiC and B_4C (Guron et al. 2008). Again here, the powders are not completely pure, although the presence of left-over carbon (the most common impurity) might be an advantage during sintering. Figure 6.8 describes the idealized ceramic conversion reactions of the PMCS and PND preceramic polymers.

FIGURE 6.7
Scanning electron micrographs of HfB$_2$ powders prepared using a carbothermal/borothermal reduction process using molar ratios of (a) 3/5 B$_4$C and 9/10 C (calcination at 1600°C for 1 h), (b) 3/4 B$_4$C and 1/10 C (calcination at 1600°C for 1 h), (c) 11/14 B$_4$C (calcination at 1600°C for 1 h), and (d) 11/14 B$_4$C (calcination at 1500°C for 1 h). (Ni D.-W. et al.: Synthesis of monodispersed fine hafnium diboride powders using carbo/borothermal reduction of hafnium dioxide. *J. Am. Ceram. Soc.* 2008. 91(8). 2709–2712. Copyright Wiley-VCH Verlag GmbH & Co. KGaA. Reproduced with permission.)

$$\left[\begin{array}{c} \text{Me} \\ | \\ \text{SiCH}_2 \\ | \\ \text{H} \end{array} \right]_X \xrightarrow{\;\Delta\;} \text{SiC} + \text{CH}_4 + \text{H}_4$$

PMCS

$$2\left[\cdots \right]_X \xrightarrow{\;\Delta\;} 5\,\text{B}_4\text{C} + \text{``C}_9\text{H}_2\text{0} + 12\,\text{H}_2\text{''}$$

PND

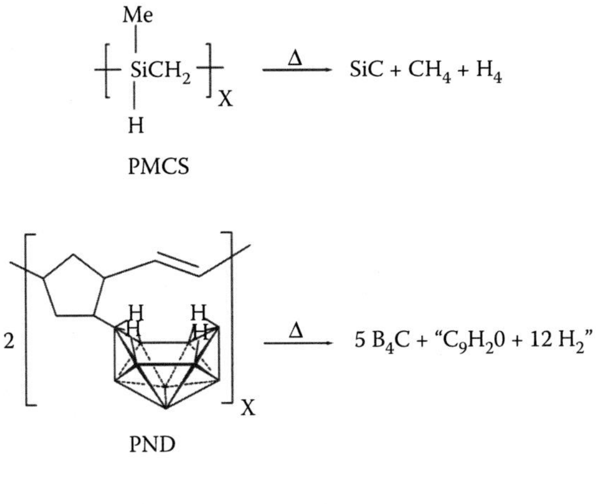

FIGURE 6.8
Idealized ceramic conversion reactions of the polymethylcarbosilane (PMCS) and poly(norbornenyldecaborane) (PND) preceramic polymers. Both precursors also retain free carbon, so the resulting products are actually SiC/C and B$_4$C/C. (Guron M.M., M.J. Kim, and L.G. Sneddon: A simple polymeric precursor strategy for the syntheses of complex zirconium and hafnium-based ultra-high-temperature silicon-carbide composite ceramics. *J. Am. Ceram. Soc.* 2008. 91(5). 1412–1415. Copyright Wiley-VCH Verlag GmbH & Co. KGaA. Reproduced with permission.)

6.3 Consolidation of Bulk Specimens

6.3.1 Tantalum Carbide: Model System for UHTC Sintering

The sintering of high-quality UHTCs is challenging because of their high melting temperatures. As an example, Table 6.5 summarizes typical densities obtained for a variety of TaC specimens, with many cases not even reaching a 90% relative density. Only recently has fully dense tantalum carbide of high purity been reported in the literature (Bakshi et al. 2011). Resistance sintering at temperatures of 1850°C and pressures of 263–363 MPa were used to obtain these samples, in comparison with a sample of 89% relative density obtained using only 100 MPa. Clearly, pressure appears to be a critical parameter for promoting higher densities in these materials and, in fact, as early as 1978, Yohe and Ruoff (1978) sintered nominally sub-micron powders using high pressures between 1.5 and 4.5 GPa and intermediate-to-high temperatures between 800°C and 1700°C. The densities and grain sizes as a function of temperature are demonstrated in Figure 6.9. Densification was complete between 1000°C and 1300°C, depending on pressure. Higher temperatures lowered the density, perhaps by initiating nondensifying sintering mechanisms or by vaporization of impurity phases. In the end, fully dense specimens were obtained using these high pressures, but were only realized after taking into consideration an estimated 11.8% tantalum oxide content. Figure 6.9 shows that a pressure of approximately 3 GPa is necessary for completion of densification and inhibition of grain growth. Densification is complete just below 1100°C, the same temperature at which grain growth is beginning to become evident.

The presence and detrimental effects of oxygen on sintering of these materials was noted in the first sintering report using nanostructured powders (Sautereau and Mocellin 1974). It was determined that the removal of oxygen to less than parts per thousand was essential for obtaining high-quality tantalum carbide of high density. However, the thermal

TABLE 6.5

Sintering Parameters and the Theoretical Densities and Grain Sizes for Nearly Stoichiometric Tantalum Carbide

Year	Author	Method	Powders	P (MPa)	T (°C)	% ρ_{th}	G (μm)
1971	Jun	HP	—	—	3150	89–95	—
1974	Sautereau	Conv	N	Ambient	1300–1600	95–96	2–20
1975	Bukatov	—	—	—	—	97.2	12
1978	Yohe	HP	N	1500–4500	800–1500	81–94	0.02–0.31
1982	Shvab	Conv	M and S	Ambient	2400–2700	< 90	1–24
2007	Zhang	HP	M	30	1900–2400	75–97	1.3–2.8
2009	Kim	IP	M and N	80	1350	96	0.94–0.03
2010	Khaleghi	RS	M	30–75	1900–2400	68–97	0.33–9.0
2010	Liu	Conv	M and S	Ambient	2200–2400	81–98	1.5–12
2010	Silvestroni	Conv	M	Ambient	1950	91	6
2011	Bakshi	RS	S	100–363	1850	89–100	0.56–5.6
2011	Silvestroni	HP	S	30	1900	85	0.8

*P = pressure, T = temperature, ρ_{th} = theoretical density, G = grain size.
**HP = hot pressing, Conv = conventional sintering, RS = resistance sintering.
***N = nanopowders, M = micropowders, S = submicron powders.

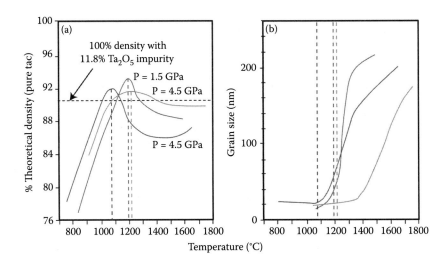

FIGURE 6.9
(See color insert.) (a) Density and (b) grain sizes of tantalum carbide sintered at temperatures between 800°C and 1700°C and pressures between 1.5 and 4.5 GPa. Dotted lines correspond to temperature of maximum density. (Adapted from Yohe W.C., A.L. Ruoff, *Am. Ceram. Soc. Bull.*, 57(12), 1978, 1123–1130.)

treatments that were necessary to remove the oxygen resulted in growth of the primary particle sizes from nanostructured to the sub-micron regime, casting some doubt on the practical interest of using nanopowders with high surface area. They also noted that liquid phases could form at temperatures as low as 1100°C as a result of binary eutectics between the carbide and transition metal impurities such as iron, nickel, cobalt, and chromium. The liquid exists intergranularly and some evidence suggests that the liquid phase is directly responsible for grain growth at the intermediate sintering temperatures used in the study (1300–1600°C). Furthermore, this is the same temperature at which densification begins, hinting at densification by liquid phase sintering as the predominant mechanism. Despite starting with nanopowders, final average grain sizes were between 2 and 20 μm.

In some cases, additions of carbon and boron carbide in small quantities have been used to enhance densification by removal of oxygen impurities (Zhang et al. 2007). The reactions between these additives and tantalum oxide produce gaseous species that can be removed prior to intermediate- to late-stage sintering, thus, increasing the purity of the sintered specimens. The reaction with carbon becomes favorable just above 1100°C, while the reaction with boron carbide is favorable at all temperatures.

It is interesting to note that the samples of highest density reported in Table 6.5, excluding the high-pressure studies previously discussed, were obtained by a conventional sintering process without pressure application (Liu et al. 2010). One possible explanation for this is a small particle size and a narrow size distribution of the starting powders, since enhanced sintering with smaller particle sizes has been reported by a variety of researchers (Kislyi et al. 1982; Kim et al. 2009; Talmy et al. 2010). In this case, the starting powders were of two sizes, 360 and 250 nm for samples B and C, respectively (Figure 6.10).

Decreasing the particle size apparently changes the mechanism for mass transport and therefore the sintering mechanism. Sub-micron powders sinter at lower temperatures (~1600°C) with a reaction order indicator of approximately $n = 1$ during the first stage of sintering, indicating diffusion/viscous flow and grain boundary sliding as the dominant sintering mechanisms. Coarser powders require higher temperatures to sinter, with a

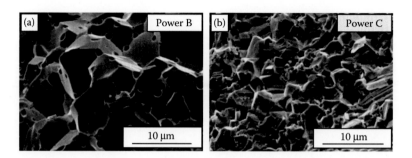

FIGURE 6.10
Microstructures observed by pressureless sintering of (a) powders with a particle size of 360 nm and (b) powders with a particle size of 250 nm at 2300°C. (Liu J.-X., Y.-M. Kan, and G.-J. Zhang: Pressureless sintering of tantalum carbide ceramics without additives. *J. Am. Ceram. Soc.* 2010. 93(2). 370–373. Copyright Wiley-VCH Verlag GmbH & Co. KGaA. Reproduced with permission.)

reaction order indicator of near $n = 3$ and indicating a surface self-diffusion mechanism. The surface diffusion mechanism is a nondensifying mode of sintering that will result in low final density.

A consequence of grain boundary sliding during the initial stage of densification is a sharp increase in grain size in the early stages of sintering (orders of magnitude) with very little linear grain growth during isothermal holding, as has been observed by multiple researchers (Shvab and Egorov 1982; Kislyi et al. 1982; Liu et al. 2011; Geguzin and Klinchuk 1976; Ashby and Verral 1974; Geguzin 1957). Zhang et al. (2007) noted that densification and grain growth in these materials have similar activation energies. The activation energy for grain growth during isothermal holding at temperatures from 2400°C to 2700°C was determined to be 380 kJ/mol, while the activation energy of viscous flow during hot pressing at 16 MPa and 3050°C was determined to be 406 kJ/mol (Shvab and Egorov 1982; Samsonov and Petrikina 1970). Thus, if viscous flow is the dominant mechanism during the initial stages of sintering, then grain growth will also be initiated at the onset of densification.

Nonetheless, the powder size might not be the only explanation for sintering behavior in these materials. It is possible that higher relative densities can be attributed to better sintering due to oxygen impurity removal by reaction of TaC with Ta_2O_3 to form sub-stoichiometric TaC and gaseous CO by-product.

In addition, these materials sometimes exhibit lamellar microstructures (Figure 6.11). Such morphologies are characteristic of sub-stoichiometric TaC, which has a higher

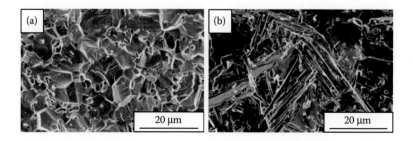

FIGURE 6.11
Microstructural differences observed for sintered specimens of (a) TaC and (b) $TaC_{0.7}$. (Hackett K. et al. Phase constitution and mechanical properties of carbides in the Ta–C system. *J. Am. Ceram. Soc.* 2009. 92(10). 2404–2407. Copyright Wiley-VCH Verlag GmbH & Co. KGaA. Reproduced with permission.)

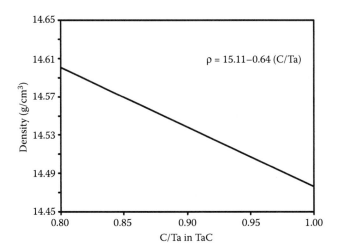

FIGURE 6.12
Variation in tantalum carbide density with carbon content. (Adapted from Samsonov G.V., R.Y. Petrikina, *Phys. Sintering*, 2(3), 1970, 1–20.)

absolute density compared to stoichiometric TaC and could contribute to the slightly higher density values obtained for some specimens (the relationship between density and stoichiometry in TaC is described in Figure 6.12 (Samsonov and Petrikina 1970). In fact, Hackett et al. have shown that while higher densities are obtained for sub-stoichiometric samples, the relative density is still low (Hackett et al. 2009).

Zhang et al. (2007) further suggested that the best sintering aids would be those that promote densification below temperatures where rapid grain growth occurs. It is unclear which additives would be most effective. It has been proposed that the combination of TaB_2 and free carbon left in the microstructure after reacting TaC and B_4C inhibits grain growth (Bakshi et al. 2011; Zhang et al. 2009), whereas additions of $MoSi_2$ do not seem to have an effect on TaC grain growth (Silvestroni and Sciti 2010). Finally, the form, reactivity, and quantity of excess carbon added to TaC have been noted to influence grain growth, thought to be due to excess carbon at the grain boundaries (Bakshi et al. 2011; Talmy et al. 2010).

Revisiting the data provided in Table 6.5, we can conclude that sub-micron grains are only maintained in sintered samples of low relative density, where densification is not adequate, or for extreme pressures or heating conditions that allow for rapid densification before grain growth can occur significantly (Yohe and Ruoff 1978; Bakshi et al. 2011; Khaleghi et al. 2010; Kim et al. 2009).

6.3.2 Sintering of Carbides and Borides

The sintering of other carbides and boride/carbide powders has been described in a variety of reports (Sonber and Suri 2011; Saito et al. 2012). In most cases, the application of pressure is necessary. For example, the fabrication of HfB_2 - 20 vol% SiC has been accomplished by hot-pressing powders of HfB_2 and SiC at 2200°C for 1 h using a moderate pressure of 25 MPa (Gasch et al. 2004). The microstructures of these materials are uniform with grain sizes in the range of 10–20 μm (Figure 6.13), although there are variations in density across the cylindrically shaped specimens that could possibly be eliminated by the application of higher pressures during sintering, thus, process improvement could still be pursued.

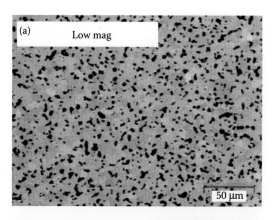

FIGURE 6.13

Scanning electron micrographs of HfB_2-20 vol% SiC composite prepared by hot-pressing HfB_2 and SiC powders at 2200°C for 1 h. (With kind permission from Springer Science+Business Media: *J. Mater. Sci.*, Processing, properties and arc jet oxidation of hafnium diboride/silicon carbide ultra-high-temperature ceramics, 3919, 2004, 5925–5937, Gasch M. et al.)

It has been found that the application of pressure during sintering is not necessary for obtaining densities less than 98% of theoretical, as evidenced by $HfC–MoSi_2$ composites obtained by pressureless sintering at 1950°C (Sciti et al. 2006). The materials have densities ranging from 96% to 98% and an average grain size of 3–4 μm (Figure 6.14). Nonetheless, the removal of the residual 2–4% porosity probably does require the application of pressure.

Another possibility is the use of reactive sintering, in which pressure is applied while the materials are undergoing some kind of chemical transformation (Anselmi-Tamburini et al. 2006; Chornobuk et al. 2009). Differences between regular sintering and reactive sintering with the application of pressure can be found in the preparation of HfB_2 by spark plasma sintering (Anselmi-Tamburini et al. 2006). Sintering of HfB_2 powders at 1900°C under a pressure of 95 MPa results in samples that are 85% of theoretical density. Reactive sintering between Hf and B at 1700°C and 95 MPa result in samples of 97% theoretical density. This still leaves the last 3% of porosity in the specimens, which is likely quite difficult to remove.

The use of sintering aids is also of great value. For the hot-pressing of HfB_2 - 20 vol% SiC composites at 1900°C, the use of Si_3N_4 greatly assists in improving density (Sciti et al.

FIGURE 6.14

Scanning electron micrographs of (a) 90HfC + 10MoSi$_2$ and (b) 80HfC + 20MoSi$_2$ composites. The brighter phase is HfC and the darker phase is MoSi$_2$. (Sciti D., L. Silvestroni, and A. Bellosi. High-density pressureless-sintered HfC-based composites. *J. Am. Ceram. Soc.* 2006. 89(8). 2668–2670. Copyright Wiley-VCH Verlag GmbH & Co. KGaA. Reproduced with permission.)

2009). For powders not containing Si$_3$N$_4$ the density reached is lower than 94%. Upon the addition of the Si$_3$N$_4$, the relative density is increased to 99% (Figure 6.15). Presumably, Si$_3$N$_4$ enables the densification of HfB$_2$ by facilitating the removal of HfO$_2$ via reactions (6.5) and (6.6).

$$Si_3N_4 + 2B_2O_3 \rightarrow 4BN + 3SiO_2 \tag{6.5}$$

$$2Si_3N_4 + 6HfO_2 \rightarrow 6SiO_2 + 6HfN + N_2 \tag{6.6}$$

The elimination of the oxide layer possibly increases the contact area between particles and accelerates transport of matter that promotes densification.

Another example of the effective use of sintering aids is for the consolidation of HfB$_2$- and HfC-based materials with the addition of MoSi$_2$ (Sciti et al. 2008a). Dense samples have been obtained using MoSi$_2$ content greater than 3 vol% at sintering temperatures between 1750°C and 1800°C. Lower amounts of the MoSi$_2$ sintering aid require higher sintering temperatures greater than 2100°C in order to obtain samples of 97% relative density. The MoSi$_2$ has the same role as described earlier (i.e., the removal of oxide from the powders surfaces) and might also promote densification by the formation of a liquid phase, as has been seen in several studies (Silvestroni and Sciti 2010; Zou et al. 2010). The use of TaSi$_2$ for the fabrication of HfB$_2$–TaSi$_2$ composites, however, still results in significant amounts of HfO$_2$, as demonstrated in Figure 6.16 (Sciti et al. 2008b).

One would expect that the formation of a liquid phase during sintering plays a major role in densification, but also does so in grain growth. During hot pressing of HfB$_2$ with HfSi$_2$ (5 vol%) at 1600°C, samples of nearly full density were obtained, but the grain size was uneven and contained faceted diboride grains up to 10 μm. In contrast, the use of B$_4$C (7 vol%) did not result in full density. In fact, the density was lower than 94%, but the grain size was smaller than 2 μm and much more uniform (Monteverde 2008). Thus, higher densities and lower grain sizes are competing effects during sintering, although exceptions can always be found. For example, TaSi$_2$ and Ir as sintering aids enhance sintering of HfB$_2$ (relative density of 99.2% versus 95.4% for the specimens not containing sintering aids), but also result in lower grain sizes (2.6 ± 0.1 μm versus 4.0 ± 0.4 μm, respectively), even though samples probably contain a liquid phase during the consolidation process (Gasch et al. 2008).

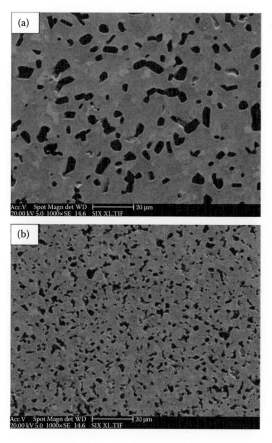

FIGURE 6.15
Scanning electron micrographs of HfB_2 hot-pressed (a) without sintering aids and (b) with Si_3N_4 as a sintering aid. (From Weng L. et al., The effect of Si_3N_4 on microstructure, mechanical properties and oxidation resistance of HfB_2-based composite, *J. Compos. Mater.*, 43(2), 2009, 113–123. Courtesy of Sage Publications.)

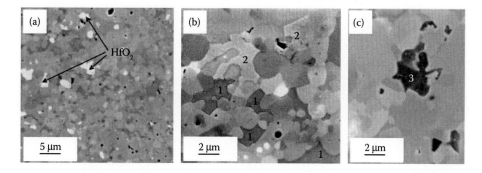

FIGURE 6.16
Scanning electron micrographs of HfB_2-15 vol% $TaSi_2$ composite (a) with significant presence of HfO_2, (b) where area "1" represents the $TaSi_2$ and area "2" represents Ta-Hf-C, and (3) where area "3" represents a SiC-based phase. (Sciti D. et al.: Sintering and mechanical properties of ZrB_2–$TaSi_2$ and HfB_2–$TaSi_2$ ceramic composites. *J. Am. Ceram. Soc.* 2008b. 91(10). 3285–3291. Copyright Wiley-VCH Verlag GmbH & Co. KGaA. Reproduced with permission.)

FIGURE 6.17
Scanning electron micrograph in back-scattered electron mode of a B_4C/Al functionally graded composite. (Reprinted from *Int. J. Refract. Met. H.*, 27(2), Hulbert D.M. et al., The synthesis and consolidation of hard materials by spark plasma sintering, 367–375. Copyright 2009, with permission from Elsevier.)

So far, we have discussed the sintering of a variety of UHTCs to intermediate or full relative densities. To finalize this section, we will describe the purposeful preparation of samples with variations in composition (i.e., functionally graded materials, FGMs), which is another materials development capability that has been explored recently. These materials have interesting thermal transport behavior and are much desired in a variety of applications in which the efficient removal of heat is important. One specific example is a composite of B_4C/aluminum (Figure 6.17), where one side is pure Al and the other is pure B_4C (Hulbert et al. 2009). For preparation of the FGM, the B_4C with porosity variations from one side to the other was prepared by spark plasma sintering using a maximum temperature of 1660°C and pressure of 53 MPa. After consolidation, one side of the compact was fully dense with the opposite side being highly porous. Afterwards, molten aluminum was infiltrated in order to fill the pores in the B_4C. Following melt infiltration, the density of the specimens varied slightly, but was found to be near 90% of the theoretical value based on the rule of mixtures. Most of the porosity was found in a region of closed porosity near the dense B_4C-rich side of the FGM.

6.4 Summary

There is a rich legacy and recent efforts in the processing of UHTCs. This chapter is a general overview that can guide the reader towards further details on processing techniques and methodologies for the preparation of UHTCs, with special emphasis on the synthesis and sintering of TaC as a model system, but also describing the preparation of a variety of carbide and boride composites.

There are many issues that need to be addressed and the field is poised for major breakthroughs with the recent development of fast sintering processes, such as spark plasma sintering, that allow one to obtain nanograined structures. The challenge with UHTCs continues to be the required high sintering temperatures for full densification, which can be difficult to achieve without promoting exaggerated grain growth.

In addition, the use of these types of materials in highly oxidative environments is a problem, as they are highly susceptible to degradation that can result in the deterioration of properties. Some UHTCs are limited by the formation (or lack of formation) of protective oxide scales on their surfaces. The general approach for obtaining a suitable oxide scale is *via* design of composite materials. Our current understanding of the oxide scale formation mechanisms is guiding the design and manufacturing of new composite UHTC materials that may result in improved performance of UHTCs.

There are two ways to design degradation resistant materials: (1) the development of a fully protective oxide scale or (2) the development of a partially protective oxide scale for which ablation can be suitably controlled. For the first to be effective, it must be demonstrated that the oxide scale is protective in the atmospheric conditions of the application and through the entire application time. The oxidation scale must slow or eliminate further oxidation to be considered protective, never leading to an accelerated and active oxidation condition in which the scale and underlying material are being rapidly removed from the UHTC surface. Maintaining a protective oxide scale provides safety and reliability that will be necessary of a UHTC component.

The development of a partially protective oxide scale is more complex, but can be beneficial. In this case, the protective oxide scale should be fully protective at low and moderate temperatures. However, in the most extreme heating conditions, controlled ablation of the material can contribute to the heat removal processes by carrying it away with the gases formed during decomposition. Ablative cooling is the currently used heat shielding technology. If UHTCs are to be used in ablative conditions, then sufficient understanding and control over the ablation behavior is necessary and there should be a demonstrated advantage over using traditional ablative materials.

To reduce costs and further promote the use of UHTCs for advanced thermal protection systems, reusability is another key aspect. From this perspective, a fully protective oxide scale is more beneficial than a partially protective oxide scale relying on ablation. To be as economically advantageous as possible, a protective scale must be able to withstand thermal cycling without losing integrity or at least require minimal refurbishment costs and time compared to already established technologies.

Acknowledgments

This chapter was written with support from the National Science Foundation under grants CMMI 0645225 and CMMI 0913373. The authors thank Eric Wuchina, Office of Naval Research, NSWCCD, W. Bethesda, Maryland; William Fahrenholtz, Missouri University of Science and Technology, Rolla, Missouri; and Yiannis Pontikes, Department of Metallurgy and Materials Engineering, KU Leuven, Belgium; for reviewing this chapter and for providing valuable technical comments and suggestions.

References

Ahlén N., M. Johnsson, A.-K. Larsson, B. Sundman, On the carbothermal vapour–liquid–solid (VLS) mechanism for TaC, TiC, and Ta$_x$Ti$_{1-x}$C whisker growth, *J. Eur. Ceram. Soc.*, 20(14–15), 2000, 2607–2618.

Alekseev N.V., Y.V. Blagoveshchenskii, G.N. Zviadadze, I.K. Tagirov, Production of fine niobium and tantalum carbide powders, *Sov. Powder Metall.*, 19(8), 1980a, 517–520.

Alekseev N.V., Y.V. Blagoveshchenskii, G.N. Zviadadze, I.K. Tagirov, Theory, production technology, and properties of powders and fibers, *Sov. Powder. Metall.*, 19(8), 1980b, 517–520.

Alfano D., L. Scatteeia, F. Monteverde, E. Bêche, M. Balat-Pichelin, Microstructural characterization or ZrB_2-SiC based UHTC tested in the MESOX plasma facility, *J. Eur. Ceram. Soc.*, 30, 11, 2010, 2345–2355.

Allison C., M. Hoffman, W.S. Williams, Electron energy loss spectroscopy of carbon in dissociated dislocations in tantalum carbide, *J. Appl. Phys.*, 53(10), 1982, 6757–6761.

Allison C.Y., R.E. Stoller, E.A. Kenik, Electron microscopy of electron damage in tantalum carbide, *J. Appl. Phys.*, 63(5), 1988a, 1740–1743.

Allison C.Y., C.B. Finch, M.D. Foegelle, F.A. Modine, Low-temperature electrical resistivity of transition-metal carbides, *Solid State Commun.*, 68(4), 1988b, 387–390.

Andrievski R.A., Nanocrystalline high melting point compound-based materials, *J. Mater. Sci.*, 29(3), 1994, 614–631.

Andrievski R.A., The state-of-the-art of high-melting point compounds, in: *Materials Science of Carbides, Nitrides, and Borides*, eds. Y.G. Gogotsi and R.A. Andrievski, Kluwer Academic Publishers, Dordrecht, The Netherlands, ISBN 0792357078, 1999, pp. 1–18.

Anselmi-Tamburini U., Y. Kodera, M. Gasch, C. Unuvar, Z.A. Munir, M. Ohyanagi, S.M. Johnson, Synthesis and characterization of dense ultra-high temperature thermal protection materials produced by field activation through spark plasma sintering (SPS): I. Hafnium diboride, *J. Mater. Sci.*, 41(10), 2006, 3097–3104.

Aronsson B., T. Lundström, S. Rundqvist, *Borides, Silicides & Phosphides*, Wiley, Hoboken, NJ, ASIN B0006BMM7W, 1965, chs. 2–5.

Aruna S.T., A.S. Mukasyan, Combustion synthesis and nanomaterials, *Curr. Opin. Solid St. M.*, 12(3–4), 2008, 44–50.

Ashby M.F., H.A. Verral, Diffusion-accommodated flow and superplasticity, *Acta Metall.*, 21(2), 1974, 149–163.

Bakshi S.R., V. Musaramthota, D.A. Virzi, A.K. Keshri, D. Lahiri, V. Singh, S. Seal, A. Agarwal, Spark plasma sintered tantalum carbide-carbon nanotube composite: Effect of pressure, carbon nanotube length and dispersion technique on microstructure and mechanical properties, *Mater. Sci. Eng. A*, 528(6), 2011, 2538–2547.

Baxter W.J., Work function of tantalum carbide and the effects of adsorption and sputtering of cesium, *J. Appl. Phys.*, 42(7), 1971, 2682–2688.

Becher P.F., Mechanical behaviour of polycrystalline TaC, *J. Mater. Sci.*, 6(1), 1971, 79–80.

Bégin-Colin S., G. Le Caër, E. Barraud, O. Humbert, Mechanically activated synthesis of ultrafine rods of HfB_2 and milling induced phase transformation of monocrystalline anatase particles, *J. Mater. Sci.*, 39(16–17), 2004, 5081–5089.

Blais J.C., G. Bolbach, Study of negative surface ionization on complex emitters, *Int. J. Mass Spectrom.*, 24(4), 1977, 413–427.

Blum Y.D., H.-J. Kleebe, Chemical reactivities of hafnium and its derived boride, carbide and nitride compounds at relatively mild temperature, *J. Mater. Sci.*, 3919, 2004, 6023–6042.

Borisova A.L., Yu.S. Borisov, A thermodynamic evaluation of the effect of nonstoichiometry of compounds on the course of their solid-phase reaction, *Sov. Powder Metall.*, 16(6), 1977, 440–447.

Bowman A.L., The variation of lattice parameter with carbon content of tantalum carbide, *J. Phys. Chem.*, 65(9), 1961, 1596–1598.

Bukatov V.G., V.I. Knyazev, O.S. Korostin, V.M. Baranov, Temperature dependence of the Young's modulus of metalline carbides, *Inorg. Mater.*, 11(2), 1975, 310–312.

Chaikovskii E.F., V.T. Sotnikov, V.A. Vlasenko, Effect of oxidation of carbon on the work function of tantalum carbide, *Sov. Phys. – Tech. Phys.*, 23(3), 1979, 360–361.

Chang Y.-H., C.-W. Chiu, Y.-C. Chen, C.-C. Wu, C.-P. Tsai, J.-L. Wang, H.-T. Chiu, Syntheses of nano-sized cubic phase early transition metal carbides from metal chlorides and n-butyllithium, *J. Mater. Chem.*, 12(8), 2002, 2189–2191.

Chen L., Y. Gu, L. Shi, Z. Yang, J. Ma, Y. Qian, Synthesis and oxidation of nanocrystalline HfB_2, *J. Alloy. Compd.*, 368(1–2), 2004, 353–356.

Chen Y.-J., J.-B. Li, Q.-M. Wei, H.-Z. Zhai, Preparation of different morphology of TaC$_x$ whiskers, *Mater. Lett.*, 56(3), 2002, 279–283.

Chornobuk S.V., A.Yu. Popov, V.A. Makara, Structure and mechanical properties of reaction-sintered ceramic composite materials based on titanium and hafnium diborides, *J. Superhard Mater.*, 312, 2009, 86–88.

Ciaravino C., F.F.P. Medeiros, C.P. De Souza, M. Roubin, Elaboration of mixed tantalum and niobium carbides from tantalite mineral (Fe,Mn)(Ta$_{1-x}$Nb$_x$)$_2$O$_6$, *J. Mater. Sci.*, 37(10), 2002, 2117–2123.

Cousin P., R.A. Ross, Preparation of mixed oxides: A review, *Mater. Sci. Eng. A-Struct.*, 130(1), 1990, 119–125.

Danil'chenko V.V., M.L. Taubin, A.S. Maskaev, I.I. Deryavko, The thermal and electrical conductivity of tantalum carbide at high temperatures, *High Temp.*, 15(3), 1977, 444–449.

Danil'yants G.I., A.V. Kirillin, S.V. Novikov, Emissivity of tantalum carbide at elevated temperatures, *High Temp.*, 261, 1988, 65–69.

Desmaison-Brut M., N. Alexandre, J. Desmaison, Comparison of the oxidation behaviour of two dense hot isostatically pressed tantalum carbide (TaC and Ta$_2$C) materials, *J. Eur. Ceram. Soc.*, 17(11), 1997, 1325–1334.

Dua A.K., V.C. George, TaC coatings prepared by hot filament chemical vapour deposition: Characterization and properties, *Thin Solid Films*, 247(1), 1994, 34–38.

Eakins E., D.D. Jayaseelan, W.E. Lee, Toward oxidation-resistant ZrB$_2$-SiC ultra high temperature ceramics, *Metall. Mater. Trans. A*, 42(4), 2011, 878–887.

Eick B.M., J.P. Youngblood, Carbothermal reduction of metal-oxide powders by synthetic pitch to carbide and nitride ceramics, *J. Mater. Sci.*, 44(5), 2009, 1159–1171.

Fahrenholtz W.G., G.E. Hilmas, A.L. Chamberlain, J.W. Zimmermann, Processing and characterization of ZrB$_2$-based ultra-high temperature monolithic and fibrous monolithic ceramics, *J. Mater. Sci.*, 3919, 2004, 5951–5957.

Fernandes J.C., F.A.C. Oliveira, B. Granier, J.-M. Badie, L.G. Rosa, N. Shohoji, Kinetic aspects of reaction between tantalum and carbon material (active carbon or graphite) under solar radiation heating, *Solar Energy*, 80(12), 2006, 1553–1560.

Fukunaga A., S. Chu, M.E. McHentry, Using carbon nanotubes for the synthesis of transition metal carbide nanoparticles, *J. Mater. Sci. Lett.*, 18(6), 1999, 431–433.

Garbe J., J. Kirschner, Spin-resolved photoemission from the (100) face of tantalum carbide, *Phys. Rev. B*, 39(9), 1989, 6115–6120.

Gasch M., D. Ellerby, E. Irby, S. Beckan, M. Gusman, S. Johnson, Processing, properties and arc jet oxidation of hafnium diboride/silicon carbide ultra high temperature ceramics, *J. Mater. Sci.*, 3919, 2004, 5925–5937.

Gasch M., S. Johnson, J. Marschall, Thermal conductivity characterization of hafnium diboride-based ultra-high-temperature ceramics, *J. Am. Ceram. Soc.*, 91(5), 2008, 1423–1432.

Gasch M., S. Johnson, Physical characterization and arcjet oxidation of hafnium-based ultra high temperature ceramics fabricated by hot pressing and field-assisted sintering, *J. Eur. Ceram. Soc.*, 30(11), 2010, 2337–2344.

Geguzin Y.E., Diffusional deformation of porous crystalline structures, *Fiz. Tverd. Tela.*, 17(7), 1957, 1950–1954.

Geguzin Y.E., Y.I. Klinchuk, Mechanism and kinetics of the initial stage of solid-phase sintering of compacts from powders of crystalline solids (sintering "activity"), *Sov. Powder Metall.*, 15(7), 1976, 512–518.

Gesheva K, E. Vlakhov, Deposition and study of CVD-tantalum carbide thin films, *Mater. Lett.*, 5(7–8), 1987, 276–279.

Gillan E.G., R.B. Kaner, Synthesis of refractory ceramics via rapid metathesis reactions between solid-state precursors, *Chem. Mater.*, 8(2), 1996, 333–343.

Gololobov E.M., N.N. Dorozhkin, B.V. Novysh, Calculation of the spectral density of the electron–phonon interaction in tantalum carbide and nitride, *Fiz. Tverd. Tela.*, 31(10), 1989, 1813–1814.

Graeve O.A., Z.A. Munir, Electric field enhanced synthesis of nanostructured tantalum carbide, *J. Mater. Res.*, 17(3), 2002, 609–613.

Graeve O.A., S. Varma, G. Rojas-George, D. Brown, E.A. Lopez, Synthesis and characterization of luminescent yttrium oxide doped with Tm and Yb, *J. Am. Ceram. Soc.*, 89(3), 2006, 926–931.

Graeve O.A., R. Kanakala, A. Madadi, B.C. Williams, K.C. Glass, Luminescence variations in hydroxy-apatites doped with Eu^{2+} and Eu^{3+} ions, *Biomaterials*, 31(15), 2010, 4259–4267.

Grossklauss W., R.F. Bunshah, Synthesis and morphology of various carbides in the Ta-C system, *J. Vac. Sci. Technol.*, 12(4), 1975, 811–813.

Gruzalski G.R., D.M. Zehner, G.W. Ownby, Electron spectroscopic studies of tantalum carbide, *Surf. Sci.*, 157(2–3), 1985, L395–L400.

Gruzalski G.R., D.M. Zehner, Defect states in substoichiometric tantalum carbide, *Phys. Rev. B*, 34(6), 1986, 3841–3848.

Guron M.M., M.J. Kim, L.G. Sneddon, A simple polymeric precursor strategy for the syntheses of complex zirconium and hafnium-based ultra high-temperature silicon-carbide composite ceramics, *J. Am. Ceram. Soc.*, 91(5), 2008, 1412–1415.

Gusev A.I., S.Z. Nazarova, N.I. Kourov, Low-temperature specific heat and magnetic susceptibility of superconducting solid solutions of niobium and tantalum carbides, *Sov. Phys.—Sol. State*, 29(6), 1987, 1082–1084.

Gusev A.I., A.A. Rempel, V.N. Lipatnikov, Magnetic susceptibility and atomic ordering in tantalum carbide, *Phys. Stat. Sol. A*, 106(2), 1988, 459–466.

Gusev A.I., Atomic ordering and the order parameter functional method, *Philos. Mag.*, 60(3), 1989, 307–324.

Gusev A.I., A.A. Rempel, V.N. Lipatnikov, Incommensurate superlattice and superconductivity in tantalum carbide, *Sov. Phys.—Sol. State*, 33(8), 1991, 1295–1299.

Gusev A.I., The evaporation of vanadium, niobium, and tantalum cubic carbides, *High Temp.*, 312, 1993, 182–187.

Gusev A.I., A.A. Rempel, V.N Lipatnikov, Incommensurate ordered phase in non-stoichiometric tantalum carbide, *J. Phys.: Condens. Matter*, 8(43), 1996a, 8277–8293.

Gusev A.I., A.A. Rempel, V.N. Lipatnikov, Heat capacity of niobium and tantalum carbides NbC_y and TaC_y in disordered and ordered states below 300 K, *Phys. Stat. Sol. B*, 194(2), 1996b, 467–482.

Gusev A.I., A.S. Kurlov, V.N. Lipatnikov, Atomic and vacancy ordering in carbide ζ-Ta_4C_{3-x} ($0.28 \leq x \leq 0.4$) and phase equilibria in the Ta-C system, *J. Solid State Chem.*, 180(11), 2007, 3234–3246.

Hackett K., S. Verhoef, R.A. Cutler, D.K. Shetty, Phase constitution and mechanical properties of carbides in the Ta–C system, *J. Am. Ceram. Soc.*, 92(10), 2009, 2404–2407.

Hassine N.A., J.G.P. Binner, T.E. Cross, Synthesis of refractory metal carbide powders via microwave carbothermal reduction, *Int. J. Refractory Met. Hard Mater.*, 13(6), 1995, 353–358.

Haussonne F.J.-M., Review of the synthesis methods for AlN, *Mater. Manuf. Process.*, 10(4), 1995, 717–755.

Hirata G.A., F. Ramos, R. Garcia, E.J. Bosze, J. McKittrick, O. Contreras, F.A. Ponce, A new combustion synthesis method for GaN:Eu3+ and Ga2O3:Eu3+ luminescent powders, *Phys. Stat. Sol. A*, 188(1), 2001, 179–182.

Hoffman M., W.S. Williams, A simple model for the deformation behavior of tantalum carbide, *J. Am. Ceram. Soc.*, 69(8), 1986, 612–614.

Hulbert D.M., D. Jiang, D.V. Dudina, A.K. Mukherjee, The synthesis and consolidation of hard materials by spark plasma sintering, *Int. J. Refract. Met. H.*, 27(2), 2009, 367–375.

Ishigaki T., S.-M. Oh, J.-G. Li, D.-W. Park, Controlling the synthesis of TaC nanopowders by injecting liquid precursor into RF induction plasma, *Sci. Technol. Adv. Mat.*, 62, 2005, 111–118.

Jang T., L.M. Porter, G.W.M. Rutsch, B. Odekirk, Tantalum carbide ohmic contacts to n-type silicon carbide, *Appl. Phys. Lett.*, 75(25), 1999, 3956–3958.

Jayaseelan D.D., R.G. de Sá, P. Brown, W.E. Lee, Reactive infiltration processing (RIP) of ultra-high temperature ceramics (UHTC) into porous C/C composite tubes, *J. Eur. Ceram. Soc.*, 31(3), 2011, 361–368.

Johnsson M., N. Ahlén, M. Nygren, Synthesis and characterization of transition metal carbide whiskers, *Key Eng. Mat.*, 132–136, 1997, 201–204.

Johnsson M., M. Nygren, Carbothermal synthesis of TaC whiskers via a vapor–liquid–solid growth mechanism, *J. Mater. Res.*, 12(9), 1997, 2419–2427.

Jun C.K., P.T.B. Shaffer, Thermal expansion of niobium carbide, hafnium carbide, and tantalum carbide at high temperatures, *J. Less-Common Met.*, 24(3), 1971a, 323–327.

Jun C.K., Shaffer P.T.B., Elastic moduli of niobium carbide and tantalum carbide at high temperature, *J. Less-Common Metals*, 23(4), 1971b, 367–373.

Kanakala R., G. Rojas-George, O.A. Graeve, Unique preparation of hexaboride nanocubes: A first example of boride formation by combustion synthesis, *J. Am. Ceram. Soc.*, 93(10), 2010, 3136–3141.

Kanakala R., R. Escudero, G. Rojas-George, M. Ramisetty, O.A. Graeve, Mechanisms of combustion synthesis and magnetic response of high-surface-area hexaboride compounds, *ACS Appl. Mater. Interfaces*, 3(4), 2011, 1093–1100.

Kecskes L.J., R.F. Benck, P.H. Netherwood, Dynamic compaction of combustion-synthesized hafnium carbide, *J. Am. Ceram. Soc.*, 73(2), 1990, 383–387.

Kelly J.P., R. Kanakala, O.A. Graeve, A solvothermal approach for the preparation of nanostructured carbide and boride ultra-high temperature ceramics, *J. Am. Ceram. Soc.*, 93(10), 2010, 3035–3038.

Kelly J.P., O.A. Graeve, Statistical experimental design approach for the solvothermal synthesis of nanostructured tantalum carbide powders, *J. Am. Ceram. Soc.*, 94(6), 2011, 1706–1715.

Kelly, J.P., O.A. Graeve, Effect of powder characteristics on nanosintering, in: *Sintering*, Eds. R.H.R. Castro and K. Van Benthem, Springer, New York, NY, ISBN 9783642310089, 2012, 57–92.

Khaleghi E., Y.-S. Lin, M.A. Meyers, E.A. Olevsky, Spark plasma sintering of tantalum carbide, *Scripta Mater.*, 63(6), 2010, 577–580.

Khusainov M.A., G.M. Demyshav, M.M. Myshlyaev, Mechanical properties and structure of carbide coatings of niobium and tantalum on graphite, *Russ. Metall.*(5), 1990, 139–142.

Khyzhun O.Y., XPS, XES, and XAS studies of the electronic structure of substoichiometric cubic TaC_x and hexagonal Ta_2C_y carbides, *J. Alloy Compd.*, 259(1–2), 1997, 47–58.

Khyzhun O.Y., E.A. Zhurakovsky, A.K. Sinelnichenko, V.A. Kolyagin, Electronic structure of tantalum subcarbides studied by XPS, XES, and XAS methods, *J. Electron Spectrosc.*, 82(3), 1996, 179–192.

Kim C., G. Gottstein, D.S. Grummon, Plastic flow and dislocation structures in tantalum carbide: Deformation at low and intermediate homologous temperatures, *Acta Metall. Mater.*, 42(7), 1994, 2291–2301.

Kim B.R., K.-D. Woo, J.-M. Doh, J.-K. Yoon, I.-J. Shon, Mechanical properties and rapid consolidation of binderless nanostructured tantalum carbide, *Ceram. Int.*, 35(8), 2009, 3395–3400.

Kislyi P.S., S.A. Shvab, F.F. Egorov, Sintering kinetics of tantalum carbide, *Powder Metall. Met. C.*, 21(10), 1982, 765–767.

Krechkivska O., M.D. Morse, Resonant two-photon ionization spectroscopy of jet-cooled tantalum carbide, TaC, *J. Chem. Phys.*, 133(5), 2010, 054309.

Kudryavtsev V.I., A.N. Khodan, V.A. Kolyagin, R.R. Chuzhko, Chemical shifts of the shell levels depending on the structure and stoichiometry of tantalum carbides, *Inorg. Mater.*, 28(1), 1992, 55–60.

Kwon D.-H., S.-H. Hong, B.-K. Kim, Fabrication of ultrafine TaC powders by mechano-chemical process, *Mater. Lett.*, 58(30), 2004, 3863–3867.

Lawson J.W., M.S. Daw, C.W. Bauschlicher, Lattice thermal conductivity of ultra-high temperature ceramics ZrB_2 and HfB_2 from atomistic simulations, *J. Appl. Phys.*, 110(8), 2011, 083507.

Lee B.I., E.J.A. Pope, *Chemical Processing of Ceramics*, Marcel Dekker, New York, NY, ISBN 0824792440, 1994, 1–554.

Lei M., H.Z. Zhao, H. Yang, B. Song, W.H. Tang, Synthesis of transition metal carbide nanoparticles through melamine and metal oxides, *J. Eur. Ceram. Soc.*, 28(8), 2008, 1671–1677.

Li P.G., M. Lei, Z.B. Sun, L.Z. Cao, Y.F. Guo, X. Guo, W.H. Tang, C_3N_4 as a precursor for the synthesis of NbC, TaC, and WC nanoparticles, *J. Alloys Compd.*, 430(1–2), 2007, 237–240.

Li P.G., M. Lei, W.H. Tang, Route to transition metal carbide nanoparticles through cyanamide and metal oxides, *Mater. Res. Bull.*, 43(12), 2008, 3621–3626.

Li X., A. Westwood, A. Brown, R. Brydson, B. Rand, A convenient, general synthesis of carbide nanofibres via templated reactions on carbon nanotubes in molten salt media, *Carbon*, 47(1), 2009, 201–208.

Lipatnikov V.N., A.A. Rempel, A.I. Gusev, Thermodynamic model of atomic ordering. IV. The order–disorder transition in non-stoichiometric tantalum carbide, *Russ. J. Phys. Chem.*, 62(3), 1986, 285–288.

Lipatnikov V.N., A.I. Gusev, A.A. Rempel, G.P. Shveikin, Effect of a structural transition on the magnetic susceptibility of tantalum carbide, *Sov. Phys. Dokl.*, 32(12), 1987, 988–990.

Lipatnikov V.N., A.A. Rempel, and A.I. Gusev, Specific heat of tantalum carbide in states with difference degress of order, *Sov. Phys.—Sol. State*, 31(10), 1989, 1818–1819.

Lipatnikov V.N., A.A. Rempel, Formation of the incommensurate ordered phase in TaC_y carbide, *JETP Lett.*, 81(7), 2005, 326–330.

Liu J.-X., Y.-M. Kan, G.-J. Zhang, Pressureless sintering of tantalum carbide ceramics without additives, *J. Am. Ceram. Soc.*, 93(2), 2010, 370–373.

Liu L., F. Ye, X. He, Y. Zhou, Densification process of TaC/TaB_2 composite in spark plasma sintering, *Mater. Chem. Phys.*, 126(3), 2011, 459–462.

Lopez O.A., J.M. McKittrick, L.E. Shea, Fluorescence properties of polycrystalline Tm^{3+} activated $Y_3Al_5O_{12}$ and Tm^{3+}-Li^+ co-activated $Y_3Al_5O_{12}$ in the visible and near IR ranges, *J. Lumin.*, 71(1), 1997, 1–11.

López-de-la-Torre L, B. Winkler, J. Schreuer, K. Knorr, M. Avalos-Borja, Elastic properties of tantalum carbide (TaC), *Solid State Commun.*, 134(4), 2005, 245–250.

Lu K., *Nanoparticulate Materials*, Wiley, Hoboken, NJ, ISBN 9781118291429, 2012, 1–497.

Nakamura K., M. Yashima, Crystal structure of NaCl-type transition metal monocarbides MC (M = V, Ti, Nb, Ta, Hf, Zr), a neutron powder diffraction study, *Mater. Sci. Eng. B*, 148(1–3), 2008, 69–72.

Ma J., Y. Du, M. Wu, M. Pan, One simple synthesis route to nanocrystalline tantalum carbide via the reaction of tantalum pentachloride and sodium carbonate with metallic magnesium, *Mater. Lett.*, 61(17), 2007, 3658–3661.

Mackie W.A., P. Carleson, J. Filion, C.H. Hinrichs, Normal spectral emittance of crystalline transition metal carbides, *J. Appl. Phys.*, 69(10), 1991, 7236–7239.

Markhasev B.I., N.C. Pioro, V.V. Klyugvant, Y.L. Pilipovskii, Y.M. Shamatov, E.I. Geshko, Conditions for the separation of ordered phases from nonstoichiometric tantalum carbide, *Inorg. Mater.*, 19(12), 1983, 1759–1761.

Martin J.L., Evidence of dislocation dissociation in nearly stoichiometric tantalum carbide using the weak-bean technique, *J. Microsc-Oxford*, 98(2), 1973, 209–213.

Modine F.A., R.W. Major, T.W. Haywood, G.R. Gruzalski, Optical properties of tantalum carbide from the infrared to the near ultraviolet, *Phys. Rev. B.*, 29(2), 1984, 836–841.

Moore J.J., H.J. Feng, Combustion synthesis of advanced materials: Part I. Reaction parameters, *Prog. Mater. Sci.*, 39(4–5), 1995a, 243–273.

Moore J.J., H.J. Feng, Combustion synthesis of advanced materials: Part II. Classification, applications and modelling, *Prog. Mater. Sci.*, 39(4–5), 1995b, 275–316.

Monteverde F., Hot pressing of hafnium diboride aided by different sinter additives, *J. Mater. Sci.*, 43(3), 2008, 1002–1007.

Munir Z.A., U. Anselmi-Tamburini, Self-propagating exothermic reactions: The synthesis of high-temperature materials by combustion, *Mater. Sci. Rep.*, 3(7–8), 1989, 277–365.

Naiki T., M. Ninomiya, M. Ihara, Epitaxial growth of tantalum carbide, *Jpn. J. Appl. Phys.*, 11, 1972, 1106–1112.

Nartowski A.M., I.P. Parkin, M. MacKenzie, A.J. Craven, I. MacLeod, Solid state metathesis routes to transition metal carbides, *J. Mater. Chem.*, 9(6), 1999, 1275–1281.

Nelson J.A., M.J. Wagner, High surface srea nanoparticulate transition metal carbides prepared by alkalide reduction, *Chem. Mater.*, 14(10), 2002, 4460–4463.

Ni D.-W., G.-J. Zhang, Y.-M. Kan, P.-L. Wang, Synthesis of monodispersed fine hafnium diboride powders using carbo/borothermal reduction of hafnium dioxide, *J. Am. Ceram. Soc.*, 91(8), 2008, 2709–2712.

Nikol'skaya T.A., R.G. Avarbé, Temperature dependence and velocity of the congruent evaporation of the tantalum carbide phase, *High Temp.*, 7(6), 1969, 1021–1025.

Opeka M.M., I.G. Talmy, E.J. Wuchina, J.A. Zaykoski, and S.J. Causey, Mechanical, thermal, and oxidation properties of refractory hafnium and zirconium compounds, *J. Eur. Ceram. Soc.*, 19(13–14), 1999, 2405–2414.

Ortman B.J., R.H. Hauge, J.L. Margrave, Chemical reactions of carbon atoms and molecules from laser-induced vaporization of graphite, TaC, and WC, *J. Quant. Spectrosc. Ra.*, 40(3), 1988, 439–447.

Pavlov I.E., S.I. Alyamovskii, G.P. Shveikin, Solubility of hydrogen in tantalum carbides and oxycarbides, *Inorg. Mater.*, 151, 1979, 58–61.

Petrov V.A., V. Ya. Chekhovskoi, A.E. Sheindlin, V.A. Nikolaeva, Total hemispherical emissivity, spectral normal emissivity at a wavelength 0.65 μm, and electrical resistivity of tantalum carbide at very high temperatures, *High Temp.-High Press.*, 1, 1969, 657–661.

Petrova I.I., V.Y. Chekhovskoi, True heat capacity of zirconium, niobium, and tantalum carbides by a pulse method, *High Temp.*, 16(6), 1978, 1045–1050.

Pierson H.O., *Handbook of Refractory Carbides and Nitrides: Properties, Characteristics, Processing and Applications*, Noyes Publications, Saddle River, NJ, ISBN 0815513925, 1996, chs. 2–6.

Pilyankevich A.N., V.N. Paderno, L.V. Strashinskaya, A.N. Stepanchuk, A.S. Pritulyak, Electron microscopical investigation of the fracture surfaces of fused carbides of some transition metals, *Sov. Powder Metall.*, 14(7), 1975, 576–579.

Preiss H., D. Schultze, P. Klobes, Formation of NbC and TaC from gel-derived precursors, *J. Eur. Ceram. Soc.*, 17(12), 1997, 1423–1435.

Price D.L., J.M. Wills, B.R. Cooper, Linear-muffin-tin-orbital calculation of TaC(001) surface relaxation, *Phys. Rev. B*, 48(20), 1993, 15301–15310.

Pulci G., M. Tului, J. Tirillò, F. Marra, S. Lionetti, T. Valente, High temperature mechanical behavior of UHTC coatings for thermal protection of re-entry vehicles, *J. Therm. Spray Technol.*, 20(1–2), 2011, 139–144.

Rafaja D., W. Lengauer, H. Wiesenberger, Non-metal diffusion coefficients for the Ta-C and Ta-N systems, *Acta Mater.*, 46(10), 1998, 3477–3483.

Ramírez-Rico J., M.A. Bautista, J. Martínez-Fernández, M. Singh, Compressive strength degradation in ZrB$_2$-based ultra-high temperature ceramic composites, *J. Eur. Ceram. Soc.*, 31(7), 2011, 1345–1352.

Rempel A.A., A.P. Drushkov, V.N. Lipatnikov, A.I. Gusev, S.M. Klotsman, G.P. Shveikin, Angular correlation of annihilation radiation in nonstoichiometric tantalum carbide, *Sov. Phys. Dokl.*, 33(5), 1988, 357–359.

Rempel A.A., V.N. Lipatnikov, and A.I. Gusev, The superstructure in nonstoichiometric tantalum carbide, *Sov. Phys. Dokl.*, 35(2), 1990, 103–106.

Ring T.A., *Fundamentals of Ceramic Powder Processing and Synthesis*, Academic Press, San Diego, CA, ISBN 9780125889308, 1996, 1–961.

Rocher M., P. Goeuriot, J. Dhers, Modelling of the growth of carbide layers in tantalum, *Key Eng. Mat.*, 206–213(1), 2002, 527–530.

Rodríguez J., D. Martínez, L.G. Rosa, J.C. Fernandes, P.M. Amaral, N. Shohoji, Photochemical effects in carbide synthesis of d-group transition metals (Ti, Zr; V, Nb, Ta; Cr, Mo, W) in a solar furnace at PSA (platforma solar de almería), *J. Sol. Energ.-T. ASME*, 123(2), 2001, 109–116.

Rowcliffe D.J., W.J. Warren, Structure and properties of tantalum carbide crystals, *J. Mater. Sci.*, 5(4), 1970, 345–350.

Rowcliffe D.J., G.E. Hollox, Hardness anisotropy, deformation mechanisms and brittle-to-ductile transition in carbides, *J. Mater. Sci.*, 6(10), 1971a, 1270–1276.

Rowcliffe D.J., G.E. Hollox, Plastic flow and fracture of tantalum carbide and hafnium carbide at low temperatures, *J. Mater. Sci.*, 6(10), 1971b, 1261–1269.

Sahnoun M., C. Daul, J.C. Parlebas, C. Demangeat, M. Driz, Electronic structure and optical properties of TaC from the first principles calculation, *Eur. Phys. J. B*, 44(3), 2005, 281–286.

Saito N., H. Ikeda, Y. Yamaoka, A.M. Glaeser, K. Nakashima, Wettability and transient liquid phase bonding of hafnium diboride composite with Ni-Nb alloys, *J. Mater. Sci.*, 47(24), 2012, 8454–8463.

Samsonov G.V., R.Y. Petrikina, Sintering of metals, carbides, and oxides by hot pressing, *Phys. Sintering*, 2(3), 1970, 1–20.

Samsonov G.V., How the defect content of the carbon sublattice influences the properties of refractory transition-metal carbides, *Inorg. Mater.*, 9(12), 1973, 1893–1898.

Sautereau J., A. Mocellin, Sintering behavior of ultrafine NbC and TaC powders, *J. Mater. Sci.*, 9(5), 1974, 761–771.

Savino R., M.D.S. Fumo, D. Paterna, A.D. Maso, F. Monteverde, Arc-jet testing of ultra-high-temperature-ceramics, *Aerosp. Sci. Technol.*, 14(3), 2010, 178–187.

Savitsky E.M., G.S. Burkhanov, Growth of single crystals of high melting metal alloys and compounds by plasma heating, *J. Cryst. Growth*, 43(4), 1978, 457–462.

Sciti D., S. Guicciardi, M. Nygren, Densification and mechanical behavior of HfC and HfB$_2$ fabricated by spark plasma sintering, *J. Am. Ceram. Soc.*, 91(5), 2008a, 1433–1440.

Sciti D., L. Silvestroni, A. Bellosi, High-density pressureless-sintered HfC-based composites, *J. Am. Ceram. Soc.*, 89(8), 2006, 2668–2670.

Sciti D., L. Silvestroni, G. Cellotti, C. Melandri, S. Guicciardi, Sintering and mechanical properties of ZrB$_2$–TaSi$_2$ and HfB$_2$–TaSi$_2$ ceramic composites, *J. Am. Ceram. Soc.*, 91(10), 2008b, 3285–3291.

Segal D.L., A review of preparative routes to silicon nitride powders, *Br. Ceram. Trans. J.*, 85(6), 1986, 184–187.

Selvaduray G, L. Sheet, Aluminum nitride: Review of synthesis methods, *Mater. Sci. Tech.*, 9(6), 1993, 463–473.

Shea L.E., J.M. McKittrick, O.A. Lopez, E. Sluzky, Synthesis of red-emitting, small particle size oxide phosphors using an optimized combustion process, *J. Am. Ceram. Soc.*, 79(12), 1996, 3257–3265.

Sheindlin A.E., I.S. Belevich, I.G. Kozhevnikov, Enthalpy and specific heat of tantalum carbide in the range 273–3600°K, *High Temp.*, 103, (1973), 581–583.

Sheindlin A.E., V.A. Petrov, A.N. Vinnikova, V.A. Nikolaeva, Integral normal emissivity of tantalum and hafnium carbides over the temperature range 1300–3000°K, *High Temp.*, 7(2), 1969, 236–238.

Shi L., Y. Gu, L. Chen, Z. Yang, J. Ma, Y. Qian, Formation of TaC nanorods with a low-temperature chemical route, *Chem. Lett.*, 33(12), 2004, 1546–1547.

Shkiro V.M., G.A. Nersisyan, I.P. Borovinskaya, A.G. Merzhanov, V.S. Shekhtman, Preparation of tantalum carbides by self-propagating high-temperature synthesis, *Sov. Powder Metall.*, 18(4), 1979, 227–230.

Shvab S.A., F.F. Egorov, Structure and some properties of sintered tantalum carbide, *Sov. Powder. Metall.*, 21(11), 1982, 894–897.

Shvab S.A., P.S. Kislyi, Some properties of tantalum carbide powder produced by synthesis from the elements, *Sov. Powder Metall.*, 13(5), 1974, 368–370.

Silvestroni L., A. Bellosi, C. Melandri, D. Sciti, J.X. Liu, G.J. Zhang, Microstructure and properties of HfC and TaC-based ceramics obtained by ultrafine powder, *J. Eur. Ceram. Soc.*, 314, (2011), 619–627.

Silvestroni L., D. Sciti, Sintering behavior, microstructure, and mechanical properties: A comparison among pressureless sintered ultra-refractory carbides, *Adv. Mater. Sci. Eng.*, 2010, 2010, 835018.

Sinha K., B. Pearson, S.R. Casolco, J.E. Garay, and O.A. Graeve, Synthesis and consolidation of BaAl$_2$Si$_2$O$_8$:Eu. Development of an integrated process for luminescent smart ceramic materials, *J. Am. Ceram. Soc.*, 92(11), 2009, 2504–2511.

Sirdeshmukh D.B., M.J.M. Rao, X-ray determination of Debye temperature of tantalum carbide, *Phys. Status Solidi A*, 17(2), 1973, K169–K172.

Sonber J.K., A.K. Suri, Synthesis and consolidation of zirconium diboride: Review, *Adv. Appl. Ceram.*, 110(6), 2011, 322–334.

Squire T.H., J. Marschall, Material property requirements for analysis and design of UHTC components in hypersonic applications, *J. Eur. Ceram. Soc.*, 30(11), 2010, 2239–2251.

Stern K.H., S.T. Gadomski, Electrodeposition of tantalum carbide coatings from molten salts, *J. Electrochem. Soc.*, 130(2), 1983, 300–305.

Sugimoto T., *Fine Particles: Synthesis, Characterization, and Mechanisms of Growth*, Marcel Dekker, New York, NY, ISBN 0824700015, 2000, 1–738.

Takahashi T., K. Sugiyami, Fibrous growth of tantalum carbide by A-C discharge method, *J. Electrochem. Soc.*, 121(5), 1974, 714–718.

Talmy I.G., J.A. Zaykoski, M.M. Opeka, Synthesis, processing and properties of TaC-TaB$_2$-C ceramics, *J. Eur. Ceram. Soc.*, 30(11), 2010, 2253–2263.

Tatarintseva M.I., N.I. Puzynina, V.N. Kulyukin, V.N. Arbekov, Crystallographic and morphological features of carbide formation in the Ta-C-Cl system, *Inorg. Mater.*, 19(11), 1983, 1630–1633.

Toth L.E., M. Ishikawa, Y.A. Chang, Low temperature heat capacities of superconducting niobium and tantalum carbides, *Acta Metall.*, 16(9), 1968, 1183–1187.

Vajeeston P., P. Ravindran, R. Asokamani, Electronic structure, bonding, and ground-state properties of AlB_2-type transition metal diborides, *Phys. Rev. B.*, 63(4), 2001, 045115.

Varma A., J.-P. Lebrat, Combustion synthesis of advanced materials, *Chem. Eng. Sci.*, 47(9–11), 1992, 2179–2194.

Veretennikov B.N., Y.P. Solodov, Y.I. Pugachev, Changes in the phase structure of tantalum carbides during prolonged heating in vacuum, *Inorg. mat.*, 13(6), 1977, 828–831.

Warner T.E., *Synthesis, Properties and Mineralogy of Important Inorganic Materials*, Wiley, Hoboken, NJ, ISBN 9780470746110, 2011, 1–289.

Weng L., X. Zhang, J. Han, W. Han, The effect of Si_3N_4 on microstructure, mechanical properties and oxidation resistance of HfB_2-based composite, *J. Compos. Mater.*, 43(2), 2009, 113–123.

Wu W.-W., Z. Wang, G.-J. Zhang, Y.-M. Kan, P.-L. Wang, ZrB_2-$MoSi_2$ composites toughened by elongated ZrB_2 grains via reactive hot pressing, *Scripta Mater.*, 61(3), 2009, 316–319.

Xiang H., Y. Xu, L. Zhang, L. Cheng, Synthesis and microstructure of tantalum carbide and carbon composite by liquid precursor route, *Scripta Mater.*, 55(4), 2006, 339–342.

Xue H., Z.A. Munir, Field-activated combustion synthesis of TaC, *Int. J. Self-Propag. High-Temp. Synth.*, 5(3), 1996, 229–239.

Yang F., X. Zhang, J. Han, S. Diu, Analysis of the mechanical properties in short carbon fiber-toughened ZrB_2-SiC ultra-high temperature ceramics, *J. Compos. Mater.*, 44(8), 2010, 953–961.

Yeh C.L., E.W. Liu, Combustion synthesis of tantalum carbides TaC and Ta_2C, *J. Alloys Compd.*, 415(1–2), 2006, 66–72.

Yohe W.C., A.L. Ruoff, Ultrafine-grain tantalum carbide by high pressure hot pressing, *Am. Ceram. Soc. Bull.*, 57(12), 1978, 1123–1130.

Zhang X., G.E. Hilmas, W.G. Fahrenholtz, Hot pressing of tantalum carbide with and without sintering additives, *J. Am. Ceram. Soc.*, 90(2), 2007, 393–401.

Zhang X., G.E. Hilmas, W.G. Fahrenholtz, Densification and mechanical properties of TaC-based ceramics, *Mater. Sci. Eng. A*, 501, 2009, 37–43.

Zou J., G.-J. Zhang, Y.-M. Kan, Pressureless densification and mechanical properties of hafnium diboride doped with B_4C: From solid state sintering to liquid phase sintering, *J. Eur. Ceram. Soc.*, 30(12), 2010, 2699–2705.

Zou J., G.-J. Zhang, S.-K. Sun, H.-T. Liu, Y.-M. Kan, J.-X. Liu, C.-M. Xu, ZrO_2 Removing reactions of groups IV–VI transition metal carbides in ZrB_2 based composites, *J. Eur. Ceram. Soc.*, 31(3), 2011, 421–427.

7

Characterization of High-Temperature Materials

Yoseph Bar-Cohen and Robert D. Cormia

CONTENTS

7.1 Introduction

Developing materials for high-temperature use as well as assuring their performance and durability requires physical and chemical characterization techniques that analyze them before, after, and possibly during various service conditions. These conditions include testing at room temperature as well as possibly during and after subjecting them to high temperatures and other conditions (pressure, chemicals, gases, and others) that mimic the environment during operation. The techniques, which are used, examine the surface and the bulk properties of the material. Some of the techniques are nondestructive like those described in this chapter, whereas the destructive test methods, and the ones that are used to remotely examine the surface nondestructively, are described in Chapter 8. Generally, image analysis is widely used to examine the physical structure of the surface of the material or structure that needs to be characterized. An easy direct accessibility of the surface allows the application of many analytical tools. Analysis of structures, below the surface, is done by techniques that provide insight and gauging of the internal characteristics that include stress and strain, and chemical composition.

There are many analytical techniques that are used to examine high-temperature materials. While these techniques are covered widely in the literature, this chapter reviews some of the key ones in order to provide completeness to this book. The methods and their underlying principles are described in the following sections.

7.2 Analyses of Materials and Processes

Materials developed for high-temperature applications require testing at various stages of their life from research and development to their retirement from use. These methods may be used to help develop the materials, for quality assurance before service, for characterization during use, or for failure analysis. Various methods of testing materials and structures are described in the following sections with the simplest being visual inspection (Figure 7.1). Visual inspection may involve unaided examination, simple imaging with photography or video viewing, or the use of optical microscope.

Testing is typically performed: (a) during fabrication, (b) prior to service, (c) during service, and also (d) after a failure—to determine the cause and prevent future similar cases. The characterization of materials includes chemical, mechanical, micro/macro, destructive as well as nondestructive analysis and testing. When the material consists of metal or metal alloy, the tests are considered to be part of the field of metallurgy. Tests may include conducting elemental analysis, determining the heat treatment condition, metallography, microstructure, and phase analysis, determination of the process by which the material was formed, and whether it conforms to certain necessary specification(s). Testing may also include failure analysis of fracture, creep, fatigue, wear, deformation, contamination (trace analysis), corrosion control, materials evaluation and development, and more. In addition, tests may include analysis of the material cross section where the sample is prepared by mounting it into a mold, and etching, grinding, and/or polishing the surface as well as possibly using replication techniques (Figure 7.2). Moreover, the tests may include microstructure evaluation, grain size determination, and analysis of the heat treatment, as well as identification of porosities or flaws, and hardness testing.

FIGURE 7.1
Charged-coupled device (CCD) microscope (model Keyence VH-8000 Digital Microscope) for visual inspection of materials. (Photographed at JPL, Pasadena, CA.)

Microscopic tests and imaging may be performed to determine the microstructure, carburization and decarburization, plating thickness, carbide precipitation, intergranular corrosion, alpha case, sensitization, surface contamination, nodularity and nodule count as well as eutectic melting.

To determine the mechanical properties of test materials, various configuration coupons are made (see an example of a dog-bone coupon in Figure 7.3, right) and are mechanically

(a)   (b)

FIGURE 7.2
(See color insert.) Surface polishing system and polished cross sections mounted in a mold. (Photographed at JPL, Pasadena, CA.)

(a)

(b)

FIGURE 7.3
Mechanical test system (made by Instron) with a test chamber of exposure to various temperatures (a) as well as a dog-bone coupon made of Al 7075 T73 (b). (Photographed at JPL, Pasadena, CA.)

tested (Figure 7.3, left). The tests are done at various conditions including heating, cooling, humidity, and so on and the sample strength vs. strain, plastic deformation, form of failure, and many other characteristics are examined. Other testing can include determination of residual stresses, thermal tests to determine durability at high temperatures and high-temperature deformation. Also, tribology measurements and testing are performed to examine friction, lubrication, and wear characteristics.

7.2.1 Surface Chemical Analysis Techniques

Surface chemical analysis is important for characterization of high-temperature materials for these reasons:

1. The surface of a material is the contact interface in mechanical systems, and surface chemistry can be different from bulk chemistry. Surface treatment and material processing can lead to significantly different surface composition, chemistry, and physical properties. Understanding surface chemistry as it relates to the processing and properties can lead to the development of better materials.

2. Surface contamination often leads to failures, especially in critical aerospace applications. Tools that provide analysis of the top 100 Å and the ability to perform compositional depth profile of the top micron layer provides insights about contamination, oxidation, corrosion, and surface segregation of elements in materials that can negatively affect the material performance.

TABLE 7.1

Surface Analysis Techniques

Acronym	Name	Analysis/Information	Strengths	Challenges
AES	Auger electron spectroscopy	Elemental and some chemical bonding, depth composition profiles	Small analysis area, depth profile capability, relatively fast, semiquantitative	Solid samples only, insulators are more difficult
XPS	X-ray photoelectron spectroscopy	Elemental and chemical bonding information, depth composition profiles	Small analysis area, depth profile capability, relatively fast, semiquantitative	Charge control on insulators, slower analysis than AES, spatial resolution to only 50 μm
SSIMS	Secondary ion mass spectroscopy	Mass spectral identification, molecular ions	Can be very surface sensitive	Data analysis and interpretation
LEED	Low-energy electron diffraction	Surface structure and geometry	Precise atomic layer analysis	Surface must be clean, single phase

3. For metals and other conductive materials, grain boundary engineering and sintering are important for developing stronger and harder materials. Surface analysis tools with high spatial resolution, such as field-emission Auger electron spectroscopy (FE-AES), can provide detailed information about the processes that are taking place at grain boundaries. This information can help develop better understanding of how processing can control parameters that influence the material performance, especially at elevated temperature.

There are many tools that are used to investigate the surface characteristics and many of them are covered in the following sections. A summary of some of the surface analysis techniques is given in Table 7.1.

7.3 Imaging and Visualization Analyzers

Visual inspection of test samples and structures is the simplest and fastest method of obtaining information about the surface, shape, anomalies, colors, and many other physical features. To enable viewing objects beyond the normal capability of the human eye, various tools are used, including magnifying glasses and various microscopes. Generally, there are three categories of microscopes: optical, electron, and scanning probe microscopy. Optical and electron microscopes use the interactions of electromagnetic radiation/electron beams with the material and collect information regarding how the specimen reflects, refracts, diffracts and/or scatters radiation. The beam can be focused onto the sample in a wide-field irradiation or in a raster scanning via a fine beam. Scanning probe microscopes use the interaction of a probe tip and the surface of the tested material, including tunneling current, magnetic attraction, and lateral force and stiction. A summary of the techniques is given in Table 7.2.

TABLE 7.2

Image Analysis Techniques

Acronym	Name	Analysis/Information	Strengths	Challenges
AFM	Atomic force microscopy	Contact and noncontact surface morphology	Straightforward analysis, digital data files	Solid samples only. Irregular surfaces can be difficult to navigate
SEM	Scanning electron microscopy	Secondary electron detectors (SED) Backscattered electrons detectors (BSD)	Fast imaging of conductive materials	Insulators are a bit more difficult to analyze. Cannot "look" at liquids
TEM	Transmission electron microscopy	High-energy electron transmission	Very high spatial resolution, some 3D information	Sample preparation can be tedious/expensive
EDX	Energy-dispersive x-ray analysis	Elemental composition	Very fast, semiquantitative	Not sensitive to low-Z-number elements
FIB	Focused ion beam	Surface topography	Fast analysis of semiconductor defects/features	Technology is more expensive than SEM
EMP	Electron microprobe	Elemental composition	Fast, sensitive, semiquantitative	

7.3.1 Optical Microscopes

Visual inspection is the most basic method of examining as well as characterizing materials and structures. Such methods are used to inspect grain boundaries and identify phases as well as to monitor the propagation of cracks, detect the presence of corrosion, and distinguish between brittle and ductile failure as well as assessing the presence of fatigue failure.

To enhance the results one can use optical methods with increasing capability starting from a simple magnifying glass, then move to recording topography, structure, integrity, and many others. To further enhance the visualization of materials, optical microscopes are used with increased magnification, resolution and stereoscopic viewing (Figure 7.4) (Murphy, 2001). Generally, optical microscopes allow detailed viewing of microstructural characteristics such as the grain boundaries, dislocations, phase structure, and others but they have a diffraction limit that is dictated by the wavelength of visible light, which is about 0.3 μm. To enhance the image contrast there are many methods that are used including contrasting colors, polarization, and phase contrasting. To increase the visualization resolution and contrast, one can use confocal microscopes, which use an aperture with a pinhole for examining sample areas that are larger than the focal plan (Figure 7.5). The pinhole light source is located in an optically conjugate plane in front of the detector, eliminating out-of-focus light (Minsky, 1961; Price and Jerome, 2011) and significantly increasing the image resolution. To ease on the scanning and viewing of samples, there are microscopes that are designed as inverted wide-field type where the sample is placed onto an XYZ stage and scanned while viewing various sections of its area. An example of such a microscope is shown in Figure 7.6.

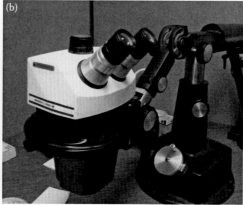

FIGURE 7.4
Collection of microscopes with various magnification levels. (a) Leica GZ4 microscope, (b) Bausch & Lomb microscope, and (c) Meiji microscope. (Photographed at JPL, Pasadena, CA.)

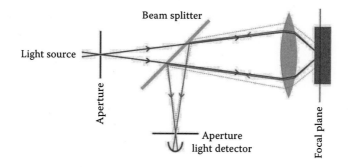

FIGURE 7.5
Schematic illustration of a confocal microscope. (From Wikimedia Commons, http://en.wikipedia.org/wiki/Confocal_microscopy.)

FIGURE 7.6
Inverted wide-field microscope. (a) Side view and (b) front view. (Model Axiovert 405M, made by Carl Zeiss). (Photographed at JPL, Pasadena, CA.)

(a)   (b)

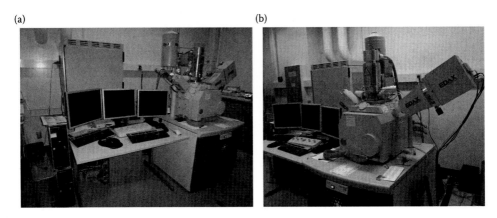

FIGURE 7.7
(See color insert.) Two views of a scanning electron microscope (SEM) model FEI Nova Nanosem 600 (made by EDAX). (a) General view and (b) close-up onto the SEM testing system. (Photographed at JPL, Pasadena, CA.)

7.3.2 Scanning Electron Microscopy

The scanning electron microscope (SEM) is one type of electron microscope (Figure 7.7) (Reimer, 1998). SEM creates images of the topmost surface of a sample by rastering a focused, collimated electron beam over the sample (Figure 7.8). The image that is generated from the interaction of the electrons with atoms on the sample surface (top 1–10 nm) contains information about the elements in the material, electrical conductivity, and surface topography (see example in Figure 7.9). This method produces secondary electrons for imaging, x-rays characteristic of elements, and backscattered electron (BSE) images that depend on atom number, Z. SEM imaging systems use multiple detectors that collect the various emitted signals.

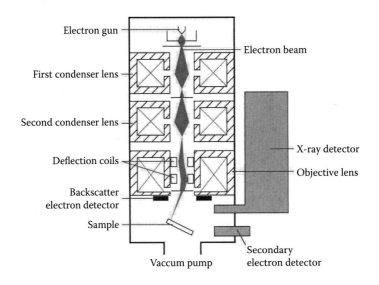

FIGURE 7.8
Schematic diagram of an SEM. (From Wikimedia Commons, http://en.wikipedia.org/wiki/Scanning_electron_microscope.)

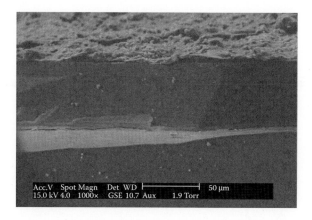

FIGURE 7.9
SEM image of corrosion on the surface of an ancient glass fragment. (From Wikimedia Commons, http://en.wikipedia.org/wiki/Scanning_electron_microscope.)

The size of the interaction volume of the electron beam depends on the electron's energy as well as the atomic number and density of the material. The magnification of an SEM can be controlled over a range of six orders of magnitude reaching as much as half a million times in field emission (FE)-SEM. The latter is three orders of magnitude higher than the best optical microscopes. Generally, the samples are mounted rigidly onto a holder. SEM imaging requires reasonable electric conductivity of the sample surface and adequate grounding to avoid accumulation of electrostatic charges. To produce a conductive surface for imaging materials that are less conductive (polymers and ceramics) analysts normally deposit an ultrathin coating (10–20 nm) of an electrically conducting material, such as gold, gold–palladium, platinum, tungsten, iridium, and even graphite.

In earlier SEM instruments, the electron beam is generated thermionically from an electron gun using a tungsten filament cathode. Tungsten is widely used in such filaments since it has a very high melting point and very low vapor pressure and it is relatively inexpensive. Other types of electron emitters include field emission guns (FEG), using a cold-cathode made of single-crystal tungsten emitter, or lanthanum hexaboride (LaB_6). The SEM that uses FEG as the emission source is known as FE-SEM. A summary of the capability and applications of SEM is given in Table 7.3.

Secondary electrons (SE): Most SEMs use a secondary electron detector (SED), which allows for secondary electron imaging (SEI). Such images have a very high spatial resolution, often less than 1 nm. Also, the very narrow electron beam of the SEM micrographs

TABLE 7.3

Summary of the Capabilities and Applications of SEM

Method Provides	Capable to Analyze	Applications
High-magnification digital images, using secondary and backscattered electrons	Solid materials, both conductive and insulating, particles, fibers, plastics and polymers. Environmental samples with special vacuum equipment	Materials characterization, failure and forensic analysis, process development and QA/QC, metallurgy

allows for a large depth of field with 3D appearance providing information about the surface structure of the tested material.

Characteristic x-rays: Exposure to the electron beam stimulates the emission of characteristic x-rays, which can be used to identify the elemental composition and relative elemental abundance of each element in the sample. Energy-dispersive spectrometers are used to measure the characteristics of the x-rays that are emitted from the test sample. Since the energy of the x-ray characterizes the difference in energy between the two shells that are involved with the emission, as well as the atomic structure of the element from which they were emitted, the elemental composition of the sample can be measured.

BSE: This is a beam of reflected electrons resulting from direct back scattering from each atom. This beam generates images that provide information about the distribution of different elements based on their different atomic numbers (Z). Since the intensity of the BSE signal is strongly related to the atomic number of the analyzed sample, it is often used to perform spectroscopy of the characteristic x-rays.

7.3.3 Energy-Dispersive X-Ray Spectroscopy

Energy-dispersive x-ray spectroscopy (EDS or EDX) is used to perform elemental analysis of test materials and involves examination of the interaction with excited x-rays (Jenkins et al., 1995). EDX systems are a commonplace addition to SEM instruments. The characterization principle is based on the fact that each element in a test material has a unique atomic structure that generates a unique set of peaks in an x-ray energy spectrum. To emit characteristic x-rays, a high-energy beam of charged particles in the form of electrons, protons, or x-rays is focused into the material. The incident beam causes electrons from various shells of the atoms creating electron vacancies. The electrons from outer shells fill the holes and release x-rays. Using an energy-dispersive spectrometer, the energies of the emitted x-rays are measured. Since the energy of the x-ray is a characteristic of the difference in energy between the two atomic shells, and the atomic structure of a specific element, this method measures the elemental composition of the material. A summary of the capabilities and applications of EDX is given in Table 7.4.

7.3.4 Focused Ion Beam

The focused ion beam (FIB) system uses an ion beam of Ga or other metals to raster scan a material's surface similar to electrons rastered on materials with SEM (Figure 7.10) (Giannuzzi and Stevie, 2004). The resulting secondary electrons (or ions) are used to create an image of the material surface. By milling small holes in the material surface using the ion beam, cross sections for imaging can be prepared. FIB systems can also be incorporated with both electron and ion beams and used to create images with much higher magnifications and resolution, as well as provide more accurate control. FIB systems are

TABLE 7.4

Summary of the Capabilities and Applications of EDX

Method Provides	Capable to Analyze	Applications
Qualitative and quantitative analysis of elements heavier than boron, spatial mapping	Solid materials, particles, fibers, and biological specimens (fixed) can be used with SEM/TEM	Materials characterization, failure and forensic analysis, process development and QA/QC, metallurgy

FIGURE 7.10
Focused ion beam (FIB) workstation. (From Wikimedia Commons, http://en.wikipedia.org/wiki/Focused_ion_beam.)

TABLE 7.5

Summary of the FIB Capabilities and Applications

Method Provides	Capable to Analyze	Applications
High-resolution imaging while ion milling through a material, ability to repair optical masks	Solid materials, metals and alloys, semiconductors, ceramics	Semiconductor failure analysis, mask repair, materials science, and preparing samples for TEM

used for such applications as failure analysis. In Table 7.5, a summary of the FIB capabilities and applications is given.

7.3.5 Transmission Electron Microscopy

Transmission electron microscopy (TEM) is an imaging technique that examines the interaction characteristics of a transmitted electron beam after passing through an ultra-thin sample of a material (Williams and Carter, 2004). The formed image is magnified and focused onto an imaging device using charged-coupled device (CCD)-based camera. Similar to SEM, it provides images with resolution that is significantly higher than optical microscopes (Figure 7.11) and it is used to examine fine details such as single column of atoms. TEM can be used to observe changes in phase and crystal orientation, electronic structure, and sample-induced electron-phase shift as well as absorption based imaging (Pennycook and Nellist, 2011). A summary of the capabilities and applications of TEM is given in Table 7.6.

7.3.6 Electron Probe Microanalyzers

Electron probe microanalyzers (EPMA) use a micro-beam of electrons to perform non-destructive chemical analysis of minute solid samples (Goldstein et al., 2003; Reed, 2005). The significance of EPMA is its ability to perform precise quantitative elemental analyses

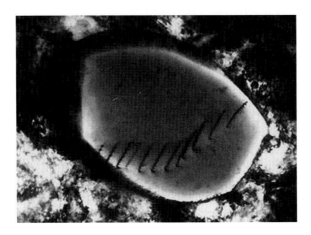

FIGURE 7.11
Transmission electron micrograph of dislocations in steel, which are faults in the structure of the crystal lattice at the atomic scale. (From Wikimedia Commons, http://en.wikipedia.org/wiki/Transmission_electron_microscopy.)

TABLE 7.6

Summary of the Capabilities and Applications of TEM

Method Provides	Capable to Analyze	Applications
Very high-resolution image analysis, some three-dimensional tomography	Solid materials that are properly prepared	Materials characterization, failure and semiconductor analysis, process development

of very small areas as small as 1–2 μm. While SEM can generate images of 3D objects, analysis by EPMA requires the use of flat polished areas that are examined in 2D. Similar to SEM, the image resolution is far higher than possible by optical imaging. The tested sample is bombarded by an accelerated and focused electron beam with energy sufficient to eject electrons and x-rays. A sample holder with three vertical axes moving stage is a standard fixture in EPMA systems. High vacuum is needed to prevent gas and vapor molecules from interfering with the electron beam. EPMA consists of various detectors that perform energy-dispersive x-ray spectrometry, wavelength-dispersive x-ray spectrometry, and visible-light microscopy. A high-powered visible-light microscope is included in order to optically observe the sample directly.

7.3.7 Scanning Probe Microscopy

Scanning probe microscopy (SPM) forms images of surfaces and surface interactions using a physical probe that scans a very small region (10 nm–20 μm) of the tested sample (Binnig et al., 1986). The image of the surface is obtained by mechanically moving the probe in a line or square raster over the surface of the sample and recording the probe–surface interaction as a function of position. Using multiple scanning probe tips allows simultaneous imaging of several surface interactions and the specific interaction that is used is called the "scanning mode." The resolution of the SPM technique depends on the specific method that is used and it can be as high as atomic levels. This capability results from the use of piezoelectric actuators that move the probe and provides atomic level precision and

TABLE 7.7

Summary of the Capabilities and Applications of SPM

Method Provides	Capable to Analyze	Applications
Family of scanning probe microscopes providing surface morphology, lateral and magnetic force, and electrostatic surface characteristics	Solid surfaces, including metals, alloys, ceramics and glasses, polymers and biopolymers	Metal finishing, process development, QA/QC

accuracy. The data are generally presented as 2D grid of points and is often presented on the computer display in false colors. A summary of the capabilities and applications of SPM is given in Table 7.7.

7.3.8 Scanning Tunneling Microscopy

Scanning tunneling microscope (STM) is a method that can be used to image surfaces at the atomic level (Binnig and Rohrer, 1986). The method is based on the concept of quantum tunneling and also involves using a probe tip. The phenomenon of quantum tunneling occurs when a particle tunnels through a barrier that classically could not be passed. STM is often operated in high vacuum using a conducting tip that is brought very close to the examined surface with bias voltage applied between tip and the sample. This system can cause electrons to tunnel between the tip and the sample (Figure 7.12).

The resulting tunneling current depends on the position of the tip, the applied voltage, and the local electron density of the sample. An image is formed by displaying the current as the tip is rastered across the material surface. An STM consists of a scanning tip, piezo-electric actuated scanner that controls the XYZ of the tip, vibration isolation system, and a computer for control, display, and data acquisition. Even though it is an imaging technique

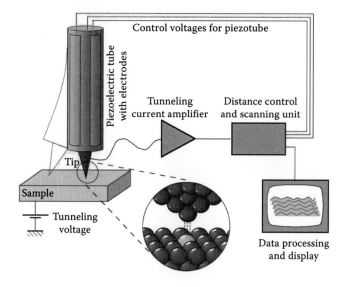

FIGURE 7.12

Schematic view of a scanning tunneling microscopy (STM). (From Wikimedia Commons, http://en.wikipedia.org/wiki/Scanning_tunneling_microscope.)

that can provide lateral resolutions of 0.1 nm resolution and depth resolution of 0.01 nm when used in UHV, it is very sensitive to the test conditions and requires extremely clean and stable surfaces, sharp tip, excellent vibration control, and sophisticated electronics. The high resolution of STM allows both imaging and manipulation of individual atoms on compliant materials. The extreme sensitivity of the tunneling current to the height of the tip is critically dependent on having effective vibration isolation. Originally, magnetic levitation was used to suppress vibrations, but today mechanical springs or gas spring systems are used. It is interesting to note that tests with STM can be performed in the air at 1 atm, as well as in various gases and liquids, with temperature that can range from close to zero Kelvins to several hundred degrees Celsius.

7.3.9 Atomic Force Microscopy

Atomic force microscopy (AFM), which is also known as scanning force microscopy (SFM), is a high-resolution SPM (Eaton and West, 2010). It normally uses a silicon or silicon nitride cantilever as the probe, which consists of a sharp tip (a tip radius of curvature on the order of 20–50 nm), to scan the sample surface. The tip is brought toward the sample surface and the atomic forces between the tip and the sample causes both attraction and deflection of the cantilever (Figure 7.13). The forces that are measured include mechanical contact forces, van der Waals forces, capillary forces, lateral forces, and electrostatic forces. The signal is produced by a laser beam that is reflected from the cantilever surface and recorded by a photodiode array having two or more position-sensitive detectors. Other methods that can be used to measure the deflection include optical interferometry, piezo-resistive, and capacitive sensing. Under ideal conditions, its resolution can reach fractions of a nanometer. It is an effective imaging, measuring, and manipulation tool for materials at the nanometer scale. The image is produced by scanning the surface with a mechanical probe and its movement is controlled by piezoelectric actuators allowing for accurate and precise movements. A summary of the capabilities of ASM and SPM is given in Table 7.8.

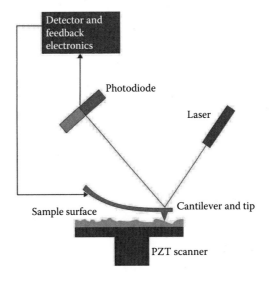

FIGURE 7.13
Schematic view of atomic force microscope (AFM). (From Wikimedia Commons, http://en.wikipedia.org/wiki/Atomic_force_microscopy.)

TABLE 7.8

Summary of the Capability of ASM and SPM

Method Provides	Capable to Analyze	Applications
Quantitative surface topology Force interaction (magnetic, electrostatic, lateral friction)	Solid materials, metals, ceramics, and glasses, and softer materials, plastics, polymers, and biopolymers	Surface finish, magnetic media, metallurgy, polymer engineering

7.4 Materials and Metallurgical Analyzers

7.4.1 X-Ray Diffraction Analysis

X-ray diffraction (XRD) analysis is used for examining the crystallography of materials as well as for obtaining information about the chemical composition and physical properties of materials (Figure 7.14) (Cullity and Stock, 2001). These properties are extracted from the scattering intensity of an x-ray beam that illuminates the material in films and powder forms as a function of the incident and scattered angle, polarization, and wavelength or energy. Given the ability to provide an absolute characterization of materials, a miniature XRD system was included in the suite of instruments on the Mars Science Lab mission's Curiosity Rover that landed on Mars in August 2012. A summary of the capabilities and applications of XRD is given in Table 7.9.

X-rays interact primarily with the electron cloud surrounding each atom, and the larger the atomic number, the higher the scattering intensity. Since a crystalline structure is a characteristic of the material being tested, a unique pattern is formed when illuminated by an x-ray beam and the resulting diffraction pattern is used to characterize the tested material. Generally, the system of XRD analysis consists of a monochromatic radiation and an x-ray detector that is encircling the sample in powder form. The detector is used to measure the diffraction intensity pattern (examples of intensity patterns are shown in Figure 7.15). An incident x-ray beam is scattered when it encounters a crystallographic

(a)   (b)

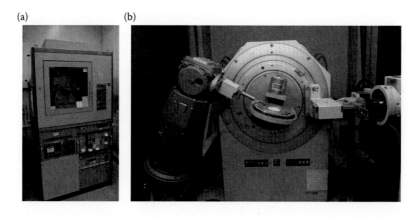

FIGURE 7.14

(See color insert.) An x-ray diffraction analysis (XRD) system Model D500 Diffraktometer (Siemens). Overall system view (a) and close-up of the sample testing source and detector system (b). (Photographed at JPL, Pasadena, CA.)

TABLE 7.9

Summary of the Capabilities and Applications of XRD

Method Provides	Capable to Analyze	Applications
Structural information, crystal lattice structure, identification of crystal and material phases	Metals, alloys, glasses and ceramics, particles, powders, polymers and biopolymers	Crystal structure and material phase determination, process development, QA/QC

(a)

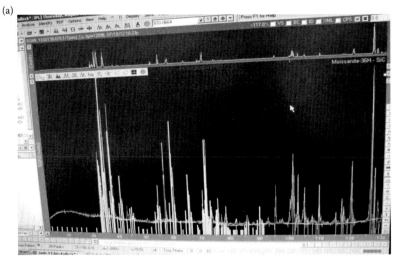

(b)

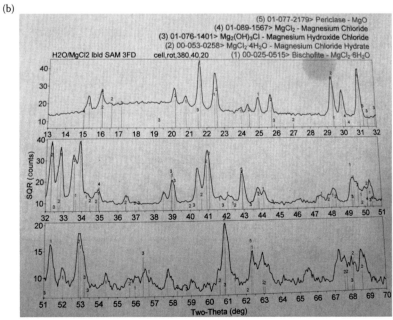

FIGURE 7.15

Diffraction patterns obtained by XRD as displayed on a computer (a) or paper (b). (Photographed at JPL, Pasadena, CA.)

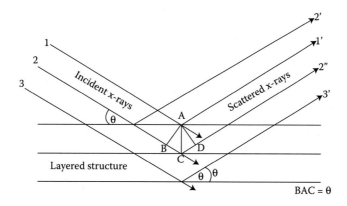

FIGURE 7.16
Incident and scattered x-rays from an atomic lattice. (Based on http://pubs.usgs.gov/of/2001/of01-041/html-docs/xrpd.htm.)

lattice and constructive interferences occur when the scattered rays from the various atomic layers are in phase (Figure 7.16). The diffraction relation is known as Bragg's law and it is as follows:

$$2d (\sin \theta) = \lambda_o \tag{7.1}$$

where d is the crystallographic lattice interplanar spacing, θ is the incident angle of the x-ray beam (Bragg's angle), and λ is the wavelength of the x-ray beam.

Generally, XRD consists of two form of scattering:

Elastic scattering: Elastic scattering of monochromatic x-rays is used to examine materials that do not have a long range order. Depending on the tested sample, the measurement is done as follows: To test structures over beam areas of nanometer through micrometer, the intensity of small-angle x-ray scattering (SAXS) is measured at angles 2θ that are close to $0°$. On the other hand, to perform measurements over 2θ angles that are larger than $5°$, wide-angle x-ray scattering (WAXS) is performed. Further, x-ray reflectivity tests of thin films are conducted to determine the thickness, roughness, and density of single and multilayers.

Inelastic scattering: Inelastic scattering modifies the phase of diffracted x-rays, and instead of producing useful data for XRD, this scattering contributes to background noise in the diffraction pattern. When the energy and angle of the inelastically scattered x-rays are monitored, one can examine the electronic band structure of tested materials. The techniques that are based on measuring such scattering are: Compton scattering, resonant inelastic, x-ray Raman scattering (XRS), and x-ray scattering (RIXS).

7.4.2 X-Ray Fluorescence Spectrometry

Bombarding materials with high-energy x-rays or gamma rays results in fluorescent emission (i.e., secondary emission) of x-rays and this emission is used to characterize materials (Jenkins, 1999). The bombardment causes emission of electrons with binding energies

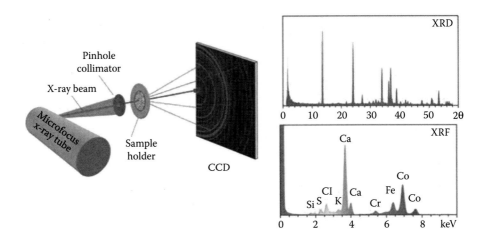

FIGURE 7.17
(See color insert.) CheMin mineralogical instrument using XRD and XRF (Blake et al., 2012. Image credit: Courtesy of David Blake, NASA Ames, and David Bish, University of Indiana.)

TABLE 7.10

Summary of the Capabilities and Applications of XRF

Method Provides	Capable to Analyze	Applications
Elemental composition of inorganic elements in a variety of materials, thin films and solids. Qualitative and semiquantitative data with the use of standards	Solid materials and thin films, ceramics and glasses, powders, particles, plastics, and polymers	Trace element quantitation, alloy composition, thin film analysis, material validation, and QA/QC

smaller than the energy of the x-rays. Following this excitement, electrons from outer, lower-energy orbitals drop to fill the generated positive holes and this leads to energy release is the form of photons that are equal to the difference in energy between of the related two electron orbitals. The emitted radiation, which has the characteristics of the atoms that are present, is measured by proportional counters or various types of solid-state detectors. X-ray fluorescence (XRF) instruments are widely used for elemental analysis of ceramics, glass, and metals. XRF and XRD instruments were used in the CheMin analyzer (Blake et al., 2012) of the Mars Curiosity suite of instruments that landed on Mars on August 5, 2012 (Figure 7.17). A summary of the capabilities and applications of XRF is given in Table 7.10.

7.4.3 X-Ray Absorption Spectroscopy

X-ray absorption spectroscopy (XAS) is used to determine the local geometric and/or electronic structure of tested materials (von Bordwehr, 1989). Tests are done using a synchrotron radiation source of intense and tunable x-rays and samples are tested in gas, solution, or solid form. The technique involves the use of a narrow parallel monochromatic x-ray beam that passes through the test material causing intensity decrease due to absorption and it depends on the types of atoms and the material density (http://www.chem.ucalgary.ca/research/groups/faridehj/xas.pdf). At certain energies, the absorption increases drastically and involves absorption edge. Each such edge occurs when the energy of the incident photons is just sufficient to cause excitation of a core electron of an absorbing atom and producing

TABLE 7.11

Summary of the Capabilities and Applications of XAS

Method Provides	Capable to Analyze	Applications
Technique for determining the local geometric and/or electronic structure of materials unique sensitivity to the local structure, as compared to x-ray diffraction	Amorphous solids and liquids, metals and alloys, solid solutions, organometallics, powders, and some plastics	Determining molecular bonding environments and fine structure, especially in amorphous and some crystalline materials

photoelectron. The energies of the absorbed radiation at these edges correspond to the binding electron energies in the K, L, M, and other shells of the absorbing elements. The emitted photoelectron wave from an absorbing atom is backscattered by neighboring atoms and the two interfere constructively and destructively and are measured as maxima and minima after the edge. There are four sections that make up an x-ray absorption spectrum:

1. Pre-edge $(E < E_0)$—This results from electron transitions from the core level to higher unfilled or half-filled orbitals.

2. X-ray absorption near edge structure (XANES) at incident x-ray beam energy of $E = E_0 \pm 10$ eV—This involves transition of core electrons to nonbound levels. Because of the high probability of such transition, it appears as a sudden absorption.

3. Near edge x-ray absorption fine structure (NEXAFS) with energies between 10 and 50 eV above the edge—The ejected photoelectrons have low kinetic energy, leading to strong multiple scattering by the first and even higher shells.

4. Extended x-ray absorption fine structure (EXAFS) starting at energies from about 50 eV up to 1000 eV above the edge—The photoelectrons have high kinetic energy and scattering that is dominated by nearest neighboring atoms.

A summary of the capabilities and applications of XAS is given in Table 7.11.

7.4.4 X-Ray Photoelectron Spectroscopy

X-ray photoelectron spectroscopy (XPS) is a surface analysis tool (Watts and Wolstenholme, 2003). It provides semiquantitative elemental composition and chemical bonding state information from the top 50 to 100 Å (Table 7.12). It differs from AES in that the composition data are more quantitative and it provides easier-to-interpret chemical state information (Table 7.13). XPS leverages the photoelectron effect where an incident x-ray (usually Al k-alpha) of sufficient energy (1486 eV) to cause the ejection of electrons with kinetic

TABLE 7.12

X-Ray Photoelectron Spectroscopy Capabilities for Materials Characterization

XPS Analysis Provides	Limitations
Semiquantitative (with calibration standards)	Cannot make measurements in liquids
Chemical bonding state information	Cannot detect hydrogen
Point analysis, line scan, and x-ray mapping	ensitive to about 0.5 atom%
Depth composition profiles (conductive and nonconductive materials)	Minimum beam area is about 50 μm

TABLE 7.13

Auger Electron Spectroscopy Method

Method Provides	Capable to Analyze	Applicable to
Surface composition	Metals and alloys	Grain boundary analysis
Some chemical bonding state information	Carbon composites	Corrosion and defect analysis
High spatial resolution	Some ceramics	Failure analysis
Line scan and X–Y elemental mapping		Semiconductors
Depth composition profiles		
Secondary electron (SED) images		
EDX (on some instruments)		

energy equal to the incident energy minus the binding energy of the element and shell and affected by the chemical bonding state. Every element (except H and He) has a set of characteristic binding energies and photoionization cross sections. The semiquantitative nature of XPS is based on a combination of theoretically calculated photoionization cross sections like the Scofield table (Scofield, 1976) and well-characterized standards.

Applications of XPS include characterization of surface passivation and conversion coatings, surface segregation and oxidation thickness, and approximate alloy composition. Contamination analysis is one of the important applications of XPS allowing for determining the nature of cleaning residues and contamination that could lead to corrosion.

7.4.5 Low-Energy Electron Diffraction

Low-energy electron diffraction (LEED) is a technique for determining the surface structure of crystalline materials by bombarding the surface with a collimated beam of low-energy electrons ranging from 20 to 200 eV and analyzing diffracted electrons as spots on a fluorescent screen (VanHove et al., 2012). The LEED technique is used to examine the diffraction pattern and provide information about the symmetry of the surface structure. Also, the intensity of the diffracted beams as a function of the energy of the incident electron beam is analyzed to obtain accurate information about the atomic positions on the surface of the tested material.

7.4.6 Neutron Diffraction

Neutron diffraction, which is also known as elastic neutron scattering, is used to determine the atomic and/or magnetic structure of materials (Lovesey, 1984; Squires, 1996). This technique is quite sensitive to particular light atoms and it can be used to distinguish between isotopes. A thermal or cold neutron beam irradiates the tested material and the formed diffraction pattern is used to determine information about the structure of the material. Similar to the XRD method, the diffraction of the scattered neutron beam follows Bragg's law and it provides complementary information. Neutrons interact with the nucleus of the atom rather than the electrons and the diffraction intensity depends on the interacted isotope. For some light atoms, the diffraction intensity is very strong even in the presence of elements with large atomic number.

The main disadvantage of neutron diffraction is the limited availability of the radiation source. Neutrons are produced by a reactor and, to select a specific neutron wavelength, such components as crystal monochromators and filters are used. Alternatively, a spallation source can be used where the energies of the incident neutrons are sorted using a series of synchronized aperture elements that filter the neutron pulses with the desired wavelength. Test samples in polycrystalline powder form are used with crystals that are much larger than the ones used in XRD. Since neutrons carry a spin, they interact with the magnetic moment of electrons allowing analyzing the microscopic magnetic structure of tested materials.

7.4.7 Auger Electron Spectroscopy

Auger spectroscopy is a surface composition analysis method and its characterization capabilities are summarized in Table 7.14 (Briggs and Seah, 1996). Auger analysis is particularly useful for grain boundary analysis and especially in conjunction with FE-SEM. AES can provide point analysis on locations as small as 50 Å, making it ideal for characterizing inclusions, defects, cracks, and so on. The Auger effect is a multielectron interaction (Figure 7.18a) starting with an incident electron that creates a core hole that is filled by an electron from higher level, and imparts transition energy onto a third electron that is

TABLE 7.14

Auger Electron Spectroscopy Capabilities for Materials Characterization

Method Provides	Capable to Analyze	Applicable to
Surface composition	Metals and alloys	Grain boundary analysis
Some chemical bonding state information	Carbon composites	Corrosion and defect analysis
High spatial resolution	Some ceramics	Failure analysis
Line scan and X–Y elemental mapping		
Depth composition profiles		
Secondary electron (SED) images		
EDX (on some instruments)		

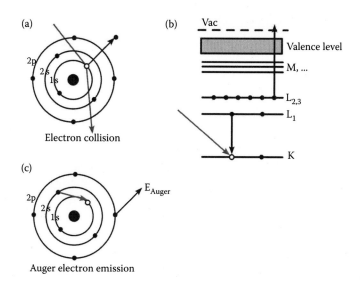

FIGURE 7.18

Atomic and spectroscopic illustrations of the Auger process: (a) Illustration of the steps that are involved with Auger effect. (b) Illustration of the same process in spectroscopic notation (K L_1 $L_{2,3}$). (From Wikimedia Commons, http://en.wikipedia.org/wiki/Auger_electron_spectroscopy.)

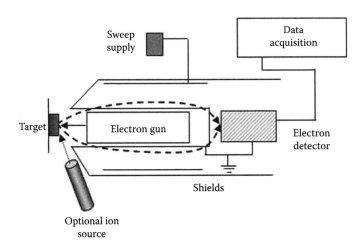

FIGURE 7.19
An Auger electron spectroscopy (AES) setup using cylindrical mirror analyzer.

emitted. An AES instruments generally consist of ionizing energy source in the form of high-voltage electron beam that is focused onto the test sample. The emitted electrons are deflected around the electron gun passing through an aperture to the back of the analyzer and are directed into an electron multiplier (Figure 7.19). Secondary electrons ejected from the surface are measured using a cylindrical mirror analyzer or hemispherical analyzer. The relatively low energy of these secondary electrons limits the information volume to the surface outer 50–100 Å layer. By varying the supplied voltage, Auger data is measured and the depth profile is measured through the use of ion gun.

The Auger effect produces a characteristic signature for each element, which is somewhat dependent on a chemical bonding state. For this reason, AES can provide semiquantitative information about surface composition using reliable standards. The spatial resolution of AES can be as low as 10–25 Å depending on the source, and typically is operated at about 50–100 Å. AES is capable of point, line, and mapping analysis and the speed of AES allows for simultaneous ion etching and analysis to provide depth composition profiles.

7.4.8 Raman Spectroscopy

Raman spectroscopy, which involves analysis of inelastic scattering (also known as Raman scattering) of monochromatic light, is used to study vibrational, rotational, and other low-frequency modes (Smith and Dent, 2005). The radiation source is a laser in the visible, near-infrared, or near-ultraviolet range. The laser beam interacts with molecular vibrations, phonons, or other excitation forms and results in the energy shift of the laser photons, providing information about the tested material. The test is done by illuminating a spot of material and the scattering is optically acquired through a monochromator. Elastic Rayleigh scattering with wavelengths that are close to the laser light is filtered out and the rest of the acquired light is dispersed onto a detector. The weak intensity of the dispersed light that results from inelastic scattering is difficult to separate from the high-intensity Rayleigh scattering. The detectors used to be based on photomultipliers, which required long acquisition times; however, in recent years, significant improvements have been made using Fourier transform (FT) spectroscopy and CCD detectors.

TABLE 7.15

Summary of the Capabilities and Applications of Raman Spectroscopy

Method Provides	Capable to Analyze	Applications
Vibrational, rotational, and low-frequency modes in a system, including molecular and crystalline bonding environments	Organic and amorphous materials, some ceramics and glasses, plastics and powders	Structural determination of nanocarbon materials, some ceramics and glasses, process development

Generally, the Raman effect occurs when light impinges on a molecule and interacts with the electron cloud to create excitation from ground state to a virtual energy state. When the molecule relaxes, it emits a photon and returns to a different rotational or vibrational state. The energy difference between the original state and the new state leads to a frequency shift from the excitation wavelength in the emitted photon. If the final vibrational state of the molecule is more energetic than the initial state, then a Stokes shift to a lower frequency takes place balancing the total energy. If the final vibrational state is less energetic than the initial state, then an anti-Stokes shift occurs and the emitted photon shifts to a higher frequency. For a molecule to exhibit a Raman effect, there is a need to have a change in the molecular polarization potential with respect to the vibrational coordinate. The level of the polarizability change determines the Raman scattering intensity, and the pattern of shifted frequencies is determined by the sample rotational and vibrational states. Raman spectroscopy is widely used to characterize materials, especially carbon, measure temperature, and find the crystallographic orientation of a sample (Gardiner, 1989). It is interesting to note that a Raman analyzer was included in the suite of instruments on the Curiosity Rover that landed on Mars in August 2012. A summary of the capabilities and applications of Raman spectroscopy is given in Table 7.15.

7.4.9 Electron Microprobe

Electron microprobes (EMP) are tools of determining the chemical composition of solid materials in small volumes (typically ≤ 10–$30~\mu m^3$). They are also known as electron microprobe analyzer (EMPA) and EPMA (Reed 1997; Goodhew et al., 2000). Similar to SEMs, they bombard the tested materials by an electron beam and the determined wavelengths of the excited x-rays are used to perform elemental analysis. Improvements in the capability of EMP instruments have led to significant accuracies in measuring minute concentration of trace elements.

7.4.10 Electron Energy Loss Spectroscopy

Electron energy loss spectroscopy (EELS) is a method of analyzing materials by exposing them to an electron beam having a narrow range of kinetic energies (Brydson, 2001). The electrons that undergo inelastic scattering lose energy and have their path deflected slightly and randomly. The energy loss can be measured by an electron spectrometer and analyzed. Inelastic interactions include phonon excitations and the inner-shell ionizations are particularly useful for detecting the elemental components of a tested material. Tests can be performed to determine the types of atoms and the atomic number of each type that is interacted by the beam. The scattering angle can also be measured and provide information about the dispersion relation of the excitation source of inelastic scattering. Generally, the EELS technique is more effective when testing materials with low atomic numbers since the excitation edges tend to be better defined.

7.4.11 Inductively Coupled Plasma Mass Spectrometry

Inductively coupled plasma mass spectrometry (ICP-MS) is a method of spectrometry, with a significant level of sensitivity, that can detect concentrations of metals and several nonmetals at levels of part per trillion (Montaser, 1998). The principle of operation of the ICP-MS method involves inducing coupled plasma onto the test material causing ionization and the ions are separated and quantified by a mass spectrometer. Generally, ICP-MS has much greater analytical precision, sensitivity, and speed than atomic absorption techniques. However, some ions interfere with the detection of others.

7.4.12 Particle-Induced X-Ray Emission

Particle-induced x-ray emission (PIXE) is another technique of elemental analysis of materials. The method is also known as proton-induced x-ray emission (Johansson et al., 1995). It involves radiating a test material by an ion beam, causing atomic interactions that release electromagnetic radiation with wavelengths in the x-ray range of the electromagnetic spectrum. The wavelengths that are emitted are related to the specific elements present in the test material. By focusing the radiating beam onto a 1-μm-diameter spot size, the method has improved significantly, allowing microscopic analysis better known as microPIXE. This method is effective in determining the distribution of trace elements in a wide range of materials.

7.4.13 Atomic Spectroscopy

Optical atomic spectroscopy is an elemental analysis method of identifying and quantifying free atoms in gas phase (Cullen, 2003). The elements that are present in the tested material are converted by atomization to gaseous atoms or elementary ions and the resulting ultraviolet and visible absorption, emission, or fluorescence of atomic species are measured. A summary of the capabilities and applications of Raman spectroscopy is given in Table 7.16.

Generally, there are three types of atomic spectroscopy: atomic absorption (AA), atomic emission (AE), and atomic fluorescence (AF).

AA: In heated gas, atoms absorb radiation with wavelengths that are the characteristic of electron transitions from ground to higher excited states. The related atomic absorption spectrum consists of resonance lines that are the result of this transition.

AE: Heat by a flame, plasma, electric arc, or spark excites atoms to higher orbitals and the excitation is accompanied by emission/radiation of a photon.

TABLE 7.16

Summary of the Capabilities and Applications of Atomic Spectroscopy

Method Provides	Capable to Analyze	Applications
Determination of elemental composition by electromagnetic or atomic or mass spectrum	Trace elements in solution, identification of some molecular species, metal and ceramic particles and some powders	Qualitative and quantitative trace element analysis, testing for impurities and contaminants in aqueous solutions/mixtures

TABLE 7.17

Static Secondary Ion Analysis Capabilities for Materials Characterization

SSIMS Is	SSIMS Information	Advantages/Disadvantages
Surface sensitive	Molecular ions	Very sensitive technique (parts per billion/trillion)
Detects positive ions very well	Inorganic ions	Fragment ions provide species identification
Analyzes fragments of organic molecules	Elemental concentration (% of volume)	Surface sensitive provides contamination analysis
Identifies contamination	Depth concentration profiles	Requires calibration for concentration profiles
Destructive technique since the surface is destroyed by the ion beam	Chemical identification (spectral libraries)	

TABLE 7.18

Summary of the Capabilities and Applications of Static Secondary Ion Analysis

Method Provides	Capable to Analyze	Applications
Molecular fingerprint information—ion masses	Organic films	Semiconductor wafer doping, characterizing thin organic films, and identifying contamination
Qualitative/semiquantitative analysis of some elements	Thin films Silicon wafers/semiconductors	

AF: Irradiation with an intense source having wavelengths that are absorbed by atoms or ions in a flame can cause fluorescence. The observed radiation mostly results from resonance fluorescence due to returning to the ground state from excited states.

7.4.14 Static Secondary Ion Analysis

Static secondary ion analysis (SSIMS) is a very sensitive surface analysis technique (Table 7.17). A summary of its capabilities and applications is given in Table 7.18. It operates using an ion beam, typically argon, which is rastered over a surface to cause fragmenting surface species and ejection into vacuum. The fragmented molecules are analyzed using time-of-flight (TOF) or quadrupole mass analyzers to obtain information about the mass of fragments and their relative intensity. SSIMS is part of an ion microprobe family used to detect trace elements in steel and other alloys; it is much more sensitive than electron microprobe techniques. Depth profiles using SSIMS can be done fairly fast, but commercially its use can be very expensive even with the use of molecular spectral libraries.

7.5 Summary/Conclusions

At elevated temperatures, unusual compounds and quasi-stable phases are formed. The resulting compositions of these compounds and their properties are unpredictable from extrapolation of low-temperature properties. The properties include the characteristics near the melting point, the quasi-stable phases, varying stoichiometry, and the generation of reactive vapor species that sometimes are corrosive. Understanding and determining the material properties require multidisciplinary knowledge and experience from materials science as well as chemical, electrical, and mechanical engineering. The methods of characterization need to account for the processing method that is used at high temperatures, including sintering, annealing, and phase transformation. The fabrication process can be multisteps that require precision control of the temperature, pressure, and processing time. Key properties that need to be characterized include particle/grain size, surface structure, chemical purity, and crystallinity. Generally, the following are the reasons for performing material analysis and the methods that used:

1. Determine the material composition and chemistry
   a. XRF
   b. EDX
   c. XPS
   d. AES
2. Determine the material phase/orientation
   a. Optical microscopy
   b. XRD
   c. SEM
   d. TEM
3. Determine the surface composition
   a. XPS
   b. AES
   c. SSIMS
4. Evaluate the grain boundary composition and chemistry
   a. AES
   b. SEM
   c. EDX
5. Properties that are needed for optimizing processes
   a. Physical properties
      i. Thermal
      ii. Electrical
      iii. Optical
      iv. Mechanical
   b. Chemical composition and chemistry
      i. XPS
      ii. AES
      iii. SEM
6. Comparing two materials (performance)
   a. Strength (to failure)

    b. Ductility (at elevated temperatures)

    c. Thermal limits and performance

    d. Electrical properties (at elevated temperature)

    e. Optical and electromagnetic properties

7. Failure analysis

    a. Optical microscopy

    b. SEM

    c. AES

    d. XPS

Combining image, surface, structural, chemical testing with physical properties measurements allows developing better materials through understanding the fundamental material structure and properties relationships. In particular, materials characterization of surfaces can be useful in understanding the mobility of atoms within a matrix at elevated temperature. Knowledge of the chemistry, kinetics, phase, and physical property changes helps materials engineers in the selection and engineering of materials. Further, grain boundary engineering and characterization help develop high-performance metal alloys, ceramics, and composites that maintain strength at high temperatures. Composite materials in particular require extensive engineering to ensure that they maintain compatibility in interfacial bonding, thermal expansion and, with all the material phases, work at the desired operating temperature. Materials characterization supports process development of composite systems, and analysis of good adhesion as well as failure analysis of poor bonding.

Structural analysis using XRD, TEM, and FE-SEM provides materials engineers with insights into the atomic and chemical bonding of crystals, what structure and chemistries make of phases, and how the macrostructure and physical properties of a material are dependent on micro/nanostructured domains.

Physical properties measurements are as important as materials characterization, and support the development and optimization of the performance of high-temperature materials. Strength, modulus of elasticity, stiffness, thermal expansion, electrical and thermal conductivity, and optical (phonon) properties are measurements made in materials laboratories and play important role in materials selection.

Acknowledgments

Some of the research reported in this chapter was conducted at the Jet Propulsion Laboratory (JPL), California Institute of Technology, under a contract with the National Aeronautics and Space Administration (NASA). Some of the material covered in this chapter is based on the nanomaterials characterization course that is taught at Foothill College, Los Altos Hills, California. The authors express their appreciation to Mark Hetzel, JPL, for allowing to photograph some of the instruments that are shown in this chapter. Also, the authors thank Kip O. Findley, Colorado School of Mines, Golden, Colorado; Vince Crist, Nanolab Technologies, Milpitas, California; and Doreen D. Edwards, North Carolina State University, Raleigh, North Carolina, for reviewing this chapter and for providing valuable technical comments and suggestions.

References

Binnig G., and H. Rohrer, Scanning tunneling microscopy, *IBM Journal of Research and Development*, 30(4), 1986, 355–369.

Binnig G., C.F. Quate, and C. Gerber, Atomic force microscope. *Physical Review Letters*, 56(9), doi:10.1103/Phys Rev Lett. 56.930, 1986, 930–933.

Blake D.F., D. Vaniman, C. Achilles, R. Anderson, D. Bish, T. Bristow, C. Chen, S. Chipera, J. Crisp, D. Des Marais, R.T. Downs, J. Farmer, S. Feldman, M. Fonda, M. Gailhanou, H. Ma, D. Ming, R. Morris, P. Sarrazin, E. Stolper, A. Treiman, and A. Yen, Characterization and calibration of the CheMin mineralogical instrument on Mars science laboratory, *Space Science Review*, 170(1–4), 2012, 341–399.

Briggs D., and M.P. Seah, *Practical Surface Analysis, Auger and X-Ray Photoelectron Spectroscopy*, ISBN-10: 0471953407, ISBN-13: 978-0471953401, John Wiley & Sons, Berlin, Germany, 1996, pp. 1–674.

Brydson R., *Electron Energy Loss Spectroscopy*, ISBN-10: 1859961347, ISBN-13: 978-1859961346, Garland Science, New York, NY, 2001, pp. 1–160.

Cullen M., *Atomic Spectroscopy in Elemental Analysis*, Blackwell, Oxford, UK, ISBN-10: 0849328179, ISBN-13: 978-0849328176, 2003, pp. 1–310.

Cullity B.D., and S.R. Stock, *Elements of X-Ray Diffraction*, 3rd Edition, ISBN-10: 0201610914, ISBN-13: 978-0201610918, Prentice Hall, Upper Saddle River, New Jersey, 2001, pp. 664.

Eaton P., and P. West, *Atomic Force Microscopy*, ISBN-10: 0199570450, ISBN-13: 978-0199570454, Oxford University Press, Cary, NC, 2010, pp. 1–288.

Gardiner, D.J. *Practical Raman Spectroscopy*, Springer-Verlag, Berlin, Germany, ISBN 978-0-387-50254-0, 1989.

Giannuzzi L.A., and F.A. Stevie (Eds.), *Introduction to Focused Ion Beams: Instrumentation, Theory, Techniques and Practice*, ISBN-10: 0387231161, ISBN-13: 978-0387231167 Springer, New York, NY, 2004, pp. 1–376.

Goldstein J., D.E. Newbury, D.C. Joy, C.E. Lyman, P. Echlin, E. Lifshin, L. Sawyer, and J.R. Michael, *Scanning Electron Microscopy and X-Ray Microanalysis*, 3rd Edition, ISBN-10: 0306472929, ISBN-13: 978-0306472923, Springer, New York, NY, 2003, pp. 689.

Goodhew P.J., J. Humphreys, and R. Beanland, *Electron Microscopy and Analysis*, ISBN: 0748409688, ISBN-13: 978-0748409686, 3rd Edition, Taylor & Francis Group; Boca Raton, FL, 2000, pp. 1–254.

Jenkins R., *X-Ray Fluorescence Spectrometry*, 2nd Edition, ISBN-10: 0471299421, ISBN-13: 978–0471299424, Wiley-Interscience, New York, NY, 1999, pp. 1–232.

Jenkins R., R.W. Gould, and D. Gedcke, *Quantitative X-Ray Spectrometry*, ISBN-10: 0824795547, ISBN-13: 978-0824795542, 2nd Edition, Marcel Dekker, New York, NY, 1995, pp. 1–484.

Johansson S.A.E., J.L. Campbell, and K.G. Malmqvist, *Particle-Induced X-Ray Emission Spectrometry (PIXE)*, ISBN-10: 0471589446, ISBN-13: 978-0471589440, Wiley-Interscience, New York, NY, 1995, pp. 1–451.

Lovesey, S.W. *Theory of Neutron Scattering from Condensed Matter; Volume 1: Neutron Scattering*. ISBN 0-19-852015-8, Clarendon Press, Oxford, United Kingdom, 1984.

Minsky M., Microscopy apparatus, US Patent 3,013,467, December 19, 1961.

Montaser A., *Inductively Coupled Plasma Mass Spectrometry*, ISBN-10: 0471186201, ISBN-13: 978-0471186205, Wiley-VCH, Berlin, Germany, 1998, pp. 1–1004.

Murphy D.B., *Fundamentals of Light Microscopy and Electronic Imaging*, ISBN-10: 047125391X, ISBN-13: 978-0471253914, Wiley-Liss, Wilmington, DE, 2001, pp. 1–360.

Pennycook S.J., and P.D. Nellist (Eds.), *Scanning Transmission Electron Microscopy: Imaging and Analysis*, ISBN-10: 1441971998, ISBN-13: 978-1441971999, Springer, New York, NY, 2011, pp. 1–774.

Reed, S.J.B., *Electron Microprobe Analysis and Scanning Electron Microscopy in Geology*, 2nd Edition, Cambridge University Press, New York, NY, 2005.

Price R.L., and W.G. Jerome (Eds.), *Basic Confocal Microscopy*, ISBN-10: 0387781749, ISBN-13: 978-0387781747, Springer, New York, NY, 2011, pp. 1–313.

Reed S.J.B., *Electron Microprobe Analysis*, ISBN-10: 052159944X, ISBN-13: 978-0521599443, 2nd Edition, Cambridge University Press, New York, NY, 1997, pp. 1–350.

Reimer L., *Scanning Electron Microscopy: Physics of Image Formation and Microanalysis*, ISBN-10: 3540639764, ISBN-13: 978-3540639763, Springer Series in Optical Sciences, New York, NY, 1998, pp. 1–541.

Scofield J.H., Hartree-Slater subshell photoionization cross-sections at 1254 and 1487 eV, *Journal of Electron Spectroscopy and Related Phenomena*, 8, 1976, 129–137.

Squires G.L. *Introduction to the Theory of Thermal Neutron Scattering*, 2nd Edition, ISBN 0-486-69447-X, Dover Publications Inc., Mineola, New York, 1996.

Smith E., and G. Dent, *Modern Raman Spectroscopy: A Practical Approach*, ISBN-10: 0471497940, ISBN-13: 978-0471497943, Wiley, New York, NY, 2005, pp. 1–222.

VanHove M.A., W.H. Weinberg, and C.-M. Chan, *Low-Energy Electron Diffraction: Experiment, Theory and Surface Structure Determination*, ISBN-10: 3642827233, ISBN-13: 978-3642827235, Springer Series in Surface Sciences, New York, NY, 2012.

von Bordwehr R.S., A history of the x-ray absorption fine structure, *Annales de Physique*, 14(4) doi:10.1051/anphys:01989001404037700, 1989, 377–465.

Watts J.F., and J. Wolstenholme, *Introduction to Surface Analysis by XPS and AES*, 2nd Edition, ISBN-10: 0470847131, ISBN-13: 978-0470847138, Wiley, New York, NY, 2003, pp. 1–224.

Williams D.B., and C.B. Carter, *Transmission Electron Microscopy: A Textbook for Materials Science* (4-vol set), ISBN-10: 030645324X, ISBN-13: 978-0306453243, Springer, New York, NY, 2004, pp. 1–703.

Internet Resources

Auger effect http://en.wikipedia.org/wiki/Auger_effect
EPMA http://serc.carleton.edu/research_education/geochemsheets/techniques/EPMA.html

8

Nondestructive Evaluation and Health Monitoring of High-Temperature Materials and Structures

Yoseph Bar-Cohen, John D. Lekki, Hyeong Jae Lee, Xiaoqi Bao, Stewart Sherrit, Shyh-Shiuh Lih, Mircea Badescu, Andrew Gyekenyesi, Gary Hunter, Mark Woike, and Grigory Adamovsky

CONTENTS

8.1 Introduction

The detection of defects and the characterization of material properties without affecting the integrity of the material or structure require the use of nondestructive testing (NDT) or nondestructive evaluation (NDE) methods. Increasingly, during service and in the field at real or near real time, nondestructive health monitoring (HM) methods are used to monitor the health of structures or determine various properties. Common NDT and NDE methods include ultrasonic, thermal, magnetic-particle, liquid penetrant, radiographic, visual inspection, eddy-current testing, and laser interferometry. These methods are widely used in industry, mechanical engineering, electrical engineering, civil engineering, systems engineering, aeronautical engineering, medicine, and even in art. Just like monitoring the health of our body, it is essential to monitor and control the material properties and integrity of critical structures at all stages of life, from cradle to retirement. A typical HM system consists of a sensor, data acquisition system, control computer or microprocessor, and software (embedded or otherwise) for analysis. While most NDE methods are applied at room temperatures, the HM methods of high-temperature materials and mechanisms need to be applied at the relevant temperatures.

The utilization of NDE for high-temperature applications require that either the sensing element be capable of withstanding the temperatures of the environment or there be a means in which to remotely monitor from a cooler location. The challenges for building a sensing system that is operational at high temperatures include the ability to develop sensing structures from proper materials as well as addressing the packaging and wiring issues. Significant efforts are made with the packaging when applying sensing systems at high temperatures since differences in coefficient of thermal expansion (CTE) and repeated temperature swings, that is, thermal cycling (can reach up to 1000°C) can cause system failure. It is typical that data acquisition systems are located in cooler environments, but this is now changing with the development of silicon carbide (SiC) and silicon-on-insulator (SOI) electronic technologies. These technologies are now allowing for more data processing and communications capability to be located in locations up to 500°C. These technologies bring the promise of smart sensor capabilities to high-temperature applications that have been until now limited to applications where traditional silicon electronics are possible.

Challenges to operating HM and NDE technologies in a high-temperature environment include material survivability and oxidation, packaging, integration, matching CTE among different material in the sensor system, thermal noise, and sensor fouling/maintainability. Fundamentally, the material that the sensor is constructed from must maintain operability over a significant amount of time in the environment which it is targeted for. In some instances, the sensor may have to routinely be able to survive thermal shocks of greater than 400°C/min. This limits the number of materials that can be combined to form a sensor, as different materials will quickly succumb to high cycle fatigue from the stresses of joined materials with different CTE and the reduced strength of materials at higher temperatures. This issue is one of the fundamental difficulties identified when packaging a sensor for utilization in a high-temperature environment. An example of a sensor that has been developed and packaged for operation up to 500°C is shown in Figure 8.1. The picture shows a packaged SiC high-temperature capacitive pressure sensor for applications in hot sections of aerospace engine environment (Chen et al., 2010). Figure 8.1 also shows a SiC pressure sensor chip attached to a ceramic packaging substrate, which is mounted inside the INCONEL seal gland. The sensor has been tested in the temperature range from room temperature to 500°C. SIC is a material with a wide bandgap that makes it very appropriate for high-temperature environments. This wide bandgap is critical as the elevated electron state in SiC is not flooded by thermally generated electrons, but instead is primarily driven by physical effects. Silicon, which has a much smaller electronic bandgap, would not be operational as a semiconductor because the higher-energy electron band would be filled by thermally energized electrons over 300°C. A ceramic material for the substrate is chosen because of the similarity of the CTE to SiC reducing the stress on the sensing element during thermal swings. The wirebonding of the sensing element is done with materials (gold and platinum) that have a good attachment over high-temperature swings and do not oxidize in the temperature regime of the sensor. All these mechanical considerations must be reviewed to develop durable sensor capability for high-temperature NDE and HM applications.

The capability and the technology readiness of the electronics is a critical part of the ability of the high-temperature system to operate at the required conditions. In an effort to enable smart sensors (sensors with integrated computation, signal processing and communications) for high-temperature HM and NDE applications, SiC and SOI electronics have been developed. For this purpose, SOI have shown capability for application at as high as 300°C, whereas SiC has been demonstrated to operate at temperatures as high as 500°C. While SiC produces stabilized transistors in this temperature regime, it is a challenge to integrate and package the related electronics into a deployable package. An example of the high-temperature packaging issue comes from the continuing development of high-temperature SiC electronics. A set of junction gate field-effect transistor (JFET) integrated circuits (ICs) that included a differential amplifier and logic gates was fabricated and run through endurance testing in a 500°C oven (Neudeck et al., 2009). A micrograph of a portion of one

FIGURE 8.1
Picture of high-temperature pressure sensor capable of operation up to 500°C. (Courtesy of Ohio Aerospace Institute and NASA GRC.)

As-fabricated chip   After prolonged 500°C test

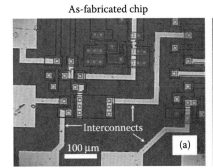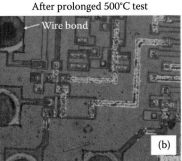

FIGURE 8.2
(See color insert.) Optical micrographs of portions of NOT gate IC chips. (a) An As-fabricated chip prior to packaging. (b) A packaged chip following failure after thousands of hours operating at 500°C. (From Lee C., S. Akbar, and C. Park, Potentiometric CO_2 gas sensor with lithium ion electrolyte. *Processing and Fabrication of Advanced Materials XI, Proceedings of the International Symposium on Processing and Fabrication of Advanced Materials*, ASM International, Materials Park, OH, 7–10 Oct. 2002, pp. 63–77.)

of the ICs before and after testing is shown in Figure 8.2. The results of this testing point to the successes and the challenges associated with developing high-temperature smart sensors for HM and NDE. The differential amplifier IC was run at 500°C and it lasted for 3900 h before a sudden failure occurred. The observed circuit failures were due to degradation of the metal/dielectric interconnect stack, but no failures in the wire bonding and packaging of the electronic system were identified. These tests demonstrated remarkably high-temperature endurance and stability necessary for a high-temperature smart sensor system.

In this chapter, NDE and HM methods are covered in relation to NDE of ceramic matrix composite (CMC), the cure monitoring of fiber-reinforced polymer matrix composites, as well as the HM of aircraft engines and steam pipes. Specifically, in the next section, recent advances in the development of NDE and HM techniques are presented with particular focus on the HM of ceramic and polymer matrix composites. Methods of HM of aircraft engines are then addressed in detail, and the current activities and future plans for the various propulsion sensor technologies in a high-temperature environment are discussed. Lastly, HM methods of steam pipe at high temperatures are presented, with emphasis on the ultrasonic pulse-echo method. Available high-temperature transducer materials and methods for analyzing received signals from steam pipe are also given in this chapter.

8.2 Nondestructive Evaluation of Composites

8.2.1 Ceramic Matrix Composites

As a follow-up to an earlier chapter describing CMC, this section reviews various NDE techniques that have been successfully utilized for assessing the damage state of woven CMC, in particular, consisting of silicon carbide fibers and silicon carbide matrices (SiC/SiC). The NDE techniques include acousto-ultrasonics (AU), modal acoustic emissions (AE), electrical resistance (ER), impedance-based structural health monitoring (IBSHM), pulse thermography (PT), and thermoelastic stress analysis (TSA). The observed damage within the composites, mostly in the form of distributed matrix cracks, was introduced using multiple experimental tactics. These included load/unload/reload uniaxial tensile tests, creep tests, and ballistic impact. Although other NDE techniques have been applied to this material system, the select NDE tools described here are limited to approaches that are of current research interest within programs at the NASA Glenn Research Center in Cleveland, Ohio.

Research is ongoing concerning the characterization of these CMCs subjected to various mechanical load and environmental scenarios in order to build a database of basic material properties and life profiles. In addition, the failure process is being closely studied to allow for better understanding and predictability (e.g., providing input for both empirical and physics-based modeling efforts), so as to build confidence and attain wide acceptance by the engineering community. As a result, there is a need for reliable NDE techniques for both laboratory-based characterization studies as well as tools/procedures for damage monitoring during service (i.e., structural health monitoring–SHM). The use of *in situ* SHM is encouraged in order to reduce the probability of catastrophic failure during service, which is a result of the statistical and brittle nature of the ceramic constituents. It is also useful for monitoring general damage progression allowing for more accurate state awareness, diagnostics, and prognostics. It should be noted that even noncritical cracks are a concern because they provide paths for interior oxidation. Early detection of this damage and the associated property changes is essential for structural applications. Accurately detecting the damage (either *in situ* or by inspecting the component during downtime) will improve vehicle safety by providing a warning to the operator when the material is damaged/compromised beyond a predefined safe limit.

Conventional NDE techniques, such as x-ray and immersion ultrasonics, have low sensitivities to the small-scale, distributed damage seen in these materials. Overall measurement/system noise (due to the complexities of the constituents as well as preexisting flaws) masks much of the new damage. In addition, the through-thickness nature of these techniques requires observations to occur parallel to the matrix crack surfaces and perpendicular to the loading direction. Furthermore, these conventional methods typically require the component to be removed from service for inspection causing excessive downtime. NDE approaches that are currently being studied at the NASA Glenn Research Center (Cleveland, Ohio, USA) include AU, modal AE, ER, IBSHM, PT, as well as TSA. The application domain ranges from a laboratory setting to the potential of service-based, *in situ* SHM. Summary descriptions are supplied for each approach. Owing to limited space and the number of NDE techniques, the portrayals are kept short and absent of specifics (e.g., equations and experimental details). The reader is encouraged to review the related references for in-depth information about the individual techniques.

8.2.1.1 Acousto-Ultrasonics

The AU method (Gyekenyesi et al., 2006) utilizes a two-transducer setup (i.e., a sender and a receiver) in order to monitor the material response to a broadband ultrasonic pulse. The ultrasonic pulse is allowed to distribute itself diffusely into the specimen. For thin specimens, multiple plate wave modes are excited. Various empirical parameters associated with the received signal are used to study the before-and-after behavior of the material of interest. The AU parameters found to correlate with distributed damage in SiC/SiC composites include the diffusion field decay rate, the mean square value (MSV) of the power spectrum, and the centroid of the power spectrum (Roth et al., 2003; Gyekenyesi et al., 2006). In-depth explanations of each of the terms are given in Ivey and Grant (2002), Roth et al. (2003), and Gyekenyesi et al. (2006) and these publications include the associated equations and the detailed steps for obtaining the values. An issue of concern that needs mentioning is the interpretation of the response related to cases where the damage or delamination takes place in layered materials, such as composites. Specifically, if the delamination occurs near the top layer of the structure on which the transducers are placed, there is a potential of false alarm as far as the severity of the effect of the damage. This delamination may not significantly affect the overall strength of the structure but may generate significant AU indications. This behavior is more of a concern for polymer composites rather than for the SiC/SiC composites described here.

Typically, the AU approach is implemented by monitoring a specimen's gage section behavior using two fixed transducers (*in situ* or interrupted). Recently, a scanning system was developed at the NASA Glenn Research Center that allows for the movement of the transducer pair, thereby, producing planar images representing localized AU behavior (Roth et al., 2002, 2003). For the fixed transducer case, AU was used to monitor damage in 0/90 woven SiC/SiC composites during load/unload/reload, uniaxial tensile tests at room temperature (Gyekenyesi et al., 2006). Figure 8.3a shows a schematic of the specimen setup as well as plots of the AU parameters versus the peak load achieved during a given

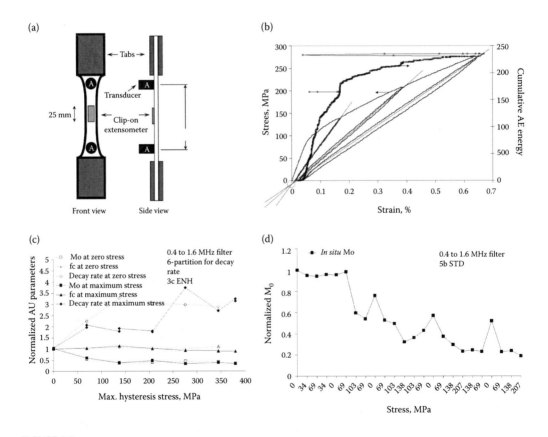

FIGURE 8.3
(a) Schematic of tensile test specimen and AU sensor alignment, (b) hysteresis stress–strain curves (also shown is gage section accumulated AE energy), (c) normalized AU parameters as a function of maximum hysteresis stress (AU captured at both the maximum stress and after unload), and (d) normalized MSV as a function of *in situ* stress with measurements taken at numerous points during the loads and unloads. (From Gyekenyesi A., G. Morscher, and L. Cosgriff, *Composites Part B*, 37B(1), 2006.)

hysteresis loop. Figure 8.3b illustrates the recurrent load/unload/reload pattern employed in many of the damage monitoring studies at NASA (the upper curve is the accumulated energy attained from AE, presented in the next section, while the lower curves are the load/unload, stress/strain hysteresis loops). The early change in the AU parameters (i.e., below 100 MPa) was correlated to initiation and accumulation of transverse matrix cracks and the associated stiffness loss (Figure 8.3c). The transverse cracks constitute the majority of the damage in these types of tests. Final specimen failure for these materials is dictated by fiber bundle fractures. In addition, the AU response was shown to be dependent on the stress state of the composites. This stress-dependent behavior was observed while unloading the specimens from the maximum stress, thereby maintaining a constant damage state (shown in Figure 8.3d, where the MSV parameter is displayed as a function of load).

The scanning approach, which provides two-dimensional gray-scale images of the specimen, was successfully applied by Roth et al. (2003) to predict the failure location in a woven carbon fiber/SiC matrix composite that was subjected to creep conditions (the AU scans were conducted during predetermined interruption points). Similarly, the scanning approach was also applied to creep-tested woven SiC/SiC specimens, again, showing a correlation between changes in the AU parameters and damage (Smith and Gyekenyesi, 2011). Figure 8.4 presents a schematic of the specimen, overlaid with the scan pattern, as

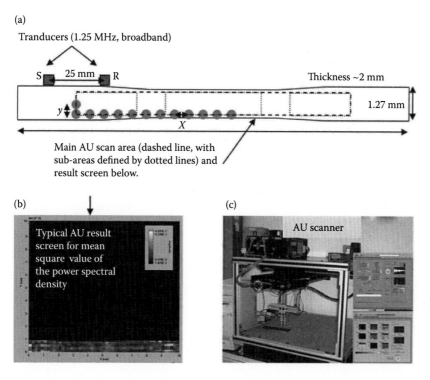

FIGURE 8.4
(a) Schematic of scan pattern, (b) typical two-dimensional, gray-scale image of scan area (MSV shown here), and (c) NASA developed AU scan system hardware. (From Roth D. et al., High frequency ultrasonic guided wave scan system for characterization of C/SiC composite creep damage, *Proceedings of Ceramamic Engineering and Science*, 2003; Smith C. and A. Gyekenyesi, Detecting cracks in ceramic matrix composites by electrical resistance, *Proceedings of SPIE: Nondestructive Characterization for Composite Materials*, Aerospace Engineering, Civil Infrastructure, and Homeland Security V, San Diego, CA, February 26–March 2, 2011.)

well as images of a typical scan data (e.g., MSV) and the scanner hardware. After analyzing the AU parameters for both the fixed sensor and scanning arrangements, the overall sensitivity of the AU technique to material change or damage was quantified and shown to correlate well with the observed damage mechanisms in SiC/SiC composites.

8.2.1.2 Modal Acoustic Emissions

Acoustic waves or AE are created when cracks are initiated or propagated due to the energy released during the creation of new crack surfaces. AE has been effectively employed to monitor and measure the degree of damage in CMC. For example, AE has proven to be an excellent technique for determining when the first matrix cracks appear, for monitoring the overall accumulation of damage during tensile tests, as well as correlating the individual AE signals to specific damage events (Morscher, 1999; Morscher and Gyekenyesi, 2002).

The captured AE data are useful for material characterization as well as SHM. The accumulated AE energy, as seen in Figure 8.3b, is well correlated to distributed transverse cracks. This includes identifying initiation, growth rate, and finally the saturation point. Furthermore, the individual AE events are utilized to assess mechanical properties. The acoustic wave speeds are slower with the accumulation of damage in a composite since the stiffness of the specimen decreases with increasing damage (Morscher, 1999).

8.2.1.3 Electrical Resistance Monitoring

Electrical resistance monitoring is a method that shows promise toward overcoming many of the deficiencies seen in other techniques. In principle, as transverse matrix cracks propagate perpendicular to the loading direction, the current-carrying capability in the longitudinal direction diminishes. The resistance can also be monitored *in situ*, eliminating the need for long downtimes due to the relative simplicity of the setup. For these reasons, correlating electrical resistance changes with damage could prove to be advantageous for these materials. Work with SiC/SiC composites has shown a direct relationship between matrix crack density and resistance change for samples subjected to either monotonic tensile tests at room temperature (Smith et al., 2008) or creep conditions (Smith et al., 2011).

In a recent study, again employing the load/unload/reload room temperature tensile tests for inducing damage, the electrical resistivity was shown to be very sensitive to transverse cracks (Smith and Gyekenyesi, 2011). At the ultimate stress, the resistance had increased by over 500%. Permanent changes in resistance were also detectable after the load was removed (even with assumed crack closure). After polishing and examining the microstructure, a parabolic relationship between resistance and crack density was observed. This relationship can be used in future HM efforts to estimate the crack density *in situ*. Current efforts include relating the actual damage to changes in resistivity by employing circuit models.

8.2.1.4 Impedance-Based Structural Health Monitoring

Impedance-based SHM uses piezoelectric patches (e.g., PZT: lead, zirconate, titanate) that are bonded onto or embedded in a structure. Each individual patch behaves as both an actuator of the surrounding structural area as well as a sensor of the structural response. The size of the excited area varies with excitation frequency, the geometry, and material composition of the structure. For a PZT patch intimately bonded to a structure, driving the patch with a sinusoidal voltage sweep, for example, deforms and vibrates the structure. In

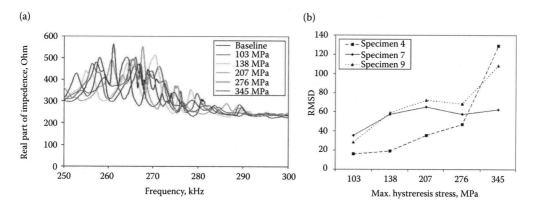

FIGURE 8.5
(See color insert.) (a) Electrical impedance response after each cycle of a load/unload/reload tensile test. (b) Root mean square deviation value for three specimens as a function of maximum stress achieved during given cycle showing sensitivity to transverse cracks accumulated during cycle. (From Gyekenyesi A. et al., *The Journal of Intelligent Material Systems and Structures*, 20(7), 2009.)

reaction to these elastic wave inputs, the structure produces a localized dynamic response. This dynamic response is transferred back to the PZT patch, which sequentially produces an electrical response that is analyzed in regard to the impedance behavior (Gyekenyesi et al., 2009). An empirical relationship is then formulated that allows for monitoring the damage state of the structure as a function of the progressive changes in the electrical impedance response of the patch throughout a structure's life.

Recently, the approach was applied to SiC/SiC composites in order to quantify the initiation and accumulation of damage during room temperature tensile tests (Gyekenyesi et al., 2009). Figure 8.5a shows the impedance curves collected after each cycle of a load/unload/reload tensile test for a single specimen. The impedance values were then compared to the baseline taken prior to damage-inducing tests using the root mean square deviation value as seen in Figure 8.5b for multiple specimens. The technique captured the initiation of transverse cracks as well as other events near failure, such as interlaminar delaminations and fiber bundle breaks. This technique also lends itself to *in situ* monitoring due to research efforts aimed at developing the electronics for low-power, autonomous sensor networks (Owen et al., 2011).

8.2.1.5 Pulsed Thermography Technique

The pulse thermography technique is based on introducing a pulse of thermal energy on a given surface of a specimen and viewing the time-based thermal response as the heat conducts into the material. Typically, the thermal energy is introduced via photographic flash lamps and the 2-D image is monitored using an infrared (IR) camera focused on the same side (note that two-sided inspections are also used although the one-sided approach may reduce the need for access and/or part disassembly). As the thermal front progresses, there is a reduction of the surface temperature. In a flawless sample, away from any geometric boundaries, the surface cools in a uniform fashion. Deviations occur when subsurface defects are present as a result of the change in material properties (e.g., thermal conductivity, density, or heat capacity). This resistance in the conductive path causes a different cooling rate at the surface directly above the defect, when compared with the surrounding defect-free material. The change in the subsurface conduction is seen as nonuniform surface temperature profile that is a function of time. Unprocessed, as-received data are typically displayed sequentially on a computer monitor in a movie fashion. Images are then visually inspected for signs of a subsurface defect by locating areas with anomalous surface temperatures (Martin and Gyekenyesi, 2002; Martin et al., 2003). In addition to simply viewing a movie of the raw image sequence, approaches exist that allow for the display of a single image aiming to isolate and locate flaws (e.g., peak contrast, peak slope, and thermal diffusivity are values captured in a single image). Further improvements are gained by representing the transient behavior at each individual pixel by a least square fit of a low-order polynomial to the logarithmic time history (Shepard, 2001). The resulting synthetic image provides increased spatial and temporal resolution, and significantly extends the range of defect depths and sample configurations to which pulsed thermography can be applied.

Recently, impacted SiC/SiC specimens were assessed using pulsed thermography. The one-sided setup was used to first capture the front side behavior (i.e., impact side), and then a second set of images was attained while viewing the backside (i.e., the exit side of the projectile). The pulsed thermography data were used in conjunction with optical assessments, modal AE attained during postimpact tensile tests, as well as electrical resistance behavior in order to predict residual tensile strength after impact (Morscher et al., 2013). Positive relationships were developed between the damage areas obtained via pulsed thermography and the observances of other NDE techniques. Furthermore, the postimpact tensile strengths showed good correlation with the pulsed thermography defined damage areas. Lastly, the data were utilized for the development of a circuit model related to the electrical resistance NDE approach described earlier.

8.2.1.6 Thermoelastic Stress Analysis

TSA is a full-field, noncontacting technique for surface stress mapping of structures. TSA is based on the fact that materials experience a temperature change when compressed or expanded (i.e., experience a change in volume). If the load causing the volumetric change is removed and the material returns to its original temperature and shape, the process is deemed reversible. This reversibility is achieved when a material is loaded elastically at a high enough rate so as to eliminate significant conduction of heat (Mackin and Roberts, 2000; Gyekenyesi and Morscher, 2010). An analytical relationship exists between the change in temperature and the change in stress. The observed cyclic change in temperature is the product of the absolute temperature, thermoelastic constant, and the cyclic change in the sum of the principal stresses. Utilizing this relationship, the method allows for a two-dimensional stress map of a loaded structure with spatial variations in the stress field showing up as spatial variations in the temperature profile.

TSA was used to study the stress fields adjacent to machined notch roots during tensile tests of woven 0/90 SiC/SiC composites in order to capture damage-induced stress relief (Gyekenyesi and Morscher, 2010). Figure 8.6 displays normalized TSA images for a SiC/SiC specimen with increasing notch lengths together with line scans representing the normalized stress between the notches. The technique showed that the stress concentration factor (SCF) at the notch roots increased with increasing notch lengths as expected. Damage-induced stress relief was not evident in these composites. In fact, the redistribution of stresses onto isolated fiber tows adjacent to the notch root caused an increase in the apparent SCF.

Multiple NDE techniques, as applied to 0/90 woven SiC/SiC composites, were described in the previous sections. This particular composite system presents multiple challenges when it comes to NDE due to its complex structure and damage behavior. To gain

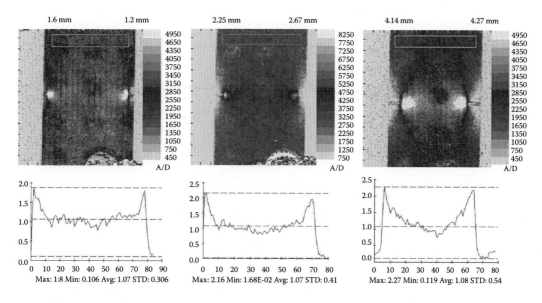

FIGURE 8.6
(See color insert.) TSA images and SCFs for SiC/SiC specimen as a function of increasing notch lengths. Notch lengths, a, are indicated above the image. TSA stress range and mean stress were 35 and 17.5 MPa, respectively. The multishade specimen images correspond to the dimensionless digital values of the TSA system while the plots represent the SCF (i.e., local stress divided by the net stress) versus pixel position along a line scan between the two notch roots. (From Gyekenyesi A. and G. Morscher, *Journal of Materials Engineering and Performance*, 19(9), 2010.)

acceptance from the design community, it is essential that reliable tools for assessing the damage state of the material be available. Hence, considerable effort is being focused on down selecting appropriate characterization tools for use in various scenarios from a laboratory setting, instruments for benchtop NDE of actual components, and lastly, capabilities for *in situ* SHM during operation.

8.2.2 Monitoring Fiber-Reinforced Polymer Matrix Composites Curing

Real-time monitoring of the cure of fiber-reinforced polymer matrix composite materials is essential for the control of their manufacturing process (Ciriscioli and Springer, 1990). The sensing for monitoring of the cure relies on measuring the physical or chemical properties that are changing during the transformation from the resin liquid state into the cured rigid form (William et al., 1985). There are many factors that affect the cure and they include the conditions under which the part is being laid up, the rate and variations in the ambient temperature, and the age of the prepreg or resin. Curing is the only time at which one can take action to possibly eliminate formation of defects as well as affect and prevent the cause of unacceptable characteristics (undercure or overcure). Real-time monitoring the cure process allows for minimizing trapped volatiles, alert of vacuum leak, prevent over-bleeding, provide information about the cure rate, and possibly help optimize the material properties.

Proper understanding of the cure process allows for establishing adequate cure model and sensors to monitor the process of new composite materials. Generally, the cure of graphite/epoxy consists of three major phases (Figure 8.7). In the first phase of the cure process, the resin behaves as a viscous liquid and at this stage it can be inflicted with defects such as voids, porosity, and inadequate fiber/resin ratio.

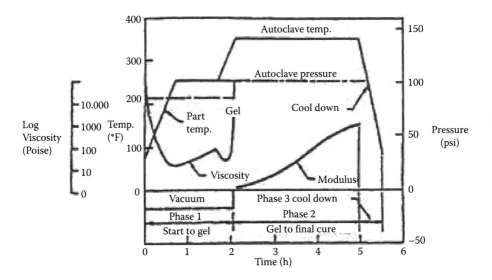

FIGURE 8.7
A typical cure cycle for graphite/epoxy laminate. (From Bar-Cohen Y., A. Chatterjee, and M. West, Sensors for cure monitoring of composite materials, *Review of Progress in Quantitative NDE*, 12A, D. O. Thompson, and D. E. Chimenti (Eds.), Plenum Press, New York, 1993, Vol. 12B, pp. 1039–1046.)

The cure cycle shown in Figure 8.7 represents a bleed-type process using bleeder cloth layer, where the resin bleeds through the laminate thickness and edges. In contrast, the no-bleeder cure does not involve the use of bleeder layers and bleeding takes place along the composite edges only. To assist in removing foreign volatiles, a vacuum is maintained on the part, whereas to prevent formation of porosity and voids, autoclave pressure is maintained on the part. During this period of heating and temperature hold, the epoxy reaches its minimum viscosity and excess resin flows into the bleeder cloth. After this period, the vacuum is removed and pressure is maintained on the part while it is in the autoclave. Then, as the curing continues, the viscosity increases. At the end of the hold period, temperature is raised again and during this resumed increase in temperature, the viscosity of the resin decreases again. If the bleeder cloth has not become saturated, the resin will continue to flow into the bleeder cloth and after passing through the second minimum, the viscosity increases dramatically, leading to gelation. The time and temperature of gelation depends on the resin chemistry, resin advancement, and the rate of temperature increase. Once gelation commences, the resin cross-links rapidly, building the physical properties that are useful in high-performance aerospace parts. After cure, the part can be fragile due to residual thermoelastic stresses, and postcure needs to be done. Postcuring takes place in an oven and it leads to an additional cross-linking and therefore more strength and environmental resistance of the laminate.

As a result of data accumulation and progress in modeling of the curing process, the field has matured significantly and many advances were made in understanding and controlling the process. Data have been accumulated to allow real-time monitoring that correlate the actual physical and/or chemical state of the resin and determine when full cross-linking has occurred for the specific part and when the cure has ended. Thus, producers of composite parts are able to shorten the cure cycle time to the minimum and fabricate parts with more predictable mechanical and performance properties. Controlling the cure process with adequate feedback requires capable sensors that provide real-time information about the conditions of the material. Several sensing methods of cure monitoring were reported and they are described herein.

8.2.2.1 Dielectric Measurement

The most widely and longest in used is the dielectric measurement method (Day, 1988; Partridge and Maistros, 2001). Such sensors measure the conductivity of ions and changes in the dielectric properties of the curing resin and they are associated to the resin viscosity and the degree of cure. As cross-linking proceeds during cure, the resin viscosity increases and the ion migration rate is reduced. The sensor consists of two electrodes with a small gap between them that is filled with resin and they are typically embedded in the uncured matrix. The sensor may also be configured as interdigitated combs in which two comb-shaped electrodes are interleaved as a pair of teeth. These sensors are either embedded in the layup tool or inside the produced part. Depending on the design, some of these sensors may be reusable.

As opposed to dielectric measurement, which is done at lower frequencies, one can monitor pairs of dipoles, which are polar molecules that rotate against each other. These dipoles are part of the resin itself and they dominate the resin's electrical response at higher frequencies. The use of a time-domain reflectometry (TDR) system to perform such measurements provides quantitative monitoring of the percent of cure at any given time measuring the dipole relaxation. For this purpose, relatively small electrodes are used that are tuned to frequencies in the microwave band (10 MHz to 10 GHz) and they are driven with very short voltage pulses (broadband frequencies).

8.2.2.2 Fiber Optics and Fluorescence Method

The combination of optical fibers and an interferometric system can be used to monitor the pressure and temperature inside a curing laminate (Wood et al., 1989; Sathiyakumar, 2006). With the aid of spectroscopy methods, such fibers can gage the changes in the resin fluorescent emission (Levy and Schwab, 1989), which is related to the resin viscosity and the degree of cure. The sensor uses optical fibers to transmit spectral radiation and stimulate the resin characteristic fluorescence emission. Optical fiber sensors can be mounted onto the curing tool and operate effectively beyond the resin-gelation stage. Also, such sensors can be used to transmit IR signals through a small portion of resin within the layup. The absorbed and released energy provides a gauge of the degree of cross-linking. Methods such as the Fourier transform IR (FTIR), Raman, and fiber Bragg grating (FBG) were reported as effective fiber-optic sensing tools (Sathiyakumar, 2006). The use of fiber optics provides an excellent gaging capability; however, they are delicate and the test systems are expensive.

8.2.2.3 Ultrasonic Method

The capability of ultrasonic waves to indicate changes in properties and presence of defects attracted researchers to explore capabilities to monitor the cure of composite materials (Huston, 1983; Stubbs and Dutton, 1996). Generally, ultrasonic cure monitoring is based on measuring the time-of-flight (TOF), in either through-transmission or pulse-echo modes. The methods are based on monitoring relationships between changes in the characteristics of propagating ultrasonic waves and the real-time mechanical properties of a cured part. In addition, acoustic velocity measurements are done using impact or laser-induced surface wave excitation in order to obtain quantitative data.

Ultrasonic information can be obtained in real time with tool-mount probes. However, it is difficult to relate specific material changes to the sensor output. For example, the TOF of an acoustic pulse traveling through a laminate is affected by changes in both the thickness and properties. Both parameters are changing simultaneously during the cure of composites and are causing a similar effect on the TOF. Also, the effects of the high temperature on the ultrasonic probe response and the requirement for acoustic coupling are hampering a greater use of ultrasonic sensors.

8.2.2.4 Thermography and Heat Flux Monitoring

Monitoring the exothermic reaction in the polymer matrix provides information that can be used to monitor cure providing thermo-kinetic information about cross-linking of thermoset resins. This method is known as the heat-flux monitoring and it measures the amount of thermal energy per unit time that is exchanged between the tool and the material being cured. The input heat to the tool and the heat produced by the exothermic reaction are measured and analyzed to determine the completion of the cure process, which is indicated by the stabilization of the thermal exchange. The sensor can be located underneath the surface of the fabrication tool and does not require direct contact with the resin and the system automatically stops the cure cycle.

8.2.2.5 Eddy-Current Measurement

Eddy current probes can be made to operate at high temperatures and perform health and process monitoring as well as various nondestructive tests (Kasuya et al., 2004; Hirsch et al., 2003).

Eddy-current monitoring of cure by measuring the change in thickness provides a test method that is not affected by the type of the tested material. The measurement is made using the liftoff effect of eddy currents (Bar-Cohen et al., 1991, 1993). A schematic view of the test system is illustrated in Figure 8.8, where the sensor is placed above the sample and a cable is connected through the wall of the autoclave to the eddy-current instrument and to a computer. The computer uses software that acquires data simultaneously from

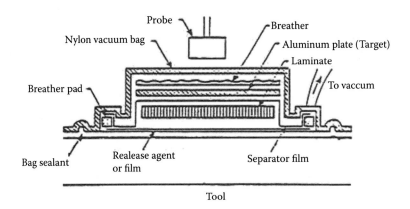

FIGURE 8.8
A schematic description of the curing bagging and the proximity sensor position. (From Bar-Cohen Y., K. H. Nguyen, and R. Botsco, Eddy currents monitor composites cure, *Advanced Materials & Processes, ASM International*, 139(4), 1991, 41–44.)

the multiple sensors of the system and displays the results on the monitor. Eddy currents are very sensitive to liftoff variations in the proximity of the eddy-current probe and the surface of a conductive material. This sensitivity to the proximity forms the basis for the thickness monitoring method. Besides measuring the change in liftoff, the instrument measures the temperature using a thermocouple. The computer controls a multiplexer, which sends both temperature and liftoff data from the individual probes and thermo-couples to a bidirectional port, which is interfaced to the computer.

Various digital signal-processing techniques and analog filtering were used to enhance the liftoff data and the instrument reading was linearized in the firmware to provide prox-imity values over a range of 0.25–0.5 mm. The digital resolution of the instrument was set to ±3 µm over the reading range. A high-temperature probe was developed to operate in the autoclave at temperatures from ambient to 204°C. To allow room for the vacuum bag above the target conductive plate, an initial gap of approximately 50 µm is used between the probe and test surface/target. An aluminum caul plate (6.4 mm thick) was used as an eddy-current target for the proximity measurements. Generally, the use of metal target simplifies the calibration of the equipment because a single setting can be used regard-less of the type of the fiber reinforcement (e.g., graphite, glass, or aramide). Also, the metal targets shield interactions with the graphite/epoxy. The measurement of the distance from such target is preferred over the direct measurement from the graphite/epoxy surface since the analysis of the eddy-current interaction with graphite/epoxy is more complex.

A rigid probe-support is attached to the tooling to allow measuring small changes in probe-to-target distance. This arrangement provides the required reference point so that, essentially, only changes in composite thickness (target movement) are detected by the probe. Errors in measurement can result from the differences in thermal expansion coef-ficients in the cure setup over the cure cycle temperature range from room temperature. The changes in electrical resistivity of the probe and target over this temperature range also contribute to measurement errors. These sources of error are minimized by determin-ing an overall system thermal expansion coefficient during an autoclave cure without a composite in the tool. The correction factor as a function of temperature is stored in the computer and automatically compensated for proximity measurement errors during the composite cure cycle.

An example of a monitored cure using four sensors for a unidirectional graphite/epoxy is shown in Figure 8.9. The agreement between the predicted final thickness and the system reading has been 3%. The first major minimum of the temporal gradient curve results from the compaction and out-gassing of trapped gasses and volatiles in the laminate. The second minimum is the result of resin flow including the flow into the bleeder cloth. Compaction and outgassing lead to a relatively small but fast reduction in laminate thickness that lasts from 5 to 10 min. Resin flow continues until gelation is reached after which there is relatively little change in thickness. Generally, a slight increase in laminate thickness is observed dur-ing the postgelation stages, which are attributed to polymer cross-linking.

8.3 Health Monitoring of Aircraft Engines

An illustration of some of the factors pushing NDE and HM utilization for high-temper-ature applications comes from aerospace. In aviation, there has been a significant push to transition from a Safe Life maintenance philosophy, where components are retired or

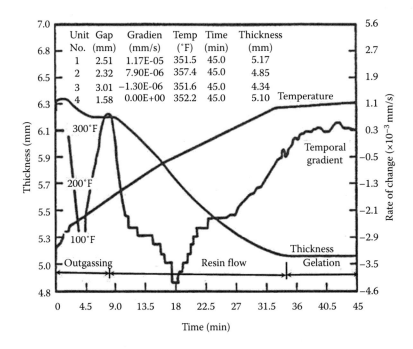

FIGURE 8.9
The average data of cure monitoring unidirectional graphite/epoxy laminate via four eddy-current sensors. (From Bar-Cohen Y., A. Chatterjee, and M. West, *Review of Progress in Quantitative NDE*, 12A, D. O. Thompson, and D. E. Chimenti (Eds.), Plenum Press, New York, 1993, Vol. 12B, pp. 1039–1046.)

changed out at a specific time whether they are damaged or not, to a retirement-for-cause philosophy where components are only changed out when a problem is discovered.

The focus of engine HM is to identify the presence of component- or system-level degradation that is a potential issue for safe operation. The technologies being researched under the aircraft engine health monitoring umbrella include multiple sensors for the identification and assessment of degradation throughout the aircraft engine. Some of the turbine engine technologies being developed for online HM include

- Propulsion health management
- Microwave blade tip clearance sensor (900°C)
- High-temperature fiber optic temperature sensors (1000°C)
- High-temperature smart sensors (500°C)
  - Pressure and emission sensors
  - Electronics for smart sensors—local processing
  - Wireless communication
- Engine emissions monitoring (high-temperature electronic nose)

In the following sections, general descriptions, current activities, and future plans for each of the propulsion sensor technologies are provided. An effort is made to provide a description of relevance as well as evaluation of sensor technique in a high-temperature environment.

8.3.1 Microwave Blade Tip Clearance/Tip Timing Sensor

The development of SHM schemes in turbine engines requires sensors that are highly accurate and can operate in a high-temperature environment. Microwave sensor technology is being investigated as a means of making noncontact structural health measurements in the hot sections of gas turbine engines. This type of sensor is beneficial in that it is accurate, it has the ability to operate at extremely high temperatures, and is unaffected by contaminants that are present in turbine engines. It is specifically being targeted for use in the high-pressure turbine (HPT) and high-pressure compressor (HPC) sections to monitor the structural health of the rotating components. It is intended to use blade tip clearance to monitor blade growth and wear, as well as blade tip timing to monitor blade vibration and deflection.

NASA has worked with Radatec (now Meggitt Inc.) through the Small Business Innovation Research Program (SBIR) for the development of microwave sensor technology for high-temperature noncontact blade tip clearance and blade tip timing measurements. The microwave blade tip clearance sensor operates essentially as a field disturbance sensor. The tip clearance probe is both a transmitting and receiving antenna. The sensor emits a continuous microwave signal and measures the signal that is reflected off a rotating blade. The sensor measures the changes in the microwave field due to the blade passing through the field. The motion of the blade phase modulates the reflected signal and this reflected signal is compared to an internal reference. Changes in amplitude and phase directly correspond to the distance to the blade. The time interval of when the blade passes through the field is measured to provide blade tip timing. (More detail on the operation of these sensors can be found in Geisheimer et al., 2004; Holst et al., 2005; Holst, 2005.) The microwave blade tip clearance probes are made of high-temperature material and are designed to operate at temperatures up to 900°C. The first-generation probes shown in Figure 8.10 operate at 5.8 GHz and can measure clearance distances up to ~25 mm (i.e., one-half the radiating wavelength). The physically smaller second-generation probes operate at 24 GHz and in theory can measure clearance distances up to ~6 mm. The 5.8 GHz sensor is targeted for use on large rotating machinery such as land-based power turbines or in the fan sections of aero gas turbine engines. The 24 GHz sensor is being targeted for use in smaller rotating machinery applications such

FIGURE 8.10
First-generation microwave blade tip clearance/timing sensor tested to 2200°F in GRC high-pressure burner rig surviving three heat/cool cycles.

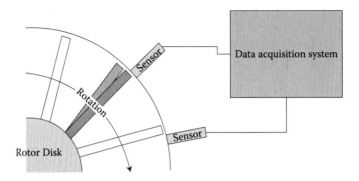

FIGURE 8.11
Illustration of how a pair of tip timing sensors would be utilized to detect the vibration frequency of a turbine rotor blade. As the blade (illustrated as a rectangle) passes each sensor, the time of arrival is recorded. Variations in between the time of arrival between the two sensors give blade vibration deflection and frequency information.

as the turbine and compressor sections of aero engines. This technology has an ultimate goal of obtaining clearance measurement accuracies approaching ±25 μm. A frequency response of up to 5 MHz is typical, with up to 25 MHz being possible with this technology. This frequency response enables the sensor to being used to make blade tip timing measurements, hence blade vibration and deflection measurements for SHM applications as shown in Figure 8.11.

Recent experimentation has shown that the microwave blade tip clearance sensor technology is a viable option for propulsion HM applications (Bencic et al., 2008; Woike et al., 2009, 2010). Both the first- and second-generation sensors have been successfully demonstrated on rotating machinery and other aero engine hardware. The first-generation, 5.8 GHz sensors were tested in a high-pressure burner facility to assess their survivability at relevant combustion temperatures. They were also used on a large axial vane fan and a NASA turbofan to acquire blade tip clearance data. The second-generation, 24 GHz sensors were used to make basic blade tip deflection and low-range clearance measurements on a simulated engine disk that had prebent blades. The second-generation sensors were also used to make blade tip clearance measurements on a small aero engine's compressor disk operated on a spin rig. The experimentation that has been accomplished to date has shown that the microwave blade tip clearance sensor technology is a viable option for propulsion HM and clearance control applications. Research will continue in the use and evaluation of microwave blade tip clearance sensor technology at NASA for SHM in gas turbine engines (Lattime and Steinetz, 2002).

8.3.2 High-Temperature Fiber Optic Sensors

The high-temperature fiber optic research is targeted toward providing expanded sensor coverage of sensed parameters through the use of fiber optic technology with the goal of providing robust and reliable measurement in hot sections of the engine. There are multiple potential advantages of optical fiber sensors. For instance, the optical devices are immune to the effects of electromagnetic interference and therefore do not need electrical insulation. This makes the sensors well suited to work in electrically charged environments and in such locations where electrical discharge may be an issue. The sensors also have significant chemical immunity to permit operations in a harsh environment. Furthermore, the

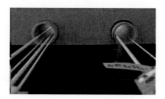

FIGURE 8.12
Fiber optic temperature sensors being tested in a high-temperature oven.

small diameter of the fiber permits easy embedding within structures. It is also anticipated that optical fiber sensors will have improved integration capability because multiple sensors can be put on a single fiber and there is no need for separate power wires.

The development of high-temperature optical thermal sensors initially focused on non-packaged devices, which simply consisted of optical fibers with FBG written into the fiber. Experiments with these high-temperature fiber optic sensors have demonstrated the long-term survivability of FBGs at temperatures up to 1000°C through many thermal cycles. In these tests, the fiber optic sensors underwent 400 h of thermal cycling up to 750°C and 700 h of thermal cycling to 1000°C.

In order to produce a high-temperature fiber optic sensor that is more robust, these FBG sensors have since been packaged in a ceramic housing. Recent research has been focused on constructing and evaluating the performance of packaged sensing devices at temperatures above 1000°C and with high thermal heating rates. A process of manufacturing high-temperature FBG-based sensors has been developed and demonstrated. The process has permitted construction of robust packaged sensing devices capable of withstanding extreme temperatures. The process starts with the construction of ceramic housings for a polyimide-coated fiber with a fiber optic Bragg grating (FBG). The structures then are placed into a furnace, shown in Figure 8.12, and heated slowly from room temperature to 1000°C. After reaching 1000°C, the sensors are kept at that temperature for periods of time from 20 to 50 h and then are allowed to cool down to the room temperature.

After the manufacturing process was completed, the sensor was tested by initially subjecting it to continuous exposure to high temperature. These packaged sensors have been tested for 500 h at 1000°C and also been subjected to 20 thermal cycles from 400°C to 800°C with various heating rates. During the tests, the gratings were able to track the temperature measured by reference s-type thermocouples within 3% accuracy. Future efforts in this area will initially focus on evaluating the durability of the FBG sensors at 1000°C for longer durations of time, up to 1000 h. Once the sensors are proven stable for such a length of time, the sensors will then be evaluated for how they respond to thermal shocks.

These sensors have the advantage that many can be placed in series in an optical fiber. The sensors can be used for either temperature measurement or strain measurement in a high-temperature environment. The sensors have significant utilization for embedded SHM and in the case of temperature measurements, can be utilized for gas path diagnostics of turbine engines and material process monitoring.

8.3.3 Smart Sensor Systems

In order for aerospace propulsion systems to meet future requirements of improved performance, increased safety, decreased maintenance, reduced emissions, and reduced cost, the inclusion of intelligence into the propulsion system design and operation are necessary (Hunter and Behbahani, 2012; Hunter et al., 2011c, 2008; Hunter, 2003; Lekki et al., 2010).

That is, a propulsion system that optimizes operational parameters based on on-board detection, diagnosis, and analysis of its own internal conditions and the environment can achieve a higher level of performance than a static system that does not take into account changing conditions. This implies the need for the development of sensor systems that will be able to operate under the harsh environments in propulsion systems to provide improved information for intelligent engine systems (Simon et al., 2004; Behbahani and Semega, 2008; DOT/FAA Report, 2008).

A fundamental priority in the implementation of any sensor system is the need for improved reliability and robustness, not only related to the operation of the sensor system itself, but also to ensure that the sensor system contributes to improving the overall vehicle reliability and safety (Kurtoglu et al., 2008). To the extent that the sensor system can internally process the data and only communicate relevant information, rather than burdening the core processing systems with large volumes of raw data, the burden on the vehicle of including new sensor technology is decreased. Further, sensor systems added to the aircraft increase the number of wires and the associated weight, complexity, and potential for failure. Thus, there is a need for high-temperature sensors and electronics, but also for high-temperature wireless technology. This implies the integration of sensors, electronics, wireless circuits, and power into a single system. This type of stand-alone sensor structure, which includes internal processing and wireless communication, is often described as a smart sensor system.

The definition of a smart sensor may vary, but typically, a smart sensor is the combination of a sensing element with processing capabilities that are provided by a microprocessor (Hunter et al. 2011b). That is, smart sensors are basic sensing elements with embedded intelligence. The sensor signal is processed by the microprocessor, which provides formatted output to an external user. A more expansive view of a smart sensor system is illustrated in Figure 8.13: a complete self-contained sensor system that includes the capabilities for logging, processing with a model of sensor response and other data, self-contained power, and an ability to transmit or display informative data to an outside user. Multiple sensor elements can be included in a smart sensor system.

The fundamental idea of a smart sensor system is that the integration of a microprocessor with sensor technology will significantly improve sensor system performance and capabilities. The embedded intelligence will continuously monitor the individual sensor elements, process the data with internal calibration tables, periodically verify sensor

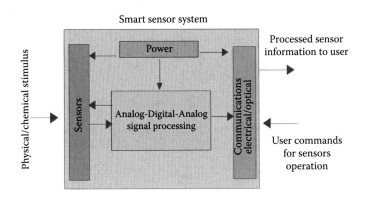

FIGURE 8.13
(**See color insert.**) A smart sensor system. The core of a stand-alone smart sensor system includes sensors, power, communication, and signal processing. (From Hunter G.W. et al. *Interface Magazine, Electrochemical Society Inc.* 20(1), Winter, 2011b, 66–69.)

calibration and health, and potentially combine various data inputs to provide the user with an assessment of the vehicle conditions. The processed data become information and can then be transmitted to external users. The user can choose the complexity of the data transmitted: from a single reading to a complete download of the sensor system's parameters. A driving goal in the development of smart sensor systems is the implementation of sensors in a nonintrusive manner and to provide the user what they need to know in order to make sound decisions. The application of such smart sensor systems includes distributed engine control and aviation safety.

A notable possible feature of a smart sensor system is the wireless capability. Currently, for almost every sensor, actuator, or processing unit that is going into an engine to improve the *in situ* monitoring of the components, communication and power wires always must follow. An A340 engine for example has 152 miles of wire weighing 1761 kg (Thompson, 2004). More sensor systems added to the aircraft increases the number of wires and the associated weight, complexity, and potential for unreliability. For example, there is a drive to minimize wiring in aircraft design since "Wiring on an aircraft is complex, difficult to route, heavy and a key source of faults which lead to delays and cancellations for passengers." Engineers thus try and minimize wiring as "every wire is a source of failure and every wire adds weight" (Thompson, 2004).

Wiring has a history of causing vehicle reliability concerns since "The problems of open and short circuits, as well as intermittent faults from degraded contacts inside connectors, caused by vibration are well known. The aging of wiring is also becoming more of an issue in older aircraft with studies identifying … numerous intermittent faults attributed to changing conductance to ground from poor airframe ground connections which are particularly difficult to isolate" (Thompson, 2004). Simply removing communication wires alone will not address the issue of the lack of reliability due to wiring. "There is no real advantage to unless the need for power wiring is also removed" (Thompson, 2004).

There has been a growing realization that "due to the increasing demand, more wireless technologies will be used for future aircraft systems, including those that impact the safety and regularity of flight" (ACP, 2008). Further, "The increased emphasis and reliance on electronic systems for modern aerospace vehicles has resulted in wiring becoming a critical safety-of-flight system. Aerospace systems now routinely use fly-by-wire technology and avionics to control and manage many of the critical vehicle sub-systems. According to a recent Air Force Research Laboratory study on Air Force mishaps, 43% of mishaps related to electrical systems were due to connectors and wiring" (Slenski and Walz, 2006; Slenski and Kuzniar, 2002).

Thus, a notable reduction in wiring has significant advantages for a range of reliability and operational reasons. However, to simply remove the wires will not provide the capabilities needed for the next generation of propulsion systems. Rather, advances in sensors and electronics are also necessary to enable intelligent engine systems. Such capabilities can be provided by the development of smart sensor systems.

One possible result of smart sensor systems is that important data can be provided to the user with increased reliability and integrity. Intelligent features can be included at the sensor level, including, but not limited to, self-calibration, self-health assessment, and compensated measurements (e.g., compensated for temperature and sensor drift). Overall, the presence of the microprocessor–sensor combination allows the design of a core system that is adaptable to a changing environment in a given application, or that can be modified to meet the needs of a wide range of different applications.

Further, smart sensors systems can also form networked systems through the communication interface, thus forming wireless smart sensor networks (WSSN). Such a network would include individual smart sensor systems that have the capability of individual network self-identification and communication allowing reprogramming of the smart sensor system as necessary. The WSSN approach allows the output from a number of sensors within a given region to be correlated not only to verify the data from individual sensors but also to provide a better situational awareness. Such communication can be between a single smart sensor and communication hub or between individual smart sensors themselves. These types of capabilities will provide for a more reliable and robust system because they are capable of networking among themselves to provide the end user with coordinated data that are based on multiple sensory inputs. Further, information can be shared in a more rapid, reliable, and efficient manner with onboard communications capability in place.

Such capabilities allow a fundamentally different approach to the vehicle monitoring. Local processing means that this infrastructure is beyond simply replacing wires with wireless for communication; it changes that nature of what is being transmitted. A major consideration for vehicle system implementation is that there is information available now related to each vehicle now; there is often a challenge as to what to do with that information. Further, if for no other reason than bandwidth restrictions, it is not feasible to transmit all of the information, all of the time, to everyone who might need it. Local processing and hierarchal approaches are needed with smart sensor systems that are conducted in such a way that any computational complexity is transparent to the user. Increased engine intelligence in this approach is established from the bottom up with integrated smart sensor diagnostics operating locally, feeding into smart nodes and subsystems, and finally across the vehicle. The approach allows an intelligent engine system enabled by integrated, local smart detection, diagnostics, and prognostics.

The ability to create complete sensor systems that can be placed wherever needed, like a postage stamp, without the need to rewire the systems ("Lick and Stick" technology) is viable in silicon (Si)-based technology for near room temperature operations (Hunter et al., 2006a,b, 2007a). Figure 8.14 shows a "Lick and Stick" smart leak detection system using silicon-based technology, including signal conditioning, data storage, power, telemetry,

(a)

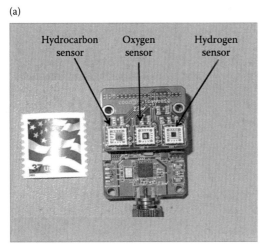

(b)

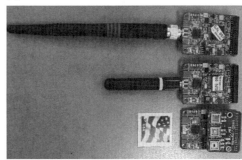

FIGURE 8.14
"Lick and Stick" leak detection system (a) with various communication configurations (b). Configurations of the system include multiple sensors, signal conditioning, power, and telemetry (Adapted from Hunter G.W. et al., Intelligent chemical sensor systems for in-space safety applications, *42nd AIAA/ASME/SAE/ASEE Joint Propulsion Conference and Exhibit*, Sacramento, CA, 9–12 July 2006b, Paper AIAA-2006-4356.).

and three sensors in one package. Multiple communication configurations are shown that can be integrated into the package. A complete "Lick and Stick" system approach allows improved sensor data and improved capability to implement the system. However, this technology is based on silicon-based electronics, which, as is discussed below, is not suitable for some propulsion system applications.

In general, highly mature smart wireless technology has been used for ground and commercial operation in other fields with notable success. However, the use of high-temperature wireless systems for aerospace applications is notably less mature. A fundamental requirement of the hardware approach is that each of the electronic components, such as the microcontroller, has the capability to operate in the environment of the engine. This includes an ability to be flight qualified and certified for long-term aircraft system operation. The potential deployment of WSSNs for engine real-time applications has to deal with many challenges, including power requirements, electromagnetic interference, space radiation, and protocol implementation (Figure 8.15).

The three key challenges related to the implementation of WSSNs in engine environments are security, communication integrity, and operation in harsh high-temperature environments (Hunter and Behbahani, 2012). For example, if smart wireless sensor systems are to be used pervasively in propulsion control systems, they need to be robust and have a long life in the harsh environment of the engine. Simon et al. (2004) provides details about these conditions that vary from more benign environments such as the cooler temperatures of the fan, to high temperature and rotating systems. The hardware configuration of the smart wireless sensor elements needs to be tailored for their function and local operational environment (Hunter et al., 2012). The implementation of wireless sensor

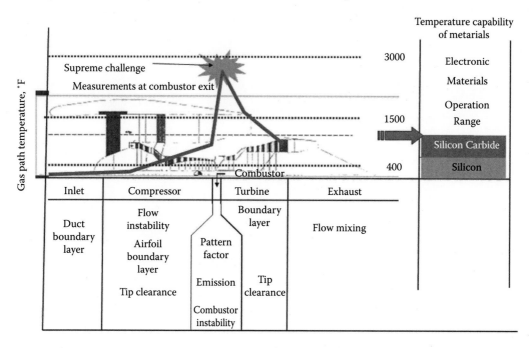

FIGURE 8.15
Correlation of position within the engine to temperature. Successful development of SiC up to 600°C has the potential to enable device operation throughout most, if not all, of the compression system and the low-pressure turbine. (From Sokolowski D., and G. Hunter, UEET Technology Forum, Intelligent Propulsion Controls, High Temperature Wireless Data Communication Technology, October 28–30, 2002.)

technology in benign parts of the aircraft can be accomplished in principle with conventional electronics and sensor technology, although flight qualification and tailoring for the aircraft environment would be necessary. Silicon-based electronics and the capabilities developed for commercial application could then be applied to engine systems depending on the part of the engine where the WSSN is being implemented.

However, there are significant parts of the engine where the use of silicon-based electronics is problematic due to the high-temperature regimes. This is illustrated in Figure 8.15, which shows the temperature distribution of an engine and the capability of two example electronic materials: Si and silicon carbide (SiC) (Sokolowski and Hunter, 2002). The operation of silicon-based electronics is limited for notable parts of the engine. In fact, "the development of high temperature electronics capability is critical for successful distributed control architecture" (Behbahani et al., 2007) and "distributed control will only find acceptance if it enables life-cycle cost reduction features through improved fault isolation made possible with embedded intelligence in system control elements. This intelligence is enabled through the capability afforded by embedded high temperature electronics" (Behbahani et al., 2007). Ideally, "Today's push for engine health management through adaptive control systems demands more robust instrumentation with inherently fail safe sensors" (Beachkofski, 2005).

Meeting the needs of engine applications to produce a complete high-temperature sensor system is a notable technical challenge. High-temperature wireless systems can be based on the basic smart sensor systems concepts being used with room temperature silicon semiconductor technology, but would require very different materials and technologies for operation at high temperatures. A high-temperature smart sensor system requires the development of a range of high-temperature, robust technologies including sensors, communication, power supplies, packaging, and electronics. Such a system is presently being developed and includes a range of components. Beyond the high-temperature sensor technologies described elsewhere (Lekki et al., 2010; Hunter et al., 2006c, 2007b), briefly such components include

- *Silicon carbide (SiC) electronics and packaging*: SiC electronics have demonstrated a significant number of world first demonstrations (Silicon Carbide Electronics website, 2013). SiC electronics has shown component operation to temperatures of 600°C (Okojie, 2007) and has the potential of meeting a range of engine application needs (Neudeck, 2000; Neudeck et al., 2002). Operation at 500°C has been demonstrated for thousands of hours; these time frames are now viable for implementation to engine conditions for extended periods (Neudeck et al., 2008; Spry et al., 2008). Further complexity in SiC electronics includes 4-bit analog/digital (A/D) circuits and binary amplitude modulation RF transmitters and they are now being developed and are planned for demonstration by 2013 (Beheim et al., 2012).

- *High-temperature wireless communications (500°C)*: NASA activities aim to use high-temperature-compatible materials to fabricate a complete wireless circuit including high-temperature passive components such as resistors and capacitors (Schwartz and Ponchak, 2005). Notable advancements include demonstration of wireless transmission at near 500°C of wireless data associated with a position transducer (Ponchak et al., 2012), and 500°C transmission of pressure data both wired and wirelessly (Hunter et al., 2011a).

- *Power scavenging*: Power scavenging using thermoelectrics or piezoelectrics (Roundy, 2003; Roundy et al., 2004) that take advantage of the energy already present within the engine have notable appeal for self-powered sensor systems. A range of relevant thermoelectric-based technologies are being investigated from thin film thermo-piles for heat flux sensors (Martin et al., 1999; Fralick et al., 2002) to environmentally durable silicide-based thermoelectric materials for 500°C operation (Dynys and Sayir, 2008; Sayir et al., 2007).

The integration of sensors, electronics, power scavenging, and wireless communication components has been demonstrated at 300°C (Hunter et al., 2010). This was considered a proving ground for technologies to allow 500°C smart system operation. In parallel, silicon on insulator (SOI) electronics has capabilities in-between the basic circuit capability in SiC, and the high maturity wireless systems in the commercial field based on Si electronics (Benson, 2009). Thus, the capabilities for smart sensor systems for engine applications vary widely, and require further development of a high-temperature capable system for full implementation.

In summary, the capability to embed operational smart sensor systems in extreme environments is critical to enabling potential revolutionary changes in propulsion systems and a key to bringing forth the next generation of complex, high-performance engines. Such smart sensor systems can enable wireless smart sensor networks that would operate at high temperatures and in harsh environments. These capabilities require a combination of high-temperature, robust technologies, including sensors, communication, power supplies, packaging, electronics, and actuators. The application of these systems ranges from enabling distributed engine controls to improving the safety of operational engine systems.

8.3.4 Engine Emissions Monitoring (High-Temperature Electronic Nose)

The detection of the chemical signature of the emissions of a propulsion system is understood to reflect the efficiency and health of the system. Rapid or sudden changes in the emissions produced by combustion indicate changes in the propulsion system combustion process or engine health state. Turbine engine exhaust emission measurements are also frequently required during engine development ground test programs (Ward et al., 2010). Exhaust emissions are quantified in performance measurements to determine combustor and/or augmenter fuel-to-air ratio and combustion efficiency, to evaluate component hardware modifications, and to optimize fuel splits, ignition characteristics, and overall engine operability. Engine exhaust emission measurements are also required to certify to the regulatory agencies that a particular engine meets environmental requirements during each phase of the landing and takeoff cycle. Aerospace Recommended Practices (ARP 1533A, 2004; ARP 1256, 1990) give standardized methodologies for the measurement of turbine engine exhaust emissions using conventional instrumentation and must be followed during environmental certification ground tests.

Ideally, an array of sensors placed in the emission stream close to the propulsion system could provide information on the gases being emitted by the propulsion system. However, there are very few sensors available commercially that are able to measure the components of the emissions *in situ* since the harsh conditions and high temperatures of the propulsion system render most sensors inoperable. Thus, in order to detect the other species present in an emission stream, the development of a high-temperature chemical sensor array technology is necessary (Hunter et al., 2006a). Such a gas sensor array would, in effect, be a *"high-temperature* electronic nose" and be able to detect a variety of gases of interest to monitor the health of the system producing those emissions.

(a)   (b)   (c)   (d)

FIGURE 8.16
Representative pictures of packaged sensors: (a) titanium oxide-based CO sensor; (b) lithium phosphate-based CO_2 sensor; (c) zirconia-based O_2 sensor; and (d) silicon carbide-based hydrocarbon sensor. (From Hunter et al., Smart sensor systems for spacecraft fire detection and air quality monitoring. In *40th International Conference on Environmental Systems*, AIAA: Portland, Oregon, 2011d; Vol. AIAA 2011-5021.)

The development of such a gas sensor array (high-temperature electronic nose) related to engine test stand evaluation systems has been ongoing for many years. Turbine engine exhaust emission measurements are frequently required during engine development ground test programs. In order to meet these needs, high-temperature emission sensor technology had been developed (Ward et al., 2010). This emission sensor technology is based on gas microsensor arrays to quantify composition of critical constituents in turbine engine exhaust products, for example, carbon monoxide, carbon dioxide, nitrogen oxide, and unburned hydrocarbons. Different chemical sensing techniques are suitable for measurement of different species (Hunter et al., 2006a,b, 2008a). By choosing the proper materials for fabrication, these sensors can withstand harsh environments. The focus is not on slight variations of similar sensor structures like many commercially available electronic nose devices, but instead on leveraging the very different sensing mechanisms to be very selective to the targeted species to be measured and reduce overall cross-sensitivity effects during sensor operation. Integration of a number of the individual high-temperature gas sensors into a single platform will enable the formation of a sensor array.

The micro-electro-mechanical-System (MEMS) chemical sensor technology being developed uses three different types of MEMS chemical sensor platforms: resistors, electrochemical cells, and Schottky diodes. Each sensor type or platform provides very different types of information on the environment and is meant to have limited cross-sensitivity (i.e., be orthogonal in its response). Integration of the information from these orthogonal sensors can provide increased whole-field information on the environment. The approach is to give quantitative readings of the species of interest. The following gives a brief discussion of the sensor technologies used for both the combustion processed of emission monitoring and fire detection (Hunter et al., 2011d). These include sensors for detection of CO, CO_2, O_2, and hydrocarbons; a representative picture of each of the sensors is shown in Figure 8.16. A more detailed description of the sensing mechanisms and operation is given elsewhere in Hunter et al. (2011d and references therein).

8.3.4.1 Carbon Monoxide Detection

The measurement of CO is an indicator of combustion inefficiency, engine health, and is a regulated air quality species. The CO detection approach uses a semiconductor oxide resistor whose electrical conductance is dependent on CO concentration in the environment (Figure 8.16a). The CO sensor for higher concentrations of CO features ~100 nm particles of anatase titanium oxide (TiO_2) as the sensing material (Akbar et al., 2006; Savage et al., 2001). In order to prevent grain growth and sensor signal drift, the anatase is coated with a surface layer of lanthanum oxide. Selectivity toward CO was obtained by using nanometer-sized CuO particles to coat the TiO_2. These anatase–lanthanum–copper (ALC) sensors showed good performance in harsh environments and an ability to detect CO in a range of concentrations, for example, above 10 ppm CO. Quantitative values of CO have been measured (Li et al., 2009; Ward et al., 2010). A modified version of this sensor without CuO is used for lower concentration. The combination of these two sensors is used to cover a range from 1 to 500 ppm. One area of research has included understanding and compensating for the effect of other species such as low concentrations of O_2, which can affect the calibration of the sensor below 10% O_2.

8.3.4.2 Carbon Dioxide Detection

The measurement of CO_2 is another indicator of combustion efficiency. The CO_2 detection approach uses an electrochemical cell with lithium (Li)-based electrolyte whose voltage output (potentiometric measurement) depends on the CO_2 concentration in the environment (Figure 8.16b) (Hunter et al., 2011d). This electrochemical cell sensor has a Li_3PO_4 electrolyte, a lithium carbonate (Li_2CO_3) sensing electrode, and a mixture of lithium titanate (Li_2TiO_3) and TiO_2 as a reference electrode (Li et al., 2009; Lee et al., 2001, 2002; Park et al., 2003; Szabo et al., 2003). The basic sensing mechanism is to measure the equilibrium potential difference between sensing and reference electrodes. The sensing electrode potential is changed depending on CO_2 concentration, while that of the reference electrode is inactive to CO_2 gas. Moreover, the use of a mixture of Li_2TiO_3 and TiO_2 enables this sensor to avoid oxygen interference. The sensor has high CO_2 specificity and sensitivity, with near-zero cross-sensitivity to species such as CO. This sensor system has also undergone engine testing as part of an engine emission sensor system development (Li et al., 2009).

8.3.4.3 Zirconia-Based Oxygen Sensor

The measurement of O_2 (oxygen) is useful as an indicator of fuel/air ratio and a good indicator of overall emissions data quality. Microfabricated oxygen sensors have been developed based on electrochemical cell technology. Commercially available O_2 sensors are typically electrochemical cells using zirconia (ZrO_2) as a solid electrolyte and platinum (Pt) as the anode and cathode. Zirconia becomes an ionic conductor of O^{2-} at higher temperatures, for example, 500°C. This property of ZrO_2 to ionically conduct O^{2-} means that the electrochemical potential of the cell can be used to measure the ambient oxygen concentration at high temperatures. The basic approach of our work is to miniaturize the sensor structure while still maintaining the capabilities of the commercially available O_2 sensors (Hunter et al., 2005, 2006b, 2008). A packaged version of a zirconia O_2 sensor is shown in Figure 8.16c. The amperometric zirconia cell oxygen sensor has been tested in multiple applications and has a nearly linear response to various concentrations of oxygen. The response is stable and sensitive to changes in ambient O_2 concentrations over a range of concentrations; other data show an ability to respond to O_2 concentration ranges beginning from 0% O_2.

8.3.4.4 Silicon Carbide-Based Hydrocarbon Sensor

The measurement of hydrocarbons is useful as an indicator of fuel/air ratio and a good indicator of engine health issues such as oil leaks. The development of hydrocarbon sensors has centered on the development of a stable silicon carbide (SiC)-based Schottky diode. The advantage of SiC over Si is its ability to operate as a semiconductor at temperatures as high as 600°C, well beyond the high temperature limits of conventional silicon semiconductor electronics. This allows SiC-based gas sensors to operate at temperatures high enough to allow the detection of hydrocarbons (Hunter et al., 2005, 2006b, 2008a, 2008b). As with the Si-based hydrogen sensor, the use of the Schottky diode structure allows the sensor to have high sensitivity. One challenge with the sensor system has been high-temperature stability for long duration. The present design uses a catalytic metal/palladium oxide/SiC structure, which has demonstrated long-term durability while still maintaining sensitivity. The present approach is shown in Figure 8.16d and it includes mounting the Schottky diode element onto a suspended platform that includes a heater and temperature detector.

A critical aspect of the harsh environment gas sensor array development is migration of high-temperature gas-sensitive materials from hand-fabricated/stand-alone lab devices to consistently manufacturable devices that can be incorporated into field supportable test packaging (Ward et al., 2010). The approach is based on individual sensor elements sampling a common emissions flow through a manifold. Combining single-element probes to a common test manifold results in lower-cost probes, more robust probes, and more flexible system operation (the end use includes only the gas sensors of interest to a test). There is a minimal impact on overall system footprint for an extractive test apparatus. Close coupling of the sensor and probe allows for the elimination of the need to transfer data over long distances. The response time of gas microsensor systems is expected to be faster than typical gas analyzers by at least the sample transport time and, in the future, this approach will possibly be integrated into the vehicle.

Figure 8.17 is an example of the general approach used for engine test stands and engine health management applications. Sensors were bonded to probe assemblies using microwelds and high-temperature packaging materials (Figure 8.17a and b), and the probe assemblies were attached to the common manifold with Aeronautical and Navy (A/N)

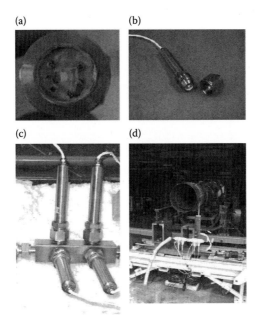

FIGURE 8.17
Emission sensor testing approach. (a) Sensor welded to the probe assembly (custom A/N fitting); (b) sensor probe assembly; (c) sensor probe assemblies in a flow-through manifold; (d) sensor manifolds mounted on test stand for J-85 turbine engine.

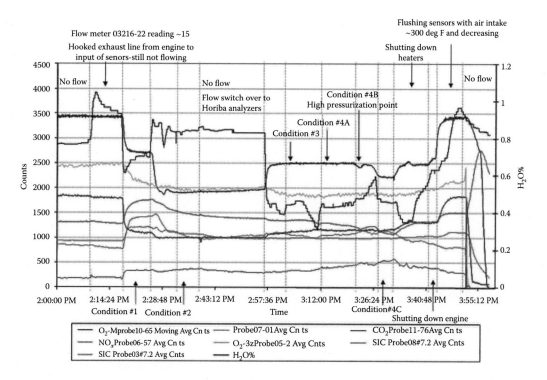

FIGURE 8.18
(**See color insert.**) Example data from emission sensor array during an engine test stand run over a range of operational conditions. (Courtesy of Makel Engineering, Inc.)

fittings (Figure 8.17c). The probe assemblies and manifold are connected to a flow sample probe that can be placed at the engine exhaust (Figure 8.17d). Samples can be taken during engine operation and the resulting emissions can be examined for trends in changes in operational parameter and health state (Figure 8.18).

Efforts are ongoing to directly demonstrate the correlation of the response of this high-temperature electronic nose to changes in engine health state. In the future, this high-temperature electronic nose system is intended to be integrated into the structure of an engine for onboard gas analysis for engine HM, combustion efficiency, and possibly even to control these emissions via feedback control.

In summary, this work highlights what is considered to be a revolutionary step change in engine test instrumentation and monitoring: a high-temperature chemical sensor array for gas analysis of CO, CO_2, NO_x, O_2, H_2, and HC into a single extractive probe system composed of miniaturized sensor technology to support test and evaluation of gas turbine engines and with the potential to be embedded in the engine structures in the future.

8.4 Health Monitoring of Steam Pipes

HM systems are used in many applications and the focus of this section is on steam pipes. Generally, steam pipes are used in major cities around the world. The system that operates in Manhattan, New York City, is managed as a district heating system, carrying steam from central power stations under the streets to support heating, cooling, or power to high-rise buildings and businesses. HM of such systems is critical to assure their safe operation. Excessive rises in the level of condensed water inside a steam pipe is a source of concern due to the possible excitation of water hammer effects that may lead to serious consequences, including damaged vents, traps, regulators, and piping. The water hammer effect is caused by accumulation of condensed water that is trapped in horizontal portions of the steam pipes. The use of ultrasonic waves has been shown to provide an effective technique of monitoring the height of the condensed water through the pipe wall while sustaining the high-temperature environment of the steam pipe system that can reach up to 250°C.

8.4.1 Methods of Water Height Monitoring

Making nondestructive measurements of the water height level through a steel pipe wall may be feasible only by an ultrasonic method. For this purpose, it is necessary to measure with good accuracy a parameter that is related to the height of the water. The author and his team considered two techniques: pulse-echo and pitch-catch (Bar-Cohen et al., 2010a,b,c). The operation principle of these techniques involves using ultrasonic waves that propagate through various materials with some of the energy reflected from the material interfaces along the path. The reflected wave amplitude and energy depend on the acoustic impedance (density × acoustic velocity) of the two materials that are involved with any interface.

Pitch-catch: The ultrasonic pitch-catch involves sending waves in an angle and receiving the reflections at the same angle at the opposite side of the surface normal. For this purpose, two separate probes are used and the piezoelectric crystals that generate and receive the wave are mounted side by side at an angle to the surface of the water. When the wave impinges onto an interface (the pipe or the water) surface, refraction occurs at an angle that is determined by the related acoustic velocities. When the wave impinges in an angle over interface with solid material, shear and longitudinal modes are generated.

Pulse-echo: In this case, the same probe is used for both transmitter and receiver of the sound waves. For the purpose of this test, the wave path is assumed to be normal to the interface surface. To measure the height of water inside a pipe, a probe is used to generate ultrasonic wave pulses and the TOF between the reflections that arrive from the bottom and the top of the water is used. The value of the height is the time multiplied by the wave velocity divided by two (taking into account the wave path back and forth through the water bulk). An illustration of the pulse-echo method using the reflections to determine the TOF inside the condensed is illustrated in Figure 8.19. To obtain high-resolution measurements, the probe needs to be broadband in relative high frequency. This enables the system to generate sharp pulse shapes in the time domain, which is needed to have sufficient time separation interval between the reflections.

In the case of pulse-echo, the probe is connected to both the transmitter (function generator), which sends high-voltage signals to generate the elastic wave pulses, and the receiver, which amplifies the attenuated reflections converting them to electric signals. To avoid damage to the receiver, the large signal from the generator is blocked by an electronic switching mechanism from reaching the receiving circuitry.

The pulse-echo and the pitch-catch methods rely on measuring the TOF from the wave reflections. Of the two methods, pulse-echo provides greater flexibility in measuring the height since there is no reliance on receiving the reflection at a specific angle. However, the numerous reflections that are received in the pulse-echo method require effective

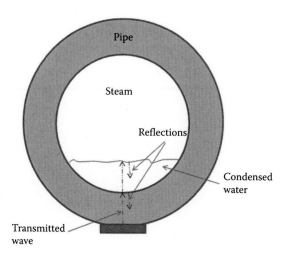

FIGURE 8.19
Illustration of the condensed water level monitoring using TOF measurements of reflected ultrasonic waves.

signal-processing technique to distinguish the reflections from the top and the bottom surfaces of the condensed water. Generally, there are several issues that need to be accounted for which include the strong reflections from the interfaces of the steel pipe and the effect of the pipe curvature that caused wave losses. Also, the measurements may involve issues associated with the interference of the multiple reflections with the pipe bottom wall and the pipe–probe interface; interference with turbulence in the condensed water; scattering from potential sediments in the bottom of the pipe inner surface along the path of the wave; and the presence of multiple reflections inside the condensed water.

8.4.2 High-Temperature Ultrasonic Probe

8.4.2.1 Design and Fabrication of Ultrasonic Probes

Given the criticality of the HT probe and the difficulties that were encountered in finding producers, ultrasonic probes were made in-house. The general configuration of an ultrasonic probe is shown schematically in Figure 8.20. Making high-temperature probes requires addressing the issues of durability, compatibility, and interfacing capability. Specifically, each of the components needs to be made of high-temperature materials. A primary consideration in the development of ultrasonic probes is the selection of the piezoelectric material. This choice determines the probe performance including the electromechanical coupling and thermal stability. For many years, it has been known that $LiNbO_3$ has a Curie temperature that is higher than 1000°C, however, since it is a single crystal, it is fragile and it has lower efficiency than piezo-ceramic transducers. Generally, the perovskite solid-solution system, $Pb(Zr,Ti)O_3$ (PZT) compositions, are preferred for most ultrasonic devices due to their high piezoelectric ($d_{33} > 300$ pC/N) and electromechanical coupling ($k_t \sim 0.50$) together with high Curie temperatures, >300°C, allowing for making ultrasonic probe with broad bandwidth and higher sensitivity over a broad temperature range. In recent years, piezoelectric ceramics that can operate at temperatures as high as 850°C have been developed (Chapter 10; Bar-Cohen, 2003; Sherrit et al., 2004; Bar-Cohen et al., 2010a, 2010d, 2011). These materials include the commercial piezo-ceramics PZ46 (made formerly by Ferroperm and currently owned by Meggitt, Denmark) and piezo-ceramics based on

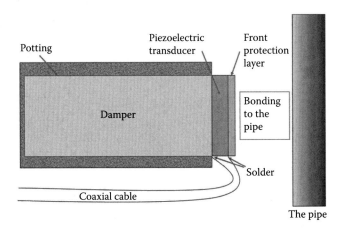

FIGURE 8.20
(**See color insert.**) A schematic diagram illustrating the probe components (not to scale).

bismuth titanate (Bar-Cohen et al., 2010d, 2011). However, the issue associated with these materials is the transducer responses are much lower than those of the perovskite solid-solution system.

In this test, Navy II type, EC-64 ceramics were selected based on their high-temperature stability up to ~250°C and high thickness electromechanical coupling, k_t ~0.46. EC-64 is made by ITT Exelis—Acoustic Systems, Salt Lake City, UT, and its properties are listed in Table 8.1.

On the back of the piezoelectric transducer, a thick and high impedance layer, referred to as backing, was attached. The purpose is to reduce the duration of the ringing in order to be able to resolve shallow water depths and have high resolution; however, the consequence is that it lowers the probe sensitivity. Therefore, the appropriate selection of backing layer is a key factor for successful and efficient design of ultrasound probes. Three different types of backing were used in this study: (A) high-impedance polymer (mixture of 20% of tungsten particles and 80% of high-temperature epoxy, Duralco 4460, Cotronics Corp., Brooklyn, NY), (B) low-impedance polymer (Duralco 4460), and (C) the transducer has no backing (backed with air). The general properties of Duralco 4460 are listed in Table 8.2.

On the front surface of a piezoelectric material, a thin layer is generally added to protect the transducer surface from the wear and corrosion when the probes are operating directly into high-impedance load, such as steel pipe (~40 MRayl). However, when the probe is used in low impedance medium, such as water or tissue (~1.5 MRayl), appropriate impedance and thickness of matching layer are required in order for efficient acoustic energy transfer between the probe and propagating medium. In general, the optimum impedance and thickness of the front layer can be obtained using the following equations:

$$Z_m = \sqrt{Z_t . Z_p}, t_m = \frac{v_m}{4 f_t} \tag{8.1}$$

TABLE 8.1

Dielectric and Piezoelectric Properties of EC-64 Piezoelectric Materials

Transducer	ρ (g/cc)	c (m/s)	k_t	ε_{33}^T	ε_{33}^S	Tan δ	c_{33}^D (GPa)
EC-64	7800	4452	0.45	$1116\varepsilon_0$	$624\varepsilon_0$	0.02	154

TABLE 8.2

Material Properties of Duralco 4460

Sample	T_m (°C)	ρ (g/cc)	c (m/s)	α (*10^5 °C)	η (centipoise)
Duralco 4460	600	1100	2200	5.4	600

Note: T_m, maximum usage temperature; ρ, density; c, longitudinal sound velocity; α, thermal expansion; η, viscosity.

FIGURE 8.21
Photographs of the produced thickness mode HT piezoelectric probe.

where Z is the acoustic impedance, f the operating frequency, t the thickness, and v the sound velocity. The subscripts are m matching layer, t transducer layer, and p propagating medium.

An ultrasonic probe was assembled with the piezoelectric transducer attached to the corrosion-resistant stainless-steel housing using an insulating commercial alumina adhesive paste (Resbond 989-FS, Cotronics Corporation of Brooklyn, NY). This ceramic adhesive can provide high bond strength and excellent high-temperature (up to 1650°C), electrical, moisture, chemical, and solvent resistance. The transducer was then electrically connected to the coaxial cable (CB-188LN-100, CD International Technology, Inc., Santa Clara, CA) using high-temperature solders (Ersin Multicore 366 Solder, Westbury, NY), where both cables and solders resist to a temperature up to 250°C. The rear face of the housing was covered with aluminum using high-temperature epoxy (Duralco 4460). The fabricated ultrasonic probes are shown in Figure 8.21.

8.4.2.2 Evaluation of Ultrasonic Probes

The performance of the fabricated ultrasonic probes was investigated using conventional pulse-echo response measurements. The fabricated probes were placed onto an aluminum plate and excited by a Panametrics pulser/receiver (model 5052PR, Panametrics Inc., Waltham, MA), with the following parameters: energy level: 1, attenuation: 20 and 2 dB step, high-pass filter: 1 MHz, damping level 4, and −20 dB amplifier gain. The signals from the receiving probes were sampled using a Tektronics model TDS2034B oscilloscope,

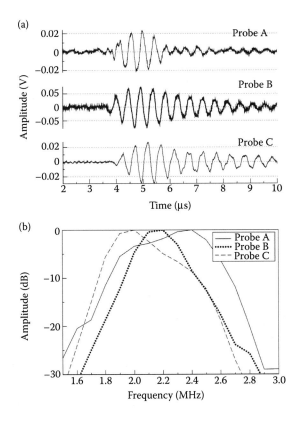

FIGURE 8.22
Time-domain (a) and frequency-domain (b) pulse-echo responses from a steel reflector in air, using probe A, probe B, and probe C.

where the sampling frequency and total sampling time were set to 25 MHz and 100 μs, respectively.

The time-domain pulse-echo waveforms and normalized frequency spectrum for prepared ultrasonic probe A, B, and C are shown in Figure 8.22. Because of a heavy backing (Z_b = 10 MRayl), the probe A shows less ringing compared to other probes B and C, leading to a broader bandwidth, on the order of 29.3%. However, the signal amplitude of probe A is significantly lower than the lightly backed (B) and the air-backed probes (C) resulting from the larger loss of power to the backing layer. As expected, the probe C exhibited the highest signal amplitudes, being on the order of 0.54 V_{pp}, which is one order of magnitude higher than that of probe A, whose signal amplitude is around 0.05 V_{pp}. However, the pulse length was increased whereas the bandwidth of the probe decreased. The measured properties of the three probes are summarized in Table 8.3.

8.4.3 Pulse-Echo Test System

To test the feasibility of the pulse-echo method, the lead author and his team used the system with the components shown as block diagram in Figure 8.23. The initial development was done at room temperature and examined the ability to transmit and receive sufficiently high signal amplitudes to allow reliable measurement. In order to examine the test system, a testbed was produced made of a steel pipe that is 91 cm (3 ft.) long with

TABLE 8.3

Measured Acoustic Performance for Various Ultrasonic Probes

	Probe A	Probe B	Probe C
Backing impedance Z_b (Rayl)	10M	2M	400
Center frequency F_C (Mhz)	2.4	2.2	2
Bandwidth at −6 dB (%)	29.3	16.9	23.2
Pick-to-pick voltage (V_{pp})	0.05	0.15	0.54

Note: Repetition rate: 4 kHz; energy level: 1; attenuation: 20 and 2 dB step; high-pass filter: 1 MHz; damping level 4; and −20 dB amplifier gain.

two end walls that simulate the steam pipe. The pipe was made of A53B steel alloy having 40.6 cm (16 in.) diameter and 0.95 cm (3/8 in.) thick wall. A plumbing was installed to allow for water entry to fill the pipe and for draining and the side walls were made of Plexiglas to allow viewing the inside of the pipe and to measure the water height. The probes were coupled to the bottom of the pipe in a pulse-echo configuration. The probes were driven by a transmitter/receiver (made by Panametrics) and, using a miniature manipulator, were aligned for maximum reflection from the water top surface.

One difficulty that may be encountered when applying this material to bond probes in the field is the rapid curing when exposed to high temperatures. The rapid curing will be addressed by the strap that was developed and by aligning the probe prior to bonding while away from the pipe surface. Once the adhesive is applied, the probe is pushed onto the surface of the pipe. In order to employ the HT probe in the field, there is a need for a mounting strap that allows for aligning and securing the probe intimate contract to the pipe surface. The design is shown schematically in Figure 8.24 and a photo of the strap mounted on a pipe is shown in Figure 8.25. The process of strapping consists of the following two steps:

Step 1: Attach the strap to the pipe and orient the probe alignment flexure to the pipe vertical direction.

Step 2: Insert the probe into the guide, apply the bonding material onto the probe face, press against the pipe, as well as preload and secure the probe backing.

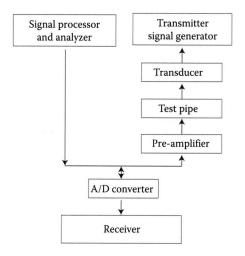

FIGURE 8.23
A schematic diagram of the test system.

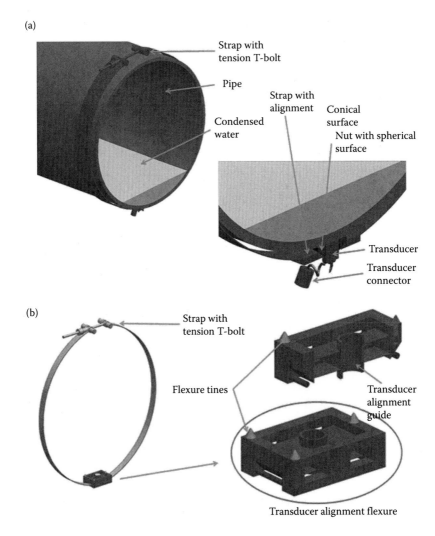

(a)

Strap with
tension T-bolt

Pipe

Strap with
alignment

Conical
surface

Condensed
water

Nut with spherical
surface

Transducer

Transducer
connector

(b)

Strap with
tension T-bolt

Flexure tines

Transducer
alignment
guide

Transducer alignment flexure

FIGURE 8.24
(**See color insert.**) Schematic view of the mounting strap for the alignment and strapping the probe to the pipe surface in the field. (a) General view of the pipe and the strap and (b) the strap components.

FIGURE 8.25
A steam pipe with the strap and the probe attached simulating operation in the field.

8.4.4 Signal Processing

Improving the reliability and performance of the computer code for the characterization of TOF requires enhancing the signal-processing algorithm. Specifically, it required modifications of the signal processing that consists of windowing, tracking, and filtering the reflected signals to obtain stable readings. For this purpose, three different approaches were introduced to characterize TOF.

8.4.4.1 Autocorrelation Method

The large number of reflections that are received from the pipe (Figure 8.26) makes it difficult to base the determination of the height on simple TOF measurements in real time and, therefore, an autocorrection technique was applied. This technique is one of the most widely used signal-processing methods to find repeating patterns or time of arrival in

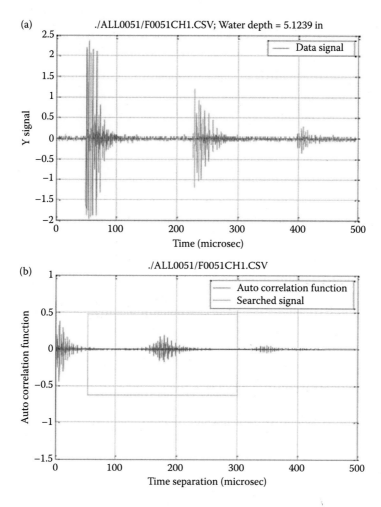

FIGURE 8.26
The TOF showing the first arrival time difference of 179.0 µs and the calculated autocorrelation time difference of 179.6 µs. (a) Time of flight data signal and (b) the auto correlation function and the searched signal.

the presence of noise (Bracewell, 1978; Cutard et al., 1994; Hertzog and Jordaan, 2004). The autocorrelation function can be defined as follows:

$$R_{xx}(\tau) = \frac{1}{T} \int_0^T x(t)x(t + \tau)dt \qquad (8.2)$$

where $x(t)$ is the signal waveform, T is the total sampling time, and τ is the time separation variable.

An example of the autocorrelation technique used is shown in Figure 8.26, where the time history of the pulse-echoed signal for testing the water level inside the pipe was recorded from the experimental setup. Note that a significant number of reflections are received from the pipe wall (the first set of reflections) and they are becoming further complicated by the multiple reflections within the water itself. This large number of reflections makes it difficult to base the determination of the height on simple TOF measurements. Figure 8.26b shows an autocorrection of the signal, where the autocorrelation leads to a first maximum group (the blue line) in the initial time stage at $t = 0$, then decays out at a certain period of time. This group of max autocorrelation is associated with the ringing signal from the pipe wall. While the second max group of the signal (the red line) appeared after a certain period of time, we obtained a local max at time $t = \tau$, when the backscattered echoes are separated by a time delay. The time, τ, thus, corresponds to a time delay between two successive echoes, corresponding to the TOF of the ultrasonic waves through the pipe and water. Since in a pulse-echo arrangement the acquired signals contain the input and output waveforms, an autocorrelation function is used as the tool to determine the TOF that is related to the water height. On the other hand, for a pitch-catch test arrangement, where the input and output signals are recorded separately, cross-correlation is used.

The water height measurements were done by both physical observation through the Plexiglas wall on the side of the tank and the autocorrelation determination in shown in Table 8.4, with a maximum difference of 6.0%. Some of the error may be attributed to the inaccuracy of physical measurement of the water height based on measuring from the side wall of the tank. It demonstrated the accuracy and feasibility of the autocorrelation method (Table 8.4).

To automate the data acquisition and analysis that determines the water height directly from the analog signals by the data acquisition system, a signal-processing computer code that includes the autocorrelation function was developed. Using the developed data acquisition and real-time signal-processing system, a test was performed to monitor the height of the water in the pipe while draining at two different rates. The results are shown in Figure 8.27. The developed algorithm and computer code were demonstrated to be relatively fast and accurate.

TABLE 8.4

Difference between the Measured and the Autocorrelation Calculations of Water Height

Water Height (cm (in.))	Calculated Height (cm (in.))	Difference%
2.5 (1.0)	2.39 (0.94)	6.0
5.1 (2.0)	4.98 (1.96)	4.0
7.6 (3.0)	7.34 (2.89)	3.7
10.1 (4.0)	10.49 (4.13)	3.3
12.7 (5.0)	12.78 (5.03)	0.6

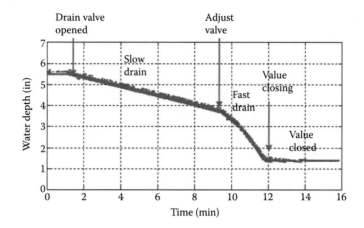

FIGURE 8.27
The determined water height for fast and slow draining rates.

8.4.4.2 Hilbert Transform

Another widely used signal-processing method for the TOF estimation is an envelope extraction method based on the Hilbert transform technique. Hilbert transform has been used to obtain an analytical signal (complex envelope) from a real signal to determine instantaneous frequency and envelop estimation (Boashash, 1992; Chen et al., 2005; Oruklu et al., 2009). The analytical signal $Z(t)$ of the echo signal $s(t)$ is defined in Equation 8.3, and the envelope of the analytical signal can be obtained with the magnitude of the signal $Z(t)$.

$$Z(t) = s(t) + jH[s(t)] = a(t)e^{-j\phi(t)} \tag{8.3}$$

where $a(t)$ is the envelop, $\phi(t)$ is phase, $j = \sqrt{-1}$, and $H[s(t)]$ is Hilbert transform of $s(t)$, defined as the Cauchy principal value of the integral:

$$H[s(t)] = \text{P.V.} \int_{-\infty}^{\infty} \frac{s(\tau)}{\pi(t - \tau)} d\tau \tag{8.4}$$

As an alternative approach, Hilbert transform-based signal processing has been developed to determine the TOF. It should be noted that the echoes generally interfere with the noises, which causes the distortion of the frequency spectrum. To overcome this problem, the signal need to be filtered, for example, in this case, a 10-order Butterworth-type high-pass filter was used with a cutoff frequency near the resonant frequency of the ultrasonic probe. This will filter out low-frequency noises due to ambient noise and interferences. The effect of a high-pass filter is shown in Figure 8.28, where the received signal and the filtered signals with their short-time Fourier transforms (STFT) are presented. It can be seen from the figures that the noise in the reconstructed signal has been greatly reduced and the echoes are clearly visible due to a high-pass filter.

One method of determining the values TOF value from the Hilbert envelope is to find the peak and threshold time of the first echo from the pipe wall, which are defined as T0 and T_1, respectively. Then, the time T_m at the local maximum of the second signal group can be found by searching above the threshold time T_1. Then the TOF can be found by the time difference $T_m - T0$ as shown in Figure 8.29.

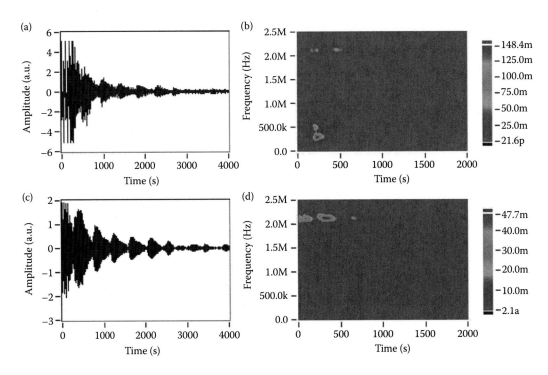

FIGURE 8.28
(See color insert.) (a) Received signal, (b) time–frequency spectrum of the (a) signal using short-time Fourier transform (STFT), (c) received signal of a high-pass filter, and (d) time–frequency spectrum of the (c) signal using STFT.

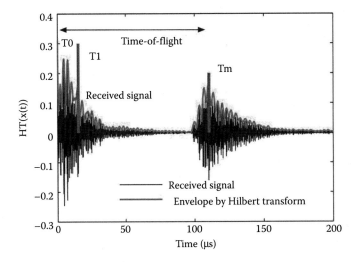

FIGURE 8.29
Received signal and the signal envelope obtained from Hilbert transform. T_0, T_1, and T_m are the peak time from the pipe wall, the threshold time, and the peak time from the water, respectively.

8.4.4.3 Shannon Energy

The normalized average Shannon energy known as Shannon envelope is also one of the widely used signal-processing methods for envelope extraction of cardiac sound signals (Liang et al., 1997; Choi and Jiang, 2008; Liu et al., 2012). The Shannon energy $S_E(t)$ and average Shannon energy $E_s(t)$ can be defined as follows:

$$S_E(t) = -x^2_{norm}(t)\log x^2_{norm}(t) \tag{8.5}$$

$$E_s(t) = \frac{1}{N}\sum_{i=1}^{N} x^2_{norm}(i)\log x^2_{norm}(i) \tag{8.6}$$

where x_{norm} is a normalized signal, N is the signal length, and $E_s(t)$ is the average Shannon energy for frame t. The normalized average Shannon energy $N(t)$, called as the Shannon envelope, is then calculated as follows:

$$N(t) = \frac{E_s(t) - M(E_s(t))}{S(E_s(t))} \tag{8.7}$$

where $M(E_s(t))$ is the mean value of $E_s(t)$, and $S(E_s(t))$ is the standard deviation of $E_s(t)$. An example of Shannon energy and Shannon envelope is shown in Figure 8.30. Note that this method emphasizes the medium intensity signal, which corresponds to the second maximum value, and attenuates the low- and high-intensity signals. Thus, the TOF can be obtained by finding the maximum intensity signal.

Figure 8.31 shows the performance comparison of the discussed methods for the determination of TOF. From Figure 8.31, the left figures show the original and processed signals

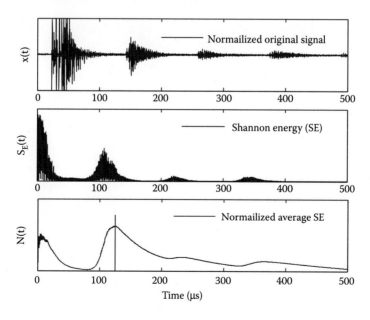

FIGURE 8.30
The normalized original signal, Shannon energy, and normalized average Shannon energy.

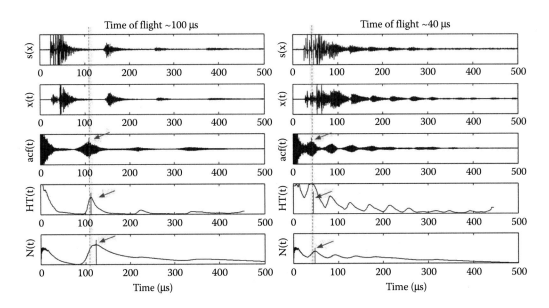

FIGURE 8.31

Signal-processing results from received signals in high (left figures) and low (right figures) oil levels. From the top down, they are original signal $s(t)$, filtered signal $x(t)$, autocorrelation of signal acf(t), Hilbert transform envelope $H(t)$, and Shannon energy envelope $N(t)$.

when TOF value was around 110 µs (approximately 7.5 cm of water height), while the right figures are those of around 40 µs, corresponding to 2.5 cm of water height. It can be seen that all processing methods provide reasonable accuracy for the detection of TOF, which is indicated by the small arrows. However, the drawback of this method is that the TOF value is generally overestimated compared to the values determined by autocorrelation and Hilbert envelop methods.

8.4.4.4 Results Discussion and Addressing the Issue of Shallow or No Water

The three different algorithms discussed above, including autocorrelation acf(t), Shannon envelope $N(t)$, and Hilbert envelope $H(t)$, have good accuracy to determine the TOF and target height under the normal operating conditions when the water lever is relative high. However, there are some cases that it is difficult to find the TOF, including the cases of shallow water or no water, water surface perturbation, and high-temperature operation. Therefore, the validation of the introduced signal-processing methods was investigated via case studies as follows.

The signal-processing results obtained from a case of shallow water that is less than 1 in. is shown in Figure 8.32. The limitation of the signal-processing methods for low height level is evident as these methods cannot resolve the overlapping echoes. The problem of the autocorrelation method arises from the fact that the noises in the received signals are not only from the ambient noise (white Gaussian noise). They also result from backscattered periodical ringing from the top and bottom of pipe wall. The consequence is the values of autocorrelation for the received echoes from pipe wall become higher than those of actual echoes from the water surface, making it difficult to determine the TOF when the target water height becomes lower, as shown in Figure 8.32. For the case of Hilbert envelope, due to the overlapping signals from the pipe wall and water surface, it may not be able to find

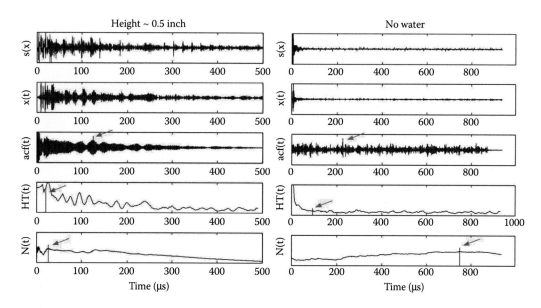

FIGURE 8.32
Signal-processing results from received signals in shallow oil. From the top down, they are original signal $s(t)$, filtered signal $x(t)$, autocorrelation of signal $acf(t)$, Hilbert transform envelope $H(t)$, and Shannon energy envelope $N(t)$.

the TOF value directly. Also, in the case of a Shannon envelope, the envelope was found to be flat, arising from the fact that the intensity of the reflected echoes are similar at low water height level, so the Shannon envelope method attenuates all signals. The results conclude that none of the above methods can be used for the low water height conditions.

In order to improve the accuracy and expand the range of the ultrasonic water level determination, a hybrid frequency analysis method was developed. This hybrid method is based on Hilbert transform (H) with a high-pass filter and followed by the fast Fourier transform (FFT). Figure 8.33 shows an example of the frequency-domain signal by

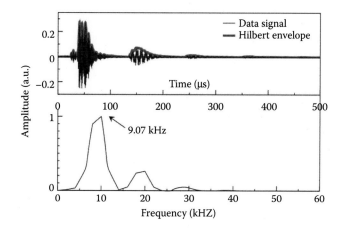

FIGURE 8.33
Reflected signal and Hilbert envelop (top). The bottom figure shows the fast Fourier transform of a Hilbert envelope.

taking the FFT of a Hilbert envelope. It can be seen that the peak frequency occurred at 9.07 kHz, whose inverse is the period of a signal envelope, which equals 110 μs. This time value corresponds to the value physically measured height level. Note that one major issue of the FFT of Hilbert envelope method is that the accuracy of echo frequency strongly depends on the search window length with respect to echo repetition period. Thus, the FFT of Hilbert envelope method was implemented only for low height of the target, with a narrow search window by cutting the signal in parts and only analyzing a small portion in time. Since the time interval between echoes is short in the low height condition, the analysis of the small part is sufficient to determine the frequency of echo repetition.

Another issue for the monitoring of steam pipe is the case of no water inside the pipe. Since there are no reflections from the water to be detected, the determination from the methods described before results in wrong TOF values. To address this issue, the Hilbert envelope energy algorithm was implemented in the data-processing system as guidance for the presence of water. The energy (E^*) can be obtained by integrating Hilbert envelope over the total sampling time T (see Equation 8.8). When there is no water or no reflections from the water, the energy that is equal to the integration of the Hilbert envelope becomes a low value. Therefore, the coherence between the energy of the echoes and the obtained TOF allows for the determination of the water level more accurately. In order to test the validity of this modified algorithm, all the water was drained from the pipe and the waveforms and water height were recorded in real time. Some of the waveforms and the corresponding water heights are shown in Figure 8.34 and this demonstrates the capability of height determination even when there is no water inside the pipe.

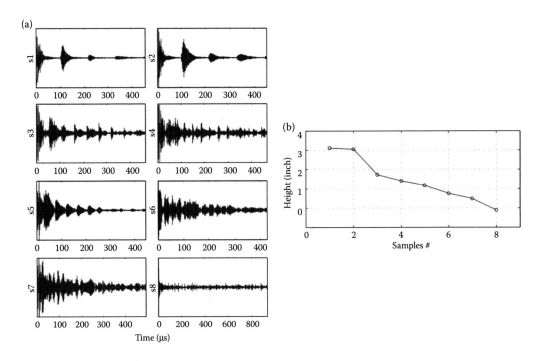

FIGURE 8.34
(a) Received waveforms (s1–s8) depending on the water height level and (b) the corresponding height level determined by the modified algorithm.

$$E^* = \frac{1}{T}\int_0^T H(t)dt \qquad (8.8)$$

8.4.5 Health Monitoring Test of Steam Pipes

8.4.5.1 Characterization of Bulk and Surface Perturbations

Using the developed algorithm for determining the height in real time, the capability to handle surface and bulk interferences was first investigated. For this purpose, surface perturbations were introduced by shaking the surface, rocking the container, and by introducing bubbles into the path of the acoustic wave. The test setup consisted of a pipe segment that was covered from its two sides by welded plates to form a container shape allowing for direct access from the top surface. A schematic view of the cross section of the test setup is shown in Figure 8.35, and Figure 8.36 shows a hose that introduced bubbles into the path of the wave inside the water.

Each test was conducted in a sequence of 1 min of water at "rest," 1 min of perturbation (bubbling or shaking), and 1 min of water at "rest" again. The data were acquired while calculating a moving average and excluding the outlier data. The bubbles were generated at the rate of ~3 bubbles per second and the surface wobbling was done at a rate of 2–3 Hz. The bubbles introduction consisted of placing an air tube 1.3 cm (0.5 in.) from the bottom of the pipe surface and the example of data obtained when generating bubbles 2.5 cm (1 in.) away from the wave path is shown in Figure 8.37. Noisy data were acquired in the window of time that the perturbation was introduced but the running average provided a reasonable accuracy of the water height measurement. Similar results were observed when introducing bubbles at various locations along the wave path as well as the direct shaking of the water surface by placing a small bowl into the water surface and raising and lowering it manually away from the water path. The result shows that the system can be used to monitor the disturbance with the correlation of the water height measurement with the further development of a desirable pattern recognition algorithm.

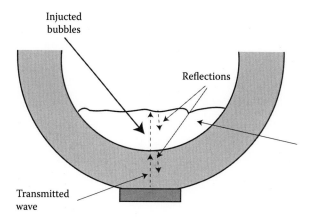

FIGURE 8.35
The cross-section illustration of the test setup.

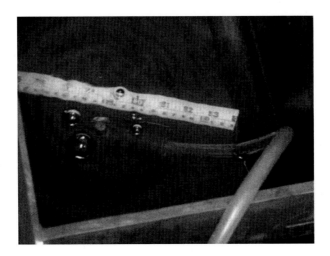

FIGURE 8.36
The bubbles that were introduced into the water via the shown hose (right).

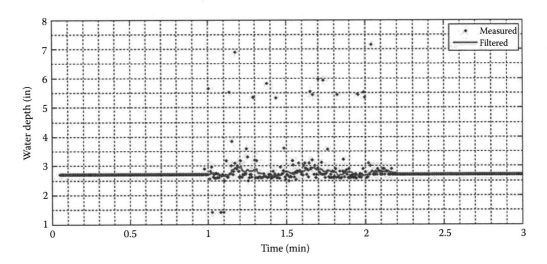

FIGURE 8.37
The water height data obtained during surface rest and perturbation from bubbles that were generated 0.5″ from the bottom of the pipe at 1″ away from the ultrasonic wave path.

8.4.5.2 High-Temperature Ultrasonic Characterization

In order to simulate the condensed water condition at 250°C, a high-temperature chamber was used consisting of oven Model 6680 (made by Blue M). The HT testbed consisted of the container shape pipe section that was mentioned earlier (see Figure 8.38). As a substitute for condensed water, the container was filled with silicone oil, where both are able to sustain such high temperatures. Using this setup, tests were performed at high temperatures while avoiding the need to deal with the hazard of high pressure that would be generated if water is placed in a closed chamber and heated.

Various ultrasonic probes were investigated to determine the optimum probe design. It was found that the reflected signals from oil using probe A and B were too low to allow

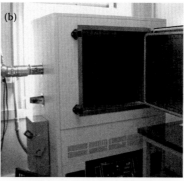

FIGURE 8.38
The high-temperature (HT) testbed with safflower oil (a) and the chamber (b) where it was subjected to 250°C.

for good height measurement accuracy. This is due to the fact that there are several effects from the steel pipe, including strong reflections from the interface of the steel pipe, the effect of the pipe curvature, and the wave losses due to scattering from a nonflat surface of the pipe. In addition, the silicone oil has higher attenuation, compared to water. Thus, only air-backed probe C with 2.25 MHz showed quite reasonable performance in terms of the sensitivity, resolution, and accuracy of the measurements. A comparison of the measured pulse-echo reflections from the steel pipe with the silicone oil using commercial custom-made probe and the in-house air-backed one is shown in Figure 8.39. It was found that the commercial probe suffered from low sensitivity and inability to receive the echoes from the oil surface at temperatures as much as 200°C and above. In contrast, the air-backed probe showed significantly higher signal-to-noise ratio with no degradation of the probe performance after the exposure to the high temperatures. This allowed for the analysis of the oil height level. However, it should be noted that when the temperature was decreased, the commercial probe was able to receive the echoes from the oil, and no performance degradation was observed during several thermal cycles from room

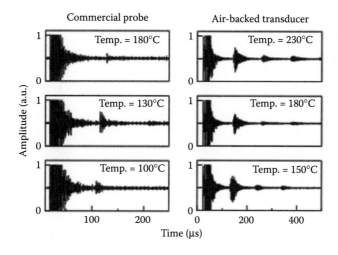

FIGURE 8.39
Pulse-echo responses with the custom-made probe (left) and air-backed probe (right) from oil containing steel pipe at various temperatures.

temperature to 250°C. Hereafter, tests were performed using the air-backed probes since they had effective capability of transmitting and receiving signal through a pipe wall at high temperatures.

One important thing that needs to be considered for ultrasonic testing is the potential effects of increasing cable length. One of the effects caused by a long cable is that it changes the transducer properties, such as frequency, bandwidth, and amplitude of pulse-echo response based on the transmission line theory. Moreover, there is a possibility that a long cable might lower the axial resolution of the probe, which prevents accurate ultrasonic distance measurements. The reason is closely associated with the extension of the excitation pulse length with increasing the cable length. For example, owing to the complex electrical impedance profile of most transducer probes, there is an electrical impedance mismatch between a cable and a probe and/or a pulser, which causes multiple reflections within the pulser and probe. When the reflected pulse from the probe reaches the probe again, it acts as a second excitation pulse that follows the original pulse. For the case of a short cable, the reflected pulse arrives very quickly after the initial excitation pulse; thus, the cable does not have a significant effect on the ultrasonic distance measurements. However, when the round trip time of reflected electrical energy becomes longer as a result of increasing the cable length, the multiple excitation pulse add to the original pulse signal by a round trip time interval, consequently increasing excitation pulse length, and limiting the axial resolution.

In order to investigate the effects of the cable length on the probe performance, two cable lengths were prepared; one is a 0.5 m (short) cable and the other is approximately a 10 m (long) cable. Figure 8.40 shows the measured electrical impedance and phase of a 2 MHz probe at room temperature with increasing the cable length. As shown, it was observed that an increase in the cable length affected the resonant characteristics of the probe, shifting the resonant and antiresonant frequencies to lower values. However, it should be noted that although the use of a long cable affected the resonant characteristics of the tested probe, the electromechanical coupling factor and the magnitude of electrical impedance at resonant frequency remained almost constant.

Figure 8.41 shows the received waveforms measured at 250°C to investigate the temperature effect on transducer performance with a long cable. It can be seen that the probe

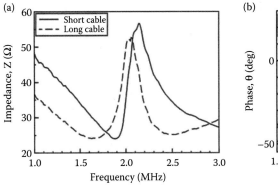
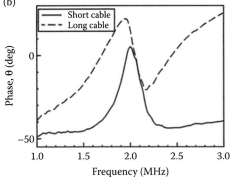

FIGURE 8.40
Effects of increasing cable length from short (0.5 m) to long (10 m) on electrical impedance and phase of a 2 MHz probe. (a) The impedance as a function of frequency and (b) the phase as a function of the frequency.

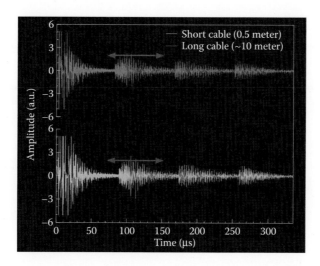

FIGURE 8.41
(See color insert.) Effects of increasing cable length from short (0.5 m) to long (10 m) on received signal waveforms measured at 250°C.

with a short cable provided slightly higher amplitude than the probe with the long cable; however, overall there is little difference between the results obtained with the short and long cables, concluding that the use of a 10 m (25.4 ft.) cable did not significantly affect the probe performance.

For the determination of the fluid height at high temperature using the ultrasonic method, the high-temperature probe was subjected to 250°C for 2.5 h, kept at 250°C for 2 h, and then cooled to room temperature. Since oil has low heat conduction, there was a need to assure that the temperature inside matches the chamber as closely as possible. For this purpose, a thermocouple was inserted into the oil and tracked the temperature as it has risen. The height of the silicone oil was measured while tracking the temperature of the chamber and the oil. Note that the thickness measurements at high temperatures require velocity recalibration as the properties of the materials change with temperature. As the sound velocities of the oil at different temperatures are not available in the literature, the sound velocity (c) of the silicone oil was estimated using the measured TOF (T_o) values that were determined by the Hilbert envelope-based signal-processing algorithm and the preestimated height (H) based on the thermal expansion coefficients of the silicone oil and the steel, which are 0.00073 cc/°C and 13 ppm/°C, respectively. The sound velocity in steel was assumed to decrease with 1 m/s per °C (Nowacki and Kasprzyk, 2009) Using the calculated velocity ($c = 2H/T_o$) and curve fitting method, the sound velocity was obtained as a function of temperature, whose result is demonstrated in Figure 8.42.

Figure 8.43 shows the calculated silicone oil height as a function of temperature with and without correction of sound velocity in steel and oil. As expected, the height increased with increasing temperature due to the thermal expansion of the silicone oil; however, note that the height was apparently overestimated without considering the temperature effect in sound velocity. Also note that the fluctuation in data point with temperature is due to the perturbations of the oil surface, which was caused by the air blown by the fan inside the HT chamber. This result demonstrates the suitability of the developed ultrasonic probe and signal-processing algorithm for the determination of the target height for high temperature up to 250°C.

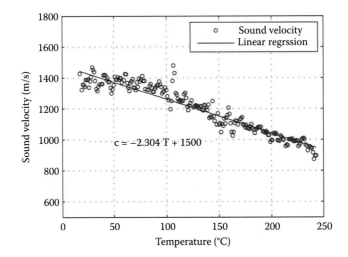

FIGURE 8.42
Estimated sound velocity in silicone oil as a function of temperature using the TOF and preestimated height of silicone oil and steel.

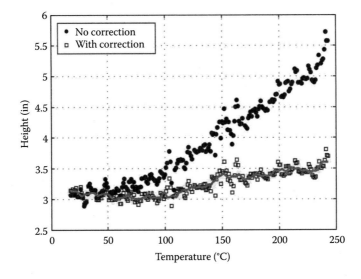

FIGURE 8.43
The measured height of the silicone oil as a function of temperature. The solid gray line shows the moving average of the determined data.

8.5 Conclusions

The detection of defects and the characterization of material properties without affecting the integrity of the material or structure require the use of nondestructive testing and evaluation methods. Increasingly, in-service monitoring methods are used to monitor the health of structures or determine various properties. Health and cure monitoring of structures at high temperatures is critical to assuring the integrity and properties compliance

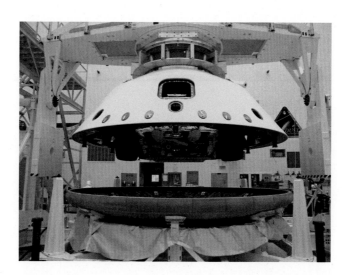

FIGURE 1.4
PICA was used in tiles form to produce the heat shield (bottom cone) of the MSL mission that landed on Mars in Aug. 2012. (Courtesy of JPL/NASA, Reference Figure No. MSL-2011-05-26-143545-IMG_0959.JPG.)

FIGURE 3.3
The Space Shuttle and its launch system require carbon–carbon leading edges and a nose-tip for the shuttle as well as carbon–carbon nozzles and exit cones for the solid rocket boosters.

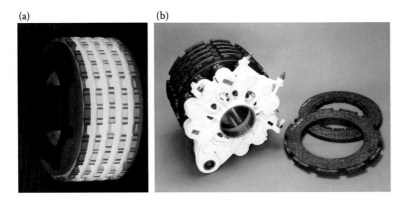

FIGURE 3.5
(a) During rejected take-off the brakes turn white hot, and the tires sometimes burn. (b) Brake assembly at room temperature along with stator and rotor shown for comparison.

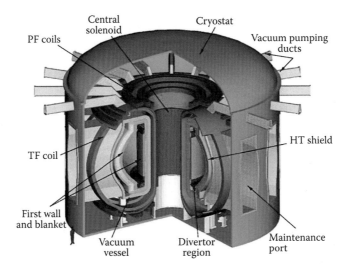

FIGURE 3.11
The ARIES-AT fusion power core. (From Najmabadi F. et al., *Fusion Eng. Design*, 80, (1–4), 2008, 3–23.)

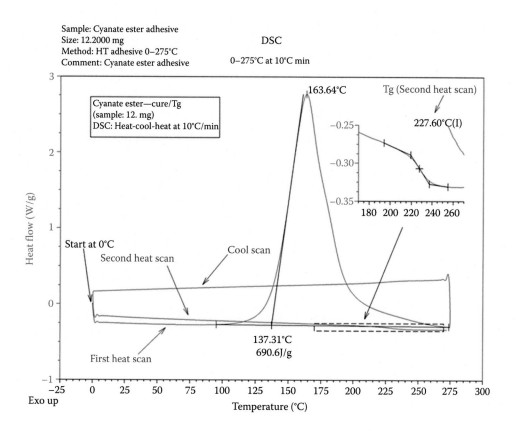

FIGURE 4.16
DSC Heat-Cool-Heat results for a cyanate ester adhesive. (Courtesy of Bill Warner, JPL Analytical Chemistry Group.)

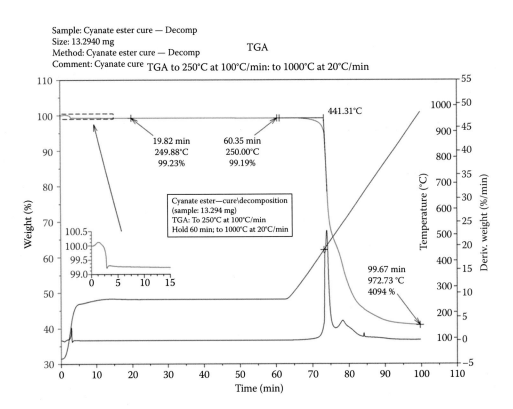

FIGURE 4.17
TGA result for a cyanate ester adhesive. (Courtesy of Bill Warner, JPL Analytical Chemistry Group.)

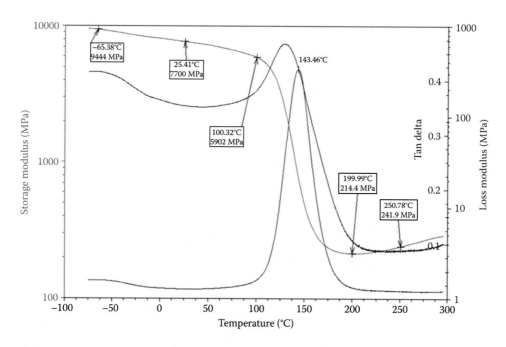

FIGURE 4.19
Example DMA thermograph for a die attach epoxy adhesive. (Courtesy of Gary Plett, JPL Analytical Chemistry Group.)

FIGURE 4.20
Die shear test coupons bonded to an ill-prepared galvanized steel pipe under conditions representative of the intended application and installation environment.

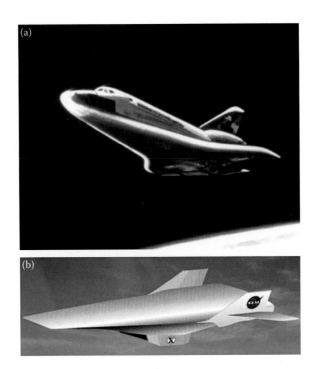

FIGURE 5.5
Hypersonic vehicles. (a) Artist's rendition of the Space Shuttle Orbiter on re-entry with its large radii leading edges and blunt nose, which leads to lower temperature, but less maneuverability. (b) Artist's rendition of the proposed Hyper X vehicle with sharp edges and more maneuverability. (From Sozer, E. 2012. Non Equilibrium Gas and Plasma Dynamics Laboratory http://ngpdlab.engin.umich.edu/static-pages/img/hypersonic-interaction-flows_hyper-x.png. Accessed December 14, 2012.)

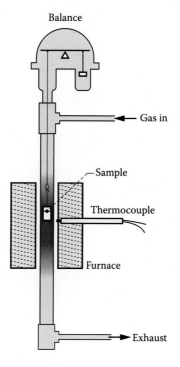

FIGURE 5.9
Thermogravimetric apparatus for studying isothermal oxidation/corrosion under controlled conditions with in-situ monitor of mass changes. These mass changes as a function of time give the kinetics of the process.

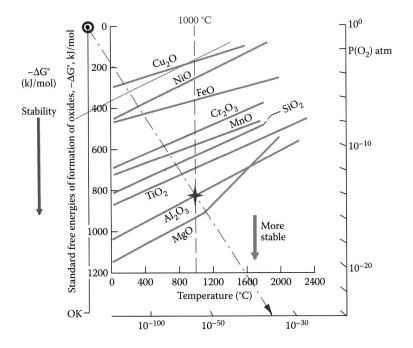

FIGURE 5.11
Modified Ellingham/Richardson diagram showing the relative thermodynamic stability (free energy of formation) of various oxides. The corresponding equilibrium gas composition can be found using the nomograph
by drawing a line from the focus point to the oxide curve at the temperature of interest and extending to the
side axes for O_2, H_2O/O_2, and CO/CO_2 equilibrium pressures. (Swalin, R. A.: *Thermodynamics of Solids*. 1962.
Copyright Wiley-VCH Verlag GmbH & Co. KGaA. Reproduced with permission.)

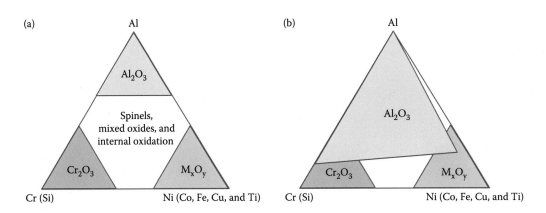

FIGURE 5.14
Notional oxide phase schematic of how alloy compositions affect scale make-up. (a) Symmetric, idealized:
single oxides near corresponding single element corners; mixed oxides near mixed alloy compositions.
(b) Asymmetric, real: regions can be highly skewed, allowing for maximum effect with minimum alloy content. Here Cr (or Si) are shown to increase the Al_2O_3 field for a base alloy of pure Ni (Co, Fe, Cu, or Ti).

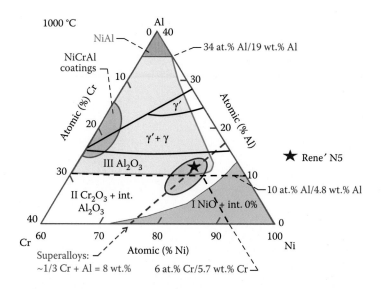

FIGURE 5.17
Actual Ni-rich corner of the Ni–Cr–Al oxide map at 1000°C showing primary regions of Al₂O₃, Cr₂O₃, and NiO. Al₂O₃ region is extended to higher Ni contents when Cr is added. NiAl and NiCrAl coating compositions are superimposed as green and blue area; superalloys follow the dashed red line; single-crystal superalloy Rene′ N5 is shown as star point. (With kind permission from Springer Science+Business Media: *J. Mat. Sci.*, Compositional factors affecting the establishment and maintenance of Al₂O₃ scales on Ni-Al-Pt systems, 44(7), 2009, 1704–1710, Gleeson, B. et al.)

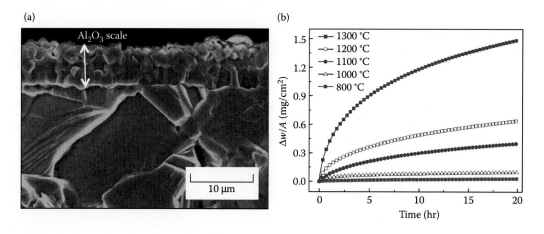

FIGURE 5.27
Protective α-Al₂O₃ scales on "312" and "211" MAX phases. (a) Fracture section showing alumina surface scale formed on 312 Ti₃AlC₂ oxidized at 1250°C for 20 h. (Reprinted from *Corr. Sci.*, 58, Wang, X. H. et al., Insights into high temperature oxidation of Al2O3-forming Ti3AlC2. 95–103, Copyright 2012, with permission from Elsevier.) (b) Isothermal oxidation kinetics of 211 Cr₂AlC oxidized at 1000–1250°C for 20 h. (Reprinted from *Acta Mat.*, 55(18), Lin, Z. J. et al., High-temperature oxidation and hot corrosion of Cr2AlC. 6182–6191, Copyright 2007, with permission from Elsevier.)

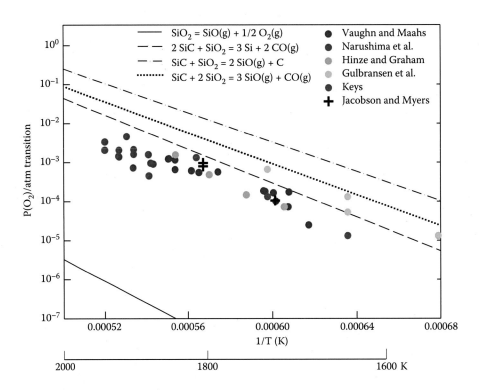

FIGURE 5.37
Plot of literature active/passive transitions a function of inverse temperature and compared to boundaries derived from the SiC/SiO_2 equilibrium and also SiO_2 decomposition.

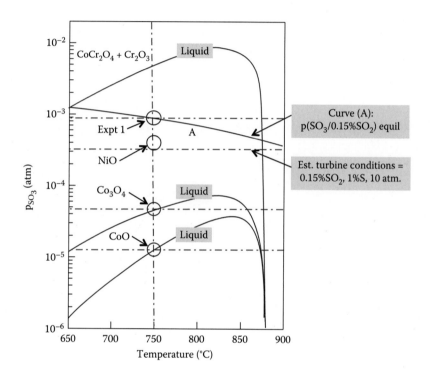

FIGURE 5.47
SO_3—Temperature reaction map for LTHC of Co. Curve A represents the equilibrium $P(SO_3)$ for an 0.15% SO_2/ O_2 environment at 1 atm. It is above that needed to form liquid $CoSO_4$–Na_2SO_4 salt solutions from CoO and Co_3O_4. Engine conditions also exceed these requirements for Co sulfation, but not Ni sulfation. (Reprinted from *Thin Sol. Films*, 119(3), Luthra, K. L. and J. H. Wood, High chromium cobalt-base coatings for low-temperature hot corrosion, 271–280, Copyright 1984, with permission from Elsevier.)

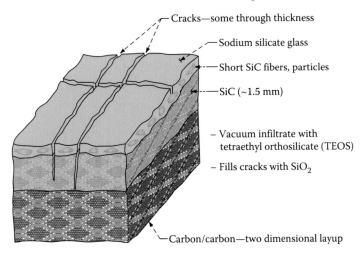

Coated reinforced carbon/carbon composite

Cracks—some through thickness

Sodium silicate glass

Short SiC fibers, particles

SiC (~1.5 mm)

– Vacuum infiltrate with tetraethyl orthosilicate (TEOS)

– Fills cracks with SiO_2

Carbon/carbon—two dimensional layup

FIGURE 5.55
Schematic of RCC.

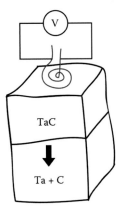

V

TaC

Ta + C

FIGURE 6.1
Schematic representation of a Ta + C self-propagating high-temperature synthesis (SHS) reaction to form TaC. The conductive coil on top (usually made of tungsten) represents one of several possible sources of heat that could initiate the SHS reaction.

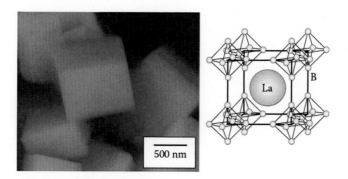

FIGURE 6.4
Scanning electron micrograph of LaB_6 powders obtained by solution combustion synthesis and the LaB_6 simple cubic unit cell. (From Kanakala R., G. Rojas-George, O.A. Graeve, *J. Am. Ceram. Soc.*, 93(10), 2010, 3136–3141; Kanakala R. et al., *ACS Appl. Mater. Interfaces*, 3(4), 2011, 1093–1100.)

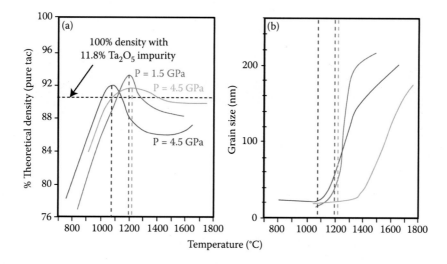

FIGURE 6.9
(a) Density and (b) grain sizes of tantalum carbide sintered at temperatures between 800°C and 1700°C and pressures between 1.5 and 4.5 GPa. Dotted lines correspond to temperature of maximum density. (Adapted from Yohe W.C., A.L. Ruoff, *Am. Ceram. Soc. Bull.*, 57(12), 1978, 1123–1130.)

FIGURE 7.2
Surface polishing system and polished cross sections mounted in a mold. (Photographed at JPL, Pasadena, CA.)

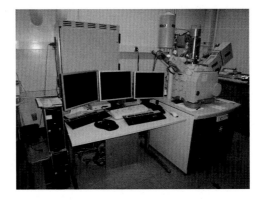

FIGURE 7.7
Two views of a scanning electron microscope (SEM) model FEI Nova Nanosem 600 (made by EDAX). (Photographed at JPL, Pasadena, CA.)

FIGURE 7.14
An x-ray diffraction analysis (XRD) system Model D500 Diffraktometer (Siemens). Overall system view (a) and close-up of the sample testing source and detector system (b). (Photographed at JPL, Pasadena, CA.)

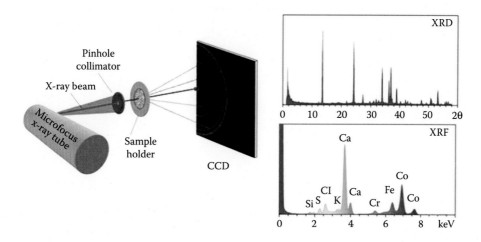

FIGURE 7.17
CheMin mineralogical instrument using XRD and XRF. (Blake et al., 2012. Image credit: Courtesy of David Blake, NASA Ames, and David Bish, University of Indiana.)

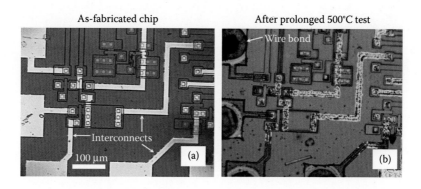

FIGURE 8.2
Optical micrographs of portions of NOT gate IC chips. (a) An as-fabricated chip prior to packaging. (b) A packaged chip following failure after thousands of hours operating at 500°C. (From Lee C., S. Akbar, and C. Park, Potentiometric CO_2 gas sensor with lithium ion electrolyte. *Processing and Fabrication of Advanced Materials XI, Proceedings of the International Symposium on Processing and Fabrication of Advanced Materials*, ASM International, Materials Park, OH, 7–10 Oct. 2002, pp. 63–77.)

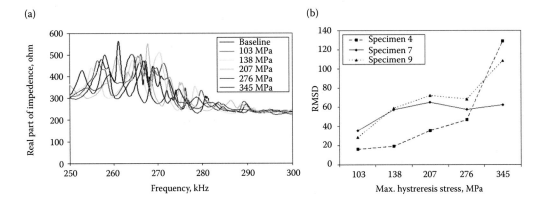

(a)

(b)

FIGURE 8.5

(a) Electrical impedance response after each cycle of a load/unload/reload tensile test. (b) Root mean square deviation value for three specimens as a function of maximum stress achieved during given cycle showing sensitivity to transverse cracks accumulated during cycle. (From Gyekenyesi A. et al., *The Journal of Intelligent Material Systems and Structures*, 20(7), 2009.)

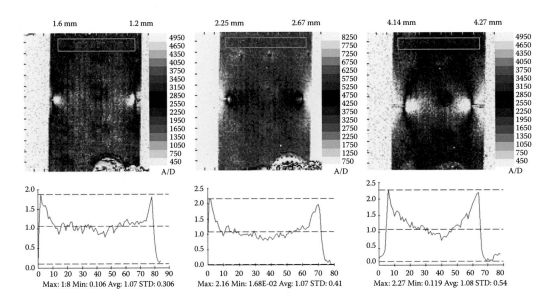

FIGURE 8.6

TSA images and SCFs for SiC/SiC specimen as a function of increasing notch lengths. Notch lengths, a, are indicated above the image. TSA stress range and mean stress were 35 and 17.5 MPa, respectively. The multishade specimen images correspond to the dimensionless digital values of the TSA system while the plots represent the SCF (i.e., local stress divided by the net stress) versus pixel position along a line scan between the two notch roots. (From Gyekenyesi A. and G. Morscher, *Journal of Materials Engineering and Performance*, 19(9), 2010.)

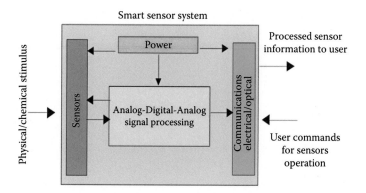

FIGURE 8.13

A smart sensor system. The core of a stand-alone smart sensor system includes sensors, power, communication, and signal processing. (From Hunter G.W. et al. *Interface Magazine, Electrochemical Society Inc.* 20(1), Winter, 2011b, 66–69.)

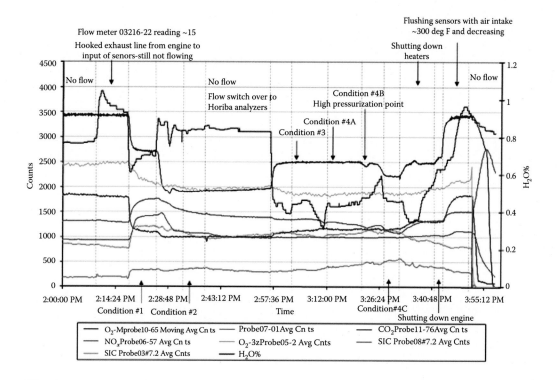

FIGURE 8.18

Example data from emission sensor array during an engine test stand run over a range of operational conditions. (Courtesy of Makel Engineering, Inc.)

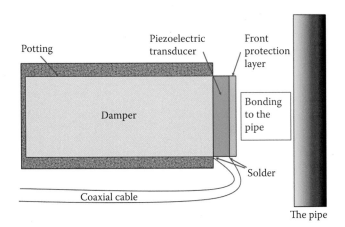

FIGURE 8.20
A schematic diagram illustrating the probe components (not to scale).

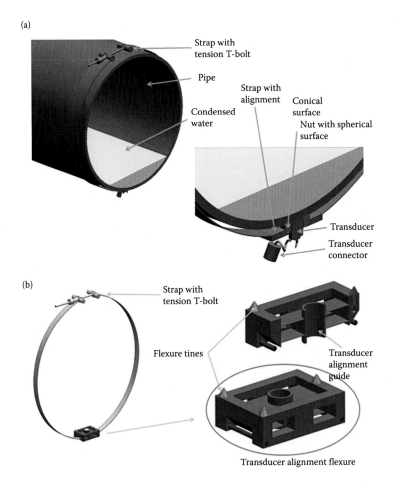

FIGURE 8.24
Schematic view of the mounting strap for the alignment and strapping the probe to the pipe surface in the field.

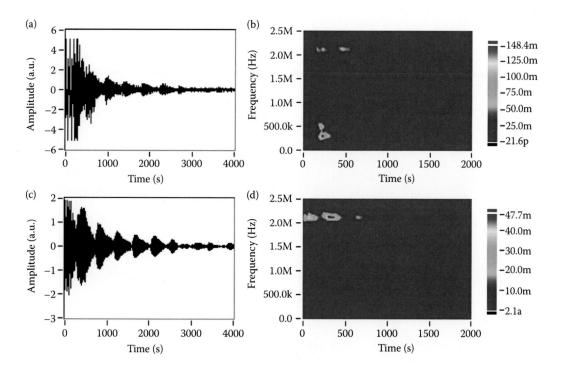

FIGURE 8.28
(a) Received signal, (b) time–frequency spectrum of the (a) signal using short-time Fourier transform (STFT), (c) received signal of a high-pass filter, and (d) time–frequency spectrum of the (c) signal using STFT.

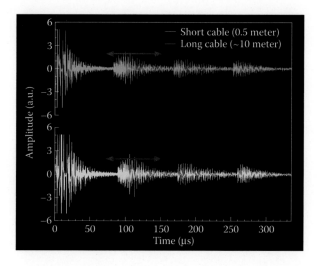

FIGURE 8.41
Effects of increasing cable length from short (0.5 m) to long (10 m) on received signal waveforms measured at 250°C.

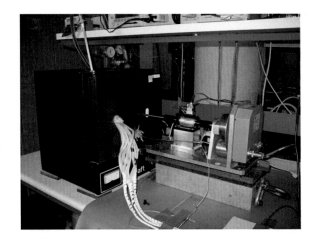

FIGURE 9.7
SRM high-temperature test setup.

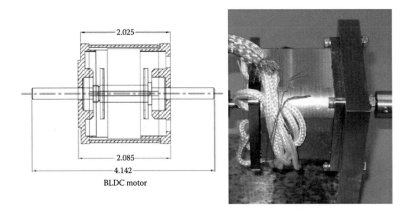

2.025

2.085
4.142
BLDC motor

FIGURE 9.12
Honeybee high-temperature BLDC motor.

FIGURE 9.14
Motor test setup.

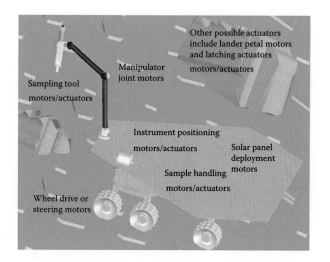

FIGURE 10.2
Schematic drawing of the possible actuator/motor locations on a rover. (Adapted from Sherrit, S., 2005, *Smart material/actuator needs in extreme environments in space*, *Proceeding of the Active Materials and Behaviour Conference, SPIE Smart Structures and Materials Symposium*, Paper #5761–48, San Diego, CA, March 7–10.)

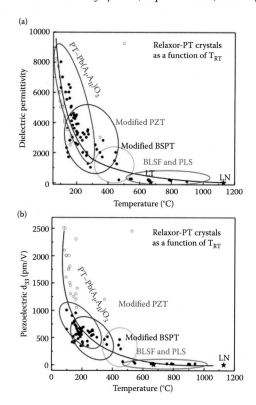

FIGURE 10.5
Temperature dependent dielectric and piezoelectric properties for various ferroelectric families. (Data obtained from Zhang, S. J., and Li, F., 2012, *Journal of Applied Physics*, 111, 031301; Zhang, S. J., and Yu, F. P., 2011, *Journal of the American Ceramic Society*, 94, 3153–3170; Sebastian, T. et al., 2010, *Journal of Electroceramics*, 25, 130–134; Sebastian, T. et al., 2012, *Journal of Electroceramics*, 28, 95–100.)

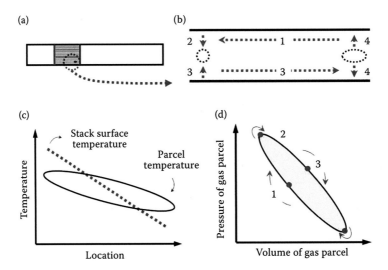

FIGURE 11.4
Thermodynamic cycle of the standing-wave thermoacoustic engine.

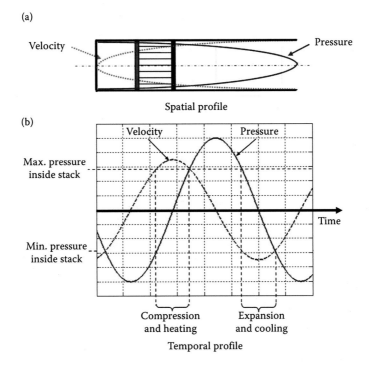

FIGURE 11.5
Spatial and temporal profiles of velocity and pressure inside the standing-wave thermoacoustic engine.

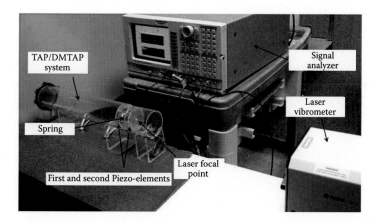

FIGURE 11.17
Experimental setup: Laser vibrometer used to scan the surface of the piezo-elements to obtain values for the transverse deflection.

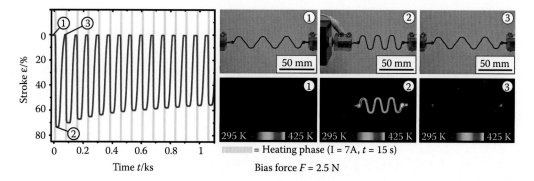

FIGURE 12.8
Additive manufactured NiTi actuator performing the shape memory effect. (From Haberland C. 2012. *Additive Verarbeitung von NiTi-Formgedächtniswerkstoffen mittels Selective Laser Melting*. Germany: Ruhr University Bochum; Aachen: Shaker Verlag GmbH.)

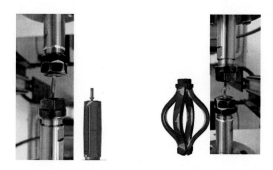

FIGURE 12.15
Left: Superelastic NiTi expandable reamer before deformation. Right: Superelastic NiTi retrograde blade after deformation. (Courtesy of Symmetry Medical Inc, New Bedford, MA.)

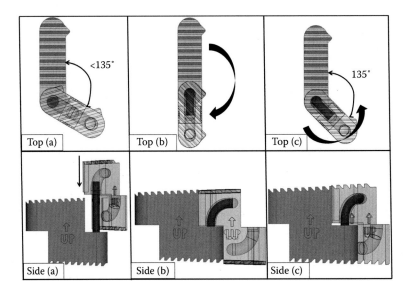

FIGURE 12.17
The assembly and deployment configuration of the intervertebral cage. (From Anderson W. 2013. Development of an Intervertebral Cage Using Additive Manufacturing with Embedded NiTi Hinges for a Minimally Invasive Deployment. University of Toledo.)

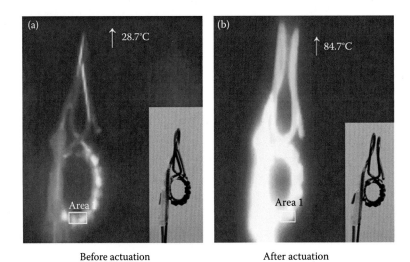

Before actuation   After actuation

FIGURE 12.20
Experimental evaluation of the smart antagonistic tissue clamp. (Courtesy of Lifewire LLC Macon GA.)

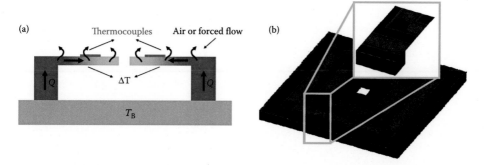

FIGURE 13.22
Simplified cross section of the proposed device with suspended membrane and central hole (a). 3D schematic of the designed micromachined generator (b). (From Dalola, S, and Ferrari, V, 2011. *Procedia Engineering*, 25, 207–210.)

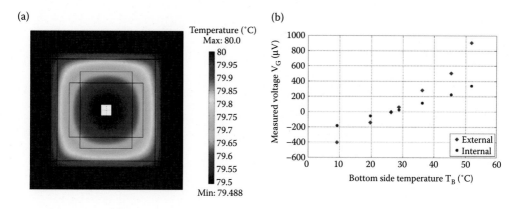

FIGURE 13.23
Simulated results of the temperature distribution for the microgenerator with a uniform temperature of the bottom side TB = 80°C and ambient temperature at 25°C (a). Measured output voltage versus the applied bottom side temperature TB. (From Dalola, S, and Ferrari, V, 2011. *Procedia Engineering*, 25, 207–210.)

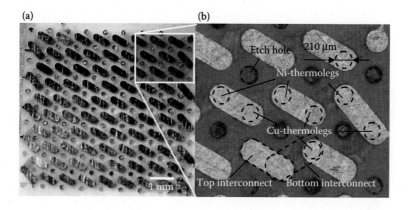

FIGURE 13.25
Pictures of an SU-8-based TEG device. (a) Complete device with 90 Cu–Ni-thermocouples in a meander shape serial connection. (b) Detailed cut out view. (Li, Y et al., Chip-level thermoelectric power generators based on high-density silicon nanowire array prepared with top-down CMOS technology, *IEEE Electron Device Letters*, 32, 674–676 © 2011 IEEE.)

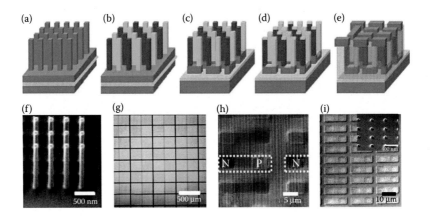

FIGURE 13.26
Schematic of fabrication. (a) SiNW formation by dry etch. (b) Ion implantation and P/N element definition with each element consisting of hundreds of SiNW. (c) P/N couples formed by dry etch. (d) SiNW top and bottom silicidation while protecting the sidewall. (e) Dielectric deposition and etch back to expose only the tip of the SiNW and top metallization. (f) SEM images of pillar formation. (g) N and P implants can be seen clearly under microscope with a different shade. (h) SEM image of SiNW after N/P implant. (i) Metallization etch showing individual N/P couples. Inset shows the tips of the SiNW exposed after oxide etch which confirms the structure of the TEG. (Li, Y et al., Chip-level thermoelectric power generators based on high-density silicon nanowire array prepared with top-down CMOS technology, *IEEE Electron Device Letters*, 32, 674–676 © 2011 IEEE.)

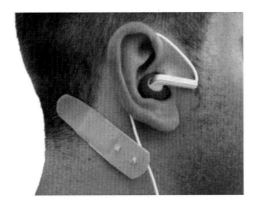

FIGURE 13.36
Demonstration of TEGs for supplying a biomedical hearing aid. (From Lay-Ekuakille, A et al., 2009. Thermoelectric generator design based on power from body heat for biomedical autonomous devices, *MeMeA 2009–International Workshop on Medical Measurements and Applications*, 1–4.)

FIGURE 14.1
Photographic view of diamond bits showing the cutters used to shear rock with a continuous scraping motion.
(a) A very early type diamond bit (surface set diamond bit). (b) 8 in. diameter PDC (polycrystalline diamond
compact) drill bit.

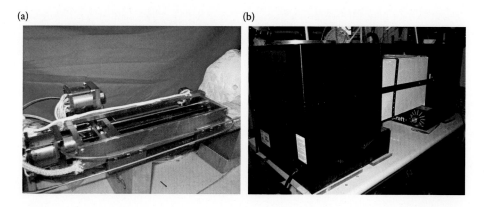

FIGURE 14.2
Left: High-temperature drill with two SR motors before the drill test. The drill was integrated and tested out-
side of the chamber first. Once the algorithm was confirmed, the drill was set up in the chamber to test at 460°C.
Right: The drill was placed inside an oven at 460°C.

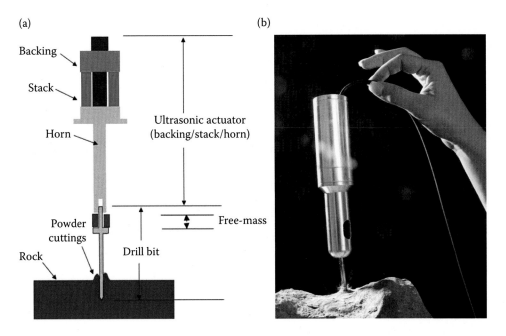

FIGURE 14.4
A schematic cross-section view (a) of the USDC and a photo showing its ability to core with minimum axial force (b).

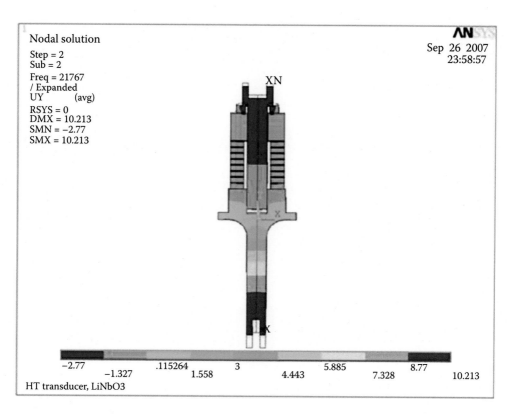

FIGURE 14.7

The first longitudinal mode of a LiNbO$_3$ shape with resonance frequency at 21.767 kHz. The color scale shows the displacement in the vertical direction.

FIGURE 14.26
1.5 in. bismuth titanate sampler on fixture after test.

(a)

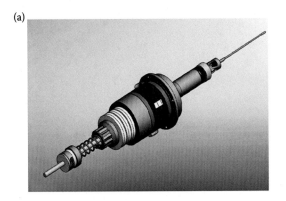

(b)

FIGURE 14.36
The components of the rotary-hammering sampler.

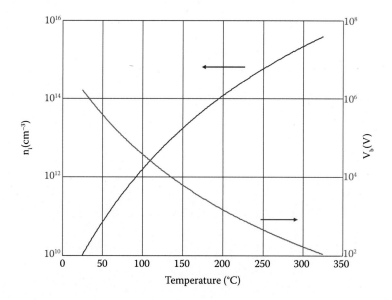

FIGURE 15.1
Intrinsic carrier concentration n_i and breakdown voltage versus temperature in Si.

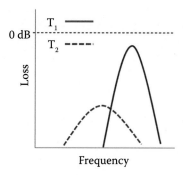

FIGURE 16.3
This illustration depicts the frequency shift resulting from the thermal expansion along the resonant dimension of a single element piezoelectric transducer and increased scattering due to material degradation. Where the difference in temperatures T_1 and T_2 can be characterized by: $T_2 \gg T_1$.

FIGURE 17.5
A synthetic image of the volcano Maat Mons on the surface of Venus that was created from Magellan orbital radar data. (From Wikimedia Commons, http://en.wikipedia.org/wiki/File:Venus_-_3D_Perspective_View_of_Maat_Mons.jpg.)

to design requirements. Controlling the processing parameters often requires operation at extreme temperature and/or pressures. Typical HM systems consist of a sensor, data acquisition, setup for processing and control as well as a microprocessor that acquires and analyzes the data and provides real-time information.

While most NDE methods are applied at room temperatures, health-monitoring methods are applied at the relevant temperatures and they may reach several hundreds of degree Centigrade. Applying HM methods at high temperatures requires sensing elements that are either capable of withstanding the environment temperatures or using remote monitoring from a cooler location. Making high-temperature sensing systems requires developing the use of sensing components from proper materials and addressing the packaging and wiring issues. Significant efforts are made when applying sensing systems at high temperatures is involved with the packaging since differences in CTE and repeated temperature swings (may reach up to 1000°C) can cause system failure. It is typical that data acquisition systems are located in cooler environments, but this is now changing with the development of SiC and SOI electronic technologies. These technologies are allowing for more data processing and communications capability to be located in locations up to 500°C. These technologies bring the promise of smart sensor capabilities to high-temperature applications that have been until now limited to applications where traditional silicon electronics are possible.

Various sensing systems were reported in this chapter, including ones for testing CMC, aircraft engines, and steam pipes.

Acknowledgments

Some of the research reported in this chapter was conducted at the Jet Propulsion Laboratory (JPL), California Institute of Technology, under a contract with National Aeronautics and Space Administration (NASA). The authors thank Edward Ecock, Josephine Aromando, and David Y. Low, Consolidated Edison Company, New York, New York, for their support of this task and for the inputs, comments, and helpful suggestions regarding the HM of the height of condensed water in steam pipes. The authors thank the students that worked at the JPL's Advanced Technologies Group, Patrick Ostlund and Nobi (Nobuyuki) Takano, Cal Poly Pomona, and Alessandro Bruno, Pisa University for their help in producing probes and fixtures for the steam pipes HM system. The authors also thank Julian Blosiu, JPL, for his helpful comments and suggestions. The authors thank Don J. Roth, NASA Glenn Research Center, Cleveland, Ohio; Ali Abdul-Aziz, NASA Glenn Research Center and Cleveland State University, Cleveland, Ohio; Subhasish (Subh) Mohanty, Argonne National Laboratory, Lemont, Illinois; Edward Ecock, Consolidated Edison Company, New York, New York for reviewing this chapter and for providing valuable technical comments and suggestions.

References

Aerospace Recommended Practice ARP 1533A, *Procedure for the Analysis and Evaluation of Gaseous Emissions from Gas Turbine Engines*, Society of Automotive Engineers, Warrendale, PA, April 2004.

Aerospace Recommended Practice ARP 1256, *Procedure for the Continuous Sampling and Measurement of Gaseous Emissions from Aircraft Turbine Engines*, Rev. B, Society of Automotive Engineers, Warrendale, PA, August 1990.

Akbar S., P.K. Dutta, and C. Lee. High-temperature ceramic gas sensors: A review. *International Journal of Applied Ceramic Technology* 3(4), 2006, 302–311.

Bar-Cohen Y., K.H. Nguyen, and R. Botsco, Eddy currents monitor composites cure, *Advanced Materials & Processes, ASM International* 139(4), 1991, 41–44.

Bar-Cohen Y., A. Chatterjee and M. West, Sensors for cure monitoring of composite materials, *Review of Progress in Quantitative NDE*, 12A, D. O. Thompson, and D. E. Chimenti (Eds.), Plenum Press, New York, 1993, Vol. 12B, pp. 1039–1046.

Bar-Cohen Y., High Temperature Technologies for Sample Acquisition and In-Situ Analysis, JPL Technical Report, Document No. D-31090, Sept. 8, 2003.

Bar-Cohen Y., S. Lih, M. Badescu, X. Bao, S. Sherrit, S. Widholm, and J. Blosiu, In-service monitoring of steam pipe systems at high temperatures. *SPIE Health Monitoring of Structural and Biological Systems IV Conference, Smart Structures and Materials Symposium*, Paper 7650-26, San Diego, CA, March 8–11, 2010a.

Bar-Cohen Y., S.-S. Lih, M. Badescu, X. Bao, S. Sherrit, S. Widholm, J. Scott, and J. Blosiu, In-service monitoring of steam pipe systems at high temperatures, NTR Docket No. 47518, Submitted on January 28, 2010. Provisional Application No. 61/312,164 filed by Caltech, Patent File No. CIT-5563-P submitted on March 9, 2010b.

Bar-Cohen Y., S-S. Lih, M. Badescu, S. Widholm, X. Bao, S. Sherrit, and Z. Chang, Task A: Development of a system for measuring condensed water levels, in Y. Bar-Cohen, J. Blosiu, Advanced Health Monitoring of Steam Pipe Systems, Final Report of the "Phase I: Feasibility Study and Proof-Of-Principle Demonstration" Contract with Con-Edison, JPL Task Plan No. 82- 12986, JPL Report No. D-68772, May 4, 2010c.

Bar-Cohen Y., X. Bao, J. Scott, S. Sherrit, S. Widholm, M. Badescu T Shrout, and B. Jones, Drilling at high temperatures using ultrasonic/sonic actuated mechanism, Invited Paper, ASCE's Earth and Space 2010 conference, Honolulu, HI on March 14–17, 2010d.

Bar-Cohen Y., M. Badescu, X, Bao, Z, Chang, S,-S, Lih, and S, Sherrit, S. Widholm, and P. Ostlund, High Temperature Piezoelectric Actuated Sampler for Operation on Venus, Final Report for the PIDDP task, NASA WBS No. 811073.02.06.55.13, Period of performance from Jan. 2007 – Jan. 2011, Report No. D-68419, Jan. 27, 2011.

Beachkofski B.K, Micro-Electro-Mechanical-System (MEMS) Requirements for Turbine Engines, 41st AIAA/ASME/SAE/ASEE Joint Propulsion Conference & Exhibit 10 - 13 July 2005, Tucson, Arizona, AIAA 2005-4372.

Behbahani A., and K. Semega, Sensing Challenges For Controls And PHM In The Hostile Operating Conditions Of Modern Turbine Engine, Structures and Controls Branch, Turbine Engine Division; AIAA-2008-5280, July 2008; AFRL-RZ-WP-TP-2008-2184; American Institute of Aeronautics and Astronautics, August 2008.

Behbahani A., D. Culley, B. Smith, C. Darouse, R. Millar, B. Wood, J. Krodel, S. Carpenter, B. Mailander, T. Mahoney, R. Quinn, C. Bluish, B. Hegwood, G. Battestin, W. Roney, W. Rhoden, and B. Storey, Status, Vision, and Challenges of an Intelligent Distributed Engine Control Architecture, AFRL-RZ-WP-TP-2008-2042, September, 2007.

Beheim G., P.G. Neudeck, and D.J. Spry, High Temperature SiC Electronics: Update and Outlook, Propulsion Controls and Diagnostics Workshop Cleveland, OH February 28–29, 2012.

Bencic T., and Woike M., Microwave Turbine-Tip-Clearance Sensor Tested in Relevant Combustion Environment in NASA Glenn Research Center Research & Technology 2007 Report, NASA/TM-2008-215054, pp. 109, 2008.

Benson D., Research Priority: Distributed Controls Architecture Propulsion Controls & Diagnostics Workshop, 9 December 2009, Cleveland, Ohio.

Bracewell R., *The Fourier Transform and its Applications* (McGraw-Hill, London, 1978).

Boashash B., Estimating and interpreting the instantaneous frequency of a signal- Part 1: Fundamentals, *Proceedings of the IEEE* 80, 1992, 520–538.

Chen L.-Y., G.M. Beheim, and R. Meredith, Packaging Technology for High Temperature Capacitive Pressure Sensors, in *Proceedings of 2010 iMAPS International High Temperature Electronics Conference* (2010 HiTEC), Albuquerque, New Mexico, May 12–15, 2010.

Chen T.-L., P.-w. Que, Q. Zhang, and Q.-k. Liu, Ultrasonic signal identification by empirical mode decomposition and Hilbert transform, *Review of Scientific Instruments* 76, 2005, 085109.

Choi S., and Z. Jiang, Comparison of envelope extraction algorithms for cardiac sound signal segmentation, *Expert Systems with Applications* 34, 2008, 1056–1069.

Ciriscioli P.R., and G.S. Springer, *Smart Autoclave Cure of Composites*, Technomic Publishing, Lancaster, PA, 1990.

Cutard T., D. Fargeot, C. Gault, and M. Huger, Time delay and phase shift measurements for ultrasonic pulses using autocorrelation methods, *Journal of Applied Physics* 75, 1994, 1909–1913.

Day D., Thermoset Cure Control fore Utilizing Microdielectrometer Feedback, 33rd SAMPE, March, 1988.

DOT/FAA Report, Engine Damage-Related Propulsion System Malfunctions, AR-08/24, December 2008.

Dynys F., and A. Sayir, *Self-Powered Wireless Sensors, Material Science & Technology Conference*, Pittsburgh, Pa., Oct. 2008.

Fralick G.C., J. Wrbanek, and C. Blaha, Thin film heat flux sensor of improved design, *Proceedings of the International Instrumentation Symposium, v 48, Proceedings of the 48th International Instrumentation Symposium* 2002, p. 409.

Geisheimer, J.L., Billington, S.A, and Burgess, D.W., A Microwave Blade Tip Clearance Sensor for Active Clearance Control Applications, AIAA-2004-3720, 2004.

Gyekenyesi A., R. Martin, G. Morscher, and R. Owen, Impedance based structural health monitoring of a ceramic matrix composite, *The Journal of Intelligent Material Systems and Structures* 20(7), 2009, 875–882.

Gyekenyesi A. and G. Morscher, Damage progression and stress redistribution in notched SiC/SiC composites, *Journal of Materials Engineering and Performance* 19(9), 2010, doi: 10.1007/s11665-010-9622-4.

Gyekenyesi A., G. Morscher, and L. Cosgriff, In-Situ monitoring of damage in SiC/SiC composites using acousto-ultrasonics, *Composites Part B* 37B(1), 2006.

Hertzog P., and G. Jordaan, Auto correlation on ultrasonic doppler signals for pregnancy determination in sheep, in AFRICON, 2004. *7th AFRICON Conference in Africa* 1, 2004, 173–178.

Hirsch T., H. Klümper-Westkamp, P. Mayr, and J. Vetterlein, Eddy Current Testing at High Temperatures for Controlling Heat Treatment Process, Open Access NDE.net NDTCE, 2003 http://www.ndt.net/article/ndtce03/papers/p022/p022.htm

Holst, T.A., T.R. Kurfess, S.A. Billington, J.L. Geisheimer, and J.L. Littles, Development of an Optical-Electromagnetic Model of a Microwave Blade Tip Sensor, AIAA-2005-4377, 2005.

Holst, T.A., Analysis of spatial filtering in phase-based microwave measurements of turbine blade tips, Master's Thesis, Georgia Institute of Technology, Atlanta, Georgia, August 2005.

Hunter G.W., *Morphing, Self-Repairing Engines: A Vision for the Intelligent Engine of the Future at AIAA/ICAS International Air & Space Symposium and Exposition*. The Next 100 Years, AIAA 2003-3045, Dayton, OH, July 2003.

Hunter G.W., C.C. Liu, and D. Makel Microfabricated chemical sensors for aerospace applications, in Gad-el-Hak, M ed. *MEMS Handbook,* 2nd edition, Design and Fabrication, CRC Press LLC, Boca Raton, Chapter 11, 2006a.

Hunter G.W., J.D. Wrbanek, R.S. Okojie, P.G. Neudeck, G.C. Fralick, L. Chen, J. Xu and G.M. Beheim, Development and application of high-temperature sensors and electronics for propulsion applications, *Proc. SPIE* 6222, 622209, 2006c.

Hunter G.W., J.C. Xu, P.G. Neudeck, D.B. Makel, B. Ward, and C.C. Liu, Intelligent chemical sensor systems for in-space safety applications, *42nd AIAA/ASME/SAE/ASEE Joint Propulsion Conference And Exhibit*, Sacramento, CA, 9–12 July 2006b, Paper AIAA-2006-4356.

Hunter G.W., J.C. Xu, and D.B. Makel, *Case Studies in Chemical Sensor Development*, Springer Press, New York, NY, Chapter 8, 2007a.

Hunter G.W., P.G. Neudeck, R.S. Okojie, G.M. Beheim, L. Chen, D. Spry, and A. Trunek, An overview of wide bandgap SiC sensor and electronics development at NASA glenn research center, *ECS Transactions*, 11(5), Editor(s): J. Wang, E. Stokes, J. Kim, H. Kuo, J. Bardwell, G. Hunter, J. Brown, 2007b, pp. 247–257.

Hunter G.W., J.C. Xu, L.K. Dungan, B.J. Ward, S. Rowe, J. Williams, D.B. Makel, C.C. Liu, and C.W. Chang, Smart sensor systems for aerospace applications: From sensor development to application testing, *ECS Transactions* 16(11), 2008, 333–344.

Hunter G.W., J.C. Xu, J.C., and D.B. Makel, Case studies in chemical sensor development, *BioNanoFluidic MEMS*, edited by P. J. Hesketh, Springer Science and Business Media, New York, 2008a, pp. 197–231.

Hunter G.W., J.C. Xu, D. Lukco, National Aeronautics and Space Administration, Washington, DC, U.S. Patent No. 7389675, June 24, 2008b.

Hunter G.W., G.M. Beheim, G.E. Ponchak, M.C. Scardelletti, R.D. Meredith, F.W. Dynys, P.G. Neudeck, J.L. Jordan, and L.Y. Chen, Development of high temperature wireless sensor technology based on silicon carbide electronics, *ECS Transactions* 33(8), 2010, 269–281.

Hunter G.W., G.M. Beheim, G.E. Ponchak, M.C. Scardelletti, R.D. Meredith, P.G. Neudeck, J.L. Jordan, L.Y. Chen, J.C. Xu, A.M. Biaggi-Labiosa, B.J. Ward, and D.B. Makel, Development of High Temperature Smart Sensor Systems presented at the 219th Meeting of the Electrochemical Society, Montreal, QC, Canada, May 1–6, 2011a.

Hunter G.W., J.R. Stetter, P.J. Hesketh, and C.C. Liu Smart sensor systems, *Interface Magazine, Electrochemical Society Inc.* 20(1), Winter, 2011b, 66–69.

Hunter G.W., L.G. Oberle, G. Baaklini, J. Perotti, and T. Hong, Intelligent sensor systems for health management applications, in *System Health Management: With Aerospace Applications*, edited by S. B. Johnson, T. Gormley, S. Kessler, C. Mott, A. Patterson-Hine, K. Reichard, and P. Scandura, Jr., John Wiley & Sons, Ltd, Chichester, West Sussex, UK, 2011c.

Hunter, G.W., J.C. Xu,, A.M. Biaggi-Labiosa, B. Ward, P. Dutta, and C.C. Liu, Smart sensor systems for spacecraft fire detection and air quality monitoring. *In 40th International Conference on Environmental Systems*, AIAA: Portland, Oregon, 2011d; Vol. AIAA 2011-5021.

Hunter G.W., and A. Behbahani, A brief review of the need for robust smart wireless sensor systems for future propulsion systems, distributed engine controls, and propulsion health management, *58th International Instrumentation Symposium*, Hyatt Regency La Jolla, San Diego, California, 4–7 June 2012.

Huston D.L., Cure monitoring of thermosetting polymers by an ultrasonic technique, *Review of Progress in QNDE*, D. O. Thompson and D. E. Chimenti (Eds.), 2B, 1983, 1711–1729.

International Civil Aviation Organization, Aeronautical Communications Panel (ACP), Eighteenth Meeting Of Working Group F, Montreal, 12–22, May 2008 http://www.icao.int/anb/panels/acp/wg/f/wgf18/acp-wgf18-wp05_avsi%20submission%20to%20icao%20may%20wg%20f%20meeting.doc

Ivey P.C., and K.R. Grant, The New Potential for Joint UK/US Studies in Sensor Technology for HCF Studies in Gas Turbine Aero-Engines, AIAA-2002-3242, June 2002.

Kasuya T., T. Okuyama, N. Sakurai, H. Huang, T. Uchimoto, T. Takagi, Y. Lu, and T. Shoji, In-situ eddy current monitoring under high temperature environment, *International Journal of Applied Electromagnetics and Mechanics* 20(3), 2004, 163–170 http://www.deepdyve.com/lp/ios-press/in-situ-eddy-current-monitoring-under-high-temperature-environment-8fzAqOTVaP

Kurtoglu T., K. Leone, M. Revely, and C. Sandifer, A Study on Current and Emerging Technologies and Future Research Requirements for Integrated Vehicle Health Management Internal NASA Report, September 2008.

Lattime S.B., and B.M. Steinetz, Turbine Engine Clearance Control Systems: Current Practices and Future Directions, NASA TM 2002-211794, AIAA-2002-3790, September 2002.

Lee C., S.A. Akbar, and C.O. Park, Potentiometric CO_2 gas sensor with lithium phosphorous oxynitride electrolyte, *Sensors and Actuators B* 80(3), 2001, 234–242.

Lee C., S. Akbar, and C. Park, Potentiometric CO_2 gas sensor with lithium ion electrolyte. *Processing and Fabrication of Advanced Materials XI, Proceedings of the International Symposium on Processing and Fabrication of Advanced Materials*, ASM International, Materials Park, OH, 7–10 Oct. 2002, pp. 63–77.

Lekki J., T. Bencic, A. Gyekenyesi, G. Adamovsky, G. Hunter, R. Tokars, and M. Woike, Aircraft engine sensors for integrated vehicle health management, *Proceedings of the 57th Joint Propulsion Meeting*, Colorado, Springs, CO, May 3–7, 2010.

Levy R.L., and S.D. Schwab, Monitoring the composite curing process with a fluorescence-base fiber-optics sensor, *SPE Proc.* 35, 1989, 1530–1533.

Liang H., S. Lukkarinen, and I. Hartimo, Heart sound segmentation algorithm based on heart sound envelogram, *Computers in Cardiology* 24, 1997, 105–108.

Li X., R. Ramasamy, and P.K. Dutta, Study of the resistance behavior of anatase and rutile thick films towards carbon monoxide and oxygen at high temperatures and possibilities for sensing applications, *Sensors and Actuators B* 143(1), 2009, 308–315.

Liu J., W. Liu, H. Wang, T. Tao, and J. Zhang, A novel envelope extraction method for multichannel heart sounds signal detection 2012, pp. 630–638.

Mackin T., and M. Roberts, Thermoelastic evaluation of damage evolution in ceramic matrix composites: A non-contacting stress mapping methodology, *Journal of American Ceramic Society.* 83, 2000, 337–343.

Martin L.C., G.C. Fralick, and K. Taylor, Advances in thin film thermocouple durability under high temperature and pressure testing conditions, *NASA Technical Memorandum*, n 208812, Jan, 1999, p. 14.

Martin R. and A. Gyekenyesi, Pulsed thermography of ceramic matrix composites, *Proceedings of SPIE's NDE and Health Monitoring of Aerospace Materials and Civil Infrastructures*, San Diego, California, March, 2002.

Martin R., A. Gyekenyesi, and S. Shepard, Interpreting the results of pulsed thermography data, *Materials Evaluation.* 61(5), 2003, 611–616.

Mohanty, S., A. Chattopadhyay, J. Wei, and P. Peralta, Unsupervised time-series damage state estimation of complex structure using ultrasound broadband based active sensing, *Structural Durability & Health Monitoring Journal* 130(1), 2010, 101–124.

Morscher G. and A. Gyekenyesi, The velocity and attenuation of acoustic emission waves in SiC/SiC composites loaded in tension, *Composites Science and Technology* 62, 2002, 1171–1180.

Morscher G., C. Baker, A. Gyekenyesi, C. Faucett, and S. Choi, Damage detection and tensile performance of various SiC/SiC composites impacted with high speed projectile, GT2013-95638, *Proceedings of ASME Turbo Expo 2013* June 3–7, 2013, San Antonio, Texas, USA.

Morscher G.N., Modal acoustic emission of damage accumulation in a woven SiC/SiC composite. *Composites Science And Technology* 59, 1999, 419–426.

Neudeck P.G., SiC technology, in *The VLSI Handbook, The Electrical Engineering Handbook Series*, W.-K. Chen, Ed. Boca Raton, Florida: CRC Press and IEEE Press, 2000, pp. 6.1–6.24.

Neudeck P.G., D.J. Spry, L.Y. Chen, G.M. Beheim, R.S. Okojie, C.W. Chang, R.D. Meredith, T.L. Ferrier, L.J. Evans, M.J. Krasowski, and N.F. Prokop, Stable electrical operation of 6H-SiC JFETs and ICs for thousands of hours at 500°C, *IEEE Electron Device Letters* 29(5), 2008, 456–459.

Neudeck P.G., R.S. Okojie, and L.-Y. Chen, High-temperature electronics- a role for wide bandgap semiconductors, *Proceedings of the IEEE* 90, 2002, 1065–1076.

Neudeck P.G., S.L. Garverick, D.J. Spry, L.-Y. Chen, G.M. Beheim, M.J. Krasowski, and M. Mehregany, Extreme temperature 6H-SiC JFET integrated circuit technology, *Physica Status Solidi A* 206, 2009, 2329–2345.

Nowacki K., and W. Kasprzyk, The sound velocity in an alloy steel at high-temperature conditions, *International Journal of Thermophysics* 31, 2009, 103–112.

Okojie R.S., Stable 600°C silicon carbide MEMS pressure transducers, *Proceedings of SPIE* 6555, 2007, 65550 V.

Oruklu E., Y. Lu, and J. Saniie, Hilbert transform pitfalls and solutions for ultrasonic NDE applications, in *Ultrasonics Symposium (IUS)*, 2009 IEEE International, Rome, Italy, September 20–23, 2009, pp. 2004–2007.

Owen R.B., A.L. Gyekenyesi, D.J. Inman, and S.H. Dong, Hardware Specific Integration Strategy for Impedance-Based Structural Health Monitoring of Aerospace Systems, NASA/CR-2011-217153, May 2011.

Park C.O., C. Lee, S.A. Akbar, and J. Hwang, The origin of oxygen dependence in a potentiometric CO_2 sensor with Li-ion conducting electrolytes, *Sensors and Actuators B* 88(1), 2003, 53–59.

Partridge I., and G. Maistros, *Dielectric Cure Monitoring for Process Control*, Chapter 17, Vol. 5, Encyclopedia of Composite Materials, Elsevier Science, London, 2001, page 413.

Ponchak G.E., M.C. Scardelletti, B. Taylor, S. Beard, R.D. Meredith, G.M. Beheim, G. W. Hunter and W.S. Kiefer, High temperature, wireless seismometer sensor for Venus, *Proceedings of the IEEE Topical Conference on Wireless Sensors and Sensor Networks Dig.*, Santa Clara, CA, Jan. 15–19, 2012.

Roth D., R. Martin, L. Harmon, A. Gyekenyesi, and H. Kautz, Development of a high performance acousto-ultrasonic scan system, *Proceedings of the 29th Annual Review of Progress in Quantitative Nondestructive Evaluation*, Western Washington University, Bellingham, Washington, July 14–19, 2002.

Roth D., M. Verrilli, R. Martin, and L. M. Cosgriff, Characterization of C/SiC composite during creep rupture tests using an ultrasonic guided wave scan system. E. Lara-Curzio and M. Readey (Eds), *Proceedings of 28th International Conference on Advanced Ceramics and Composites B: Ceramic Engineering and Science Proceedings*, The American Ceramic Society, Westerville, OH, Cape Canaveral Port/Cocoa Beach, FL, January 25–30, 2003, pp. 267–274.

Roundy S.J., Energy scavenging for wireless sensor nodes with a focus on vibration to electricity conversion, Ph.D. Dissertation, University of California at Berkeley, 2003.

Roundy S., D. Steingart, L. Frechette, P.K. Wright, and J. Rabaey, Power sources for wireless networks, *Proceedings of the First European Workshop on Wireless Sensor Networks (EWSN '04)*, Berlin, Germany, Jan. 19–21, 2004.

Sathiyakumar S., Cure monitoring of composite materials using optical techniques, PhD Thesis, Victoria University, Melbourne, Australia, 2006.

Savage N.O., S.A., Akbar, and P.K. Dutta, Titanium dioxide based high temperature carbon monoxide selective sensor, *Sensors and Actuators B* 72(3), 2001, 239–248.

Sayir A., F. Dynys, A. Sehirlioglu, and T. Caillat, Thermoelectric properties of CrSi2-Si(Ge) eutectic for oxidizing environment, *Material Science & Technology Conference*, Detroit, MI., Sept. 2007.

Schwartz Z.D., and G.E. Ponchak, High temperature performance of a SiC MESFET based oscillator, *2005 IEEE MTT-S International Microwave Symposium Dig.*, Long Beach, CA, June 11–17, 2005, pp. 1179–1182.

Shepard S., Advances in pulsed thermography, in *Thermosense XXIII-Proceedings Vol. 4360*, editors A. Rozlosnik and R. Dinwiddie, Orlando, Florida, April 16–19, 2001.

Sherrit S., X. Bao, Y. Bar-Cohen, and Z. Chang, Resonance analysis of high temperature piezoelectric materials for actuation and sensing, *Proceeding of the Industrial and Commercial Applications of Smart Structures Technologies Conference*, SPIE Smart Structures and Materials Symposium, Paper #5387-58, San Diego, CA, March 15–18, 2004, pp. 411–420.

Silicon Carbide Electronics http://www.grc.nasa.gov/WWW/SiC/discoveries.html, visited on Jan 29, 2013.

Simon D.L., S. Garg, G.W. Hunter, T.-H. Guo, and K.J. Semega, *Sensor Needs for Control and Health Management of Intelligent Aircraft Engines*, U.S. Army Research Laboratory, Glenn Research Center, Cleveland, Ohio, ARL-TR-3251, GT2004-54324; NASA/TM-2004-213202; August 2004.

Slenski G.A., and M.F. Walz, Novel technologies for improving wire system integrity, *9th Joint FAA/DOD/NASA Conference on Aging Aircraft*, March 2006, Atlanta, GA. http://www.agingaircraft-conference.org/all_files/29/29a/70_doc.pdf

Slenski G., and J. Kuzniar, Aircraft wiring system integrity initiatives- a government and industry partnership, paper at, *6th Joint FAA/DOD/NASA Conference on Aging Aircraft*, San Francisco, CA, September 2002.

Smith C. and A. Gyekenyesi, Detecting cracks in ceramic matrix composites by electrical resistance, *Proceedings of SPIE: Nondestructive Characterization for Composite Materials*, Aerospace

Engineering, Civil Infrastructure, and Homeland Security V, San Diego, CA, February 26–March 2, 2011.

Smith C.E., G.N. Morscher, and Z.H. Xia, Electrical resistance as a Nondestructive evaluation technique for SiC/SiC ceramic matrix composites under creep-rupture loading, *International Journal of Applied Ceramic Technology* 8, 2011, 298–307.

Smith C.E., G.N. Morscher, and Z.H. Xia, Monitoring damage accumulation in ceramic matrix composites using electrical resistivity, *Scripta Materialia* 59, 2008, 463–466.

Sokolowski D., and G. Hunter, UEET Technology Forum, Intelligent Propulsion Controls, High Temperature Wireless Data Communication Technology, October 28–30, 2002.

Spry D.J., P.G. Neudeck, L.-Y. Chen, G.M. Beheim, R.S. Okojie, C.W. Chang, R.D. Meredith, T.L. Ferrier, and L.J. Evans, Fabrication and testing of 6H-SiC JFETs for prolonged 500°C operation in air ambient, in *Silicon Carbide and Related Materials 2007, Materials Science Forum*, T. Kimoto, Ed. Trans Tech Publications, Switzerland, 2008.

Stubbs D.A., and R.E. Dutton, An ultrasonic sensor for high-temperature materials processing, *Journal of Materials (JOM), The Minerals, Metals & Materials Society* 9, 1996, 29–31. http://www.tms.org/pubs/journals/jom/9609/stubbs-9609.html

Szabo N., C. Lee, J. Trimboli, O. Figueroa, R. Ramamoorthy, S. Midlam-Mohler, A. Soliman, H. Verweij, P. Dutta, and S. Akbar, Ceramic-based chemical sensors, probes and field-tests in automobile engines, *Journal of Materials Science* 38(21), 2003, 4239–4245.

Thompson H.A., Wireless and internet communications technologies for monitoring and control, *Control Engineering Practice* 12, 2004, 781–791.

Ward B.J., K. Wilcher, and G. Hunter, Gas Microsensor Array Development Targeting Enhanced Engine Emissions Testing, AIAA Infotech@Aerospace 2010, AIAA, Reston, VA, April 20–22, 2010, AIAA 2010-3327.

William J.G., T.M. Donnellan and R.E. Trabocco, A Predictive Model for Resin Flow Behavior During Composite Processing, Naval Air Development Report No. NADC-85164-60, 1985.

Woike M., J. Roeder, C. Hughes, and T. Bencic, Testing of a Microwave Blade Tip Clearance Sensor at the NASA Glenn Research Center, NASA TM 2009-215589, AIAA-2009-1452, 2009.

Woike M., A.A. Aziz, and T. Bencic, A Microwave Blade Tip Clearance Sensor for Propulsion Health Monitoring, AIAA-2010-3308, AIAA Infotech@Aerospace 2010, April 2010.

Wood R.L., A.K. Tay and D.A. Wilson, Design and fabrication considerations for composite structures with embedded fiber-optic sensors, *Proceedings of the SPIE Vol. 1170*, Fiber Optic Structures and Skins II, Boston, MS, 4–8, Sept. 1989, pp. 150–159.

Internet Resources

Cure monitoring http://en.wikipedia.org/wiki/Cure_monitoring

Monitoring the cure itself http://www.compositesworld.com/articles/monitoring-the-cure-itself

9

High-Temperature Motors

Nishant Kumar

CONTENTS

9.1 Introduction

Actuators and sensors are basic devices for producing and controlling motion in robotic and manipulation mechanisms. While sensors convert energy into electrical signals, actuators reverse this process by converting electrical signals into mechanical forces. A sensor provides a measure of the physical quantities of the surrounding environment, whereas an actuator produces a motion in response to these physical quantities. Together, both types of devices play the important role of interfacing between the mechanism and the physical world.

Technological developments have enabled actuators and sensors to be very small, precise, and fast. While these developments were intended for operation on Earth, recent interest in planetary exploration has placed unprecedented demand on their operation. Planetary environments have posed challenges that include operation at high temperatures.

In this chapter, recent technological breakthroughs in the development of high-temperature actuators and sensors are discussed. The actuators and sensors were made capable of operation at temperatures above 460°C and pressure in the range of 90 bars, akin to ambient condition on Venus. The developmental work done on these were part of an NASA SBIR effort to enable high-temperature technology for exploration on Venus.

9.2 Actuators

9.2.1 Actuation Mechanisms and Power Sources

Actuation occurs when one or more forms of energy are converted into mechanical motion of an object. In robotic mechanisms, this can be as simple as a direct drive or as complicated as a multilevel, interlinked mechanical system. In multilevel systems, the complexity in the mechanism is required to achieve a particular motion trajectory. The complexity also helps to create mechanical advantage that cannot be achieved with just the driver in the system.

All actuators need an input signal and a power source. The input signal controls the flow of power to the actuator and is usually in the form of an electric signal. The choice of actuator type varies based on the choice of power source. Common sources of power are electric, pneumatic, and hydraulic. Other less common sources are thermal, chemical, and light. An example of a thermal actuator is a shape memory wire that significantly expands, contracts, or bends when the temperature crosses the designed threshold of the shape memory alloy. Of all the choices of power sources, electrical power is the most readily accessible and can be generated and stored by converting other forms of energy. Common actuators utilizing this are motors, solenoids, and piezoelectric actuators (Chapters 10 through 13).

9.2.2 Actuation Performance in Different Environment

The physical properties of a material vary according to environmental factors like temperature, pressure, and humidity. The physical properties that can be affected include mechanical properties like size, shape, and strength; electrical properties like conductivity and dielectric strength; and magnetic properties like permeability and retentivity. Changes in these properties alter the performance characteristics of an actuator. For example, in a mechanical system at increased temperature, parts made with different materials expand differently due to differences in the coefficient of thermal expansion. High temperature also causes oxidation and dust generation on interacting surfaces. If actions are not taken to mitigate the effects of these changes, they can cause increased levels of friction, thus requiring more energy for the actuator to operate. In electrical systems, resistance increases linearly with temperature due to the temperature coefficient of resistance, and this causes a proportional increase in power consumption.

Motors used at high-temperature face these challenges and can therefore easily fail. This section aims to describe a new type of high-temperature motor (developed at Honeybee Robotics) specially designed, through selection of compatible materials and design choices, so as to withstand and operate continuously at temperatures above 460°C.

9.3 Motors

Motors are actuator devices that convert electrical energy into rotating mechanical motion. Motors are made in different sizes and forms and can be broadly classified as brushed or brushless. While brushed types have brushes to carry current to their coils, brushless motors are characterized by the absence of any electrical connection between the stator and the rotor. Some examples of this class of motors are AC (alternating current) induction motors, stepper motors, switched reluctance motors (SRM) and brushless direct current (BLDC) motors. Owing to the absence of contacting commutation brushes, and associated oxide formation, brushless motors are favorable for high-temperature applications.

9.3.1 Switched Reluctance Motors

A SRM does not have permanent magnets on the rotor. In this type of motor both the stator and the rotor have salient poles, and only the stator has windings. The stator and rotor are built using a stack of laminated sheets of magnetic alloy. The stack lamination reduces the eddy current losses as the motor operates.

The rotary motion in a SRM is produced by the rotor pole moving to a position so as to minimize the reluctance in the stator coil (Ehsani et al., 1992). In other words, the torque produced by the attraction of the energized pole moves the rotor so that the inductance increases to the maximum value. This movement maximizes the flux-linkage between the two poles, as shown in Figure 9.1. The increase in the inductance is proportional to the degree of alignment of the rotor pole to the stator. This direct correlation of inductance to the rotor position also provides the opportunity to sense the position by estimating the inductance of the rotor. The direction of the current in the energized pole is immaterial (Miller and Hendershot 1993) for the rotor's direction of rotation. The commutation of the stator coils is achieved through electronic circuits.

Because magnets tend to de-magnetization at high temperature, SRM is a good brushless motor option for high-temperature applications due to the absence of magnets. The simple design and low part count drives it toward low-cost machine construction. It has high fault tolerance because the coils are independent of each other and, in the event that one coil fails, the motor can continue to operate at a lower torque level.

As illustrated in Figure 9.2, the motor's torque is produced only in the direction of rising inductance. The torque is produced from the rate of change of coenergy in the motor magnetic circuits.

The controller for a typical SMR, with eight stator-poles and six rotor-poles consists of four branches of metal–oxide–semiconductor field-effect transistors (MOSFETs) where two MOSFETs are connected in series with each phase of the four phases of the motor. An example of the controller is shown in Figure 9.3 below. The series configuration causes the current to be unipolar, and not alternating, which has an advantage of causing low core losses in the motor. Switching of the field-effect transistors (FETs) is controlled through a digital signal processor (DSP) controller that runs an appropriate algorithm for the switching sequence.

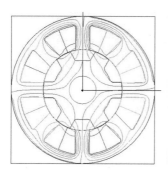
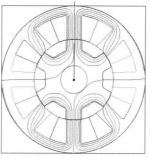

FIGURE 9.1
SRM internal structure and flux distribution.

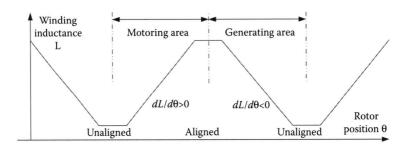

FIGURE 9.2
Inductance profile of SRM motor phase.

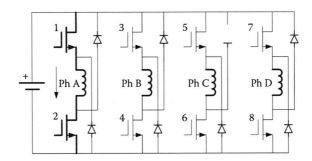

FIGURE 9.3
Four-phase SRM driver circuit.

9.3.1.1 High-Temperature SRM Design

The SRM (made by Honeybee Robotics) is a four-phase motor with an eight stator-pole and a six rotor-pole. It is based on a National Electrical Manufacturers Association (NEMA) 23 motor frame and measures 5.4 cm (2 inches) in diameter and 5.4 cm (2 inches) in length. The selection of the size is based on the following:

1. Since NEMA 23 is a standard size frame, there are many motors available with which the characteristics of the motor can be compared.
2. Since the specific size is widely used, this motor can easily replace motors already being used.
3. The size is relatively simple for fabrication in a lab without the need for special tools and assembly equipment.

The SRM design (Figure 9.4) utilizes high-temperature materials and components and they were selected based on the requirement for the motor to survive 460°C in a Venusian environment for extended durations of time. The motor is made of the following:

- Motor body
- Stator
- Rotor
- Motor shaft
- Stator winding

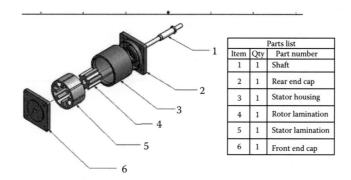

Parts list		
Item	Qty	Part number
1	1	Shaft
2	1	Rear end cap
3	1	Stator housing
4	1	Rotor lamination
5	1	Stator lamination
6	1	Front end cap

FIGURE 9.4
The structure of the SRM motor (made by Honeybee Robotics).

- Bearing
- Adhesive

The components used in the construction of the motor body meet the requirement for it to perform without degradation at temperatures near 500°C. The drive electronics of the SRM consist of a bank of switches that are activated in sequence to allow current to flow in the selected motor coil. The direction of the flow of current is immaterial, as the operation of the motor is based on the amount of flux linkage, not the direction of the flux. Literature shows a number of configurations of switches that one can use for the drive circuit (for example, Miller and Hendershot, 1993). One commonly used configuration consists of two switches and two diodes per phase. An eight-pole motor, using two opposite poles in a phase, would need four sets of switches to complete the drive circuit.

One of the challenges in a brushless motor is the detection of the shaft position with respect to the stator (Hossain et al., 2003). This is required to commutate the motor by exciting the specific motor phase to a specific rotor position. Traditionally, in a brushless DC motor this is achieved by electronic position feedback devices such as hall sensors, resolvers, and optical encoders. Although there have been breakthroughs in operating these devices at higher temperatures, progress that would allow them to operate at temperatures as high as 460°C has fallen short. Hence, a high-temperature SRM cannot utilize traditional positional feedback devices.

As mentioned earlier, one of the characteristics of a SRM is the varying inductance of a stator phase with varying rotor position as shown in Figure 9.5. Sampling and comparing the inductance of different phases is a strategy that is being used to control the SRM that is described herein. The phase circuit can be represented by the formula

$$iR + L(di/dt) + Vd = 0 \tag{9.1}$$

where
Vd is voltage drop across the diode (Figure 9.3),
i is the current, and
di/dt is the rate of change of current

The slope of the data group in Figure 9.6 represents di/dt.

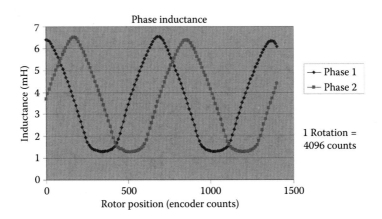

FIGURE 9.5
Measured phase inductance of the SRM.

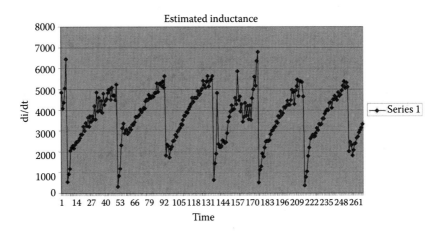

FIGURE 9.6
Estimated di/dt through voltage and current sampling (Sample points are rotor positions).

9.3.1.2 Test Results

The prototype SR motor was tested inside a furnace with a protruding shaft connected to a dynamometer (Figure 9.7). The furnace temperature was increased slowly until the desired temperature was reached. The chamber was kept at positive pressure with CO_2 gas so as to minimize the surface oxidization at higher temperature. The test was performed at various temperatures and voltage levels. At 460°C, the no-load speed of the motor was observed to be 7500 rpm and the maximum torque that it achieved was in the range of 25 oz-in before it stalled, as shown in Figure 9.8. The high current observed at the maximum torque region (Figure 9.9) can be attributed to the combined increase in friction and decrease in the permeability of the motor core.

Maxon motors have been used previously in mechanisms on Mars rovers. Hence, Maxon RE-25 motor, which is representative of the class, was selected as a base for comparative study of our motor. The properties of the SRM prototype were compared to this motor and showed promising results (Table 9.1).

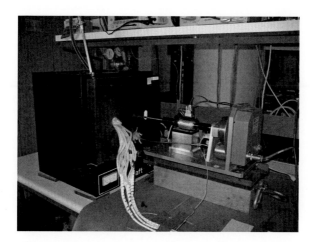

FIGURE 9.7
(**See color insert.**) SRM high-temperature test setup.

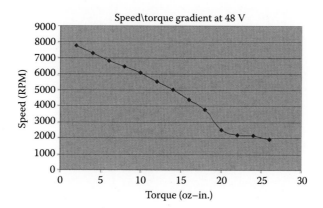

FIGURE 9.8
Motor speed-torque gradient Test Data at 48 V and 460°C.

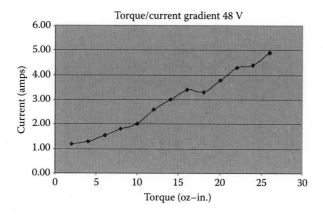

FIGURE 9.9
Motor torque-current gradient Test Data at 48 V and 460°C.

TABLE 9.1

Comparison of the Maxon RE-25 Motor and the SRM Motors

Characteristics	Units	Maxon RE-25 Range at 25°C	SRM Prototype Range at 460°C
Applied voltage	V	4.5–48	20–48
Maximum speed	rpm	5500	7500
No-load speed	rpm	4790–5500	7000–7500
No-load current	mA	7–80	1000–1200
Stall torque	mN-m	119–144	200–250

9.3.2 Brushless DC Motor

A BLDC motor consists of a rotating permanent magnet and a set of current-carrying coils on the stator, whose polarity is sequentially changed to produce a unidirectional torque. Just like in a brushed DC motor where the sequential change in polarity of the coil is achieved by a set of mechanical brushes and a commutator, in a BLDC motor this is achieved by an electronic position-sensing device. This device could be equipped with a set of hall sensors or an optical encoder. The flow of current in the coils is controlled by set of switches consisting of transistor or MOSFET.

In a BLDC motor, there are three basic arrangements of the magnets and the coil, interior-rotor motor, exterior-rotor motor and axial-gap rotor motor as shown in Figure 9.10.

The interior-rotor of a BLDC motor consists of a shaft mounted on two bearings, which holds the rotating permanent magnets. The magnets are held on the shaft by different methods, each accommodating unique characteristics of the motor. To prevent the magnets from falling apart due to extreme rotational forces at high rpm, care is taken in retaining them on the shaft. The stator holds the current-carrying conductor in the form of coils that are placed in slots distributed along the inside of the stator. As the stator and conductor are on the exterior, cooling is achieved easily through convection.

In an exterior-rotor motor, magnets and a rotating drum form the exterior of the motor. The stator consists of a stack of lamination with salient poles extending out from the center of the motor, similar to a DC motor rotor armature. The cost of manufacturing motors with this stator-rotor arrangement is low. The stator winding can be easily

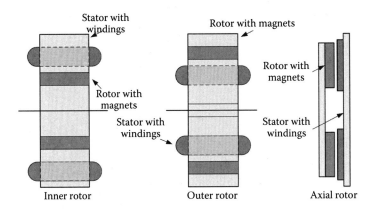

FIGURE 9.10
Three basic BLDC motor configurations.

mechanized using a fly-winding method, and the magnets are held in the inside of the drum with epoxy, assisted by centrifugal force during rotation. Simple design and low cost of production has made this motor popular in cooling fan applications where long life is desirable.

An axial-gap BLDC motor, also known as a pancake motor or disc-type motor, Figure 9.11, consists of a rotating disc of magnets, magnetized axially, and is held on a circular steel disc. The stator constitutes a circularly arranged set of coils mounted on a printed circuit board with all necessary connections made on the PCB. This type of motor is favored where there is a flat constraint on the space, as on a rotating disc or in a flywheel application.

The BLDC motor design begins with design requirements and specifications, which may include maximum power, rated speed, rated torque, etc. Constraints may include bus voltage, maximum temperature, or envelope dimension. The design starts with a rough estimate of the correct magnet grade. Magnet choice is an important step in BLDC motor design, since magnet flux determines the airgap flux. The number of phases, the number of magnet poles, the number of stator slots, and the winding configuration must then be selected (Hendershot and Miller, 1994; Hanselman, 1994). The rotor and magnet configuration is designed, and then the stator winding is determined. Parameters like material-hoop stress, RMS current density and rotor torque density are selected in the next step.

The general guideline for the basic design (Hanselman, 1994; Momen, 2004) of a BLDC motor follows. The torque of a motor with one slot per pole per phase is

$$\tau_{mean} = 2n_s N_m R_g B_g L_{stk} I_{coil} \tag{9.2}$$

where
n_s is number of phases
N_m is the number of magnet poles
B_g is airgap flux density
R_g is airgap radius
L_{stk} is stack length
I_{coil} is coil rms current

This equation suggests that increasing the number of magnet poles increases the torque generated by the motor. A basic rule of thumb is that the number of poles should be

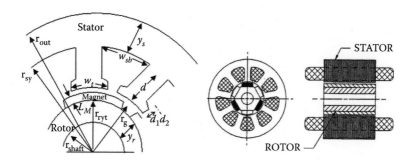

FIGURE 9.11
BLDC motor geometry. r_{out} is stator outer radius, r_{sy} is stator yoke internal radius, r_g is rotor radius, r_{ry} is rotor yoke external radius, r_{shaft} is shaft radius, y_s is stator back iron thickness, y_r is rotor yoke thickness, d_s is stator pole height, w_t is stator tooth width, w_{sb} is the stator slot internal width.

inversely proportional to the maximum speed of rotation, which would limit the commutation frequency to avoid excessive switching losses in the transistors and iron losses in the stator. For very high speeds, two- and four-pole motors are preferred. If smooth torque is required at low speed, such as in a DC torque motor, a larger number of poles should be selected (Hendershot, 1994). Yet another consequence of increasing the number of magnet poles is that the required rotor and stator back iron thickness decreases since the amount of flux to be passed by the back iron decreases. As a result, the overall diameter can be reduced by increasing the number of poles.

One of the primary constraints on Permanent Magnet BLDC stator design is that the total number of stator slots must be an even integer multiple of the number of phases. There are many combinations of slot and pole numbers that can be used effectively. The number of stator slots and rotor poles depends upon many factors, including

- Magnet material and grade
- Configuration
- Mechanical assembly of the rotor and magnets
- Speed of rotation
- Inertia requirements

With a large numbers of slots/pole, the cogging torque is inherently reduced by the fact that the relative reluctance variation seen by the magnet is reduced as it successively covers and uncovers the slots one at a time.

9.3.2.1 Design of BLDC (Honeybee Robotics) Motor

The design of the BLDC motor is a progression of the design of the SRM described above. The two main differences between any SRM and a BLDC motor are the structure of the rotor and the coil arrangement on the stator. In an SRM, the rotor has salient poles and is made of a stack of lamination, whereas a BLDC motor has rotating permanent magnets on a shaft. The SRM stator is a double salient, and phases are singly excited, whereas in the BLDC motor conductors are distributed evenly and uniformly to yield highest utilization of the active conductors that produce torque.

Based on NEMA 23 motor frame size, the motor measures were 5.4 cm (2 inches) by 5.4 cm (2 inches). The BLDC motor designed is a three-phase, six-slot and four-pole configuration interior-rotor motor. The stator is similar to that of the 3-phase induction motor and the magnets are on the rotor. The cross section of the prototype BLDC motor is shown in Figure 9.12. All components in the motor were validated for continuous operation at 460°C.

9.3.2.2 Control of BLDC Motor

In the BLDC motor, commutation is achieved by reading the position feedback from sensors arranged in a configuration that enables them to sense the pole position sequentially as the motor shaft moves. By utilizing the position information, the controller excites the coils in sequence to produce a rotational motion of the motor shaft. The most common way to get the position feedback in a BLDC motor is through hall sensors attached on the axial side of its stator. These hall sensors are triggered sequentially by the magnetic field of the rotor magnets. The operational temperature range for existing hall devices is limited to temperatures less than 180°C.

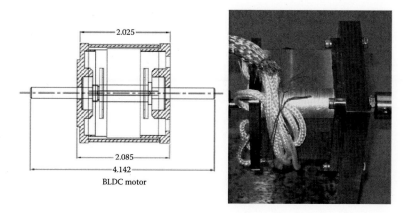

FIGURE 9.12
(**See color insert.**) Honeybee high-temperature BLDC motor.

In the description of the high-temperature SRM design above, a method is mentioned of detecting the position of the rotor by utilizing a sensorless control. This method, combined with new rotor geometry, can be used to construct a position-sensing device that provides the position information of the high-temperature BLDC motor.

The design of the high-temperature position sensor involves a proprietary rotor design and an algorithm that is implemented through a high-speed DSP. Built using the same construction method used to make the high-temperature SRM and BLDC motors, this device is capable of working continuously at 460°C. The algorithm makes it possible to nullify the effects of parametric changes such as wire resistance, permeability and air gap, due to increased temperature.

Similar to the hall sensor, the output of this position sensor provides shaft position feedback to the high-temperature BLDC motor that commutates the motor. Figure 9.13 shows the typical switching circuit of a BLDC motor controller. It consists of three half bridges of MOSFETs connected in parallel to the power source and each branch of half bridge connected to one of the three terminals of the motor. The excitation of the element of the half bridges is facilitated by the DSP controller which executes the algorithm of choice.

9.3.2.3 High-Temperature BLDC Motor Testing

The high-temperature BLDC motor was actuated at temperatures up to 460°C, at speeds up to 1000 rpm, and under load conditions from no-load to an applied torque load of 80 mN-m. Operational limits of the motor were obtained by increasing the motor's speed and loading at temperature until stalling occurred. The motor was also run continuously at different speeds and under different load cases to determine effects on the motor over time.

The BLDC motor was tested in a furnace at 460°C with a one atm CO_2 atmosphere to simulate Venus's atmospheric conditions without the pressure. The motor was placed in the furnace (Figure 9.14), and the output shaft was rigidly coupled to a rod extending through the furnace door to the ambient environment. The other end of the rod was attached via a flexible coupling to a dynamometer that applied and measured torque loads to the BLDC motor.

The motor temperature was raised from room temperature to 460°C in increments of 100°C, using a set torque of 10 oz-in (approximately 0.5 × room temperature stall) at speeds from 200 to 1200 rpm.

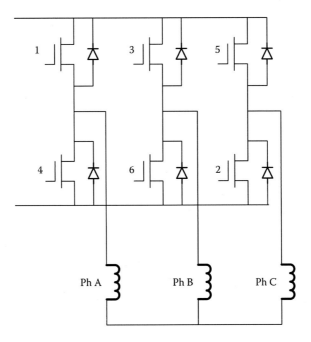

FIGURE 9.13
Three-phase BLCD bridge circuit.

The results of these tests are summarized in Figure 9.15. The family of curves shows current versus speed at 5 different temperatures under a constant load of 10 oz-in. These plots clearly show a trend towards increased current as both speed and temperature increase.

The motor with position-sensor (Figure 9.16) was also tested at 460°C at 1200 psi with CO_2 at JPL's Venus Test Chamber (Figure 9.17). Test data and post-test inspection demonstrated the survivability of the HT Motor/Sensor at Venus' atmosphere and verified that pressures beyond one atmosphere do not affect the performance of the system. These reported results show great promise in terms for operating motors at temperatures as expected on Venus.

FIGURE 9.14
(**See color insert.**) Motor test setup.

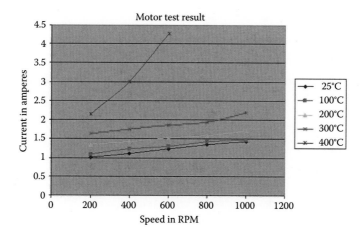

FIGURE 9.15
BLDC motor test data.

FIGURE 9.16
Motor with position sensor.

9.4 Conclusion

High-temperature technologies are crucial in the exploration of some of the planets in our solar system. The development of a high-temperature motor and position-sensing device, described in this chapter, was aimed to enable actuation capability. Mechanisms designed to utilize these devices will not only facilitate interaction with the in-situ environment, but also increase the survivability of the related mechanism.

Two main motor types and their associated position sensing methods were described. The BLDC motors are widely in use, and SRM motors are becoming popular as new control methodologies are developed. The test results of these motors are promising and are comparable to those of commercially available motors. While developmental effort has mainly focused on resolving technical problems, future effort would include

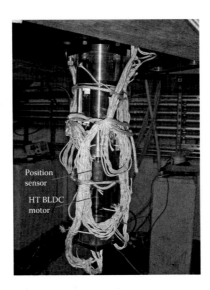

FIGURE 9.17
High-temperature and high-pressure testing at JPL.

addressing the economic challenges of producing these devices in large quantity and at low cost.

Acknowledgments

The author thanks Iqbal Husain, North Carolina State University, Raleigh, North Carolina; Ted Iskenderian, Yoseph Bar-Cohen, and Robert Troy, Jet Propulsion Laboratory (JPL)/Caltech; and Jason Herman, Honeybee Robotics Inc., New York, for reviewing this chapter, and for providing valuable technical comments and suggestions. The author thanks Roopnarine, Jerri Ji, and Sase Singh for their innovation and dedication toward the development of the high temperature technology at Honeybee Robotics. Also, the author expresses his appreciation to Louisa Avellar, who grammatically edited the chapter during her internship at the NDEAA Lab of JPL, Pasadena, California, Lastly, the author thanks his wife, Navneet, and son, Vihaan, for their time and patience during the writing of this chapter.

References

Ehsani M., I. Husain, and A. Kulkarni, Elimination of discrete position sensor and current sensor in switched reluctance motor drives, *IEEE Trans. Ind. Appl.*, 28(1), 128–135, 1992.
Hanselman D., *Brushless Permanent-Magnet Motor Design*, McGraw-Hill, New York, 1994.
Hendershot J.R., and T.J.E. Miller, *Design of Brushless Permanent Magnet Motors*, Magna Physics Publishing, Hillsboro, OH and Oxford University Press, Oxford, 1994.

Hossain S.A., I. Husain, H. Klode, B.P. Lequesne and A.M. Omekanda, Four quadrant and zero speed sensorless control of a switched reluctance motor, *IEEE Transactions. on Industry Applications*, 39(5), 1343–1349, 2003.

Miller T.J.E. and Hendershot J.R, *Switched Reluctance Motors and Their Control*, Magna Physics Publishing, Hillsboro, OH, 1993.

Momen M.F., Thermal analysis and design of switched reluctance and brushless permanent magnet machines, PhD Dissertation, University of Akron, August 2004.

10

High-Temperature Electromechanical Actuators

Stewart Sherrit, Hyeong Jae Lee, Shujun Zhang, and Thomas R. Shrout

CONTENTS

10.1 Terrestrial and Extraterrestrial Extreme Environments

10.1.1 Introduction

Electromechanical materials including piezoelectric and electrostrictive single crystal and ceramic materials have the potential to operate at extremely high temperatures. For example, although it has a relatively small piezoelectric constant, $LiNbO_3$ has a Curie temperature >1100°C. This material and others have the potential to produce useful work at elevated temperatures. This chapter looks at the potential applications of these electromechanical materials for high-temperature applications. The majority of this chapter discusses recent developments in high-temperature electromechanical materials and potential actuator design approaches that can produce useful work. A small section on competing technologies that can operate at high temperatures is also included. Some of the potential negative issues and pitfalls that can occur when designing mechanisms that have to operate at high temperatures are also discussed. Finally, we summarize the advantages and disadvantages of these materials over competing actuator technologies.

10.1.2 Applications

Actuation materials are an essential component of mechanisms that can operate in extreme environments. Because they are embedded in the mechanism, there are a variety of mechanical and electrical interfaces that need to be considered when designing mechanisms that can withstand and operate at extreme temperatures to ensure that the system operates as intended. It is not enough to prove that the actuator material at the component level will operate effectively at elevated temperatures. There could also be combined effects that need to be addressed such as high-pressure and high-radiation environments. Also, an environment that is benign at room temperature may become highly reactive to materials in the actuator or mechanism at high temperature. Electrical properties, such as conductivity for conductors and the resistivity for insulators, need to be evaluated over the whole range of operational temperatures. In many applications the temperature might be cycled so one needs to thermally cycle the actuator to ensure that the increase and decrease in temperatures does not produce a critical stress causing a fracture, rupture, de-poling, de-lamination or a critical change of friction within the device. All of the same issues that can affect the life of the actuator must also be considered when designing the mechanism. Although we have concentrated our investigation in the actuation, one needs to also invest in appropriate gearing and rotary and linear bearings in order to produce efficient actuators.

A variety of applications exist where actuators and sensors that operate at elevated temperatures (T > 100°C) would be extremely beneficial. These include high-temperature fans and turbines, motors for valves for the oil and natural gas industries (Hooker et al. 2010) and kiln automation (Baldor 2012) and actuators for automotive engines such as fuel injectors (Schuh et al. 2000) and cooling system elements (Stoeckel 1990). The majority of industrial actuator applications are at or below the 250°C temperature limit. Sensors that determine a variety of properties, however, have been developed to operate at much higher temperatures (Lloyd Spetz et al. 1997, Göpel et al. 2000, Omega 2013).

Another realm that requires high-temperature actuators and sensors is extraterrestrial exploration. A series of future NASA missions will require a variety of new technologies in order to meet the mass/power/volume envelopes and extreme conditions found on interesting future exploration sites in the Solar System (Gershman and Wallace 1999). Examples of the types of environments that may be encountered and typical temperature and pressure ranges are shown in Figure 10.1. As can be seen from the graph, the range in expected pressures is high vacuum to 1000 bar and the temperature range is 460°C on Venus to −215°C on Pluto. In addition to the thermal and pressure environments experienced at various places in the solar system, spacecraft and instruments have to be designed to function in gravity fields that can range from μg on comets and asteroids to 2.5 g on Jupiter. Another issue is the large radiation dose that can be acquired outside the safety of the Earth's atmosphere and magnetosphere. Metals, ceramics, and carbon/carbon composite materials can withstand extremely high dosage of the order of 10^{12} Rads without noticeable degradation in mechanical properties (Dolgin 1991).

Electronic properties, however, are generally not as resilient and a large area of research has been dedicated to understanding space environmental effects of an assortment of materials (Haymes 1971, Hastings and Garrett 1996). One important technology area that is required is advanced actuators and technology development for in-situ missions under

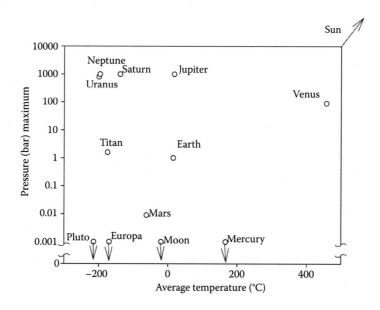

FIGURE 10.1

Estimated pressure maximums vs. temperature for a selection of proposed mission destinations. (Adapted from Planets 2012 http://en.wikipedia.org/wiki/Planets#Accepted_planets, or http://solarsystem.nasa.gov/planets/or http://www.nineplanets.org/ [downloaded Dec 6, 2012]; Sherrit, S., 2005, Smart material/actuator needs in extreme environments in space, *Proceeding of the Active Materials and Behaviour Conference, SPIE Smart Structures and Materials Symposium*, Paper #5761–48, San Diego, CA, March 7–10.)

extreme environmental conditions is required. These actuators need to exert sufficiently high strokes, torques, and forces while operating under relatively harsh conditions. A particular challenge from a temperature perspective is Venus, which is one of the planets that has been included in the NASA Decadal Planning effort for future missions.

The environment on Venus poses a challenge to the existing actuation technology and there are no commercially available motors that can operate at its ambient temperature of 460°C. Electromechanical materials (e.g., piezoelectrics and electrostrictive) have the potential to operate at these temperatures and produce the stroke/velocity and force/torque that is required to operate the needed mechanisms. The actuators that are needed for robotic exploration include lander petal motors (Jandura 2004), drive/steering motors, manipulator joint motors, latching and deployment motors and sampling tool motors (Sherrit 2005). The requirements on these actuators are very restrictive due to the mass, volume, and power envelopes imposed by space applications. A schematic diagram showing some of these actuators, which may be used in space exploration missions, is shown in Figure 10.2. Except for the extreme high-pressure region, of these restrictions (pressure, static g, dynamic g, temperature, radiation hardiness) high-temperature operation in particular is the most restrictive. Operating at 460°C, as is found on the surface of Venus, is an extreme challenge. In addition, Venus has a pressure of 9 MPa (primarily CO_2) at the surface and opaque sulfuric acid clouds in the atmosphere. Shielding, damping, sealing, heating or cooling can be incorporated in designs to account for many of these environmental factors, however, working in a high-temperature environment for extended times is a very difficult technical challenge. Active cooling using one-shot chemical cooling or powered refrigeration techniques can ameliorate the problem for drive electronics and batteries; however, the actuators, bearings, cabling/insulation, solders, and control sensors may have to be

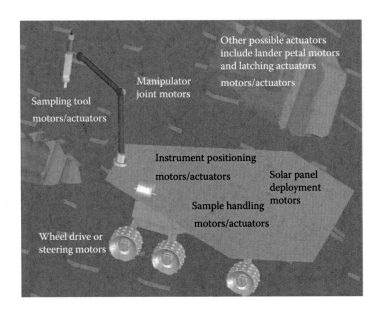

FIGURE 10.2
(See color insert.) Schematic drawing of the possible actuator/motor locations on a rover. (Adapted from Sherrit, S., 2005, Smart material/actuator needs in extreme environments in space, *Proceeding of the Active Materials and Behaviour Conference, SPIE Smart Structures and Materials Symposium*, Paper #5761–48, San Diego, CA, March 7–10.)

located external to any environmentally controlled space. It is feasible that some of these components may be located intermediately between the high- and controlled-temperature regions and operated at a temperature below 460°C.

Ultrasonic drills, corers and rock abrasion tools have been designed, fabricated and tested at JPL at −60°C and in simulated Mars environments ($\approx$1 kPa CO_2). Piezoelectric rotary motors have also been developed for composite manipulators (Schenker et al. 1997) for a Mars environment. In order to determine whether these materials can be used to build actuators that can operate at or above 460°C, we have studied an assortment of ceramics to determine the feasibility of these emerging high-temperature piezoelectric materials (Sherrit et al. 2004a) in support of this NASA need. In addition to the increase in operational temperatures of standard motors and actuators, one future area of interest is in high-temperature MEMS research which can be used for high-temperature, high-pressure valving for instrumentation and sample processing (Chen et al. 2009).

10.2 High-Temperature Electromechanical Materials

10.2.1 Piezoelectric Materials

Piezoelectricity: The phenomenon of piezoelectricity is responsible for two distinct but related relationships called the direct and converse effects. The direct effect describes the generation of charge on a crystal face as a result of an application of a stress, while the converse effect describes the generation of a mechanical strain as a result of the application of an electric field. At low values of stress and electric field, the relationship between charge and stress, strain, and field, is observed to be linear for a piezoelectric material. The charge reverses sign going from tension to compression and the strain switches from dilation to contraction under a change in the polarity of the electric field. Macroscopic descriptions of the piezoelectric effect under small fields can be described using a set of coupled linear equations:

$$S = s^E T + dE$$
$$D = \varepsilon^T E + dT$$

(10.1)

where d is the piezoelectric coefficient, and S, T, D, and E are the strain, stress, electric displacement and electric field, respectively. s^E and ε^T are elastic compliance and dielectric permittivity, respectively. The superscripts, T and E, in dielectric permittivity and elastic compliance indicate particular boundary conditions for the coefficients; constant stress (mechanically free) T and constant electric field (short circuit) E, respectively. Other representations of the linear equations of piezoelectricity derived from possible thermodynamic potentials are given in the various standard references in the field (IEEE Std. 176 – 1987; Damjanovic 1998, Jaffe and Berlincourt 1965, Curie and Curie 1880, Cross et al. 1972).

A common feature of various materials that permit piezoelectricity is that they have no center of symmetry. Depending on the crystal structure, the centers of positive and negative charges may not coincide without the application of an external electric field. In the 32 crystallographic point groups, only 20 point groups allow piezoelectricity. Of these 20 point groups, 10 are polar materials that possess a temperature-dependent spontaneous polarization (pyroelectrics). Ferroelectric materials are a special class of materials in which the spontaneous polarization can be permanently reoriented between two or more distinct directions with respect to the crystal axes through the application of an electric field (Nye 1985, Herbert 1982, Moulson and Herbert 2003). A polarization develops once a ferroelectric material is cooled through the Curie temperature, from the prototype paraelectric phase to ferroelectric phase with low symmetry. In general, a uniform alignment of polarization only occurs in certain regions of a material, while in other regions of the material, the spontaneous polarization may be in another direction. Such regions with uniform polarization are called ferroelectric domains. The interface between two domains is called the domain wall. Ferroelectric materials can be characterized by measuring their polarization P (D) as function of applied ac electric field (E), that is, P–E hysteresis loop (Xu 1991, Zhang and Li 2012).

Ferroelectrics are strongly nonlinear in their dielectric behavior and the variation of polarization with an electric field is generally hysteretic as a consequence of domain wall motion. As shown in Figure 10.3, at low electric fields, a linear relationship between P and E is observed, since the field is not sufficient enough to switch any domains and the material behaves as a normal dielectric. As the applied field increases, a number of negative domains switch to the positive direction and the polarization increases rapidly until most of the domains are aligned along the applied field direction, whereupon the polarization becomes saturated. As the field strength decreases, the polarization decreases, when the field is reduced to zero, some of the domains will remain aligned in the positive direction and the material will exhibit a remnant polarization P_R (poled state). The extrapolation of the linear segment BC of the curve back to polarization axis (at the point of E) represents the spontaneous polarization P_S. The remnant polarization cannot be removed till the applied field in negative direction reaches a certain value, the field required to reduce the polarization to zero is referred to as the coercive field E_C. The corresponding strain loop with typical butterfly shape is shown in the figure. Generally, P_S, P_R, and E_C decrease as the material approaches the Curie temperatures T_C, above which, the values drop to zero

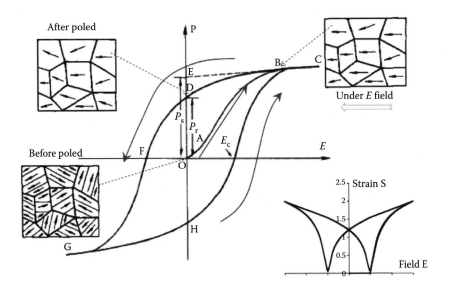

FIGURE 10.3
A typical polarization electric field hysteresis loop for a ferroelectric material, the small insets show domain reversal (polarization rotation).

in the absence of an external field. The piezoelectric coefficient of a ferroelectric material is closely associated with the spontaneous polarization (Shrout et al. 2002)

$$d = 2\varepsilon_r \varepsilon_0 Q P_S \tag{10.2}$$

where Q is the electrostrictive coefficient and ε is the permittivity, Q is temperature independent while permittivity and P_S are strongly related to temperature. For the shear coefficient, there is no two in Equation 10.2 (Haun et al. 1989).

The electromechanical coupling factor k is another critical parameter in the selection of piezoelectric materials, which is defined as the energy coupling between stored mechanical (or strain) energy and input electrical energy or vice versa. The magnitude of electromechanical coupling is directly related to the elastic constants, dielectric permittivity and piezoelectric coefficients, for example:

$$k_{ij}^2 = \frac{d_{ij}^2}{\varepsilon_0 \varepsilon_{ii}^T s_{ij}^E} \tag{10.3}$$

where the subscripts take account of the directional dependencies while the superscripts correspond to the defining boundary conditions. Note that the electromechanical coupling can also be determined by measuring the change in permittivity or elastic constant due to the change in mechanical (free and clamped) or electrical boundary conditions (open and short circuit), respectively. A common definition of the effective coupling can be expressed as the resonance f_r vibrating under short-circuit conditions and the antiresonance f_a vibrating under an open-circuit condition, as given in the following equation:

$$k_{ij}^2 = 1 - \frac{\varepsilon_{ij}^S}{\varepsilon_{ij}^T} = 1 - \frac{c_{ij}^E}{c_{ij}^D} = 1 - \frac{f_r^2}{f_a^2}, \tag{10.4}$$

where superscripts S and D refer to as constant strain (mechanically clamped) and constant dielectric displacement (open circuit), respectively. Thus, the definition indicates that high electromechanical coupling allows effective energy conversion between electrical and mechanical energy, improving overall transducer performance, such as transducer efficiency, sensitivity, and bandwidth (Berlincourt et al., Jaffe 1964, Jaffe et al. 1971, Sherman and Butler 2007).

Other important considerations for the selection of piezoelectric materials are loss parameters. Elastic, dielectric, and piezoelectric coefficients possess phase angles (therefore coefficients exhibit imaginary parts) that in the pure elastic and dielectric relationships correspond to an energy loss per cycle. In the elastic case and in the literature on piezoelectricity, the loss is usually designated by the mechanical quality factor Q_m, which is defined as the real part of the elastic constant divided by the imaginary part of the elastic constant (i.e., inverse of $\tan\gamma$, where γ is the phase angle mentioned above). In the dielectric case, the loss coefficient is the dissipation, or $\tan\delta$ and is the ratio of the imaginary part of the dielectric permittivity to the real part. For piezoelectric actuators, the loss coefficients can be as important for the efficient design of piezoelectric horns or motors as the real parts of the elastic, dielectric and piezoelectric coefficients. This is because high loss tangent results in excessive heat generation as the portion of electrical input power is dissipated as heat, limiting output power (Uchino 2000, Uchino and Giniewicz 2003, Uchino et al. 2006).

10.2.2 High-Temperature Piezoelectrics

In order to use piezoelectric materials for a specific high temperature application, many aspects need to be considered along with piezoelectric properties, such as phase transition, thermal aging, thermal expansion, electrical resistivity, chemical stability (decomposition and defect creation), and the stability of properties at elevated temperatures. Among them, property degradation and phase transitions at elevated temperatures are the critical limitations for use of piezoelectric materials at high temperatures. For example, the essential feature of ferroelectrics is that the spontaneous polarization can be reversed by the application of a large poling field. This can lead to a net spontaneous polarization along the poling field (called remanent polarization), enabling piezoelectricity (as shown in Figure 10.3). However, the aligned spontaneous polarization can revert back to a random orientation with increasing temperature, resulting in a decrease in electromechanical properties. At a certain temperature, the material transforms from a ferroelectric phase to a high symmetry non-ferroelectric phase, and this transition temperature is referred to as the Curie temperature (T_C). Above this temperature, the materials are permanently depolarized and cannot be used for piezoelectric applications. Therefore, the maximum operating temperature of ferroelectric materials is determined by their respective Curie temperature. In practice, the operating temperature must be substantially lower than the Curie temperature in order to minimize thermal aging and property degradation. As a general rule of thumb, it is not recommended to exceed half of the Curie temperature of ferroelectric materials (Zhang and Yu 2011, Damjanovic 1998, Kazys et al. 2008, Turner et al. 1994, Zhang et al. 2005a,b,d, Buchanan 1991, Shrout and Zhang 2007, Panda 2009, Messing et al. 2004, Fraser 1966, Smith and Welsh 1971, Venet et al. 2005, Damjanovic 2008, Newnham 2005).

There are a great number of ferroelectric materials that are candidates for high-temperature piezoelectrics, including polycrystalline ceramics with perovskite, tungsten bronze, bismuth layer (bismuth layer structured ferroelectric, BLSF—Aurivillius) or double perovskite layer structures (PLSs). Single crystals, including $LiNbO_3$ (LN) and $LiTaO_3$ (LT), with the corundum structures are ferroelectric materials, with Curie temperatures of 1150°C and 720°C, respectively. As presented in Table 10.1, ferroelectric materials are found to be diverse, with high sensitivity achieved for piezoelectrics with the perovskite structure, but with poor temperature stability, while ferroelectrics with the PLS exhibit low piezoelectric coefficient, with enhanced temperature stability.

Ferroelectric polycrystalline ceramics with the perovskite structure, such as $Pb(Zr,Ti)O_3$ (PZT), offer high piezoelectric coefficients $d_{33} > 350$ pm/V and electromechanical coupling factors $k_{33} > 0.7$ (Berlincourt 1971). The perovskite systems have been extensively studied for more than 60 years for various applications. These high-performance ferroelectric ceramics are based around a specific set of compositions that have two phases in equilibrium over a relatively temperature independent range, corresponding to a morphotropic phase boundary (MPB). The usage temperature range for PZT ferroelectric ceramics is limited by their respective Curie temperatures, T_C, ranging from 160 to 350°C. The piezoelectric properties, however, are found to be degraded prior to the Curie temperatures due to aging, limiting their use to about $1/2T_C$ as previously noted (Shrout et al. 2002, Zhang and Yu 2011). It should be noted that bismuth based perovskite polycrystalline ceramics, for example, $BiScO_3$-$PbTiO_3$ (BSPT) and $Bi(Mg_{1/2}Ti_{1/2})O_3$ based systems with a MPB composition, show comparable piezoelectric properties to PZT based materials, but with higher Curie temperatures (~100°C higher than PZT ceramics), expanding the temperature usage range (Eitel et al. 2002, Zhang et al. 2005a,c, Sebastian et al. 2010, 2012).

The tungsten bronze family of oxygen octahedral ferroelectrics possess the general formula $(A1)_2(A2)_4C_4(B1)_2(B2)_8O_{30}$. Lead metaniobate $PbNb_2O_6$ (PN) belongs to the tungsten bronze family, in which five out of the available six A sites are occupied by Pb^{2+} and the B sites by Nb^{5+}, while the C sites are empty. Lead metaniobate is widely used in nondestructive evaluation (NDE) type transducers, due to its ultralow mechanical $Q_m \sim 20$ (wide bandwidth) and high d_{33} to d_{31} ratio (high degree of anisotropy). Commercial PN compositions are modified to enhance specific electrical characteristics but at the cost of the lower T_C. A commonly used composition contains about 10% Ba (PBN) and has a T_C of about 400°C, being further limited by high electrical conductivity above 300°C (Zhang and Yu 2011, Ferroperm 2012).

Another piezoelectric family for high-temperature applications are bismuth layer (Aurivillius) structured ferroelectrics (BLSFs). The general formula for BLSFs is $(Bi_2O_2)^{2+}(A_{m-1}B_mO_{3\,m+1})^{2-}$ where A is a mono-, di-, or trivalent ion or a mixture of the three, allowing dodecahedral coordination, B is a combination of cations well suited for octahedral coordination. m is the number of octahedral layers in the perovskite slab, which ranges from 1 to 6, such as $Bi_4Ti_3O_{12}$, $SrBi_2Nb_2O_9$, $CaBi_2Nb_2O_9$, Bi_3TiNbO_9, $CaBi_4Ti_4O_{15}$, and $SrBi_4Ti_4O_{15}$. In fact, the m value can also be fractional, as in $Na_{0.5}Bi_{4.5}Ti_4O_{15}$ and $K_{0.5}Bi_{4.5}Ti_4O_{15}$ compounds. The bismuth titanate family has high Curie temperatures (e.g., T_C ~650°C for $Bi_4Ti_3O_{12}$ and T_C ~940°C for Bi_3TiNbO_9), low aging effects, low dielectric loss, and high mechanical quality factor Q_m up to 10,000, make them promising candidates for various high temperature applications. However, the dielectric and piezoelectric properties are relatively low, being 120 for ε_r and 18 pm/V for d_{33}, respectively, and low electrical resistivity at high temperatures limits their use to well below 600°C. (Zhang and Yu 2011, Damjanovic 1998b, Panda 2009, Messing et al. 2004, Takenaka and Nagata 2005, Moure et al. 2009). Piezoelectric activity in BLSFs can be further improved by suitable doping, while enhancing the electrical resistivity and achieving a compromise between good polarizability and a high Curie temperature, demonstrating piezoelectric d_{33} values and Curie temperatures on the order of 10 ~ 30 pm/V and 500 ~ 900°C, respectively (Zhang and Yu 2011). The schematic structure of perovskite, tungsten bronze, and BLSF materials are compared in Figure 10.4 (Trolier-McKinstry 2008).

TABLE 10.1

Room Temperature Electrical and Physical Properties of Various High Temperature Ferroelectric Materials (* Free Relative Dielectric Permittivity)

	T_m/T_C (°C)	Structure	Dielectric Properties	Coupling Factors	Piezoelectric Coefficients (pm/V)	Physical Properties	Ref.
Lead zirconate titanate PZT5A	365	Perovskite	$\varepsilon_r \sim 1700$, $\tan\delta \sim 2\%$			$\rho = 7.9$ g/cc, $s_{33}^E = 18.8$ pm²/N, $Q_m = 75$	Berlincourt (1971)
Lead titanate PbTiO₃	400	Perovskite	$\varepsilon_r \sim 210$, $\tan\delta \sim 1.4\%$	$k_t \sim 0.4$, $k_{31} \sim 0.05$, $k_{33} \sim 0.40$		$s_{33}^E \sim 7$ pm²/N, $Q_m > 500$	Ferroperm Piezo (2012)
BiScO₃-PbTiO₃	450	Perovskite	$\varepsilon_r \sim 2010$, $\tan\delta \sim 5\%$	$k_t \sim 0.49$, $k_{31} \sim -0.22$, $k_{33} \sim 0.69$	$d_{33} \sim 400$, $d_{15} \sim 520$	$\rho = 7.9$ g/cc, $s_{33}^E \sim 24$ pm²/N, $Q_m = 50$	Eitel et al. (2002)
BiScO₃-PbTiO₃-Mn	442	Perovskite	$\varepsilon_r \sim 1450$, $\tan\delta \sim 1\%$	$k_t \sim 0.49$, $k_{31} \sim -0.33$, $k_{33} \sim 0.69$	$d_{33} \sim 360$, $d_{15} \sim 520$	$\rho = 7.9$ g/cc, $s_{33}^E \sim 21$ pm²/N, $Q_m = 100$	Zhang et al. (2005c)
BMT-BF-BS-PT	450	Perovskite	$\tan\delta \sim 3.6\%$	$k_p \sim 0.44$	$d_{33} \sim 328$	/	Sebastian et al. (2010 & 2012)
BiInO₃-PbTiO₃-Nb	542	Perovskite	$\varepsilon_r \sim 250$, $\tan\delta \sim 1.4\%$	$k_t \sim 0.38$, $k_{15} \sim 0.38$,	$d_{33} \sim 60$, $d_{15} \sim 85$	/	Zhang et al. (2005a)
Lead metaniobate	>400	Tungsten Bronze	$\varepsilon_r \sim 220$, $\tan\delta \sim 0.6\%$	$k_t \sim 0.34$	$d_{33} \sim 100$, $d_{15} \sim 50$	$\rho = 5.6$ g/cc, $Q_m = 15$–25	Ferroperm Piezo (2012)
Bi₄Ti₃O₁₂	650	Bismuth layer	$\varepsilon_r \sim 120$, $\tan\delta \sim 0.4\%$	$k_t \sim 0.2$, $k_{31} \sim 0.02$, $k_{33} \sim 0.09$	$d_{33} \sim 18$, $d_{15} \sim 16$	$\rho = 6.55$ g/cc, $s_{33}^E \sim 44$ pm²/N, $Q_m > 600$	Ferroperm Piezo (2012)
CaBi₄Ti₄O₁₅-Mn	800	Bismuth layer	$\varepsilon_r \sim 148$, $\tan\delta \sim 0.2\%$	$k_{15} \sim 0.055$, $k_{33}' \sim 0.084$	$d_{33} \sim 14$, $d_{15} \sim 9$	$Q_m > 4300$	Zhang et al. (2006)
La₂Ti₂O₇ (textured)	1461	Perovskite layer	$\varepsilon_r \sim 46$, $\tan\delta \sim 0.2\%$	/	$d_{33} \sim 2.6$	$Q_m > 3000$	Zhang and Yu (2011)
LiNbO₃ crystal	1150	Corundum	$\varepsilon_r \sim 25$, $\tan\delta \sim 0.5\%$	$k_t \sim 0.17$ (z cut), 0.49 (y/36° cut)	$d_{31} = -1$, $d_{33} = 6$, $d_{15} = 68$, $d_{22} = 21$	$\rho = 4.65$ g/cc, $s_{33}^E \sim 5$ pm²/N, $Q_m = 10{,}000$	Newnham (2005)
LiTaO₃ crystal	665	Corundum	$\varepsilon_r \sim 43.4$, $\tan\delta \sim 0.5\%$	$k_{33} \sim 0.14$	$d_{31} = -3$, $d_{33} = 9.2$, $d_{15} = 26$, $d_{22} = 8.5$	$\rho = 7.45$ g/cc, $s_{33}^E \sim 4.4$ pm²/N, $Q_m = 10{,}000$	Smith and Welsh (1971)
PMNT crystal	135	Perovskite	$\varepsilon_r \sim 5000$, $\tan\delta \sim 0.2\%$	$k_{33} \sim 0.90$	$d_{33} > 1500$, $d_{15} > 2000$	$\rho = 8$ g/cc, $s_{33}^E \sim 60$ pm²/N, $Q_m = 100$	Zhang et al. (2008c)

PMNT: Pb(Mg$_{1/3}$Nb$_{2/3}$)O₃-PbTiO₃; BMT-BF-BS-PT: Bi(Mg$_{1/2}$Ti$_{1/2}$)O₃-BiFeO₂-BiScO₃-PbTiO₃.

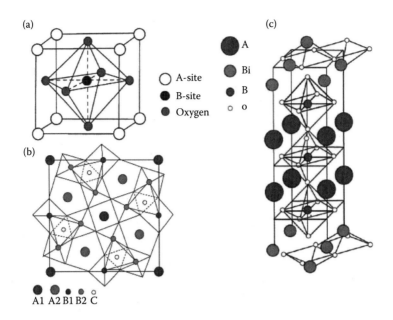

FIGURE 10.4
Schematic structures for perovskite (a), tungsten bronze (b) and bismuth layer (c) materials. (With kind permission from Springer Science+Business Media: *Piezoelectric and Acoustic Materials for Transducer Applications*, Crystal chemistry of piezoelectric materials, 2008, 39–56, Trolier-McKinstry, S.)

PLS ferroelectrics have the general formula $A_2B_2O_7$ and possess an anisotropic layered structure similar to the Aurivillius family, which is actually a derivative of the perovskite structure. In the PLS, the perovskite slabs consist of vertex-sharing BO_6 octahedrons separated by additional O atoms. In this family, $Sr_2Nb_2O_7$ and $La_2Ti_2O_7$ are the two most recognized compounds, possessing the highest known ferroelectric Curie temperatures, 1342°C and 1500°C, respectively (Wang and Wang 2008, Fuierer and Newnham 1991). Of particular significance is that high electrical resistivity was achieved at elevated temperatures, being on the order of 10^7 Ohm · cm at 700°C. Recently, textured $La_2Ti_2O_7$ ceramics were achieved by spark plasma sintering, exhibiting a Curie temperature of ~1460°C and piezoelectric d_{33} about 2.6 pm/V, with electrical resistivity being on the order of 3×10^8 Ohm · cm at 500°C (Zhang and Yu 2011, Yan et al. 2009).

It should be noted that in addition to ferroelectric polycrystalline ceramics, relaxor-ferroelectric single crystals have also been widely studied and utilized in various applications, where the term relaxor ferroelectrics is used for solid solutions between relaxors and ferroelectrics. The relaxors have a common characteristic of two or more cations occupying equivalent crystallographic sites, for example, $A(B_1,B_2)O_3$, which cause different behaviors compared to normal ferroelectrics, such as a broad Curie peak range in the ferroelectric-paraelectric phase transition, strongly frequency dependent dielectric permittivity, and Curie-Weiss square behavior above the dielectric permittivity (Cross 1987, Smolenskii et al. 1961, Randall et al. 1993). Relaxor ferroelectric crystals with the perovskite structure, including $Pb(Mg_{1/3}Nb_{2/3})O_3$-$PbTiO_3$ (PMN-PT) and $Pb(In_{1/2}Nb_{1/2})O_3$-$Pb(Mg_{1/3}Nb_{2/3})O_3$-$PbTiO_3$ (PIN-PMN-PT), were reported to possess very high piezoelectric coefficients, being >1500 pm/V, but their usage temperature range is limited by a low ferroelectric—ferroelectric phase transition temperature, due to their strongly curved MPBs (~90–130°C) (Zhang and Li 2012). In contrast, $LiNbO_3$ (LN) and $LiTaO_3$ (LT) belonging to a *3m* symmetry with the corundum structure, exhibit very high Curie temperatures, being on the order of ~1150°C and 720°C, respectively; however, the piezoelectric coefficients d_{33} are about 6 pm/V for LN and 9 pm/V for LT crystals. The thickness shear vibration of LN crystals has been extensively studied for transducer applications due to the relatively higher piezoelectric d_{15}, being on the order of 68 pm/V. Though LN possesses a high Curie temperature, the usage temperature is limited to <600°C, due to its low resistivity ($10^4 \sim 10^6$ Ohm · cm at 500°C) and increased attenuation at elevated temperatures. Furthermore, LN crystals suffer from chemical decomposition at modest temperatures (Zhang and Yu 2011).

General observations of various ferroelectric polycrystalline ceramics as a function of Curie temperature are summarized in Figure 10.5. For comparison, the room temperature values of dielectric permittivity and piezoelectric coefficient of relaxor-PT single crystals and LN crystals are also added. As expected, the values of dielectric permittivity and piezoelectric coefficients were found to decrease with increasing Curie temperatures for

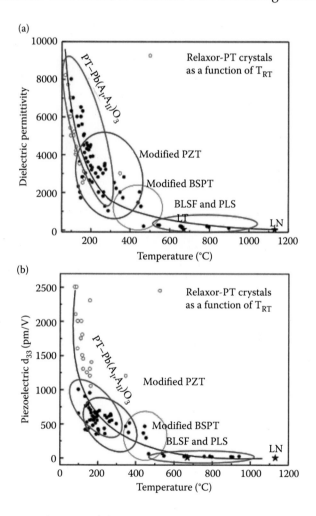

FIGURE 10.5

(See color insert.) Temperature dependent dielectric (a) and piezoelectric properties (b) for various ferroelectric families. (Data obtained from Zhang, S. J., and Li, F., 2012, *Journal of Applied Physics*, 111, 031301; Zhang, S. J., and Yu, F. P., 2011, *Journal of the American Ceramic Society*, 94, 3153–3170; Sebastian, T. et al., 2010, *Journal of Electroceramics*, 25, 130–134; Sebastian, T. et al., 2012, *Journal of Electroceramics*, 28, 95–100.)

ferroelectric polycrystalline ceramics. The highest piezoelectric properties and lowest Curie temperature were observed for relaxor $Pb(A_IA_{II})O_3$ based polycrystalline materials with the perovskite structure, while the lowest piezoelectric properties and highest Curie temperatures were found for materials with the double PLS. In contrast, the dielectric permittivity and piezoelectric coefficients of relaxor-PT ferroelectric single crystals exhibit a similar trend but as a function of rhombohedral to tetragonal phase transition temperature T_{RT}, instead of the Curie temperature, due to the strongly curved MPB (Zhang and Li 2012).

In addition to high piezoelectric properties and transition temperature, ferroelectrics with high electrical resistivity are also desired so that a large field can be applied during the poling process without breakdown or excessive charge leakage. High insulation resistance "R" is also required during device operation. For example, in the case of sensors, the piezoelectric material must not only develop a charge for the applied stress or strain but must also maintain the charge for a time long enough to be detected by the electronic system, while for an actuator, the electric field must be maintained for the duration of the desired strain or deformation (Shrout et al. 2002). Figure 10.6 gives the electrical resistivity as a function of temperature for various ferroelectric materials (Zhang and Yu 2011), including polycrystalline ceramics with perovskite, bismuth layer and PLSs, and lithium niobate single crystals. The highest resistivities are observed for ceramics with the PLS, being on the order of 10^8 ~10^9 Ohm · cm at 500°C, decreasing to 10^7 Ohm · cm for BLSF ceramics, and further reduced to 10^6 Ohm · cm for PZT ceramics and LN crystals at the same temperature. According to the Arrhenius law, the activation energies were found to be on the order of 1.3–1.5 eV for all the ceramics, with lower values of 1.1 eV observed for LN crystals, reflecting that the oxygen vacancies dominate the conductivity.

In contrast to ferroelectric materials, non-ferroelectric piezoelectric crystals, such as α-quartz and wurtzite-structured compounds (AlN and ZnO), have distinct advantages over ferroelectric materials, including minimal aging behavior and high mechanical

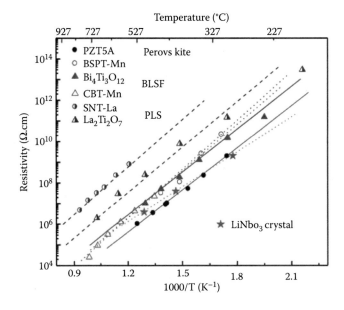

FIGURE 10.6
Electrical resistivity as a function of temperature for various ferroelectric materials. (After Zhang, S. J., and Yu, F. P., 2011, *Journal of the American Ceramic Society*, 94, 3153–3170.)

quality factors, due to the absence of ferroelectric domains. In addition, these piezoelectric crystals exhibit high electrical resistivity and exhibit no Curie temperature prior to their respective melting temperatures, which results in an expanded usage temperature range. However, the properties of non-ferroelectrics are considerably low in comparison with ferroelectrics; for example, the piezoelectric coefficients of α-Quartz and AlN are only about ~2 pm/V and ~5 pm/V, respectively. The detailed properties of non-ferroelectric piezoelectric single crystals are listed in Table 10.2 (Zhang and Yu 2011).

Quartz is the best known and the earliest piezoelectric material used in electronic devices, such as oscillators, resonators and filters (Mason 1950). Quartz possesses excellent electrical resistivity ($>10^{17}$ Ohm · cm at room temperature) and ultralow mechanical loss (high mechanical Q_m), and is highly temperature-stable, making it the material of choice in telecommunication. However, disadvantages of quartz, include a relatively small piezoelectric coefficient (d_{11} ~ 2.3 pm/V) and a low α – β phase transition temperature at 573°C, being further limited by mechanical (ferrobielastic) twinning which occurs at 300;°C (Zhang and Yu 2011). GaPO$_4$ shares many of the positive features of quartz, such as high electrical resistivity and mechanical quality factor at temperatures up to 970°C, where a α–β phase transition occurs. GaPO$_4$ crystal has been reported to possess high mechanical Q_m, about 20,000 at room temperature, decreasing significantly at temperatures above 700°C, due to the increase in structural disorder (Haines et al. 2002). GaPO$_4$ has been extensively studied for high temperature gas, temperature and pressure sensors for a number of challenging environmental applications (Zhang and Yu 2011; Zarka et al. 1996).

AlN belongs to group IIIA nitrides and possesses the wurtzite structure with point group *6 mm*. In air, surface oxidation occurs above 700°C, and even at room temperature, surface oxide layers of 5–10 nm have been detected. Thus, the usage temperature is about 700;°C without protection from oxidation, while going up to ~1100°C in non-oxidizing atmosphere (Kazys et al. 2005). AlN thin films have already been widely used, such as film bulk acoustic resonators. The piezoelectric properties of AlN thin films were measured and the bulk values were derived and reported to be d_{33} ~5.6 pm/V and d_{14} ~9.7 pm/V (Wright 1997). Bulk AlN crystals can be grown, however, with poor quality and low electrical resistivity due to defects (Zhang and Yu 2011).

In the early 1980s, the search for new piezoelectric compounds intensified, particularly CGG (Ca$_3$Ga$_2$Ge$_4$O$_{14}$) had been found to comprise an entire group of crystal materials with at least 40 members (Zhang and Yu 2011). Various materials were obtained by incorporating different cations in CGG, referred to as the langasite family. These crystals are found to be readily grown by the Czochralski pulling or Bridgman growth method. Of particular significance is the absence of phase transitions prior to their respective melting temperatures, usually in the range of 1200–1550°C. Langasite family crystals belong to the trigonal system (point group 32) and therefore are not pyroelectric, analogous to quartz, with general formula of A$_3$BC$_3$D$_2$O$_{14}$. The structure of langasite—La$_3$Ga$_5$SiO$_{14}$, and its isomorphs (such as langatate—La$_3$Ga$_{5.5}$Ta$_{0.5}$O$_{14}$), however, are disordered, where the structural disorder affects the material uniformity and results in incoherent phonon scattering, which increases the acoustic friction. The piezoelectric coefficients of the disordered langasite materials were reported to be in the range of 6–7 pm/V, with electromechanical coupling factors being on the order of 0.15–0.17 (Zhang and Yu 2011). Recent developments on LGS-type compounds, based on stringent structural and charge compensation rules, have led to the identification of a group of totally "ordered" langasite–structure crystals, such as Ca$_3$TaGa$_3$Si$_2$O$_{14}$ (CTGS) (Zhang and Yu 2011, Zhang et al. 2009, Chai et al. 2000, Chou et al. 2001). The reported ordered structure crystals are expected to give lower acoustic loss and higher acoustic velocity. As listed in Table 10.2, the piezoelectric coefficients

TABLE 10.2

Room Temperature Electrical and Mechanical Properties of Various High-Temperature Piezoelectric Crystals

Material Family	T_m/T_C (°C)	Symmetry	Dielectric Properties	Coupling Factors	Piezoelectric Coefficients (pm/V)	Physical Properties	Ref.
α-SiO_2	573 (α-β)	32	$\varepsilon_r \sim 4.5$, $\tan\delta \sim 0.4\%$	$k_t = 0.09$	$d_{11} = 2.3$, $d_{14} = -0.67$	$\rho = 2.65$ g/cc, $Q_m = 100{,}000$	Zhang and Yu (2011)
$GaPO_4$	970 (α-β)	32	$\varepsilon_{r11} \sim 6.1$	$k_{eff} = 0.16$	$d_{11} = 4.5$, $d_{14} = 1.9$	$Q_m = 20{,}000$	Zhang and Yu (2011)
ZnO	$T_m \sim 1975$	6 mm	$\varepsilon_r \sim 11$	$k_{33} \sim 0.408$, $k_{31} \sim -0.189$, $k_t \sim 0.282$	/	/	Jaffe, Berlincout (1965)
AlN	$T_m \sim 2200$	6 mm	$\varepsilon_r^S \sim 8.5$	$k_t \sim 0.24$, $k_{33} \sim 0.30$	$d_{31} = -2$, $d_{33} = 5$, $d_{15} = 4$	$\rho = 3.26$ g/cc, $s_{33} = -2.8$ pm²/N	Jaffe, Berlincout (1965)
$La_3Ga_{5.5}Ta_{0.5}O_{14}$ (disordered)	$T_m \sim 1450$	32	$\varepsilon_{r11} \sim 19.6$	$k_{12} = 0.16$	$d_{11} = 6.4$	$s_{11}^E = 9.2$ pm²/N, $Q_{11} = 4000$	Zhang and Yu (2011)
$Ca_3TaGa_3Si_2O_{14}$ (ordered)	$T_m \sim 1350$	32	$\varepsilon_{r11} \sim 18.2$	$k_{12} = 0.12$	$d_{11} = 4.6$	$s_{11}^E = 9.3$ pm²/N, $Q_{11} = 19{,}000$	Zhang and Yu (2011), Zhang et al. (2009)
$YCa_4O(BO_3)_3$	$T_m \sim 1510$	m	$\varepsilon_{r22} \sim 12$	$k_{26} = 0.19$	$d_{26} = 7.8$	$s_{66}^E = 16.3$ pm²/N, $Q_{26} = 9000$	Zhang and Yu (2011), Yu et al. (2010)

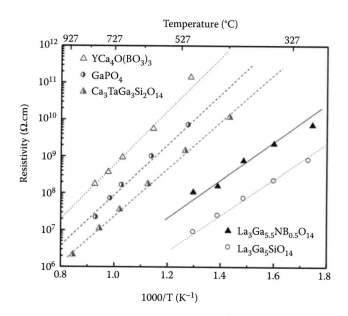

FIGURE 10.7
Electrical resistivity as a function of temperature for selected non-ferroelectric piezoelectric single crystals. (After Zhang, S. J., and Yu, F. P., 2011, *Journal of the American Ceramic Society*, 94, 3153–3170.)

and electromechanical coupling factors were found to be slightly lower when compared to their disordered counterparts, being on the order of 4–5 pCN and 0.12–0.16, respectively, while exhibiting much higher mechanical quality factor and electrical resistivity (as shown in Figure 10.7) (Zhang and Yu 2011).

Oxyborate crystals with the general formula $ReCa_4O(BO_3)_3$ (Re = rare earth element, abbreviated as ReCOB) have been extensively studied for nonlinear optical applications. The crystals can be readily grown from the melt using the Czochralski (CZ) pulling or Bridgman techniques at around 1500°C. Analogous to langasite crystals, no phase transitions occur prior to their melting temperatures, which result in an expanded usage temperature range (Zhang and Yu 2011, Zhang et al. 2008a,b). Recently, oxyborate crystals have attracted great attention for piezoelectric applications, due to their reasonable piezoelectric properties (d_{26} ~7–16 pm/V, being 4–8 times of the value of α-quartz and $k_{26} = 0.19$) (Yu et al. 2010), ultra-high electrical resistivity and mechanical quality factors at elevated temperatures (Zhang and Yu 2011).

Figure 10.7 shows the electrical resistivity as a function of temperature for piezoelectric single crystals, where it is observed that langasite crystals with a disordered structure possess relatively low resistivities, with the lowest values of ~8×10^6 Ohm · cm at 500°C for LGS crystals, while ordered langasite crystals, for example, CTGS, were found to have higher resistivity, being 100 times the value of its disordered counterparts. The $GaPO_4$ and YCOB crystals were found to possess the highest resistivities, being 10^{10} and 10^{11} Ohm · cm at 500°C, respectively. Of particular significance is that the resistivities of YCOB, $GaPO_4$, and CTGS crystals were on the order of >10^6 Ohm · cm at 1000°C, significantly expanding the potential usage range (Zhang and Yu 2011).

Figure 10.8 summarizes the sensitivity versus maxima usage temperature range for various piezoelectric materials. The relaxor-PT ferroelectric single crystals with the perovskite structure were found to possess the highest piezoelectric properties with d_{33} values being on the order of >2000 pm/V, however, their usage temperature range is limited by low

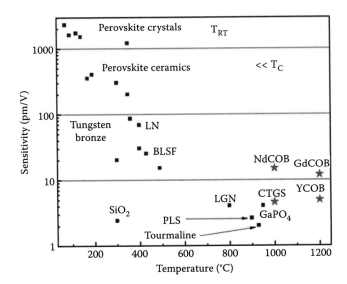

FIGURE 10.8
Potential operational temperatures and sensitivity (piezoelectric coefficient) range for various piezoelectric materials.

ferroelectric phase transitions. Perovskite polycrystalline ceramics, such as PMNT, PZT, and BSPT, have actuations in the range of 100 to –1000 pm/V, with a usage temperature range of 100–350°C, depending on their Curie temperatures, and subsequently limited to 1/2 to 2/3 T_Cs by thermally activated aging behavior. Ferroelectric materials with tungsten bronze and Aurivillius structures possess medium piezoelectric properties, ranging from 10 to 100 pm/V, with operational temperatures up to 550°C, and limited by the low electrical resistivity at elevated temperatures. It should be noted that though LN crystals possess a Curie temperature of 1150°C, their low resistivities and oxygen loss at elevated temperatures restrict application to less than 600°C (Sweeney and Halliburton 1983, Damjanovic 1998b, Zhang and Yu 2011). Generally, nonferroelectric piezoelectric single crystals possess low piezoelectric coefficient, falling in the range of 1–20 pm/V, with usage temperatures in the range of 350 ~ >1000°C, depending on their phase transitions, melting temperature or electrical resistivity. Thus, to evaluate a material for high-temperature applications, the following material parameters are suggested: piezoelectric coefficients and electromechanical coupling factors, usage temperature range (phase transition, melting temperature, electrical resistivity, chemical stability, etc.) and temperature stability of the properties. Furthermore, the mechanical quality factor Q_m is also very important for resonance-based sensing and actuation applications. In addition, radiation and corrosion resistant properties are desirable for applications under harsh environments.

10.2.3 Electrostrictive Materials

Electrostriction is used to describe the electric-field induced strain, which is present in all dielectric materials. A key difference between piezoelectricity (converse piezoelectric effect) and electrostriction is that piezoelectricity displays a linear relationship between strain S and electric field E, while the electrostrictive strain shows a quadratic dependence on the electric field E, or electric polarization P (Cross et al. 1980, Uchino et al. 1981). Electrostrictive strain S is simply expressed as a function of electric field E or electric polarization P, that is, $S = M \cdot E^2$ or $Q \cdot P^2$, where M and Q are forth rank tensors representing electrostriction coefficients. The general description of electrostriction can be written in the following tensor notation:

$$S_{ij} = M_{ijkl}E_kE_l,$$
$$S_{ij} = Q_{ijkl}P_kP_l, \tag{10.5}$$

Note that electrostrictive strain generally deviates from the quadratic relation (E^2) and saturates at a high electric field level because of dielectric nonlinearity; thus, the Q electrostriction coefficient is preferably used to describe the electrostrictive effect, which is linearly proportional to the square of electric polarization (P^2) (Newnham 2005).

All dielectric materials possess electrostriction; however, the practical applications for electrostrictive actuators are limited to the perovskite ferroelectric family that has relatively high dielectric permittivities. The usefulness of electrostrictive materials over piezoelectric materials for actuator applications arises from the absence of macroscale domains as they are generally operated in the para-electric state above the temperature of dielectric permittivity maximum (T_{max}). The absence of macroscale domains results in low hysteresis loss, low aging, improved reproducibility, reduced creep effects and a high-response speed (<10 µs). These features are particularly suitable for actuator devices, such as servo transducers and micro-positioning devices (Uchino 1996, 2000). Electrostrictive materials can also induce large piezoelectric coefficients with a biasing field, allowing for more linearized strain field response.

Lead magnesium niobate (PMN) and modified PMN compositions, known as relaxors, are the most widely studied electrostrictive materials, offering a relatively broad operating temperature range due to the diffuseness of maximum dielectric permittivity ($\varepsilon_{max} = K_{max}\varepsilon_o$). The temperature at which the dielectric permittivity reaches maximum value (T_{max}) is around −10°C for PMN, whose value can be shifted upward with increasing PbTiO$_3$ (PT) content. Materials with a range of maximum temperatures (T_{max}) from −5°C to 80°C and higher have been produced with controlling the PT content up to 30%. However, the diffuseness of the transition decreases with increasing PT content. Above 30% of PT, the material is transformed into a normal ferroelectric, exhibiting a sharp dielectric transition. The values of electrostriction coefficients of PMN relaxors are reported to be $Q_{11} \sim 0.025$ and $Q_{12} \sim 0.0096$ m^4/C^2 (Uchino et al. 1980). The maximum relative permittivities range from 20,000 to 30,000 at 1 MHz with large diffused dielectric permittivity maxima ($\Delta T = 20$–30°C) as a result of the relaxor characteristic.

Lanthanum-doped lead zirconate-lead titanate Pb$_{1-x}$La$_x$(Zr$_{1-y}$Ti$_y$)$_{1-x/4}$O$_3$ (PLZT) is another example of relaxor ferroelectrics. For PLZT relaxors, most investigations have been focused on compositions of Zr/Ti ratio 65/35, that is, PLZT (x/65/35). With increasing lanthanum content, T_{max} shifts toward a lower temperature as increasing the diffuseness of maximum dielectric permittivity; however, the maximum dielectric permittivity decreases with increasing lanthanum content. PLZT (x/65/35) relaxors with $x = 0.07$–0.11 exhibit a broader temperature region ($\Delta T = 100$–140°C) than PMN based relaxors with maximum relative permittivities ($K_{max} = 6000$–8000), implying the potential for a large E-field induced strain over a broad temperature range. The longitudinal electrostriction coefficients Q_{11} are between 0.018 and 0.022 m^4/C^2 depending on lanthanum content (Meng et al. 1985). Note that the electrostriction coefficients of relaxor ferroelectrics are generally lower than those of normal ferroelectrics; for example, Q_{11} of PMN and PLZT relaxors are ~0.022 m^4/C^2, while those of normal ferroelectrics are 0.08–0.11 m^4/C^2. However, the strain levels of electrostrictive materials are on the order of 10^{-3}, two orders higher than those of the linear piezoelectrics because of much higher dielectric permittivities of relaxor ferroelectrics compared to those of normal ferroelectrics.

Other electrostrictive materials include strontium doped barium titanate $(Ba_{1-x}Sr_x)$ TiO_3 (BST) and Sn-doped barium titanate $Ba(Ti_{1-x}Sn_x)O_3$ (BTS) ferroelectrics, whose compositions possess comparable dielectric permittivities to relaxor ferroelectrics. The maximum relative permittivities of BST and BTS are reported to be 15,000–20,000 and 28,000–35,000, respectively. However, these materials have a low degree of diffuseness of the dielectric permittivity maximum, limiting the working temperature range. The properties of these electrostrictive materials are summarized in Table 10.3. For comparison, the properties of piezoelectric materials are also included (Fielding 1993, Sundar et al. 1995, Newnham et al. 1997).

For high-temperature actuation, electrostrictive materials with large diffused dielectric permittivity maxima at high temperature are desired. However, the implementation of electrostrictive materials has been limited below 125°C due to the decreased diffuseness of the maximum dielectric permittivity. In recent years, guided by the crystal chemistry arguments based on the perovskite tolerance factor, a new MPB piezoelectric material, $BiScO_3$-$PbTiO_3$ (BS-PT), has been developed. The BS-PT materials possess relative permittivity maxima greater than 40,000 with a T_{max} of around 450°C, whose temperature is ~100°C higher than traditional PZT ceramics (Eitel et al. 2002, Zhang et al. 2005b). This allows BSPT to operate at 450°C as an electrostrictor, by applying an ac drive field superimposed on a dc bias to induce piezoelectricity. Of particular significance is that the mechanical Q_m was found to be very high, with saturated values of over 1000 at 450°C under dc bias (Hackenberger et al. 2004). However, one drawback of this composition is the operating temperature window is narrow due to the sharp phase transition as found in normal ferroelectrics.

New bismuth and lead-based perovskite compositions have been proposed and developed for high-temperature relaxor ferroelectric materials; specifically $xBiScO_3$-$yPb(Mg_{1/3}Nb_{2/3})$ O_3-$zPbTiO_3$ (BS-PMN-PT) compositions. The maximum relative permittivities K_{max} range from 10,000 to 24,000 at 1 kHz, with the temperatures of the maximum relative permittivity

TABLE 10.3

Dielectric and Electrostriction Properties for Various Piezoelectric and Electrostrictive Materials

Family	Composition	T_{max} (°C)	K (RT, 1 kHz)	Q_{11}/Q_{12} (10^{-2} m⁴/C²)
Normal Ferroelectrics	$PbTiO_3$	400	270	8/−2.9
	$BaTiO_3$	120	1070	11/−4.5
	PZT	365	1700	-
Family	Composition	T_{max} (°C)	K_{max} (1 MHz)	Q_{11}/Q_{12} (10^{-2} m⁴/C²)
Relaxor (Dispersive)	(1-x)PMN-xPT	27 (x = 0.07) (ΔT ~20–30)	20–30,000 (8800)*	2.52/−0.96
Relaxor (Dispersive)	PLZT (x/65/35)	39 (x = 11) (ΔT ~100–140)	6–8,000 (8500)*	2.1/−0.91
Normal Ferroelectrics	$(Ba_{1-x}Sr_x)TiO_3$	32 (x = 0.3) (ΔT ~0)	15–20,000 (5400)*	-
Normal Ferroelectrics (Pinched)	$Ba(Ti_{1-x}Sn_x)O_3$	8 (x = 0.13) (ΔT ~0)	28 ~ 35,000 (7000)*	-

Note: *Dielectric properties are E-field dependent values at ~10 kV/cm.

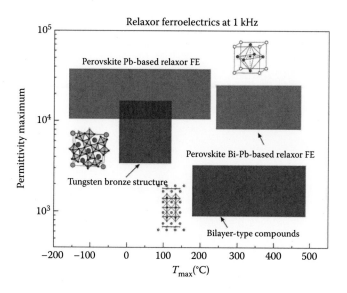

FIGURE 10.9

Comparisons of various dielectric materials in terms of relative permittivity maximum (K_{max}) and the temperature of the relative permittivity maximum (T_{max}) for Pb-based perovskites, Bi-Pb relaxors, tungsten bronze structured relaxors, and Bi-layered compounds. (Adapted from Stringer, C. J., 2006, Structure property performance relationship of new high temperature relaxors for capacitor applications, PhD Dissertation, Pennsylvania State University.)

T_{max} in the range of ~250–350°C, depending on compositions. Figure 10.9 shows the comparison of various dielectric materials in terms of K_{max} and T_{max} with Bi-Pb-based relaxor ferroelectrics, exhibiting that Bi-Pb-based relaxor compositions possess both high K_{max} and T_m, making these materials candidates for high temperature electrostrictive applications (Stringer 2006).

10.2.4 Electromechanical Actuators

Electromechanical materials can deform or move under an external electrical stimulus. We can apply these materials in a variety of configurations to produce actuators to do useful work. These can be small stroke high force systems for micro displacements or high torque rotating mechanisms for motors. Linear actuators can be classified by the stroke, force, and frequency of operation whereas and rotary actuators can be specified by the speed and the torque. In terms of sizing the actuator, one also has to consider whether it is going to be activated continuously (Power) or intermittently (Energy per activation) and their respective densities (Force ρ_F, Energy ρ_E, Torque ρ_τ, Power ρ_P). Another useful specification of an actuator is the efficiency η which is the work done divided by the input energy per cycle. In this chapter, we will look at the available materials that can be used to actuate at high temperatures either directly through the application of current and/or voltage or indirectly by generating a local heat through the dissipation of electrical power.

One of the limitations of solid state electromechanical material is that the strain in the material is small compared to electromagnetic actuators even though the generated stresses can be very large. A wide variety of techniques have been developed to increase the strain for a given applied voltage as is shown schematically in Figure 10.10. One of the most common electroceramic actuator configurations is the stack, which as the name

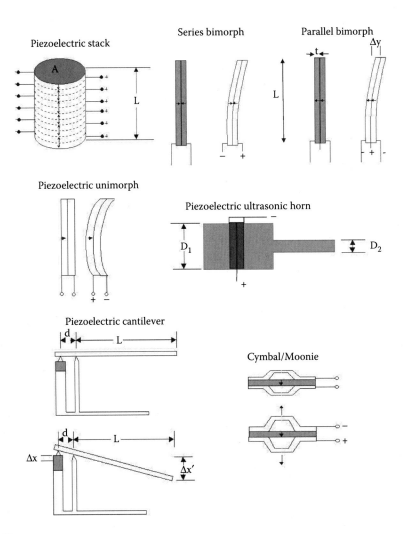

FIGURE 10.10
Schematic diagrams of a variety of techniques used to increase the strain in electromechanical actuators.

implies, is a number of thin alternately poled piezoelectric layers connected mechanically in series and electrically in parallel. The total displacement of a stack of reasonable length is limited to approximately 100 microns at room temperature without the use of other amplification techniques. There has been much interest in precision positioning for optical instrumentation and recent studies have looked at the life issues for these devices (Sherrit et al. 2011). Another strain amplification technique is to use bending strain as is found in bi-morphs. Piezoelectric bi-morphs can be manufactured in both series and parallel configurations. It is manufactured from two plates of piezoelectric material poled in opposite directions and bonded together. Under the application of electric field one piezoelectric layer contracts in the thickness direction while the other expands.

Owing to the contraction and expansion in the thickness direction, one layer expands along the length direction and the other contracts inducing bending in the layers. For a series arrangement with a free length L and total thickness t and a transverse piezoelectric coefficient d_{31}, the tip displacement is seen to be proportional to the voltage V and is found to be

$$\Delta y = \frac{3d_{31}VL^2}{2t^2} \qquad (10.6)$$

For the parallel arrangement, the displacement is twice that of the series configuration for the same voltage and is

$$\Delta y = \frac{3d_{31}VL^2}{t^2} \qquad (10.7)$$

Tip displacements of the order of a centimeter can be induced in these actuators. A variety of papers have looked at the response of these actuators to tip force, pressure, moment, voltage etc. (Ballato and Smits 1991, Smits and Dalke 1989, Smits and Ballato 1993, Smits et al. 1997).

Another interesting bender material that has been developed includes the Rainbow (reduced and internally biased oxide wafers) and Cerambow actuators developed at Clemson by Heartling (Sherrit et al. 1994, Barron et al. 1996, Haertling 1997) and Thunder (thin-layer composite-unimorph ferroelectric driver and sensor) developed by NASA Langley—see Face International (1999). These unimorph materials have at least one active layer and an inactive layer. Under expansion in the poling direction the strain in the plane perpendicular to the poling direction undergoes a contraction. Since this strain occurs only on the active side of the interface of the piezoelectric/electrostrictive material, a bending moment is induced in the structure. Large deflections can be induced in these materials and the size of the deflection can be shown to increase with an increase in the lateral dimensions.

Another strain amplification technique occurs in the flextensional actuator materials known as the Moonie and Cymbal developed at Penn State (Dogin et al. 1997). Flextensional composites use endcaps that convert the transverse strain to longitudinal strain. A variety of companies make these actuators for normal operating temperatures (Claeyssen et al. 2001; Dynamic Structures and Materials LLC 2012). Cantilever designs can also be used as a lever to increase the strain. A piezoelectric/electrostrictive material is connected mechanically to a lever arm, which is pivoted a distance d from the sample. From the pivot the lever arm extends out to a distance L. For a displacement Δx in the sample the displacement at the other end is $\Delta x' = (L/d)\,\Delta x$. We have a strain amplification of (L/d). For example, if a piezoelectric stack of maximum displacement of 100 microns is used and a strain amplification factor $(L/d = 10)$, then displacements of the order of 1 mm can be achieved with an equivalent reduction in force. For a discussion and comparison of the cantilever, multilayer, flextensional, rainbow, and bimorph strain amplification techniques see the paper by Near (Near 1996). Another piezoelectric multilayer-stacked hybrid actuation/transduction system was recently reported that uses synergetic contributions from positive-strain and negative-strain piezoelectric multilayer-stacks to give displacements of about 3.5 times those of the same-sized piezoelectric flex-tensional actuator/transducer (Xu et al. 2011).

A technique that is used in high-power applications to increase the strain in a material for a given field is to drive the material at its resonance frequency. In general, for any mechanical system it can be shown that at resonance the strain in the material is amplified by a factor called the mechanical Q_m. The mechanical Q_m is a measure of the mechanical losses in the material. For a high Q_m material the mechanical losses are small. For PZT the mechanical Q_m can range from 50 to 1000, which means that for an ideal system that strain amplification can be as high as a 1000 times. It should be noted, however, for materials with large Q_ms it is very easy to drive the sample into a nonlinear regime, which would reduce the overall amplification and/or damage the sample. Another problem with resonance amplification is that the resonance frequency for a piezoelectric material is determined from the velocity of sound and the dimension in which the material is resonating. This means that in practice for most piezoelectric materials the frequency is fixed and is limited to a range in the 100s of kHz and above.

An alternative resonance amplification technique that can be designed to operate at a lower resonance frequency is the ultrasonic horn. A variety of horn designs have been conceived for different applications, however the stepped horn (Belford 1960, Sherrit et al. 1999, 2004b) typically produces the largest stroke amplification, which is proportional to the area ratios of the base and the horn tip. If a strain wave is induced on the large end of the horn at the horn resonance frequency, the displacement at the small end is amplified by a factor $M = (D_1/D_2)^2$ due to the conservation of the wave momentum. This amplification is in addition to the amplification due to the mechanical Q_m of the material. In previous work we developed a horn that was demonstrated to drill at 500°C (Bao et al. 2012). Other horn configurations include the exponential ($M = D_1/D_2$) (Mason 1958) and linear horns ($M = 4.61$ max).

One advantage of the ultrasonic horn is one can adjust the resonance frequency by adjusting the relative length of each section. The frequency can be adjusted by a factor of 2 without changing the overall length. It should be noted that the amplification is limited in practice by the critical strain of the tip material. Other methods of stroke amplification techniques include producing rotary and linear motors from microscopic displacements in electromechanical materials (Watson et al. 2009, Ueha and Tomikawa 1993, Sashida and Kenjo 1993).

10.3 Competing Actuation Technologies

10.3.1 Magnetostrictive Materials

Magnetostrictive materials have been known to exhibit strain as a function of the change in the magnetic state of a material for over a century and were originally described by James Joule (Joule 1842). Materials such as Terfenol D and Galfenol (Clark and Wun-Fogle 2002) have been used to produce high frequency high power sonar projectors, optics (Apollonov et al. 1992) and ultrasonic actuators (Hudson, Busbridge and Piercy 1998). A variety of magnetostrictive compounds have been shown to have reasonable magnetostrictive coefficients at elevated temperatures. The family of RFe2 alloys (R = Tb, Dy, Sm) have demonstrated very high magnetostriction (greater than 1000 ppm) at temperatures as high as 200°C (Clark and Crowder 1985). Piezomagnetism is a phenomenon related to magnetostriction but rather than a change in strain that is proportional to the square of the change in magnetic field, it is linear with the magnetic field change. Like electrostriction, a magnetostrictive material can be forced to a linear piezomagnetic regime by DC biasing and adding an AC field with amplitude below the DC bias field. At present, magnetostrictive materials with operational temperatures in the 500°C range are available but have magnetostrictive coefficients that are small (Belson 1967) compared to the giant magnetostrictive materials (Terfenol D and Galfenol).

10.3.2 Shape Memory Alloys

Shape memory alloys (SMA) have been implemented in a variety of low frequency applications where high force is required. The materials are alloys that have the ability to recover a permanent strain when heated above a certain temperature (e.g., Nitinol NiTi) (Butera 2008, Gilbertson and Busch 1996). They exhibit a stable high- and low-temperature phase that has different crystal strains in a given direction. High-temperature SMA materials NiTiHf (Meng et al. 2000) have been developed that can actuate above 400°C with reduced strain and increased aging. A recent review on the high temperature SMA by Ma and colleagues (Ma et al. 2010) suggests a variety of materials are available in the 100–400°C range. These materials have been used in a variety of applications where simplicity and high force are required and where a large number of cycles and frequency of operation are not critical (Badescu et al. 2012, Wu and Schetky 2000).

10.3.3 Phase Change Materials

New classes of materials that have been used in actuators for space where large stroke and force are required include phase change actuators, which have a volume expansion during a liquid to solid phase transformation. One particular substance that exhibits upwards of 10% stroke is the paraffin actuators produced by StarSys Inc. (Sierra Nevada corp 2012; Caltherm 2012). A high-temperature version of this approach has been developed by Therm-Omega-Tech Inc. (Thermoloid 2012). These actuators use the material volume expansion during phase change to drive a piston. They can be designed to produce large strokes of the order of a few centimeters and forces up to 500N or more.

One of the advantages of these actuators is they can be driven with a heater or by local environmental changes to balance thermal environments or act as thermal switches to activate when the local temperature exceeds the phase transition temperature. A variety of the Thermo-Omega-Tech actuators are shown in Figure 10.11.

10.3.4 Conventional Electromagnetic Actuators

The availability of conventional motors to build high temperature rotary and linear actuators is limited to a few suppliers who have targeted the oil and gas industries. These actuators are designed to handle large pressures as well as high temperature. A variety of companies have motors and actuators that can operate up to the intermediate temperature range of 200–300°C. These include Maxon (Maxon 2012, Wittenstein 2012, Ducommun 2012), photographs of these motors are shown in Figure 10.12 along with some of their operational specifications in the figure caption.

These motors can use a variety of high-temperature magnetic materials such as Samarium Cobalt SmCo or Alinco that have high remnance, coercivity, and energy density and can operate up to 350°C. Other groups who are interested in extraterrestrial

FIGURE 10.11
Various commercial Thermo-Omega-Tech actuators that can operate up to 149°C or greater. (Courtesy of Therm-Omega-Tech Inc.)

environments such as Venus have designed motors that have operated for a relatively short time at 500°C, including Honeybee Robotics and a high temperature motor developed at JPL and discussed at length in Chapter 9 of this book. Another issue for motors operating at 500°C is the availability of high-performance high-temperature gearing (Wright et al. 2010; Townsend 1994). A variety of other issues including winding material, insulation and structural magnetic steels have to be considered when pushing motor designs to 500°C (Moser 2003).

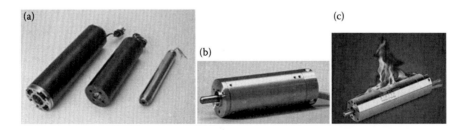

FIGURE 10.12
Various commercial motors that will operate up to 200°C or greater. (a) Ducommun 30,000 psi. (1.5" – 3.7" OD), 1.8" with 60 oz-in torque and 2800 RPM, 120 Watts, 200°C. (Courtesy of Ducommun LaBarge Technologies.) (b) Wittenstein HT motor—Up to 30,000 psi, (up to 300°C). (Courtesy of Wittenstein Cybermotor.) (c) Maxon Inc. DC brushless motors—Up to 25000 psi, (200°C) (0.95 "Diameter). (Courtesy of Maxon Inc.)

10.4 Actuator Life

10.4.1 Thermal Expansion

Thermal expansion is a strain in a material due to a change in temperature. The strain is usually described by a linear relationship with temperature

$$l = l_o[1 + \alpha(T - T_o)], \tag{10.8}$$

where T_o and l_o are the reference temperature and the length at T_o and l is the length at temperature T.

The coefficient α is the linear thermal expansion coefficient and is only valid over a limited temperature range. There is also a volume thermal expansion and lateral thermal expansion (area) which may also need to be considered. When designing mechanisms that are required to work at high temperature, one may need to consider the change of dimensions between the assembly temperature and the operational temperature and ensure that sufficient margin is built in to the design to allow for dimensional changes and resultant stress build up. In mechanisms that have to be assembled with a given pre-stress, soft spring mechanisms such as Belleville washers or flexures (Sherrit et al. 2010) can be used to apply the load. Since they are soft, small changes in the bolt dimension can be accommodated by the spring without increasing or decreasing the stress considerably. Other things to consider are structures that feed through another structure for relative movement. The gap must be sufficient at all temperatures between the assembly temperature and operational temperatures.

10.4.2 Thermal Aging and Degradation

Many of the electromechanical mechanisms used for producing stroke including piezo-electric, piezomagnetic, electrostrictive, and magnetostrictive are based on cooperative microscopic behavior in the material (e.g., domains) and since domain wall mobility and stability by their nature are thermally activated, operation at elevated temperatures can lead to the reduction in the magnetization or polarization in the material over time, resulting in aging. Note that unlike degradation this phenomenon may be reset by the application of large electric or magnetic fields. Typically at room temperature, the aging is described by logarithmic time-dependent properties relative to an initial time period.

Property degradation in the material can also occur on a larger timescale due to slow chemical reactions that may be caused by possible impurities or mobile dopants in the material. In addition, reactions with the ambient environment may cause severe problems if not limited. Examples of the latter include oxidation in an oxygen rich environment or rusting in humid environment or hydrogen embrittlement. In some materials, high-temperature operation leads to the grain growth, which can change both the elastic properties of structural materials (e.g., ductility) and the electromechanical properties of active materials.

10.4.3 Creep

Another major source of failure under load at high temperature for structural materials required by many mechanisms is creep. Creep is defined as the continuing time dependent deformation at constant temperature of a structure under constant load or stress.

It is characterized by a typical initiation behavior and then by Primary, Secondary, and Tertiary creep (Webster and Ainsworth 1994). In order to produce a housing or structure that preloads another structure, one has to make sure that the critical creep levels for the material are not enhanced at elevated temperatures. A variety of materials can operate easily above the 500°C mark, but one has to insure that the material will not behave poorly at the operational temperature since creep is thermally activated. Creep should be investigated for a specific application when the structure is operating at about 30% its melting temperature (Kelvins). Structural ceramics can usually operate at up to 40 to 50% of the melting temperature due to their resistance to creep fatigue (Craig 2005).

10.5 Summary

There are many technologies that have been used to produce actuation in research mechanisms up to 500°C. It is apparent that many applications in the aerospace industry like the adjustment of turbine blades or active deformation of an aeroshell could benefit substantially from robust commercial actuators that can operate up to and above these temperature ranges. Currently, a variety of commercial off the shelf actuators and motors are available up to 300°C and have been used successfully in the oil and gas industry.

We present a table summarizing the capabilities of the various actuation materials and mechanisms in Table 10.4. In general these types of comparisons suffer from a lack of consistency for each of the properties under comparison. For example, the operation temperature, response time, the drive source, aging, and hysteresis are fundamental properties that exist independent of the actuator design. However, the importance of each of these properties may depend on a specific design (e.g., hysteresis and closed loop control) and therefore can be only used as a general guide.

It should be noted that only piezoelectric materials have critical temperatures with measurable piezoelectric response at temperatures exceeding 600°C. This suggests that for extremely high-temperature applications, piezoelectrics may be the only active material that can potentially be used to produce reusable actuators in this temperature range. The search for new and better materials that can operate at elevated temperatures is ongoing. It

TABLE 10.4

Comparison of Actuator Metrics for Different Actuation Mechanisms

Actuation Method	Max Critical Temperature of Class	Demonstrated Operating Temperature	Response Time (s)	Drive Source	Aging	Hysteresis
Piezoelectric	>600°C	>600°C	<10 μs	Electric Field	Moderate	Moderate
Electrostrictive	>450°C	450°C	<10 μs	Electric Field	Small	Small
Magnetostrictive	Up to 500°C	500°C	<1 μs	Magnetic Field (Current)	Large	Large
Shape Memory Alloy	Up to 400°C	400°C	≈1 s	Voltage (VI Heat)	Large	Large
Phase Change	>150°C	150°C	≈10 s	Voltage (VI Heat)	Large	Large
Electromagnetic	600°C	−80 to 300°C	<ms	Voltage Current	Small	Moderate

is apparent that in order to produce robust high temperature actuators for commercial use, developments in the active and structural materials, conductive materials and insulation materials will be required to enable this technology.

Acknowledgments

Some of the research reported in this chapter was conducted at the Jet Propulsion Laboratory (JPL), California Institute of Technology, under a contract with the National Aeronautics and Space Administration (NASA). The authors thank Dragan Damjanovic, Institute of Materials, School of Engineering, EPFL, Lausanne, Switzerland; David Cann, Oregon State University, Corvallis, Oregon; Troy Y. Ansell Oregon State University, Corvallis, Oregon; and Xiaoning Jiang, North Carolina State University, Raleigh, North Carolina, for reviewing this chapter and for providing valuable technical comments and suggestions.

References

Apollonov, V. V., Aksinin, V. I., Chetkin, S. A., Kijko, V. V., Muraviev, S. V., Vdovin, G. and Popo, V., 1992, Magnetostrictive actuators in optical design, *Proceedings of SPIE 1543, Active and Adaptive Optical Components*, pp. 313–324.

Badescu, M., Sherrit, S., Bar-Cohen,Y., 2012, Compact, low-force, low-noise linear actuator, *NASA Tech Briefs*, 36(9), 70–71.

Baldor/Dodge/Reliance, 2012, http://www.reliance.com/pdf/motors/data_sheets/raps869.pdf [downloaded June 6, 2012].

Ballato, A., and Smits, J. G., 1991, Network representation for piezoelectric bimorphs, *IEEE Transactions on Ultrasonics, Ferroelectrics and Frequency Control*, 36, 595–602.

Bao, X., Bar-Cohen, Y., Sherrit, S., Badescu, M., and Shrout, T. R., 2012, High temperature piezoelectric drill. *Proc. SPIE 8345, Sensors and Smart Structures Technologies for Civil, Mechanical, and Aerospace Systems*, San Diego, CA.

Barron, B. W., Li, G., and Haertling, G. H., 1996, Temperature dependent characteristics of CERAMBOW actuators, *Proceedings of the 10th IEEE International Symposium on the Application of Ferroelectrics*, East Brunswick, N.J., vol. 1, pp. 305–308.

Belford, J. F., 1960, The stepped horn, *Proceedings of the National Electronics Conference*, pp. 814–822, Chicago.

Belson, H. S., 1967, High-temperature magnetostriction in alloys, *J. Appl. Phys.*, 38, 1327–1328.

Berlincourt, D. A., Curran, D. R., and Jaffe, H., 1964, Piezoelectric and piezomagnetic materials and their function in transducers, *Physical Acoustics*, 1, edited by Mason W.P., Academic Press, New York.

Berlincourt, D., 1971, Piezoelectric crystals and ceramics. Chapter 2. *Ultrasonic Transducer Materials*, E. b. Mattiat, Ed., ed: London: Plenum, pp. 63–124.

Buchanan, R. C., 1991, *Ceramic Materials for Electronics: Processing, Properties and Applications*, Marcel Dekker, Inc., New York.

Butera, F., 2008, Shape Memory Actuators for the Automotive Industry, Advanced Materials & Processes, March, pp. 37–40.

Caltherm thermal phase change actuators 2012, http://www.caltherm.com/thermostatic-solutions/thermal-actuators.php [downloaded Oct 10, 2012].

Chai, B. H. T., Bustamante A. N. P., and Chou, M. C., 2000, A new class of ordered langasite structure compounds, *IEEE/EIA Internaltional Frequency Control Symposium*, pp. 163–168, Kansas City, Missouri.

Chen, Y. M., Sheppy, M., Yen, T.-T., Vigevani, G., Lin, G.-M., Kuypers, J., Hopcroft M. A., and Pisano, A. P., 2009, Bi-chevron aluminum nitride actuators for high pressure microvalves, *Procedia Chemistry* 1, 706–709.

Chou, M. M. C., Jen, S., and Chai, B. H. T., 2001, Investigation of crystal growth and material constants of ordered langasite structure compounds, *IEEE Internaltional Frequency Control Symposium*, pp. 250–254, Seattle, WA.

Claeyssen, F., Le Letty, R., Barillot, F., Lhermet, N., Fabbro, H., Guay P., Yorck M., and Bouchilloux, P., 2001, Mechanisms based on piezoactuators, *Proceedings of SPIE 4332, Smart Structures and Materials 2001: Industrial and Commercial Applications of Smart Structures Technologies*, pp. 225–233, June 14, 2001 [see also http://www.cedrat-technologies.com/en/mechatronic-products/actuators/apa.html]

Clark, A. E., and Crowder, D. N., 1985, High temperature magnetostriction Of $Tbfe_2$ and $Tb_{.27}dy_{.73}fe_2^*$, *IEEE Transactions On Magnetics*, Mag-21(5), 1945–1947.

Clark, A. E., and Wun-Fogle, M., 2002, Modern magnetostrictive materials—Classical and non-classical alloys, 4699, 421–436.

Craig, B. D., 2005, Material failure modes, part I-A brief tutorial on fracture, ductile failure, elastic deformation, creep, and fatigue, *J Fail. Anal. and Preven.*, 5, 13–14.

Cross, L. E., Jang, S.J., Newnham, R. E., Nomura, S., and Uchino, K., 1980, Large electrostrictive effects in relaxor ferroelectrics, *Ferroelectrics*, 23, 187–191.

Cross, L., Hench, L.L., and Dove, D.B., 1972, *Thermodynamic Phenomenology of Ferroelectricity in Single Crystal and Ceramic Systems*, LL Hench and DB Dove. Dekker, New York.

Cross, L. E.,1987, Relaxor ferroelectrics, *Ferroelectrics*, 76, 241–267.

Curie, P. and Curie, P., 1880, Development by pressure of polar electricity in hemihedral crystals with inclined faces, *Bulletin de la Société minéralogique*, 3, 90–93.

Damjanovic, D., 2008, Lead-based piezoelectric materials, in *Piezoelectric and Acoustic Materials for Transducer Applications*, A. Safari and E. K. Akdoğan, eds. Springer, pp. 59–79.

Damjanovic, D., 1998b, Materials for high temperature piezoelectric transducers, *Current Opinion in Solid State & Materials Science*, 3, 469–473.

Damjanovic, D., 1998, Ferroelectric, dielectric and piezoelectric properties of ferroelectric thin films and ceramics, *Reports on Progress in Physics*, 61, 1267–1324.

Dogin, A., Uchino, K., and Newnham, R. E., 1997, Composite piezoelectric transducer with truncated conical endcaps 'Cymbals', *IEEE Transactions on Ultrasonics, Ferroelectrics and Frequency Control*, 44, 597–605.

Dolgin, B. P., 1991, Radiation Effects on Non-Electronic Materials Handbook, JPL Document D5312.

Ducommun, 2012, Product specification sheet [downloaded Sept 20, 2012], http://www.ducommun.com/dti/products/motorsresolvers/downHoleApplication.aspx

Dynamic Structures and Materials LLC, 2012, see http://www.dynamic-structures.com/blog/2010/9/21/piezoelectric-actuation-mechanisms-flextensional-piezo-actua.html [downloaded October 10, 2012]

Eitel, R. E., Randall, C. A., Shrout, T. R., and Park, S. E., 2002, Preparation and characterization of high temperature perovskite ferroelectrics in the solid-solution $(1-x) BiScO_3-xPbTiO_3$, *Japanese Journal of Applied Physics*, 41, 2099.

Face International, 1999, THUNDER Technical Specifications, Face International Corp. 427 35th St. Norfolk, VA, 23508 [downloaded 2012] http://www.faceinternational.com

Ferroperm Piezoceramics A/S. Available: 2012 http://www.ferroperm-piezo.com

Fielding Jr, J. T., 1993, Field-induced piezoelectric materials for high frequency transducer applications, Ph.D. Thesis, Pennsylvania State University.

Fraser, D. B., 1966, Lithium niobate: A high-temperature piezoelectric transducer material, *Journal of Applied Physics*, 37, 3853.

Fuierer, P. A., and Newnham, R. E., 1991, $La_2Ti_2O_7$ ceramics, *Journal of the American Ceramic Society*, 74, 2876–2881.

Gershman, R., and Wallace, R. A, 1999, Technology needs of future planetary missions, *Acta Astronautica*, 45, Nos. 4–9, 329–335.

Gilbertson, R. G., and Busch, J. D., 1996, A survey of micro-actuator technologies for future spacecraft missions, *The Journal of The British Interplanetary Society*, 49, 129–138.

Göpel, W., Reinhardt, G., and Rösch, M., 2000, Trends in the development of solid state amperometric and potentiometric high temperature sensors. *Solid State Ionics*, 136, 519–531.

Hackenberger, W., Alberta, E., Rehrig, P., Zhang, S. J., Randall, C.A., Eitel, R., and Shrout, T. R., 2004, High temperature electrostrictive ceramics for a Venus ultrasonic rock sampling tool, *14th IEEE International Symposium on Applications of Ferroelectrics*, pp.130–133, Montreal, Canada.

Haertling, G., 1997, Rainbow acuators and sensors: A new smart technology, *Proceeding of the SPIE conference on Smart Materials, In Smart Material Technologies*, 3040, 81–92.

Haines, J., Cambon, O., Keen, D. A., Tucker M. G., and Dove, M. T., 2002, Structural disorder and loss of piezoelectric properties in quartz at high temperature, *Applied Physics Letters*, 81, 2968–2970.

Hastings, D., and Garrett, H. B., 1996, Spacecraft-environment interactions, *Atmospheric and Space Science Series*, Ed. A. J. Dessler, Cambridge University Press, Cambridge, England.

Haun, M. J., Furman, E., Jang, S. J., and Cross, L. E., 1989, Thermodynamic theory of the lead zirconate-titanate solid solution system, part I: Phenomenology, *Ferroelectrics*, 99(1), 13–25.

Haymes, R. C., 1971, *Introduction to Space Science*, John Wiley & Sons Inc, New York, NY.

Herbert, J. M., 1982, *Ferroelectric Transducers and Sensors*, Gordon and Breach, London.

Hooker, M. W, Hazelton, C. S., Kano, K. S., Adams, L. G., Tupper, M. L. and Breit, S., 2010, Novel High-Temperature Materials Enabling Operation of Equipment in Enhanced Geothermal Systems, presented at Energy, Cocoa Beach, FL, February 2010.

Hudson, J., Busbridge, S. C., and Piercy, A. R., 1998, Electromechanical coupling and elastic moduli of epoxy-bonded Terfenol-D composites, *Journal of Application Physics*, 83, 7255–7257.

IEEE Standard on Piezoelectricity: ANSI/IEEE Standard, NY, 1987.

Jaffe, B., Cook, W. R., and Jaffe, H., 1971, *Piezoelectric Ceramics*, Academic Press, New York

Jaffe, H., and Berlincourt, D., 1965, Piezoelectric transducer materials, presented at the *Proceedings of the IEEE.* 53(10), 1372–1386.

Jandura, L., 2004, Brake failure from residual magnetism in the mars exploration rover lander petal actuator, *Proceedings of the 37th Aerospace Mechanism Symposium*, Johnson Space Center, May 19–21, pp. 221–235.

Joule, J. P., 1842, On a new class of magnetic forces, *Annals of Electricity, Magnetism, and Chemistry*, 8, 219–224.

Kazys, R., Voleisis, A., and Voleišienė, B., 2008, High temperature ultrasonic transducers: Review, *Ultragarsas (Ultrasound)*, 63, 7–17.

Kazys, R., Voleisis, A., Sliteris, R., Mazeika, L., Nieuwenhover, R. V., Kupschus, P., and Abderrahim, H., 2005, High temperature ultrasonic transducers for imaging and measurements in a liquid Pb/Bi eutectic alloy, *IEEE Trans. Ultrason. Ferroelectric, Ferquency Control*, 52, 525–537.

Lloyd Spetz, A., Baranzahi, A., Tobias, P., and Lundström, I., 1997, High temperature sensors based on metal–insulator–silicon carbide devices. *Physica Status Solidi (a)*, 162(1), 493–511.

Ma, J., Karaman, I., Noebe, R. D., 2010, High temperature shape memory alloys, *International Materials Reviews* 55(5), 257–315.

Mason, W. P., 1958, *Physical Acoustics and the Properties of Solids*, D. Van Nostrand Co Inc., Princeton, NJ.

Mason, W. P., 1950, *Piezoelectric Crystals and their Application to Ultrasonics*, D. Van Nostrand, NY.

Maxon, 2012 Product specification sheet, http://test.maxonmotor.com/docsx/Download/Product/Pdf/EC_22_HD_09_10_en.pdf [downloaded Sept 20, 2012].

Meng, X.L., Zheng, Y.F., Wang, Z., and, Zhao, L.C., 2000, Effect of aging on the phase transformation and mechanical behavior of $Ti_{36}Ni_{49}Hf_{15}$ high temperature shape memory alloy, *Scripta Mater.* 42, 341–348.

Meng, Z. Y., Kumar, U., and Cross, L. E. 1985, Electrostriction in lead lanthanum zirconate-titanate ceramics, *J. Am. Ceram. Soc.*, 68(8), 459–462.

Messing, G., Trolier-McKinstry, S., Sabolsky E., Duran, C., Kwon, S., Brahmaroutu, B., Park, P. et al., 2004, Templated grain growth of textured piezoelectric ceramics, *Critical Reviews in Solid State and Materials Sciences*, 29, 45–96.

Moser, J., 2003, Electromagnetic Devices for operation in Ambient High Temperatures of (1000°F) or 500°C. Industrial Paper found at "http://www.firstmarkaerospace.com/pdf/VHTmotors.pdf [downloaded Oct 2, 2012]

Moulson, A. J., and Herbert, J. M., 2003, *Electroceramics: Materials, Properties, Applications*, John Wiley & Sons Ltd, New York.

Moure, A., Castro, A., and Pardo, L. 2009, Aurivillius type ceramics, a class of high temperature piezoelectric materials: Drawbacks, advantages and trends, *Progress Solid State Chemisty*, 37, 15–39.

Near, C., 1996, Piezoelectric actuator technology, *Proceeding of the SPIE Smart Structures and Materials Conference*, vol. 2717, pp. 246–258, San Diego, CA.

Newnham, R. E., Sundar, V., Yimnirun, R., Su, J., and Zhang, Q. M., 1997, Electrostriction: Nonlinear electromechanical coupling in solid dielectrics, *Journal of Physical Chemistry B*, 101, 10141–10150.

Newnham, R. E., 2005, *Properties of Materials: Anisotropy, Symmetry, Structure*, Oxford University Press, USA.

Nye, J. F., 1985, *Physical Properties of Crystals, Their Representation by Tensors and Matrices*, Clarendon Press, Oxford.

Omega Temperature products 2013 http://www.omega.com/temperature/tsc.html. [downloaded March 2013]

Panda, P., 2009, Review: Environmental friendly lead-free piezoelectric materials, *Journal of Materials Science*, 44, 5049–5062.

Planets 2012 http://en.wikipedia.org/wiki/Planets#Accepted_planets, or http://solarsystem.nasa.gov/planets/or http://www.nineplanets.org/ [downloaded Dec 6, 2012]

Randall, C. A., Hilton, A., Barber, D., and Shrout, T. R., 1993, Extrinsic contributions to the grain size dependence of relaxor ferroelectric Pb $(Mg_{1/3}Nb_{2/3})O_3$: $PbTiO_3$ ceramics, *Journal of Materials Research*, 8, 880–884.

Sashida, T., and Kenjo, T., 1993, *An Introduction to Ultrasonic Motors*, Oxford Univ. Press, Oxford.

Schenker, P. S., Baumgartner, E. T., Lee S., Aghazarian, H., Garrett, M.S., Lindemann, R. A., Brown, D. K., Bar-Cohen, Y., Lih, S., Joffe, B., Kim, S. S., Hoffman, B. D., and Huntsberger, T. L., 1997, Dexterous robotic sampling for Mars in-situ science, *Proceedings of SPIE*, 3208, 170–185, Intelligent Robots and Computer Vision XVI: Algorithms, Techniques, Active Vision, and Materials Handling; David P. Casasent; Ed.

Schuh, C., Steinkopff, T., Wolff, A., and Lubitz, K., 2000, Piezoceramic multilayer actuators for fuel injection systems in automotive area, *Proc. SPIE* 3992, Smart Structures and Materials 2000: Active Materials: Behavior and Mechanics, pp. 165–175.

Sebastian, T., Sterianou, I., Reaney, I. M., Leist, T., Jo, W., and Rodel, J., 2012, Piezoelectric activity of (1-x)[0.35BMT-0.3BF-0.35BS]-xPT ceramics as a function of temperature, *Journal of Electroceramics*, 28, 95–100.

Sebastian, T., Sterianou, I., Sinclair, D. C., Bell, A. J., Hall, D. A., and Reaney, I. M., 2010, High temperature piezoelectric ceramics in the BMT-BF-BS-PT system, *Journal of Electroceramics*, 25, 130–134.

Sherman, C. H., and Butler, J. L., 2007, *Transducers and Arrays for Underwater Sound*, vol. 124: Office of Naval Research, Springer.

Sherrit, S., 2005, Smart material/actuator needs in extreme environments in space, *Proceeding of the Active Materials and Behaviour Conference, SPIE Smart Structures and Materials Symposium*, Paper #5761–48, San Diego, CA, March 7–10.

Sherrit, S., Bao X., Bar-Cohen, Y., and Chang, Z., 2004a, Resonance analysis of high temperature piezoelectric materials for actuation and sensing, *SPIE Smart Structures and Materials Symposium*, Paper #5388–34, San Diego, CA, March 15–18.

Sherrit, S., Badescu, M., Bao X., Bar-Cohen, Y., and Zhang, Z., 2004b, Novel horns for power ultrasonics, *Proceedings of the IEEE International Ultrasonics Symposium*, Vol. 3, pp. 2263–2266, UFFC, Montreal, Canada.

Sherrit, S., Wiederick, H. D., Mukherjee, B. K., and Haertling, G.H., 1994, The dielectric, piezoelectric, and hydrostatic properties of PLZT based rainbow ceramics, *Proceedings of the 9th International Symposium on the Application of Ferroelectrics*, University Park, Pennsylvania, pp. 390–393.

Sherrit, S., Dolgin, B. P., Bar-Cohen, Y., Pal, D., Kroh, J., and Peterson, T., 1999, Modeling of horns for sonic/ultrasonic applications, *Proceedings of the IEEE Ultrasonics Symposium*, pp. 647–651, Lake Tahoe.

Sherrit, S., Bao, X., Badescu, M., Bar-Cohen, Y., and Allen, P., 2010, Monolithic rapid prototype flexured ultrasonic horns, *Proceedings of the IEEE International Ultrasonics Symposium*, pp. 886–889, San Diego, CA.

Sherrit, S., Bao, X., Jones, C. M., Aldrich, J. B., Blodget, C. J., Moore J. D., Carson, J. W., and Goullioud, R., 2011, Piezoelectric multilayer actuator life test, *IEEE Trans. Ultrasonics, Ferroelectrics and Frequency Control*. 58(4), 820–828.

Shrout, T. R., Eitel, R., and Randall, C., 2002, High performance, high temperature perovskite piezoelectric ceramics, *Piezoelectric Materials in Devices*, ed. N. Setter, Switzerland pp. 413–432.

Shrout, T. R., and Zhang S., 2007, Lead-free piezoelectric ceramics: Alternatives for PZT?, *Journal of Electroceramics*, 19, 111–124.

Sierrra Nevada Corp, 2012, High Output Parafin Actuators - http://www.spacedev.com/ss_space_technologies.php

Smith, R. T., and Welsh, F. S., 1971, Temperature dependence of the elastic, piezoelectric, and dielectric constants of lithium tantalate and lithium niobate, *Journal of Applied Physics*, 42, 2219–2230.

Smits, J. G., Choi, W. S., and Ballato, A., 1997, Resonance and antiresonance of symmetric asymmetric cantilevered piezoel.ectric flexors, *IEEE Transactions on Ultrasonics, Ferroelectrics, and Frequency Control*, 44(2), 250–258.

Smits, J. G., and Ballato, A., 1993, Dynamic behavior of piezoelectric bimorphs, *Proceedings of The IEEE International Ultrasonics Symposium*, pp. 463–465, Baltimore, MD.

Smits, J. G., and Dalke, S. I., 1989, The Constitutive equations of Piezoelectric Bimorphs, *Proceedings of the IEEE International Ultrasonics Symposium*, pp. 781–784, Montreal, Canada.

Smolenskii, G. A., Isupov, V. A., Agranovskaya, A. I., and Krainik, N. N., 1961, New ferroelectrics of complex composition, *Soviet Physics, Solid State*, 2, 2651–2654.

Stoeckel, D., 1990, Shape memory actuators for automotive applications, *Materials & Design*, 11(6), 302–307.

Stringer, C. J., 2006, Structure property performance relationship of new high temperature relaxors for capacitor applications, PhD Dissertation, Pennsylvania State University.

Sundar, V., WaGachigi, K., McCauley, D., Markowski, K. A., and Newnham, R. E., 1995, Electrostriction measurements in diffuse phase transition materials and perovskite glass ceramics, *Proc. ISAF*, pp. 353–356, State College, PA.

Sweeney, K. L. and Halliburton, L. E., 1983, Oxygen vacancies in lithium niobate, *Applied Physics Letters.*, 43, 336–338.

Takenaka, T., and Nagata, H., 2005, Current status and prospects of lead free piezoelectric ceramics, *Journal of European Ceramic Society*, 25, 2693–2700.

Thermoloid, 2012, Thermal actuators up to 150°C or higher, [downloaded Sept 24, 2012], http://www.thermomegatech.com/pages/Thermal-Actuators.html

Townsend, D. P., 1994, Surface fatigue life of high temperature gear materials, *30th Joint Propulsion Conference cosponsored by the AIAA*, ASME, SAE, and ASEE, Indianapolis, Indiana, June 27–29, Also found on NASA Technical Report Server - http://ntrs.nasa.gov/archive/nasa/casi.ntrs.nasa.gov/19940025722_1994025722.pdf

Trolier-McKinstry, S., 2008, Crystal chemistry of piezoelectric materials, Chapter 3, Safari, A. and Akdogan E.K., *Piezoelectric and Acoustic Materials for Transducer Applications*, Springer Science, New York, pp. 39–56.

Turner, R. C., Fuierer, P. A., Newnham, R. E., and Shrout, T. R., 1994, Materials for high temperature acoustic and vibration sensors: A review, *Applied Acoustics*, 41, 299–324.

Uchino, K., 1996, *Piezoelectric Actuators and Ultrasonic Motors* vol. 1, Boston, Kluwer Academic Publishers.

Uchino, K., 2000, *Ferroelectric Devices* vol. 16. Marcel Dekker, Inc, New York, NY.

Uchino, K., and Giniewicz, J. R. 2003, *Micromechatronics*, Marcel Dekker, New York.

Uchino, K., Nomura, S., Cross, L. E., Jang, S. J., and Newnham, R. E., 1980, Electrostrictive effect in lead magnesium niobate single crystals, *Journal of Applied Physics*, 51(2), 1142.

Uchino, K., Nomura, S., Cross, L. E., Newnham, R. E., and Jang S. J., 1981, Electrostrictive effect in perovskites and its transducer applications, *Journal of Materials Science*, 16, 69–578.

Uchino, K., Zheng, J. H., Chen, Y. H., Du, X. H, Ryu, J., Gao, Y., Ural, S., Priya, S., and Hirose, S., 2006, Loss mechanisms and high power piezoelectrics, *Journal of Materials Science*, 41, 217–228.

Ueha, S., and Tomikawa, Y., 1993, *Ultrasonic Motors*, Claredon Press, Oxford.

Venet, M., Vendramini, A., Zabotto, F., Guerrero, F., Garcia, D., and Eiras, J., 2005, Piezoelectric properties of undoped and titanium or barium-doped lead metaniobate ceramics, *Journal of the European Ceramics Society*, 25, 2443–2446.

Wang, M., and Wang, J. F., 2008, Äurivillius phase potassium bismuth titanate $K_{0.5}Bi_{4.5}Ti_4O_{15}$, *Journal of the American Ceramic Society*, 91, 918–923.

Watson, B., Friend, J., and Yeo, L., 2009, Review: Piezoelectric ultrasonic micro/milli-scale actuators, *Sensors and Actuators A*, 152, 219–233.

Webster, G. A., and Ainsworth, R. A., 1994, *High Temperature Component Life Assessment* 2010, Published by Blackman and Hall, London, UK, 1st edition.

Wittenstein, 2012, See http://www.wittenstein-us.com/motors/extreme-servo-motors/extreme-temperature-motors.html [downloaded June 6, 2012]

Wright, A. F., 1997, Elastic properties of zinc blende and Wurzite AlN, GaN and InN, *Journal of Applied Physics*, 82, 2833–2839.

Wright, J. A., Sebastian, J. T., Kern, C. P., and Kooy, R. J, 2010, "Design, Development and Application of New, High–Performance Gear Steels, Gear Technology, pp. 46–53.

Wu, M. H., and Schetky, L. McD., 2000, Industrial applications for shape memory alloys, *Proceedings of the International Conference on Shape Memory and Superelastic Technologies*, Pacific Grove, California, pp. 171–182.

Xu, Y. H., 1991, *Ferroelectric Materials and Their Applications*, North-Holland, NY.

Xu, T., Jiang, X. N., and Su, J., 2011, A piezoelectric multilayer-stacked hybrid actuation transduction system, *Applied Physics Letters*, 98(24), 062124.

Yan, H. X., Nign, H. P., Kan, Y. M., Wang, P. L., and Reece, M. J., 2009, Piezoelectric ceramics with super high Curie points, *Journal of the American Ceramic Society*, 92, 2270–2275.

Yu, F. P., Zhang, S. J., Zhao, X., uan, D. R., Wang, C. M. and Shrout, T. R, 2010, Characterization of neodymium calcium oxyborate piezoelectric crystal with monoclinic phase, *Crystal Growth and Design*, 10, 1871–1877.

Zarka, A., Capelle, B., Detaint, J., Palmier, D., Philippot, E., and Zvereva, O., 1996, Studies of $GaPO_4$ crystals and resonators, *IEEE Int. Freq. Contr. Symp.*, pp. 66–71, Honolulu, Hawaii.

Zhang, S. J., and Li, F., 2012, High performance ferroelectric relaxor-$PbTiO_3$ single crystals: Status and perspective, *Journal of Applied Physics*, 111, 031301.

Zhang, S. J., and Yu, F. P., 2011, Piezoelectric materials for high temperature sensors, *Journal of the American Ceramic Society*, 94, 3153–3170.

Zhang, S. J., Alberta, E. F., Eitel, R. E., Randall, C. A., and Shrout, T. R., 2005a, Elastic, piezoelectric and dielectric characterization of modified $BiScO_3$-$PbTiO_3$ ceramics, *IEEE Trans. Ultrason. Ferroel. Freq. Control*, 52, 2131–2139.

Zhang, S. J., Eitel, R. E., Randall, C. A., and Shrout, T. R., 2005b, Maganese modified $BiScO_3$-$PbTiO_3$ piezoelectric ceramic for high temperature shear mode sensor, *Applied Physics Letters*, 86, 262904.

Zhang, S. J., Frantz, E., Xia, R., Everson, W., Randi, J., Snyder D. W., and Shrout, T. R., 2008b, Gadolium calcium oxyborate piezoelectric single crystals for ultrahigh temperature (>1000°C) applications, *Journal of Applied Physics*, 104, 084103.

Zhang, S. J., Kim, N., Shrout, T. R., Kimura, M., and Ando, A., 2006, High temperature properties of manganese modified $CaBi_4Ti_4O_{15}$ ferroelectric ceramics, *Solid State Communications*, 140, 154–158.

Zhang, S. J., Lee, S. M., Kim, D. H., Lee, H. Y., and Shrout, T. R., 2008c, Elastic, piezoelectric and dielectric properties of 0.71PMN-0.29PT crystals obtained by solid state crystal growth, *Journal of the American Ceramic Society*, 91, 683–686.

Zhang, S. J., Xia, R., Lebrun, L., Anderson D., and Shrout, T. R., 2005d, Piezoelectric materials for high power, high temperature applications, *Materials Letters*, 59, 3471–3475.

Zhang, S. J., Xia, R., Randall, C. A., Shrout, T. R., Duan, R. R., and Speyer, R. F., 2005c, Dielectric and piezoelectric properties of niobium-modified $BiInO_3$-$PbTiO_3$ perovskite ceramics with high Curie temperature, *Journal of Materials Research*, 20, 2067–2071.

Zhang, S. J., Zheng, Y. Q., Kong, H., Xin, J., Frantz, E., and Shrout, T. R., 2009. Characterization of high temperature piezoelectric crystals with an ordered langasite structure, *Journal of Applied Physics*, 105, 114107.

Zhang, S. J., Fei, T. T., Chai, B. H. T., Frantz, E., Snyder, D. W., Jiang X. N., and Shrout, T. R., 2008a, Characterization of piezoelectric single crsytal $YCa_4O(BO_3)_3$ for high temperature applications, *Applied Physics Letters*, 92, 202905.

11

Thermoacoustic Piezoelectric Energy Harvesters

Mostafa Nouh, Osama Aldraihem, and Amr Baz

CONTENTS

11.1 Introduction

11.1.1 Overview

This chapter presents the new development in the area of thermoacoustic energy harvesting whereby the output electric transducer of a conventional thermoacoustic harvester is coupled with an elastic structure in the form of a simple spring-mass system to amplify the deflection and strain experienced by the electric transducer in order to amplify the harnessed output power.

Furthermore, the emphasis is placed here on this class of harvesters with piezoelectric transducers because of their ability to convert thermal energy, such as solar or waste heat energy, directly into electrical energy without the need for any moving components in an environmentally friendly manner.

Therefore, this chapter is organized in six sections. In Section 11.1, a brief history of thermoacoustics and thermoacoustic devices is presented. In Section 11.2, the phenomenon of thermoacoustic is introduced. Section 11.3 outlines the concept of conventional thermoacoustic-piezoelectric (TAP) energy harvesters. The fundamentals and the basic equations governing the operation of dynamically magnified thermoacoustic-piezoelectric (DMTAP) harvesters are outlined in Section 11.4 along with comparisons with the TAP systems. Experimental behavior of both the TAP and DMTAP systems are discussed in Section 11.5 in a comparative manner. Section 11.6 summarizes the basic concepts presented in the chapter as well the merits and limitations of both the TAP and DMTAP systems.

11.1.2 History of Thermoacoustics

The *thermoacoustics* phenomenon, as the name implies, is a phenomenon resulting from the interactions and mutual conversions between thermal and acoustic energies that take place inside specially configured acoustic resonators. The *thermoacoustics* phenomenon was first observed in the mid-nineteenth century; glassblowers noticed that a hot glass bulb when connected to a cooler tube, starts radiating acoustic waves. This observation became known as the Sondhauss oscillations after the German physicist who in 1850 attempted to quantitatively describe the underlying physics contributing to the sound waves radiation.

In 1859, another major development in the area of thermoacoustics was discovered by Rijke (1859), whereby heat was used to sustain acoustic oscillations in a cylindrical tube open at both ends. The oscillations occur spontaneously if the combustion progresses more rapidly or efficiently during the compression phase of the pressure oscillation than during the rarefaction phase.

The foundation for theoretical thermoacoustics was initiated, in 1868, by Kirchhoff in an attempt to study the acoustic attenuation in a duct due to oscillatory heat transfer between the isothermal tube wall and the gas inside the tube.

However, in his notable monograph of 1878, Lord Rayleigh (John William Strutt) introduced the qualitative Rayleigh's criterion for describing heat-driven oscillations, which applies equally to the Soundhauss oscillations and the Rijke tube. In his criterion, Lord Rayleigh indicated that: "If heat be given to the air at the moment of greatest condensation, that is, greatest density, or be taken from it at the moment of greatest rarefaction, the vibration is encouraged."

In his classical book on *The Theory of Sound*, Lord Rayleigh also discussed the potential of generating temperature differences using acoustic oscillations (1887). This discussion has laid the foundation for the invention of thermoacoustic refrigerators.

Since then, the topic of thermoacoustics has remained nearly untouched until 1949, when thermoacoustic oscillations and acoustic "singing" were reported while operating liquid helium dip-sticks to monitor the level inside cryogenic storage vessels. These oscillations are called Taconis oscillations after the scientist who encountered these oscillations while studying the properties of $^3He/^4He$ mixtures at the Kamerlingh Onnes Laboratory at Leiden Institute of Physics in the Netherlands.

In 1969, serious analytical work has been carried out by Rott (1969, 1980) resulting in a series of publications describing acoustic oscillations in a gas in a channel with an axial temperature gradient. His work signaled a revival of thermoacoustic research and established its theoretical foundations based on the linear thermoacoustic theory. Rott's work was inspired by the Taconis oscillations (1949), and has departed from the boundary-layer approximation considered by Kirchhoff (1868). Instead, he formulated the mathematical framework for small-amplitude damped and excited oscillations in wide and narrow tubes with an axial temperature gradient.

In all the above-mentioned developments, the thermoacoustic systems were of the standing-wave configurations. In 1979, a radical departure to the more efficient traveling wave design has been reported by Ceperley, which was a realization based on the Stirling cycle. The design was improved by Yazaki et al. (1998) and perfected by Backhaus and Swift (2000).

In the 1980s, extensive efforts were exerted at the Los Alamos National Laboratory by Swift (1988), and that resulted in the development of powerful theoretical predictive tools (*DeltaEC*), and in the design and building of a wide variety of practical thermoacoustic devices. The most prominent of these efforts resulted in the invention of a standing-wave thermoacoustic refrigerator by Hofler et al. (1988).

Apart from all the above-mentioned milestones in the discovery and development of thermoacoustics, the field has a very rich history with "many roots, branches, and trunks intricately interwoven, supporting and cross-fertilizing each other" as indicated by Swift (2002).

Figure 11.1 displays only the main milestones in the discovery and development of the field of thermoacoustics.

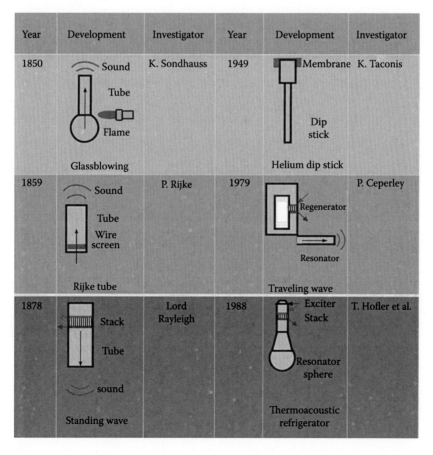

FIGURE 11.1
Milestones of discovery and development of the field of thermoacoustics.

11.2 Thermoacoustic Phenomenon

The *thermoacoustics* phenomenon manifests itself in one of the following two effects that are namely the forward and reverse effects as displayed in Figure 11.2. In the forward effect, steady thermal energy is converted into self-sustained acoustic oscillations and then to electricity resulting in a *thermoacoustic engine* as indicated in Figure 11.2a. The conversion from acoustic to electrical energy is achieved using electromagnetic or piezoelectric transducers. In the reverse effect, electric energy is used to drive an acoustic speaker in order to generate persistent acoustic oscillations which produce steady temperature gradient yielding a *thermoacoustic refrigerator* as depicted in Figure 11.2b.

In this chapter, the emphasis is placed on thermoacoustic engines in order to utilize this unique class of systems in harvesting waste heat energy. The basic configuration of thermoacoustic engines is the standing-wave configuration shown in Figure 11.3.

A typical thermoacoustic engine consists of a series of small parallel channels, referred to as the stack. The stack is placed between two heat exchangers that act as a heat source and sink as shown in Figure 11.3. These three components are located inside an acoustic tube called the "resonator." The engine can be thermally driven by any source of heat, such as waste heat or concentrated solar energy. The temperature gradient across the stack

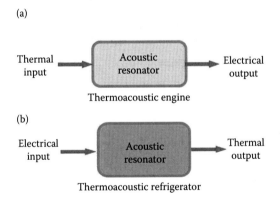

FIGURE 11.2
Forward and reverse thermoacoustic effects.

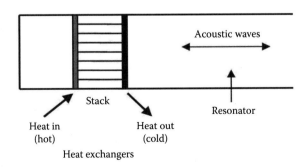

FIGURE 11.3
Schematic drawing of a standing-wave thermoacoustic engine.

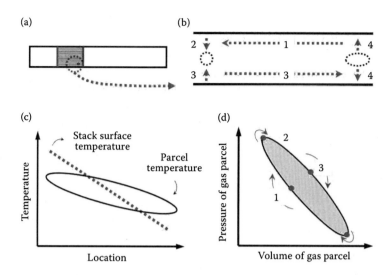

FIGURE 11.4
(**See color insert.**) Thermodynamic cycle of the standing-wave thermoacoustic engine.

generates self-sustained acoustic oscillations which can be converted into electrical energy using various types of transducers such as electromagnetic or piezo-membranes.

The operation of a thermoacoustic engine is typically based on exciting a working fluid in the presence of a temperature gradient that exceeds a certain critical value as shown in Figures 11.4a and b. An acoustic wave travels up the temperature gradient while working gas parcels are in intimate thermal contact with the adjacent solid boundaries of the stack. In a standing-wave engine, the pressure and velocity fluctuations in the stack are such that heat is given to the oscillating gas at high pressure and removed at low pressures so as to satisfy Rayleigh's criterion (1878) as displayed in Figure 11.4c. The working gas undergoes a thermodynamic cycle where energy is added to the acoustic wave as indicated in Figure 11.4d.

Figure 11.5 displays the spatial and temporal variation of velocity and pressure inside a standing-wave thermoacoustic engine. Note that the pressure and velocity are 90° out of phase. Note also that the gas starts moving toward the hot end of the stack when its velocity is positive and toward the cold end as its velocity becomes negative. It is important also to emphasize that as the compression and heating occur simultaneously so is the expansion and cooling. This can also be ascertained from the thermodynamic cycle shown in Figure 11.4d. In order to achieve such a cycle, the stack must have poor heat transfer characteristics to delay the heating so that it follows the compression process and the cooling follows the expansion. Such poor heat transfer and time delays result in irreversibility and inefficient energy conversion (Ceperley 1979).

11.3 TAP System

This section presents the concept and principle of operation of the conventional TAP harvester. A schematic drawing of this class of harvesters is shown in Figure 11.6. The harvester consists of four sections: heat cavity, stack, resonator tube, and piezoelectric diaphragm. The

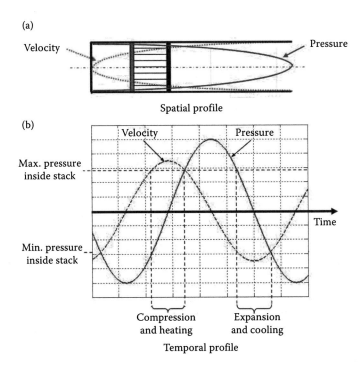

FIGURE 11.5
(See color insert.) (a) Spatial and temporal profiles of velocity and (b) pressure inside the standing-wave thermoacoustic engine.

heat source generates a temperature gradient along the stack that, in turn, produces standing acoustic waves in the resonator tube. The oscillation energy of the acoustic waves is harnessed by the piezoelectric diaphragm, which converts the incident pressure pulsations directly into electrical energy, to power the load Z_L, without the need for any moving parts.

It is important here to mention that the technology of using piezoelectric alternators dates back to 1974, when Martini et al. utilized a piezoelectric regenerator to convert the acoustic oscillations of a Stirling engine into electric energy. Examples of more recent attempts include the work of Keolian and Bastyr (2006), Symko et al. (2004), Symko and Abdel-Rahman (2007), and Matveev et al. (2007). In the work of Keolian and Bastyr (2006),

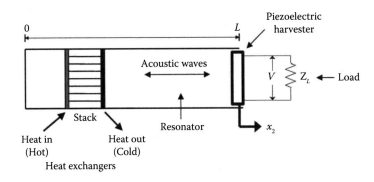

FIGURE 11.6
Schematic drawing of a standing-wave thermoacoustic energy harvester.

the emphasis was placed on the development of large-scale thermoacoustic engines and the proposed system included heavy moving masses communicating with arrays of piezo-electric alternators. This is contrast of the work of Symko et al. (2004), Symko and Abdel-Rahman (2007), and Matveev et al. (2007), where focus was on the development of small engines for thermal management in microelectronics. Note that the Symko et al. (2004), and Symko and Abdel-Rahman (2007) was primarily experimental in nature whereas the work of Matveev et al. (2007) was limited to theoretical analysis.

In all the above-mentioned work on conventional thermoacoustic engines with piezoelectric transducers, it is observed that the maximum conversion efficiency from acoustical to electrical energy is relatively low. Peak efficiencies of about 15% are reported by Matveev et al. (2007) and Smoker et al. (2012). Furthermore, the critical temperature differences necessary for producing self-sustained acoustic oscillations is fairly high reaching about 500–600°C (Matveev et al. 2007 and Smoker et al. 2012). Such low conversion efficiency and high critical temperature differences, limit considerably the practicality of the conventional TAP harvesters. It is therefore essential to seek other viable approaches to the design of these harvesters.

11.4 DMTAP versus Conventional TAP System

In this section, the serious limitations of conventional thermoacoustic engines with piezoelectric have been addressed by considering a radically different approach whereby the conventional TAP harvester is coupled with an elastic structure in the form of a simple spring-mass system to amplify the strain experienced by the piezo-element. The proposed system is referred to as a dynamic magnifier and has been shown in different areas to amplify significantly the deflection of vibrating structures (Cornwell et al. 2005, Ma et al. 2010, Aldraihem and Baz 2011).

DMTAP systems can be advantageous when the appropriate properties of the magnifier are chosen. The DMTAP can be designed to achieve a higher efficiency than a conventional TAP of the same size, and/or a lower temperature gradient across the stack ends.

The characteristics and the behavior of the DMTAP systems are presented in the following sections in comparison with those of the conventional TAP of the same size.

Figure 11.7 shows a schematic drawing of the DMTAP system, which consists of a conventional TAP harvester coupled with an elastic structure in the form of a simple spring-mass system to amplify the strain experienced by the piezo-element.

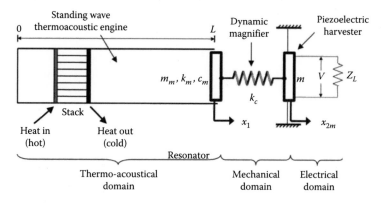

FIGURE 11.7
Schematic of standing-wave thermoacoustic engine integrated with a dynamic magnifier system (DMTAP).

11.4.1 TAP versus DMTAP

The equations governing the dynamics of the TAP and DMTAP configurations are summarized in Table 11.1. Detailed derivations of these equations are given by Nouh et al. (2012). The table includes: the equations of motions of the mechanical components of the two systems, the equation defining the harvested voltage, the mechanical impedance at the resonator end, and the equations necessary to compute the self-sustained frequencies of oscillations inside the two harvesters. Note that for effective and efficient operation of the harvesters, the mechanical impedance is matched with acoustic impedance at end of the resonator to ensure maximum energy conversion of the acoustical energy into mechanical energy. The table includes also the equations for computing the generated acoustical power, the output electrical power, and the conversion efficiency from acoustical power into electrical power.

In Table 11.1, m, b, and A_p denote the effective vibrating mass, damping coefficient, and the cross-sectional area of the piezo-element with k_s representing the ratio of the tube to the piezo-element cross-sectional areas. In addition, k_m and c_m are the internal stiffness and internal damping of magnifier mass m_m respectively with k_c denoting the stiffness of the spring coupling the mass m_m with the piezo-element. Note that while the x_2 (displacement of the piezo-element) is the only mechanical degree of freedom (DOF) of the TAP system, x_1 (the displacement of m_m and x_{2m} (displacement of the piezo-element) are the DOFs of the DMTAP system respectively. Also, Z_R is the impedance at resonator end, which is equal to that of the piezo-element for the TAP configuration and that of the dynamic magnifier for the DMTAP configuration. Furthermore, $P(L)$ is the pressure at the resonator end where L is the resonator length. In the electrical parts of the harvesters, s is the piezo-element internal stiffness, ω is the angular frequency, d_{33} is the piezoelectric strain coefficient, C_p is the piezoelectric clamped capacitance, V is the voltage across the load impedance Z_L, and ψ is the reciprocal of the piezoelectric coupling factor which is equal to $(d_{33}s)$.

11.4.2 Wave Forms

In this section, analysis of the wave forms that are encountered in both of the TAP and DMTAP configurations are presented in a comparative manner to illustrate the inherent and salient features of the two systems. In order to achieve such a goal, numerical examples are presented using the design parameters of the TAP system obtained from Nouh et al. (2012) and the parameters of the DMTAP system are as listed in Table 11.2. The magnifier is assumed to have only a mass and stiffness m_m and k_c. Also, the working gas is assumed to be air with a mean pressure P_m of 10^5 Pa and an average temperature T_m of 400 K. An electric load Z_L of a 1000 Ω is attached to the piezo-element. The resonator length is considered as a variable parameter, which is varied in the range from 0.5 to 4 cm.

The system frequencies calculated from Table 11.1 are plotted against the resonator length as shown in Figure 11.8. The vertical axis represents a normalized frequency equal to $\omega L/c$, with c denoting the sound speed inside the resonator. Also, the solid line represents the natural frequency of the piezo-element alone while the dashed and the dash-dotted lines represent the closed-open and closed-closed tube frequencies for comparative purposes. Closed-open tubes are ideally quarter wavelength resonators (i.e., $\lambda = 4L$) with a resonant frequency given by $\omega = 2\pi c/\lambda$. Combining these two facts, the normalized frequency of closed-open tubes would be a constant ($\omega L/c = \pi/2$). In the case of closed-closed tubes, the tubes are half wavelength resonators yielding a normalized frequency of $\omega L/c = \pi$.

TABLE 11.1

Main Equations of the TAP and DMTAP Harvesters

Type of Equation	TAP	DMTAP
Equations of motion	Piezo-element $m\ddot{x}_2 + b\dot{x}_2 + sx_2 - d_{33}sV - \dfrac{SP(L)}{k_s} = 0$	Magnifier mass $m_m\ddot{x}_1 + c_m\dot{x}_1 + k_mx_1 + k_c(x_1 - x_{2m}) - \dfrac{SP(L)}{k_s} = 0$ Piezo-element $m\ddot{x}_{2m} + b\dot{x}_{2m} + sx_{2m} + k_c(x_{2m} - x_1) - d_{33}sV = 0$
Output voltage equation	$sd_{33}\dot{x}_2 + C_p\dot{V} + \dfrac{V}{Z_L} = 0$	$sd_{33}\dot{x}_{2m} + C_p\dot{V} + \dfrac{V}{Z_L} = 0$
Equation of mechanical impedance at resonator's end	$Z_R = \dfrac{k_s}{S}\left[iom + b + \dfrac{s}{io} + \dfrac{\psi^2 Z_L}{1 + ioZ_LC_p}\right]$	$Z_R = \dfrac{k_s}{S}\dfrac{\left[iom_m + c_m + \dfrac{k_c}{io} + \dfrac{k_m}{io}\right]\left[iom + b + \dfrac{s}{io} + \dfrac{k_c}{io} + \dfrac{\psi^2 Z_L}{1 + ioZ_LC_p}\right] + \dfrac{k_c^2}{\omega^2}}{\left[iom + b + \dfrac{s}{io} + \dfrac{k_c}{io} + \dfrac{\psi^2 Z_L}{1 + ioZ_LC_p}\right]}$
Equation for computing frequency of self-sustained oscillations	$f_{TAP} = i(\rho c)\cot(kL) - Z_R = 0$	$f_{DMTAP} = i(\rho c)\cot(kL) - Z_R = 0$
Acoustical power	$\dot{E}_T = \dfrac{1}{2}S\,\mathrm{Re}\{P(L)\mathrm{conj}[U(L)]\}$	
Output electrical power	$\dot{E}_L = \dfrac{1}{2}\mathrm{Re}\left\{\dfrac{V\,\mathrm{conj}(V)}{Z_L}\right\}$	
Efficiency	$\eta_e = \dfrac{\dot{E}_L}{\dot{E}_T}$	

TABLE 11.2

TAP and DMTAP System Parameters

Parameter	Value
S	$7.85e-5\ M^2$
k_s	4
m, m_m	$3.46e-7\ \text{kg}$
b	$3.88e-5\ \text{kg/s}$
s	$574\ \text{N/m}$
ψ^2	$9.44e-9\ \text{kg/(s.\ \Omega)}$
C_p	$2.76e-8\ \text{F}$
k_c	$229.6\ \text{N/m}$

The results displayed in Figure 11.8 suggest that adding the dynamic magnifier to the thermoacoustic tube, results in reducing the frequency of the self-sustained oscillations. More interestingly, it can be noticed that the behavior of the resonator approaches that of a half wavelength resonator at increasing lengths for the TAP case. For the DMTAP case, the behavior of the resonator tends to fall somewhere in between the quarter and the half wavelength tubes at increasing lengths.

Figures 11.8a and b provide displays of the acoustic waveforms along the tube for a resonator length of 1.5 and 4 cm respectively. The plots show the variation of the real component of pressure and the imaginary component of velocity along the length of the resonator. These are the dominating components of both pressure and velocity expressions in their complex form.

It is evident here that, as suggested by Figure 11.8, the behavior of the TAP with a length of 1.5 cm resembles that of a closed-open tube. Under this condition and at the open end

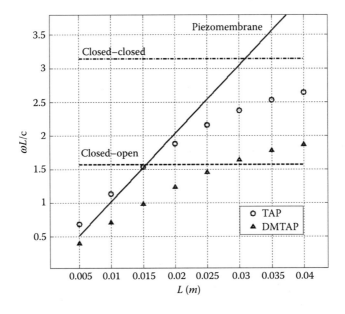

FIGURE 11.8
Variation of dimensionless frequencies with resonator length for TAP, DMTAP, piezo-element, closed-closed tube, and closed-open tube.

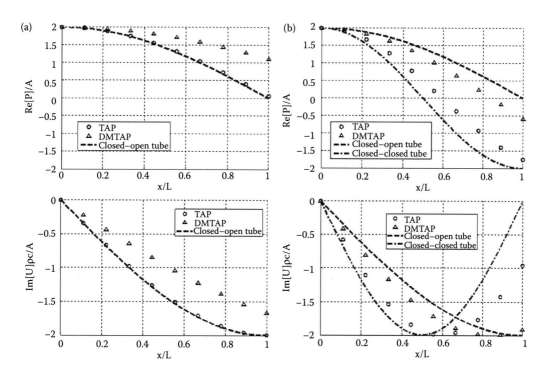

FIGURE 11.9
Pressure and velocity waveforms for TAP and DMTAP systems in comparison with closed-closed and closed-open tubes for resonator lengths of (a) 1.5 cm and (b) 4 cm.

of the tube, the pressure amplitude eventually decays to zero to match the outside mean pressure and the velocity reaches its antinode (peak) point as confirmed by the results shown in Figure 11.9a.

At that same length of the resonator, the DMTAP is relatively close to a closed-open tube behavior, but is expected to fully imitate it at around 3 cm of tube length. On the other hand, both the TAP and the DMTAP resonators of 4 cm long are expected to fall somewhere in between the closed-closed and the closed-open tube behaviors as implied by Figure 11.8. It is also confirmed here that the TAP is closer to a closed end behavior where the working gas parcels are ideally at rest and a pressure node is detected halfway through the tube. The DMTAP in this case is relatively closer to the open end tube behavior as illustrated by Figure 11.9b.

11.4.3 Magnification Ratio

To examine the effect of using the dynamic magnification concept, it is useful to investigate the displacement x_2 of the piezo-element in the TAP case in relation to the displacement x_{2m} of the DMTAP case. If the ratio x_{2m}/x_2 exceeds unity, then this indicates that more strain is experienced by the piezo-element and, hence, more power output is expected. The mass m_m is taken to be the same as m as shown in Table 11.2, and while c_m and k_m are neglected for simplicity. The ratio x_{2m}/x_1 in the DMTAP can be an acceptable approximation of the magnification ratio under these assumptions. Appropriate selection of k_c that would ensure the ratio exceeds 1 makes the use of DMTAP advantageous.

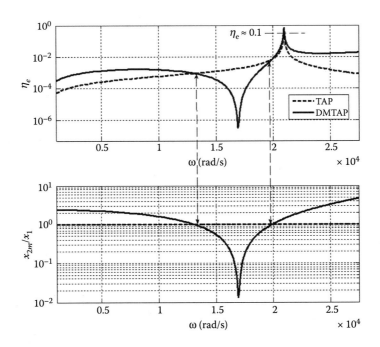

FIGURE 11.10

Frequency response of conversion efficiency η_e and corresponding magnification ratio x_{2m}/x_1 for a TAP and a DMTAP at $m_m = m$, $k_c = 0.11$s (double headed arrows indicate frequencies at which magnification ratio is equal to one).

Figure 11.10 shows a preferred scenario for using a DMTAP under the above simplified conditions. The plot shows the effect of frequency on the efficiency η_e of conversion from acoustic to electric energy at a load Z_L of 1000 Ω. In the plot, the tube lengths of the TAP and the DMTAP are set to achieve the same resonate frequencies.

It is interesting to note that the efficiency of the DMTAP starts exceeding that of the TAP when the ratio x_{2m}/x_1 starts exceeding 1. This is indicated by the black arrows. Even though there is a bandwidth of lower frequencies where the DMTAP shows better efficiency than the TAP, it is of greater interest to have η_e of the DMTAP higher than that of the TAP at the resonant frequency. This is more important to look for since only around resonance the efficiency becomes significantly high.

11.4.4 Power and Efficiency

Figure 11.11 provides a comparison between the TAP and the DMTAP configurations in terms of the harvested piezoelectric power and conversion efficiency. Output power used here is normalized using system parameters, namely gas density ρ, speed of sound c, resonator cross section S and the wave amplitude squared A^2. This quantity is convenient for comparing useful amounts of electricity generated by the piezo-element for given values of sound pressure amplitude. The comparison is made for electric load of 1000 Ω, and is based on the parameters provided in Table 11.3.

From the obtained results, it is noticed first that the strain amplification that takes place in the piezo-element due to the addition of the magnifier significantly increases the amount of useful electric energy harvested and enhances the efficiency as well. At a load

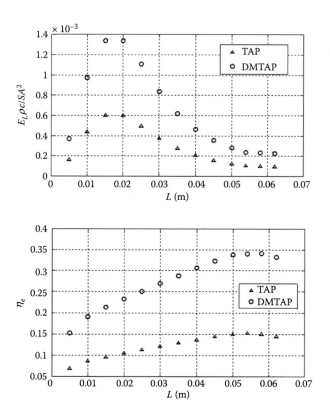

FIGURE 11.11
Dimensionless harvested electric power output and corresponding acoustic to electric energy conversion efficiency for an electric load of 1000 Ω.

resistance of 1000 Ω, the efficiency η_e peaks at around 34% for the DMTAP system with a resonator length of 5.5 cm. Therefore, it can be seen that with appropriate selection of the added mass and spring stiffness, a DMTAP can contribute to raising the overall efficiency η_o of a thermoacoustic standing-wave engine by improving the efficiency of its energy harvester component η_e.

TABLE 11.3

Comparisons between Characteristics of the TAP and DMTAP Harvesters

Feature	TAP	DMTAP
Dimensionless frequencies of resonator ($\omega L/c$)	Closer to closed-closed resonator $= \pi$	Closer to a closed-open resonator $= \pi/2$
Acoustic waveforms	Approaches behavior of a closed-closed resonator	Approaches behavior of a closed-open resonator
Displacement of the piezo-element	Small	Larger
Acoustic to electric energy conversion efficiency	Small	Larger
Harvested electric power output	Small	Larger
Temperature difference required to onset acoustic oscillations	High	Lower
Optimum location of stack	Away from piezo-element	Closer to piezo-element

11.4.5 Onset Temperature of Self-Sustained Oscillations

According to the analysis presented by Nouh et al. (2012), the onset temperatures necessary for achieving self-sustained oscillations are displayed in Figure 11.12 for both of the TAP and DMTAP configurations.

Figure 11.12 indicates that for the shorter resonator length of 1.5 cm, the DMTAP requires a lower temperature difference for almost any position of the stack along the tube length when compared with the TAP. Such temperature difference may reach values as low as 200°K at $x_s/L = 0.425$. This feature is indicative of an important performance enhancement resulting from the addition of the dynamic magnifier. With this small temperature difference across the stack, lower thermal input is needed to initiate the self-sustained oscillations.

For longer resonators, the comparison is more critical and is sensitive to the placement of the stack. Note that these resonators have pressure waves with a node close to the middle of the tube. Consequently, there exists a point where the temperature difference required becomes negative. In physical terms, that requires a heat input to the right end of the stack instead of its left end. In turn, this means switching the locations of the hot and cold heat exchangers. In that domain, the TAP seems to require a less temperature difference than the DMTAP as shown in Figure 11.12. For example, for a 4-cm-long resonator the temperature difference becomes almost 200°K for the TAP and 500°K for the DMTAP when the stack is placed at $x_s/L = 0.95$.

Accordingly, it should be emphasized here that the optimal stack placement in standing-wave engines should be in the left quarter of the resonator, to compromise between better acoustic power output and better efficiency. Stacks should be typically located whereas the magnitude of gas velocity is relatively small to reduce any viscous dissipation losses that might affect the conversion efficiency, yet simultaneously at a location where the pressure-velocity product is reasonably high to generate more acoustic power. Taking the above factors into consideration, it stands that DMTAP systems would be potentially more useful to use given the optimal stack location.

Table 11.3 presents a comparison between the main characteristics of the TAP and DMTAP harvesters in view of the discussions presented in Section 11.4. This comparison highlights the favorable attributes of the DMTAP as compared to the conventional TAP harvester.

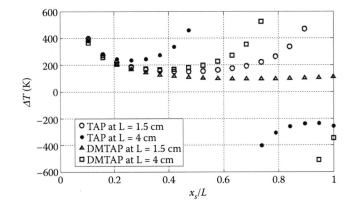

FIGURE 11.12
Temperature difference required to onset acoustic oscillations for TAP and DMTAP of tube lengths 1.5 and 4.5 cm.

11.5 Experimental Validation

11.5.1 Experimental Setup

Experiments are carried out to demonstrate the feasibility of the concept of the DMTAP discussed in Section 11.4. The experimental setup used is shown in Figure 11.13. The acoustic oscillations produced by the stack are simulated by a speaker placed at the beginning of the resonator. The resonator is 2.75″ (6.985 cm) in diameter. A circular buzzer piezo-element placed on a 0.008″ (0.203 mm) thick aluminum sheet of a diameter equal to that of the tube is attached to the other end of the resonator. The piezo-element is manufactured by Digi-Key Corp., part no. AB4113B. This piezo-element is then connected to a similar one through a mechanical spring of known stiffness. Details of the experimental setup and results are reported by Nouh (2013).

The second piezo-element is supported by a separate stand than the rest of the resonator. Furthermore, the section between the two piezo-elements is open to the air. This way, by detaching the coupling spring and the second piezo-element, the system at hand is simply a speaker-driven cavity with one piezo-element at one end. The piezo-element converts the incoming acoustic energy from the speaker into an electrical output, thus simulating a TAP-like system. When the second piezo-element is reattached to the first one using the coupling spring, the system in effect acts like a DMTAP. The first piezo-element in this case acts as the dynamic magnifier mass.

11.5.2 Voltage Output from Piezo-Elements

For different values of the stiffness k_c of the spring connecting the two piezo-elements, the system is found to have a first natural frequency in the range starting 580–750 Hz when

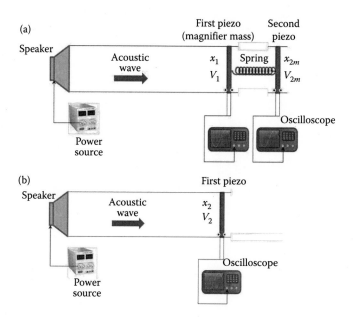

FIGURE 11.13
Schematic diagram of the experimental setup when used as (a) DMTAP-like system and (b) TAP-like system.

operating as a TAP- or a DMTAP-like system. Thus, using the speaker as the source of the acoustic energy and with no electrical loads used, a sine sweep over the domain 0–800 Hz is carried out while monitoring the voltage output from both piezo-elements (V_1 and V_{2m}) over this range of frequencies. Having values of V_{2m} less than V_1 would mean that the TAP still operates as a better energy harvester than the DMTAP. This would eliminate the need for having to compare V_{2m} with the voltage V_2 from the first piezo when no springs are attached (i.e., the TAP case). However, having values of V_{2m} higher than V_1 is an indication that the power output in the second piezo-element is more than that obtained from the first one. Even though this can be taken as a valid approximation for the magnification in most cases, the voltage output from the second piezo V_{2m} should still be compared with the voltage V_2 to confirm that the proposed system does serve as a magnifier of the power harvested.

The variable parameters in such an experiment are mainly the spring constant k_c and the masses of the two piezo-elements. Both piezo-elements are similar and have the same weight. Both elements are supported by an aluminum backing of the same thickness, thus their total masses are equal. Small masses in the order of 1–5 g can still be attached to the piezo-elements as a way of varying the mass of the DMTAP.

Several combinations of k_c and added weights are attempted while monitoring the voltage output over the sine-swept frequency range on a signal analyzer. Figure 11.14 shows an example of the case where the energy traveling from the first to the second piezo through the coupling spring is not magnified; hence, V_1 values are higher than V_{2m} over the considered frequency range. The spring used here has a constant of k_c equals to 17,800 N/m. In this case, the added spring-mass structure serves as means of dissipating or absorbing the energy being conveyed to the second piezo-element instead of amplifying it. The figure displays also the nature of the dominant modes in the frequency spectrum. Notably are the two modes, at 490 and 580 Hz, resulting from combining the harvester with the dynamic magnifier. However, because of the weak nature of the coupling between the harvester and magnifier, the contribution of the second mode to the output voltage is also weak.

Upon the addition of small masses to the piezo-elements, the voltage obtained from the second piezo begins improving in comparison with the case with no masses added.

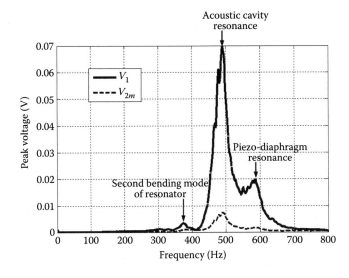

FIGURE 11.14
Sine-swept frequency response of voltage outputs V_1 and V_{2m} of the two piezo-elements ($k_c = 17,800$ N/m).

However, it is the combination of the masses and the proper spring constant that decides the performance of the DMTAP. The best results are obtained using a spring of k_c equal to around 29,180 N/m with no added masses. In this case, a strong coupling exists between the piezo-element and the dynamic magnifier resulting in enhanced performance.

Figure 11.15 shows the response of the voltages: V_1, V_2 and V_{2m} for this case. Note that V_1 in this case rises from its peak of 70 mV, in Figure 11.14, to about 160 mV, while V_{2m}

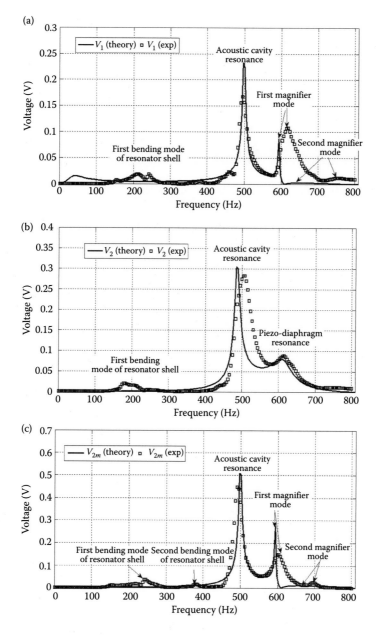

FIGURE 11.15
Sine-swept frequency response of voltage outputs V_1, V_2, and V_{2m} of the two piezo-elements ($k_c = 29{,}180$ N/m).

dramatically jumps from a peak of 8 mV to almost 450 mV. It is evident here that the energy transferred through the spring to the second piezo-element is magnified. This is manifested clearly by comparing the performance with that of the case without a magnifier where the output voltage V_2 peak at about 290 mV and, hence, are much lower than the 450 mV achieved by the second piezo in the DMTAP case.

The equations listed in Table 11.1 are used to model the system at hand and compare with the experimental results. A MATLAB code is developed to simulate the experiment and predict values of V_1, V_2, and V_{2m} over the frequency range 0–800 Hz for the cases presented in Figure 11.15. The obtained theoretical characteristics and the corresponding experimental frequency responses are displayed in Figure 11.16. The figure shows, to a great extent, a good agreement between theoretical predictions and experimental results.

Displayed on the figures are the clearly identified structural and acoustic modes of the resonator as well as the structural modes of the piezo-diaphragm and magnifier

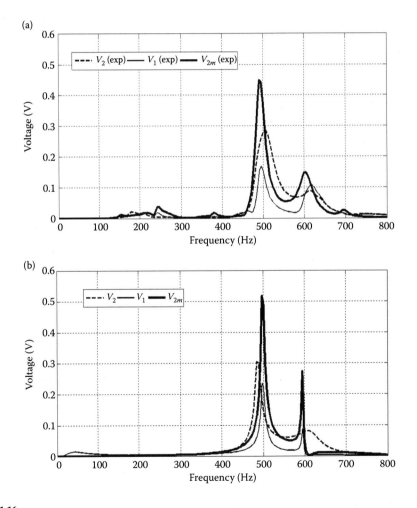

FIGURE 11.16
(a) Sine-swept frequency response of voltage outputs V_1, V_2, and V_{2m} of the two piezo-elements compared with (b) the theoretical predictions ($k_c = 29{,}180$ N/m).

system. Note that in the case of a harvester without a dynamic magnifier, the output voltage V_2 shows only on distinct peak at 500 Hz to indicate the acoustic resonance of the resonator cavity and another peak at 615 Hz to quantify the resonant frequency of the piezoelectric diaphragm. However, in the case of a harvester with a dynamic magnifier, the frequency spectrum of the output voltage V_{2m} shows two bending modes of the resonator shell at 230 and 380 Hz, acoustic resonance of the resonator cavity at 500 Hz, combined resonant frequencies of the piezoelectric diaphragm and the magnifier at 610 and 700 Hz.

11.5.3 Vibrometer Scanning of Piezo Surface

To verify that the voltage measurements presented earlier reflect and give an indication of the corresponding piezo deflection and for the purpose of voltage-displacement calibration as well, the surface of the piezo-element in the TAP and the DMTAP case is scanned using a laser vibrometer during operation. The contours are obtained using the PSV200 scanning Laser Doppler Vibrometer from Polytec-PI, Hopkinson, MA (used in previous experiments). The setup for this experiment is shown in Figure 11.17 and the scanning patterns of the vibrometer are shown in Figure 11.18.

Figure 11.18 shows results from a vibrometer scanning of the first piezo-element when the spring and the second piezo are detached (the TAP case) and a scanning of the second piezo-element after placing it back and attaching it to the first piezo using the spring (the DMTAP case). The plots shown are at 510 and 775 Hz.

The contours of transverse velocity show clearly the difference in the deflection pattern between first mode and the second mode. It is also evident that the measurements of the voltage output are confirmed as the DMTAP case shows to have a significantly higher transverse velocity than the TAP in both cases. It is also shown that operating at the first mode is not only favorable to avoid energy cancellation, but also because the piezo-elements experience much higher strain compared to the secondary modes.

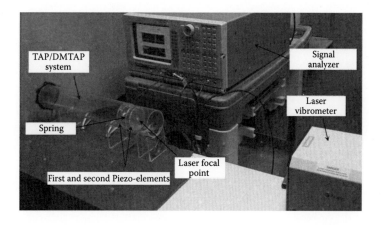

FIGURE 11.17
(See color insert.) Experimental setup: Laser vibrometer used to scan the surface of the piezo-elements to obtain values for the transverse deflection.

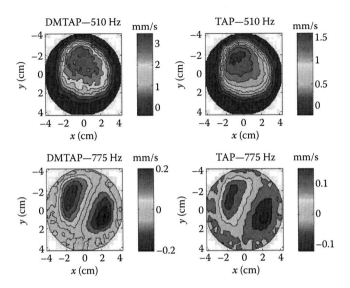

FIGURE 11.18
Contours of transverse velocity of DMTAP and TAP showing the first and second deflection modes.

11.6 Summary/Conclusions

This chapter has presented an attempt to enhance the performance of conventional standing-wave thermoacoustic engines integrated with piezo-elements by augmenting the system with a dynamic magnifier aimed at enhancing the piezo strain. The theory governing the operation of the DMTAP has been introduced in comparison with the TAP dynamics. Numerical examples were presented to illustrate the performance characteristics of both systems.

It was shown that with appropriate selection of the dynamic magnifier parameters, a consequent amplification of the harvested electric energy can be significantly observed. The efficiency of energy conversion using the DMTAP was almost doubled for short-length resonators. It is also shown that the DMTAP can simultaneously reduce the critical temperature difference, necessary for self-sustained oscillations, particularly for short resonators with optimal stack location.

The obtained results demonstrate the feasibility of the DMTAP as an effective means for improving the performance of standing-wave thermoacoustic harvesters. The performance of the DMTAP can be further enhanced by looking into optimization schemes to provide a methodology for calculating the optimal design parameters of the thermoacoustic harvesters with dynamic magnifiers.

Acknowledgments

The research reported in this chapter was conducted at the University of Maryland and supported by funding from King Saud University (Visiting Professors Program) and the and National Plan for Science and Technology (NPST).

The authors also thank Wael Akl, Faculty of Engineering, Ain Shams University, Cairo, Egypt; Hyeong Jae Lee, Jet Propulsion Laboratory (JPL), Pasadena, California; and Ji Su, NASA Langley Research Center (LARC), Virginia, for reviewing this chapter and for providing valuable technical comments and suggestions.

Nomenclature

A_p	cross-sectional area of piezo-element
B	piezo-element damping coefficient
C	speed of sound in gas medium
c_m	damping of magnifier mass
C_p	piezoelectric clamped capacitance
d_{33}	piezoelectric strain coefficient
$\dot{E}_L$	electric power output from piezo-element
$\dot{E}_T$	acoustic power reaching piezo-element
K	complex wave number
k_{33}	electromechanical coupling factor
k_c	stiffness of spring coupling magnifier mass with piezo-element
k_m	stiffness of magnifier mass
k_s	ratio of resonator to piezo-element cross-sectional areas
L	resonator length
M	piezo-element effective vibrating mass
m_m	magnifier mass
P	spatial component of working gas pressure
P_m	mean pressure of working gas
S	piezo-element stiffness
S	resonator cross-sectional area
T_m	mean temperature of working gas
U	spatial component of working gas x-velocity
V	voltage across electric load
X_1	displacement of magnifier mass
x_2	displacement of piezo-element in TAP
x_{2m}	displacement of piezo-element in DMTAP
X	direction of acoustic wave propagation
y_o	half plate spacing in stack
Z_L	impedance of electric load
Z_R	impedance of system attached to the right end of the resonator

Greek Symbols

η_e	acoustic to electric energy efficiency
η_o	overall efficiency of thermoacoustic device
Λ	wavelength

Ψ reciprocal piezoelectric coupling factor
Ω angular frequency

References

Aldraihem O., and A. Baz, Energy harvester with a dynamic magnifier, *Journal of Intelligent Material Systems and Structures*, 22(6), 521–530, 2011.

Backhaus S., and G. Swift, A thermoacoustic-stirling heat engine: Detailed study, *Journal of the Acoustical Society of America*, 107(6), 3148–3166, 2000.

Backhaus S., and G. Swift, A thermoacoustic-stirling heat engine, *Nature*, 399, 335–338, 1999.

Ceperley P. H, A pistonless stirling engine–the traveling wave heat engine, *Journal of the Acoustical Society of America*, 66, 1508–1513, 1979.

Cornwell P. J., J. Goethal, J. Kowko, and Damianakis M., Enhancing power harvesting using a tuned auxiliary structure, *Journal of Intelligent Material System Structures*, 16, 825–834, 2005.

Hofler T., Wheatley J. C., Swift G. W., and Migliori A., Acoustic cooling engine, US Patent No. 4,722,201, 1988.

Keolian R. M. and K. J. Bastyr, Thermoacoustic Piezoelectric Generator, US Patent, Patent No. 7,081,699, 2006.

Kirchhoff G., Ueber den Einfluss der W¨armteleitung in einem Gas auf die Schallbewegung (On the influence of heat conduction in a gas on sound movement), *Annalen der Physik (Annals of Physics)*, 134, 177, 1868.

Ma P. S., Kim J. E., and Kim Y. Y., Power amplifying strategy in vibration powered energy harvesters, Active and Passive Smart Structures and Integrated Systems 2010, *Proceedings of SPIE Conference*, Vol. 7643, Paper #7643–23, San Diego, CA, 2010.

Matveev K., Wekin A., Richards, C., and Shafrei-Tehrany N., On the coupling between standing-wave thermoacoustic engine and piezoelectric transducer, IMECE2007–41119, *Proceedings of IMECE2007*, Seattle, WA, 2007.

Nouh M., Thermoacoustic-piezoelectric systems with dynamic magnifiers, *Ph.D. Dissertation*, University of Maryland, College Park, MD, 2013.

Nouh, M., Aldraihem O., and Baz A., Energy harvesting of thermoacoustic-piezo systems with a dynamic magnifier, *Journal of Vibration and Acoustics*, 134, 061015, 2012.

Rijke P. L., On the vibration of the air in a tube open at both ends, *Philosophical Magazine*, 17, 419–422, 1859.

Rott N., Damped and thermally driven acoustic oscillations in wide and narrow tubes. *Journal of Applied Mathematics and Physics (Zeitschrift für Angewandte Mathematik und Physik)*, 20, 230–243, 1969.

Rott N., Thermoacoustics, *Advances in Applied Mechanics*, 20, 135–175, 1980.

Symko, O.G., Abdel-Rahman E., Kwon Y. S., Emmi M., and Behunin R., Design and development of high-frequency thermoacoustic engines for thermal management in microelectronics, *Microelectronics Journal*, 35, 185–191, 2004.

Symko O. G., and E. Abdel-Rahman, High Frequency Thermoacoustic Refrigerator, United States Patent # 7,240,495, 2007.

Strutt J. W. (Lord Rayleigh), The explanation of certain acoustical phenomena, *Nature*, 18, 319–321, 1878.

Sondhauss K., On acoustic oscillations of the air in heated glass tubes and in closed pipes of non-uniform width, *Pogendorff's Annals of Physics and Chemistry (Pogendorff's Annalen der Physik und Chemie)*, 79, 1–34, 1850.

Swift G., *Thermoacoustics: A Unifying Perspective for some Engines and Refrigerators*, Acoustical Society of America, American Institute of Physical Press, NY, 2002.

Swift G. W., Thermoacoustic engines, *Journal of the Acoustical Society of America*, 84, 1145–1180, 1988.

Smoker J., Nouh M., Aldraihem O., and Baz A., Energy harvesting from a standing wave thermoacoustic piezoelectric resonator, *Journal of Applied Physics*, 111, 104901–1:11, 2012.

Taconis K. W., Measurements concerning the vapour-liquid equilibrium of solutions of He3 in He4 below 2.19°K, *Physica*, 15, 733, 1949.

Yazaki T., Iwata A., Maekawa T., and Tominaga A., Traveling wave thermoacoustic engine in a looped tube, *Physical Review Letters*, 81, 3128–3131, 1998.

Internet Resource

UMD/Baz: http://www.baz.umd.edu/labs/thermoacoustic.html

12

Shape Memory and Superelastic Alloys

Mohammad Elahinia, Masood Taheri Andani, and Christoph Haberland

CONTENTS

12.1 Introduction

Shape memory alloys (SMAs) can recover large amounts of deformation (axial strain of 10%) under specific thermo-mechanical conditions. The key feature of these alloys is their ability to undergo large seemingly plastic strains and subsequently recover the deformation when the load is removed or the material is heated. This unique capability is the basis for various applications for these materials ranging from automotive actuators to medical devices.

In 1932, the first alloy to exhibit the shape memory effect (SME) (AuCd) was discovered by Arne Ölander (Ölander 1932a, 1932b). This alloy could only survive small stresses or strains and was never developed into an actuator. In 1961, while searching for high corrosion resistant materials a group of researchers stumbled upon another SMA. The alloy was composed of Nickel and Titanium (NiTi) and was discovered at the US Naval Ordinance Laboratory (NOL), thus, the original name for the material became NiTiNOL (Elahinia 2004). Consequently, several other metal combinations have been discovered to exhibit the same effect. Alloys such as AgCd, AuCd, CuSn, InTi, NiAl, NiTi, and MnCu all exhibit shape memory tendencies. Despite the subsequent discoveries, nickel–titanium (NiTi) has proven to be the most promising alloy for many applications (Duerig et al. 1999; Mabe et al. 2004; Es-Souni et al. 2005; Saadat et al. 2002). The superiority of Nitinol is due to its high ductility, large recoverable motion, excellent corrosion resistance, stable transformation temperatures, its biocompatibility and its ease of electrical heating. In addition to the distinctive properties of SME and superelasticity (SE) as will be discussed later, SMAs exhibit good resistance to wear and corrosion, have superior energy absorption capacity, and are often biocompatible. These properties make NiTi SMAs good candidates for a variety of applications from aerospace and automotive to biomedical applications. SMAs have been used for actuation, energy absorbing, and sensing. As actuators, SMAs offer several advantages for system miniaturization such as excellent power to mass ratio, maintainability, reliability, and clean and silent actuation. There are disadvantages such as low energy efficiency due to conversion of heat to mechanical energy and difficulties in control due to hysteresis, nonlinearities, parameter uncertainties, and in the challenge of measuring state variables such as temperature.

12.2 Phase Transformation in SMAs

The underlying reason for the unique properties of SMAs lays in their crystalline structure. Their crystal structure undergoes a solid–solid phase transformation when cooled from its stiff, high-temperature austenite (A) phase to its softer, low temperature martensite (M) structure. This inherent phase transformation can be stress or temperature induced.

12.2.1 Phase Transformation

The stress–temperature transformation plot is a schematic representation of the transformation regions for SMAs (Bekker and Brinson 1998). The lines in the plot show the phase boundaries that separate the two solid phases of an alloy. Usually a stress–temperature transformation plot shows the temperature along the abscissa and stress along the ordinate. The mechanism in which the crystalline structure of the austenite phase transforms to martensite phase is called lattice distortion and the transformation is called martensitic transformation. The martensite crystal can also be formed as a twinned, detwinned, or reoriented form. The reversible transformation of the material from austenite, which is known as the parent phase to martensite makes the special thermomechanical behavior of SMAs. A widely accepted stress–temperature transformation plot of SMA materials is shown in Figure 12.1. As described earlier, SMAs can exhibit a SME or pseudoelasticity. The type of the effect depends on the chemical composition of the alloy, on the processing history and of course on the ambient temperature. Both effects and the special thermo-mechanical properties of SMAs are described in the following sections.

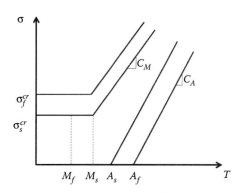

FIGURE 12.1
Stress-temperature-transformation plot of a shape memory material depicts the stable area for each crystalline structure. Crystal transformation takes place as the result of variation in stress and temperature ("s" refers to transformation start and "f" to transformation finish, respectively).

12.2.2 Shape Memory Effect

The SME is the ability of these alloys to recover a certain amount of unrecovered strain upon heating. This phenomenon takes place when the material is loaded such that the structure reaches the detwinned martensite phase and then unloaded while the temperature is below the austenite start temperature (A_s). Heating up the material at this stage into austenite will lead to strain recovery of the material and the material will regain its original shape. This phenomenon can be better understood in the combined stress–strain temperature diagram as shown in Figure 12.2.

Starting from point A the material is initially in the austenite phase. Cooling down the alloy to a temperature below its martensite finish temperature (M_f) will result in the twinned martensite crystal, point B. At this point loading the alloy at the same temperature will lead the crystal to transform to the detwinned martensite phase at point C. This stress–strain path is nonlinear because of the transformation phenomenon. Unloading the applied stress at same temperature will result in linear strain recovery to point D and remains in the detwinned martensite phase and a residual strain. By heating the

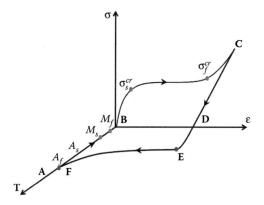

FIGURE 12.2
Shape memory effect path in stress-strain-temperature space.

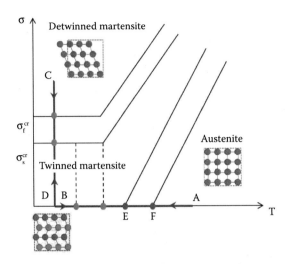

FIGURE 12.3
Shape memory effect in transformation diagram.

alloy above the austenite start temperature (A_s) at point E, the transformation from detwinned martensite to austenite phase starts. This transformation recovers the residual strain, which will be fully recovered at point F where the alloy passes the austenite finish temperature. The stress–temperature crystalline structure pattern during SME is depicted on the transformation diagram (Figure 12.3) for more clarity.

12.2.3 Pseudoelasticity

Pseudoelastic or superelastic behavior is the ability of SMAs to recover large amount of strains through mechanical loading/unloading. Figure 12.4 shows a typical superelastic stress–strain response of SMAs. The superelastic behavior starts from temperatures above austenite finish temperature (A_f) where the material is fully austenite (point A) and continues in loading by an applied force to make the detwinned martensite crystal form (point B). During this (forward) transformation from austenite to martensite the transformation strain is generated. Upon unloading, the generated strain is fully recovered in the (backward) transformation and the original form is achieved (point C). The

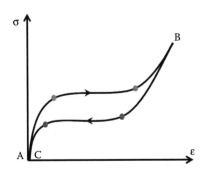

FIGURE 12.4
Superelastic stress–strain response of SMAs.

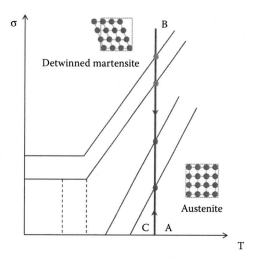

FIGURE 12.5
Superelasticity in transformation diagram.

stress–temperature crystalline structure cycle of such behavior is shown on the transformation plot in Figure 12.5.

12.3 High-Temperature Shape Memory Alloys

Current practical uses for SMAs are limited to temperatures below 100°C. This is the transformation temperature limit of the two most commercially successful SMA systems: the near equiatomic NiTi binary and Cu-based SMAs. Naturally, this limits the application of SMAs at higher temperatures, and necessitates design modifications for SMA containing components in order to reduce operating temperatures to below 100°C, or completely abandon their use. Several studies (Fristov et al. 2006) have been conducted on possibility of developing SMAs with elevated transformation temperatures. This class of materials is referred to as high-temperature shape memory alloys (HTSMAs).

Several alloy systems can be selected for HTSMAs with stable transformation temperatures above 100°C. Ni–Ti–Pd and Ni–Ti–Pt systems are the two well-known HTSMAs with reasonable balance of properties and good work output for use between 100 and 300°C. However, the incomplete shape recovery is the major drawback of these alloys. Adding extra elements such as boron, gold, cesium, and scandium can address this issue to some extent although they also deteriorate the desirable transformation temperatures. The Ni–Ti–Hf and Ni–Ti–Zr systems are possible alternatives to the expensive Ni–Ti–Pd and Ni–Ti–Pt systems, in which the less expensive ternary alloying elements (Hf and Zr) are substituted at the expense of Ti. In addition to these systems, there are few alloys such as Ti–Pd and Ti–Au with characteristic temperatures between 400°C and 700°C. These materials are of particular interest for engine and fuel systems. However, poor ductility and cyclic stability as well as undesirable phase decomposition and recrystallization are among the challenges that have limited the development of these high-temperature systems. No HTSMAs with transformation temperatures above 300°C have yet manifested an acceptable SME or SE.

Owing to the lack of machinablitity, quality standards for stability, ductility, functional behavior and reliability, no successful applications have been realized so far for HTSMAs. In addition, because of the high working temperature and the possibility of growing large local stresses in the material, creep can be a significant issue, limiting the processing and functionality possibilities of HTSMAs. Precipitation (during heat treatment and aging) in conventional SMAs usually causes stable phases, may be useful for improvement of SE and provides the best hardening performance. But aging may naturally occur in HTSMAs and affects the transformation temperatures. It is well known for NiTi based SMAs that aging can lead to precipitation of Ni-rich phases and thus the matrix composition shifts to higher Ti contents. This will result in an increase of the transformation temperatures (Eggeler et al. 2005; Khalil-Allafi et al. 2002). Oxidation is another important issue in HTSMA systems. Generally, when exposed to high temperatures the pick-up of oxygen leads to the formation of Ti-rich phases (e.g., $Ti_4Ni_2O_x$), which are stabilized by oxygen. This results in a shift of the stoichiometry to higher Ni contents which cause a decrease of the transformation temperatures (Nevitt 1960; Olier et al. 1997).

More research is needed to identify the operational limits of HTSMAs including the effects of temperature and stress on strain recovery, work output, and fatigue life. Furthermore, models that accurately describe the engineering aspects of HTSMAs should be developed. The exposure to high temperature creates a series of new variables, such as creep and visco-plasticity, and the effects of extensive plasticity, must be accounted for in modeling. A review work by Ma et al. (2010) covers these issues in more details.

12.4 Modeling and Characterization

Conducting experiments on SMAs is expensive and time consuming. Therefore, in the design of SMA devices it is essential to have reliable modeling platforms to avoid unnecessary trials and errors. In a general case, the behavior of an SMA-based device could be demonstrated through the following interconnected sub-models: constitutive model, phase transformation kinetics, heat transfer model, and kinematics and dynamics of the device (Figure 12.6). Among these sub-models, the first three as shown in Figure 12.6 are aimed to capture the thermo-mechanical response of SMA and will be discussed briefly in the following sections.

12.4.1 Constitutive Model

An SMA's material behavior depends on stress, temperature, and crystallographic phase, all of which are mutually dependent on each other. Because of this interdependency the formulation of adequate macroscopic constitutive laws is necessarily complex (Brinson and Huang 1996). There have been many attempts to mathematically model the SMA features over the past three decades. The resulting models can be divided into two categories: micromodels and phenomenological macromodels. In general, micromechanical-based models (Lexcellent et al. 1996; Levitas and Ozsoy 2009) utilize information about the microstructure of the SMA to predict the macroscopic response. Micromechanical models are useful in understanding the fundamental phenomena of SMA, although they may not be easily deployed for engineering applications. On the other hand, phenomenological models (Liang and Rogers 1990; Brinson 1993; Boyd and Lagoudas 1996; Arghavani et al. 2010;

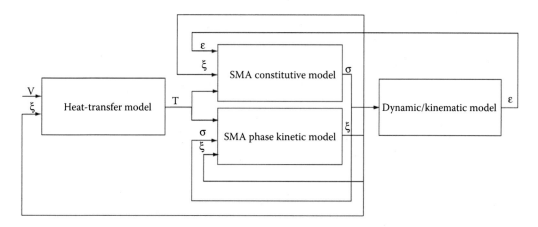

FIGURE 12.6
Modeling of the SMA systems; four sub-models are inter-connected. (From Elahinia M H and Ahmadian M. 2005. *Smart Materials and Structures* 14(6): 1309.)

Saleeb et al. 2011; Lagoudas et al. 2012; Mehrabi and Kadkhodaei 2013) use the principles of continuum thermodynamics to describe the material response. They are calibrated by a limited number of parameters measured at the macroscopic scale through experimental observations, and thus, they are more computationally efficient.

A common 1-D phenomenological thermomechanical based model is discussed in the following. This model is developed by Liang and Brinson (Liang 1990; Brinson 1993) and has been used for the design and control by many researchers (Williams et al. 2010; Elahinia 2004). The constitutive law can be written as

$$\sigma = D(\xi)(\varepsilon - \varepsilon_1 \xi_s) + \Theta(T - T_0) \tag{12.1}$$

Where σ is the Cauchy stress, D is the Young's modulus of the SMA which depends on the martensite volume fraction ξ, ε is the Green Strain, ε_1 is the maximum residual strain, ε_s is the stress-induced martensite volume fraction, Θ is the thermal coefficient of expansion, T is the current temperature and T_0 is the reference temperature.

In this model, the martensite volume fraction (ξ) is decomposed into temperature-induced (ξ_t) and stress-induced (ξ_s) martensite components and only the stress-induced part is accompanied by a macroscopic strain. By this separation, the model is capable of capturing the full range of SMA behavior, including SME and pseudoelasticity. In this model, the Reuss scheme is applied to calculate the value of the SMA Young's modulus,

$$D(\xi) = \left(\frac{1}{D_M} + \frac{1 - \xi}{D_A} \right)^{-1} \tag{12.2}$$

where D_M and D_A are the pure martensite and austenite Young's modulus, respectively.

12.4.2 Phase Transformation Kinetics

The martensite volume fractions, ξ_s and ξ_t are determined by transformation kinetic equations which relate the evolution of the martensite fractions with stress and temperature.

The phase diagram depicted in Figure 12.1 is used and the transformation kinetics proposed by Brinson (Brinson 1993) are presented as follows:

For $T > M_s$ and $\sigma_s^{cr} + C_M(T - M_s) < \sigma < \sigma_f^{cr} + C_M(T - M_s)$:

$$\xi_s = \frac{1 - \xi_{s0}}{2} \cos\left\{\frac{\pi}{\sigma_s^{cr} - \sigma_f^{cr}}[\sigma - \sigma_f^{cr} - C_M(T - M_s)]\right\} + \frac{1 + \xi_{s0}}{2} \qquad (12.3)$$

$$\xi_T = \xi_{T0} - \frac{\xi_{T0}}{1 - \xi_{s0}}(\xi_s - \xi_{s0}) \qquad (12.4)$$

For $T < M_s$ and $\sigma_s^{cr} < \sigma < \sigma_f^{cr}$:

$$\xi_s = \frac{1 - \xi_{s0}}{2} \cos\left\{\frac{\pi}{\sigma_s^{cr} - \sigma_f^{cr}}(\sigma - \sigma_f^{cr})\right\} + \frac{1 + \xi_{s0}}{2} \qquad (12.5)$$

$$\xi_T = \Delta_{T\xi} + \frac{\Delta_{T\xi}}{1 - \xi_{s0}}(\xi_s - \xi_{s0}) \qquad (12.6)$$

where if $M_f < T < M_s$ and $T < T_0$

$$\Delta_{T\xi} = \frac{1 - \xi_{s0} - \xi_{T0}}{2} \cos\left\{\frac{\pi}{M_s - M_f}(T - M_f)\right\} + \frac{1 - \xi_{s0} + \xi_{T0}}{2} \qquad (12.7)$$

otherwise $\Delta_{T\xi} = \xi_{T0}$ \qquad (12.8)

For $T > A_s$ and $C_A(T - A_f) < \sigma < C_A(T - A_s)$:

$$\xi = \frac{\xi_0}{2}\left\{\cos\left[\frac{\pi}{A_f - A_s}(T - A_s - \frac{\sigma}{C_A})\right] + 1\right\} \qquad (12.9)$$

$$\xi_s = \xi_{s0} - \frac{\xi_{s0}}{\xi_0}(\xi_0 - \xi) \qquad (12.10)$$

$$\xi_T = \xi_{T0} - \frac{\xi_{T0}}{\xi_0}(\xi_0 - \xi) \qquad (12.11)$$

12.4.3 Heat Transfer Model

Because of the heat driven phase transformation, the transition temperatures of SMAs are very important factors when determining the stress. An accurate model of heat transfer is necessary to model the behavior of the actuator in different climates and environments and particularly over the wide range of ambient temperatures in which a device may operate. The heat transfer model uses the applied current and the ambient temperature to determine the active wire temperature. The phase transformation temperatures vary with stress (Taheri Andani et al. in press).

The assumed SMA wire heat transfer equation consists of electrical heating and natural convection:

$$mc_p \frac{dT}{dt} = \frac{V^2}{R} - h(T)A_c(T - T_\infty) - m\Delta H\dot{\xi} \tag{12.12}$$

where m is mass per unit length, c_p is the specific heat, R is resistance per unit length, h(T) is the heat convection coefficient, A_c is the circumferential area of the SMA wire. Also, V is the applied voltage, T_∞ is the ambient temperature, ΔH is the latent heat, and $\dot{\xi}$ is the phase transformation rate.

When capturing the SMA response in transformation, all the three sub-models presented above should be solved simultaneously. Different finite element and analytical platforms have been developed in recent years to employ these models to design and analyze the SMA devices (Qidwai and Lagoudas 2000; Mirzaeifar et al. 2011; Taheri Andani et al. in press).

12.5 Fabrication Processes

To be successful any engineering component must be durable, compact and reproducible. Owing to the important properties and characteristics mentioned in the previous section, NiTi SMAs are gaining attention for many applications.

The SME, SE, damping, and impact absorbing capabilities of near equiatomic Nitinol, as well as its thermo-mechanical properties are strongly dependent on the stoichiometry and thermal/mechanical treatment of the sample. Very small atomic variation of the Ni content on the order of 0.1% is shown to significantly change the transformation temperatures on the order of 10°C (Frenzel et al. 2010). Impurities should also be avoided from the NiTi matrix since the transformation temperatures, hysteresis, and ductility of the material are very sensitive to these impurities. In other words, the properties of Nitinol are significantly affected by the manufacturing processes. There are abundant amounts of processing data available in the industry to process the material; however, they are often kept proprietary and not released in the public domain.

12.5.1 Conventional Methods

Conventional fabrication methods of SMA components include arc or induction melting followed by a hot working process and lastly machining to the final shape. Melting in an inert atmosphere limits contamination and also makes it possible to achieve a good mixture of the constituent elements resulting in material homogeneity and uniformity of the characteristic properties (Funakubo 1987; Frenzel et al. 2004; Morgan and Broadley 2004). Two major methods of melting are Vacuum Induction Melting (VIM) and Vacuum Arc Remelting (VAR). The cost of production by these methods is comparable and they both provide material suitable for current engineering devices. Both VAR and VIM processes have major drawbacks including extreme reactivity of the melt and undergoing segregation due to the density difference of the reacting melts. Especially when using graphite crucibles the melt will pick-up carbon. But by using optimizied melting, procedures that pick-up can be reduced significantly (Frenzel et al. 2004). Additionally, rapid grain growth occurs as a result of high-temperature working which leads to poor fatigue properties. Although VAR and VIM melting processes have basic differences, VAR and VIM/VAR products appear to have similar mechanical and fatigue properties (Reinoehl et al. 2000). However, using water-cooled copper crucibles VAR allows for the productions of highest purity NiTi (Frenzel et al. 2010).

In addition to conventional casting processes, electron beam melting (EBM) process is also being used to make NiTi products. In this process melting is performed in a water-cooled copper crucible and under a high vacuum. It is difficult to control the nominal chemical composition in EBM because of the vacuum operation and high heating temperature. This causes some constituent elements to evaporate, changing the martensitic transformation temperatures. In spite of this shortcoming, this method is often used to prepare such SMAs that do not require precise control of the transformation temperature (Otubo et al. 2003).

Machining of NiTi is also challenging due to its characteristic properties and its resistance to deformation, both of which cause severe tool wear. Abrasive methods such as grinding and abrasive saws are therefore preferred for this material. However, they do not allow for final shaping and they may induce significant heat. The machinability of NiTi significantly depends on the cutting speed and feed rate, both of which are typically very high. Machining NiTi components with high cutting feeds and speeds will help to extend the tool life and improve product quality although these elevated speeds result in an increased work hardening on the subsurface zones of the part (Weinert and Petzoldt 2004). Laser cutting, water jet cutting, electro discharge machining (EDM), photo-chemical etching and are a few of the alternative processes to manufacture finished NiTi products. Laser cutting has the advantage of causing no mechanical stresses, no tool wear, and high lateral resolution. Unfortunately, laser-cut or EDMed parts must undergo further processing to remove the heat-affected zone. Nevertheless, laser cutting represents the current state of art especially for stent production. Modern laser cutting machines, using a pulsed Nd:YAG laser and equipped with a CNC motion control system, offer high speed, high accuracy, and the capability for rapid prototyping (Wu 2002).

When producing NiTi components for industrial purposes, near net shape fabrication processes are preferred. Powder metallurgy (PM) is a well-known process for its ability to provide semi-finished and net-shaped products. In addition, another benefit of PM is that it is an effective method for manufacturing porous NiTi. Porous SMAs have gained the researchers' attention for many engineering applications due to their unique mechanical properties that can be adjusted via manipulating the production procedure to that of the required application (Elahinia et al. 2012).

Different PM processes have been experimentally developed for NiTi. In conventional sintering (CS), a green compact of Ni and Ti powders or pre-alloyed NiTi powders is prepared and sintered at near melting temperatures to produce binary NiTi. Conventional sintering requires long heating times and samples are limited in shape and size. The porous structure in the product of conventional sintering is also shown to be of small size and irregular shape. A maximum porosity of 40% can be achieved with this procedure (Li et al. 1998; Yuan et al. 2005). Another PM process is self-propagating high-temperature synthesis (SHS). This process initiates by a thermal explosion ignited at one end of the specimen that propagates through the specimen in a self-sustaining manner (due to the exothermic reaction between Ni and Ti). Hot iso-static pressing (HIP) is a high pressure sintering technique that can be utilized to manufacture very dense products. A powder mixture of elementary powders or pre-alloyed powders is encapsulated in an evacuated, air-tight welded canister and undergoes simultaneous iso-static pressure and elevated temperature. Alternatively, Argon can be used as an inert environment without the need for the air-tight canister of elemental compact. In this case, HIP can be used to compress and trap Ar gas bubbles in between the elemental powders. A subsequent high-pressure diffusion stage will lead to Ar-filled pores. The HIP process has several advantages such as a shorter solid state diffusion time, good control over the pore size, the ability to manipulate various geometries, and finally having a more stable and manageable reaction compared to SHS process. Metal injection molding (MIM) is another powder procedure for manufacturing near-net-shape NiTi products. MIM operates on the same principles as the injection molding of plastics. This process was developed for the manufacturing of small and complex shaped parts in a cost-effective manner. By using space holders MIM allows for the production of porous NiTi parts, too (Bram et al. 2011).

12.5.2 Additive Manufacturing

The general term Additive Manufacturing (AM) describes the process of making a part by adding successive layers of material. AM is an attractive manufacturing method as it adds material and hence AM wastes less material than more traditional subtractive manufacturing methods such as mill or lathe work. Each layer is melted according to a geometry defined by a three-dimentional CAD model. Building parts with very complex geometries without any sort of cutting tools or fixtures is the primary advantage of AM process. It is also a fast production route from CAD to physical part with very high material utilization and does not require expensive castings or forgings. Therefore, it is a very cost-effective, energy efficient, and environmentally friendly manufacturing process. The most common and known processes for making shapes from metal powders are the powder-bed-based technologies Selective Laser Sintering, Selective Laser Melting (SLM), Direct Metal Laser Sintering, LaserCusing and EBM as well as the nozzle-based technologies Laser Engineered Net Shaping (LENS) and Direct Metal Deposition (DMD). AM technologies lead to today's freeform fabrication machines that build parts using a wide variety of metal powder particles (Gibson et al. 2010).

These computer-controlled techniques were originally developed for producing plastic prototypes but have subsequently been extended to metals, ceramics, and composite powders. In fact, AM can produce parts from a relatively wide range of commercially available powder materials (Gibson et al. 2010). As shown in Figure 12.7, powder-bed-based AM is an iterating process. A powder layer (typically 25 μm to 0.1 mm thick) is deposited by a blade or a rotating roller. A high power laser or electron beam locally melts the powder. After solidification, solid structures remain which are surrounded by loose powder. Afterwards another powder layer is deposited on top of the previous layer. This procedure is repeated until a desired three-dimentional shape is produced. Usually, all the process is done in an enclosed chamber filled with argon or nitrogen to minimize oxidation.

Although AM is a comparatively young processing method, it is predicted to gain in importance for processing SMAs (Gausemeier et al. 2012). Most of the recent publications emphasize that AM seems to be a promising method for near-net-shape processing shape memory components. Clare et al. (2008) used SLM to build micro NiTi cantilever beams that were used to make a two-way trained actuator. Meier et al. (2011) showed that SLM makes the manufacturing of high-quality NiTi parts possible. They also experimentally studied the effect of SLM processing on microstructure and characteristic temperatures of

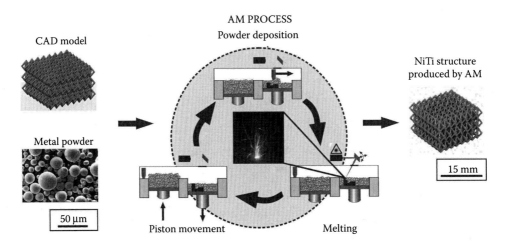

FIGURE 12.7
Schematic of powder-bed based AM process.

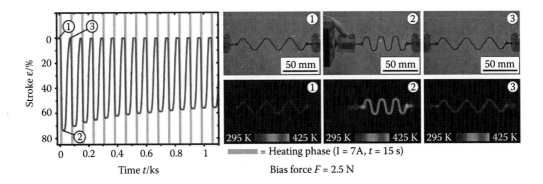

FIGURE 12.8
(**See color insert.**) Additive manufactured NiTi actuator performing the shape memory effect. (From Haberland C. 2012. *Additive Verarbeitung von NiTi-Formgedächtniswerkstoffen mittels Selective Laser Melting*. Germany: Ruhr University Bochum; Aachen: Shaker Verlag GmbH.)

NiTi and showed that SLM parts have unique shape memory properties. Figure 12.8 exemplarily shows the functionality of a shape memory actuator produced by SLM.

12.5.3 Heat Treatment, Shape Setting and Training

NiTi products (e.g., bars, wires, ribbons, and sheets) are normally finished with cold working in order to achieve tight dimensional tolerance control and enhanced surface quality. Cold working suppresses the shape memory response of these alloys. It also raises the strength and decreases the ductility. Cold working does not, however, raise the stiffness of the material. Heat-treating after cold working reduces the effects of cold working and restores the shape memory response of SMAs. Therefore, in order to optimize the physical and mechanical properties of a NiTi product and to achieve desired shape memory and/or pseudoelasticity properties, the material is cold worked and heat treated. Both superelastic and shape memory properties can be optimized through cold working and heat treatment.

 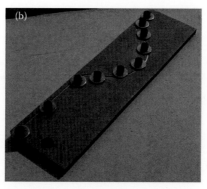

FIGURE 12.9
A shape setting fixture for NiTi for creating a helical shape for an expandable pedicle screw (Data from Tabesh M. 2010. Finite Element Analysis of Shape Memory Alloy Biomedical Devices. University of Toledo, Toledo, OH (Master's Thesis)) and for creating a curved beam for an organ positioner (Data from Kanjwal K et al. 2011. *Journal of Interventional Cardiac Electrophysiology* 30(1): 45–53).

This thermomechanical process is applied to all NiTi alloys although differing amounts of cold work and heat treatment may be used for different alloys and property specifications (Lagoudas 2007).

Shape setting is accomplished by deforming the Nitinol part to the shape of the desired component, constraining the Nitinol by clamping, and then heat treating. This is normally done on the materials after cold working, for example with cold drawn wires. However, annealed wires may also be shape set with a subsequent lower temperature heat treatment. In the case of shape setting SMA wires, a tooling fixture made of stainless steel is typically used to hold the wire in a taut position (see Figure 12.9). The effect of duration and temperature of the shape setting process will affect the shape recovery quality, transformation temperatures, and mechanical behavior of NiTi SMA as investigated in Liu et al. (2008).

Training an SMA refers to the process of repeatedly loading/unloading the material response becomes consistent. Once trained the hysteretic response of the material stabilizes and the inelastic strain saturates and disappears. Figure 12.10 shows the cyclic mechanical loading of a superelastic NiTi wire. At the end of the first cycle, a permanent plastic strain remains in the material. The additional permanent strain associated with each consecutive cycle begins to gradually decrease until it practically ceases to further accumulate. Training is a recommended process to be conducted on every off-the-shelf NiTi component before practical application to avoid an inconsistent strain response.

12.6 SMA Actuation

SMA actuators are attractive due to their many potential applications in mechanical, electrical, civil, medical, and aerospace systems (Hartl and Lagoudas 2007; Lagoudas et al. 2000; Nespoli et al. 2010; Rofooei and Farhidzadeh 2011; Saadat et al. 2002; Song et al. 2006). SMAs have many advantages over conventional actuation methods. Their significantly reduced weight, size, complexity, and noiseless operation make them suitable as actuators

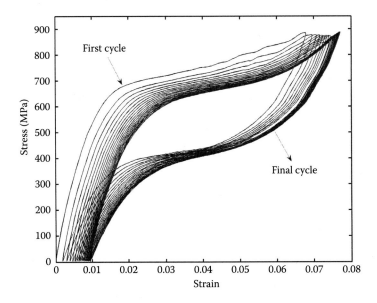

FIGURE 12.10
Training is required to be conducted on SMA to stabilize its cyclic response. (From Taheri Andani M. 2013. Constitutive modeling of superelastic shape memory alloys considering rate dependent non-Mises tension-torsion behavior. University of Toledo, Toledo, OH (Master's Thesis).)

for a wide range of applications from micro robotics to medical application and aerospace systems. Their benefits over other smart materials such as piezoelectric materials, electrostrictive materials, and magnetostrictive materials include their high-force-to-weight ratio, and large displacement capabilities.

SMA actuators could replace conventional actuators thanks to the large mechanical strains induced by heating and cooling. Large mechanical stress is consequently produced when SMA elements are heated or cooled beyond critical temperatures. SMA actuators can apply up to 600 MPa. As an example, a 0.1-mm diameter Nitinol wire can therefore apply a force of 4.7 N. This means that the wire can lift up an object about 100,000 times its own weight. Another advantage that SMA actuators can have over conventional actuators is that they can be made compact and simplistic. There are many advantages to employ SMA wires as actuators in various applications. In the following, possible loading and actuation modes of SMA devices are discussed. Example(s) are also presented for each mode to better show the mechanism of the actuation.

12.6.1 One-Way and Two-Way Actuation

One-way SME refers to the ability of an SMA deformed at a low temperature to recover the deformation when heated to a higher temperature. In this behavior, the austenite shape is memorized due to self-accommodated structure of martensite variants. One-way shape memory is the most frequently utilized SMA behavior in practice. In actuation applications, however, the shape memory behavior is never used in this fashion. Instead, an external biasing force always exists on SMA actuators during thermal cycling. During transformation, the associated shape change causes SMA to push against the biasing force, thus doing mechanical work.

With two-way SMA actuators, the SMAs can exhibit repeatable shape changes without a biasing mechanical load. Instead the SMA will alternate between memorized shapes when subjected to a cyclic thermal load. In this case the SMA memorizes a martensite shape different from the austenite shape. This behavior is called the two-way SME. In this mode, when the SMA is cooled from austenite to martensite, instead of adapting to a self-accommodated structure, some variants of the martensite are favored and the martensite adopts a shape different from that of the self-accommodated structure. Two-way SME is caused by internal stresses that develop in the SMA from plastic deformation in martensite usually through aging for precipitation under stress and/or under constraint, which is called training.

Most of the SMA actuators utilize these alloys in the wire form. The SMA actuators are designed to use the SME in creating motion and force. When an SMA wire is heated, the material applies a large amount of force and displacement while returning to a memorized length. Joule heating is an effective and simple way for actuating SMA components.

So far, most of developed SMA actuators operate in simple axial loading with small displacements. However, it is possible to overcome these drawbacks through design optimization. Figure 12.11 shows a simple mechanism used by Elahinia and Afsharioun (2002) to create a rotational motion by actuating an SMA wire. The arm mechanism converts the small axial deformation of the wire to a large rotational displacement and the bias spring makes the cyclic motion possible.

Utilizing SMAs in the form of springs is another solution to overcome the deformation limitations of these materials. Spring-type SMA actuators provide large deformation in addition to smooth, silent, and clean motion compared to traditional types of actuators. This kind of actuator is particularly appropriate for robotic applications where size and weight are critical. Moghaddam et al. (2011) designed a light-weight biped using two SMA based actuators. Figure 12.12 shows a configuration of this device. The experiments conducted on a prototype of the device demonstrated a promising capability of the SMA springs in providing a flexible motion under unwanted turbulences.

It is well established that using SMA actuators can drastically reduce the number of moving mechanical parts, simplify the assembly procedure and reduce manufacturing costs. These benefits have caused researchers to use SMAs to design less expensive and compact actuators. The automotive is a large potential market for these types of actuators. Williams et al. (2010) designed an SMA external side mirror actuator. As shown in Figure 12.13, the mirror consists primarily of a modified ball and socket joint and four SMA wires.

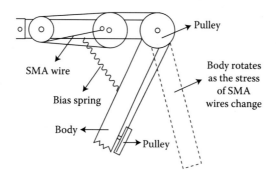

FIGURE 12.11
SMA rotary actuator. (From Elahinia M H and Ashrafiuon H. 2002. *Journal of Vibration and Acoustics* 124(4): 566–575.)

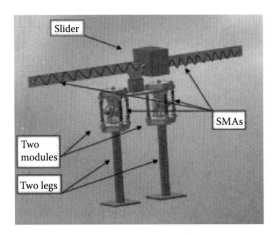

FIGURE 12.12
Configuration of a biped actuated by SMA springs. (From Moghaddam M M et al. 2011. *Journal of Intelligent Material Systems and Structures* 22(13): 1489–1499.)

The spring-loaded joint provides mirror stability between actuation. At the time of actuation, the force of the SMA wires disengages the joint to provide friction-free motion. This joint passively provides a larger range of motion with smaller actuation force. The actuation force, unlike in the first generation mirror, is not used to overcome friction. This feature of the joint, additionally, mitigates the disturbances such as aerodynamic forces and variation in environment temperature. The proposed actuator provides a two degree of freedom by rotating the mirror about its two axes.

By developing the modern manufacturing methods, the interest in porous structures as actuators is also increasing. Figure 12.14 depicts an SMA porous beam, which is built through AM and is supposed to operate in a one-way shape memory regime. The porous structure gives the flexibility of having very light-weight actuator with desirable stiffness.

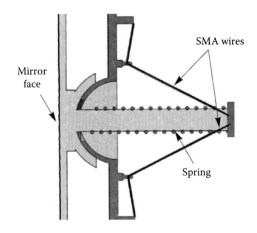

FIGURE 12.13
Configuration of an SMA mirror actuator. (From Williams E, Shaw G and Elahinia M H. 2010. *Mechatronics* 20(5): 527–534.)

FIGURE 12.14
A shape memory porous beam. (From Haberland C. 2012. *Additive Verarbeitung von NiTi-Formgedächtniswerkstoffen mittels Selective Laser Melting.* Germany: Ruhr University Bochum; Aachen: Shaker Verlag GmbH.)

12.6.2 Superelasticity

Superelastic SMAs are beneficial in passive applications where large recoverable deformations are required. SMAs can provide almost constant level of force (stress) over a large range of strain (up to 10%). Dental archwires, brassieres, springs, cell phone antennae, and many other applications use the pseudoelastic property of SMA to increase the durability and performance of consumer goods such as those aforementioned (Haga et al. 2005).

Nitinol expandable reamer aimed to cut the bone is an example of a superelastic SMA device. As shown in Figure 12.15, the blades undergo large bending deformations without considerable residual deformation.

A third application is an intervertebral cage, developed using superelastic Nitinol hinges under torsional loading (Figure 12.16). An intervertebral cage is a spacer that is used when two vertebrae are fused together to alleviate low back pain. The cage was built using AM out of medical grade titanium. Elliptically shaped rods were employed so that residual torque at closure is possible. In the Figure 12.17, two individual cage segments are shown being assembled. The two segments are "over-rotated" (panel a) and the two parts are made to come together (panel b). The cage is then straightened and put in its pre-deployment condition (panel b). After deployment, the SE of the Nitinol causes the cage to go from the constrained straight form to the oval closed form (panel c).

FIGURE 12.15
(**See color insert.**) Left: Superelastic NiTi expandable reamer before deformation. Right: Superelastic NiTi retrograde blade after deformation. (Courtesy of Symmetry Medical Inc, New Bedford, MA.)

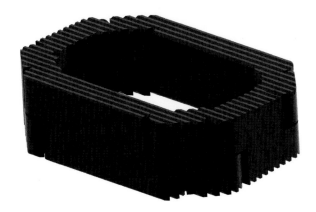

FIGURE 12.16
An intervertebral cage was developed using superelastic NiTi elliptical shape rods as hinges. (From Anderson W. 2013. Development of an Intervertebral Cage Using Additive Manufacturing with Embedded NiTi Hinges for a Minimally Invasive Deployment. University of Toledo.)

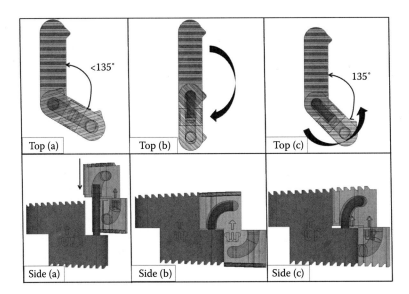

FIGURE 12.17
(**See color insert.**) The assembly and deployment configuration of the intervertebral cage. (From Anderson W. 2013. Development of an Intervertebral Cage Using Additive Manufacturing with Embedded NiTi Hinges for a Minimally Invasive Deployment. University of Toledo.)

12.6.3 Antagonistic Superelastic-Shape Memory

A distinct actuation approach is possible based on combining the two distinct properties of SMAs: SME and SE. A SM member in conjunction with a SE element would make an antagonistic manner actuator, such that they force each other in the opposite directions. The SM component would control the actuation while the SE element provides the required actuation force and large actuation stroke.

The function of the actuator is as follows: assuming that the actuator is supposed to work between two temperatures; low temperature (T_l) and high temperature (T_h). The SE member is always in austenite phase either at low temperature or high temperature. This means that its austenite finish temperature is lower than both low and high temperatures ($A_f^{SE} < T_1$ and $A_f^{SE} < T_h$). On the other hand, the SM element is initially in its martensite phase, but transforms to austenite at high temperature. To this end, the SM material should be selected such that its martensite finish temperature (M_f^{SM}) is higher than the low temperature and its austenite finish temperature (A_f^{SM}) is lower than high temperature ($T_1 < M_f^{SM}$ and $T_h > A_f^{SM}$). The SE and SM elements' original shapes are set such that in low temperature, the SE element is stronger and the assembly stays toward the SE element, which is set to be the inactivated form of the actuator. At high temperature, the SM element passes its austenite temperature and moves the assembly toward its original shape that is set to be the activated form of the device.

The described antagonistic actuator could be assembled in many different ways. An interesting configuration is having the SM element as a at wire and SE element as a ring shaped wire. By wrapping the flat wire around the round member, a smart antagonistic actuator would be resulted. Figure 12.18a and b, respectively, display the initial shape set or memorized forms of the SE and SM elements before assembly.

When the two members are assembled at a low temperature, they apply force to each other in a way that the SM at wire tries to enclose the ring while the SE element tends to keep the ring open. Depending on the material properties of the two components, an equilibrium position will be obtained. Figure 12.19a displays the equilibruim or neutral position of the device at low temperature. By heating the SM element either by changing the thermal environmental condition or by resistant heating, the assembly tends toward the shape set form of the SM member, which likes to enclose the ring as shown in Figure 12.19b. By cyclic heating and cooling these two configurations can be perfectly repeated. It's worth mentioning that the actuation stroke and shape generally depend on the material properties and shape set forms of the SM and SE elements and thus there are numerous possible actuation trends. This is a very beneficial point in design.

Tissue clamps are widely used to hold the body organs during the surgery. The conventional devices are usually large, heavy and hard to control. A smart tissue clamp is designed based on the introduced antagonistic actuation method. This novel device is light and tiny, with a large actuation stroke and the ability of remote releasing control. A prototype of the smart tissue clamp was fabricated and experimentally evaluated in the

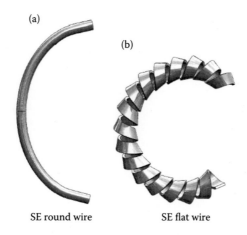

(a)

(b)

SE round wire   SE flat wire

FIGURE 12.18
Initial memorized shapes of the members before assembly in an antagonistic actuator. (Courtesy of Lifewire LLC Macon GA.)

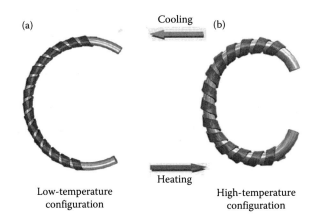

FIGURE 12.19
Smart antagonistic actuator assembly. (Courtesy of Lifewire LLC Macon GA.)

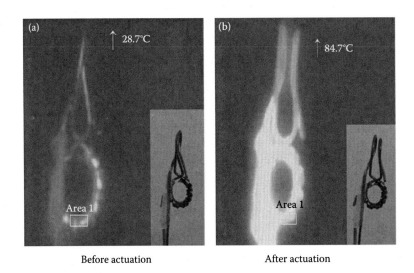

FIGURE 12.20
(**See color insert.**) Experimental evaluation of the smart antagonistic tissue clamp. (Courtesy of Lifewire LLC Macon GA.)

Dynamic and Smart Systems Laboratory at the University of Toledo. The shape memory wire was heated and the temperature of the device was measured by an infrared camera. Figure 12.20a shows the device before the actuation of the SM wire at room temperature. As soon as the SM wire is heated, it starts to release the clamp as shown in Figure 12.20b. By cyclic heating and cooling, the clamping and releasing configurations would be repeated.

12.6.4 Control

Owing to nonlinear hysteresis behavior of SMAs, the control of SMA actuators is a challenging task. The hysteresis properties existing in SMA involve the relationships between displacement and temperature, Young's modulus and temperature, displacement and voltage, strain and temperature as well as martensite fraction and temperature. Another issue in controlling SMA is its challenging actuation impetus (adding and removing heat). The SMA actuation is mostly initiated by applying a current to the element, which is therefore subject to Joule heating and natural convection. The SMA system may experience a large wait time for the element to cool. Heat removal techniques are usually in the form of using heat sink, water immersion, and/or forced convection. The temperature and type of environment, the convection of the environment and the surface to volume ratio of the SMA wires could all influence the heating and cooling time. These time delays for SMA actuators to respond to control signals make another nonlinear problem and present a challenge in controller design.

In the design of an SMA controller it is important to avoid overheating and/or overloading. If an SMA component is overheated it will prolong the cooling procedure, decreasing the speed of the actuation. Overloading might causes a malfunction or decreases the fatigue life of the system.

Several methods of controlling SMA actuators have been investigated; most of these control systems, however, are non-model based with limited possibility for stability and robustness analysis. Different variations of linear Proportional Integral Derivative controls have been explored by several (Arai et al. 1994; Gorbet and Wang 1995; Shameli et al. 2005; Silva 2007; Gédouin et al. 2011; Esfahani and Elahinia 2010) while some others have used Pulse Width Modulation (Gharaybeh and Burdea 1994; Honma et al. 1984). Several nonlinear control schemes such as fuzzy logic, neural networks, feedback linearization, sliding mode control, and variable structure control have also been explored by researchers (Choi and Cheong 1996; Dickinson and Wen 1998; Kilicarslan et al. 2010). Song et al. (2003) have applied sliding mode feedback control in conjunction with a feed forward neural network to an SMA actuator with linear motion. Neural networks showed great possibility in mapping and adaptation for capturing the hysteresis behavior of the SMA. However, neural networks need training to operate and therefore the training data and training methods are critical. This may require a large amount of data to secure the statistical accuracy. Furthermore, other controllers have also been designed that include an additional open-loop part for improving the performance (Raynaerts and Van Brussel 1991; Van der Wijst et al. 1997). Elahinia et al. (2005) developed a model-based back stepping controller for SMA actuators. The same group developed a stress-based and temperature-based controller that demonstrated enhanced tracking performance for rotary SMA actuators (Elahinia et al. 2004b; 2004a). A comprehensive review of control methods of SMAs is presented in Elahinia et al. (2010).

12.7 Summary

This chapter introduced the unique properties of SMAs such as the SME and SE. The underlying microstructural phenomena corresponding to such behaviors were briefly discussed and the phase diagram was introduced as a beneficial tool to schematically represent the associated phase transformations. As mentioned, although Nitinol is the most common SMA, other shape memory materials are also available for specific applications. In general, processing SMAs is challenging. However, different manufacturing techniques are applied to process SMAs nowadays. Among these conventional techniques new and innovative manufacturing techniques are gaining importance. For example, AM is a fast and less expensive method. Furthermore, it may open up new potential for SMA applications because it allows for a high freedom in fabrication, which is not provided by conventional techniques. A well-known phenomenological modeling approach was presented and shown to be necessary in the design and analysis processes of the SMA systems. SMAs provide several interesting methods to design compact and efficient actuators for various applications. Control of SMA systems on the other hand is a challenging topic due to their highly non-linear response and has been the focus of many extensive works by several researchers in the last decade.

Acknowledgments

The authors acknowledge the financial support of the National Science Foundation (CBET-0731087), Ohio Department of Development (WP 10-010 Nitinol Commercialization Accelerator), deArce Memorial Fund, UT Innovation Enterprise, Lifewire LLC, Jacobson Center for Clinical & Translational Research, and Ohio Board of Regents. The authors also thank Alexander Czechowicz, Ruhr-University, Bochum, Germany; Eugenio Dragoni, University of Modena and Reggio Emilia, Italy; and Osman E. Ozbulut, University of Virginia, Charlottesville, Virginia, for reviewing this chapter and for providing valuable technical comments and suggestions.

References

Anderson W. 2013. *Development of an Intervertebral Cage Using Additive Manufacturing with Embedded NiTi Hinges for a Minimally Invasive Deployment.* Toledo, OH: University of Toledo.

Arai K, Aramaki S and Yanagisawa K. 1994. Continuous system modeling of shape memory alloy (SMA) for control analysis. In *Proceedings of the 5th International Symposium On Micro Machine and Human Science*, Nagoya, Japan, IEEE.

Arghavani J, Auricchio F, Naghdabadi R, Reali A, and Sohrabpour S. 2010. A 3-D phenomenological constitutive model for shape memory alloys under multiaxial loadings. *International Journal of Plasticity* 26(7): 976–991.

Bekker A and Brinson L C. 1998. Phase diagram based description of the hysteresis behavior of shape memory alloys. *Acta Materialia* 46(10): 3649–3665.

Boyd J G and Lagoudas D C. 1996. A thermodynamical constitutive model for shape memory materials. Part I. The monolithic shape memory alloy. *International Journal of Plasticity* 12(6): 805–842.

Bram M, Köhl M, Buchkremer H P and Stöver. 2011. Mechanical properties of highly porous NiTi alloys. *Journal of Materials Engineering and Performance* 20(12): 522–528.

Brinson L C. 1993. One-dimensional constitutive behavior of shape memory alloys: Thermomechanical derivation with non-constant material functions and redefined martensite internal variable. *Journal of Intelligent Material Systems and Structures* 4(2): 229–242.

Brinson L C and Huang M S. 1996. Simplifications and comparisons of shape memory alloy constitutive models. *Journal of Intelligent Material Systems and Structures* 7(1): 108–114.

Choi S-B and Cheong C-C. 1996. Vibration control of a flexible beam using shape memory alloy actuators. *Journal of Guidance, Control, and Dynamics* 19(5): 1178–1180.

Clare A T, Chalker P R, Davies S, Sutcliffe C J and Tsopanos S. 2008. Selective laser melting of high aspect ratio 3D nickel-titanium structures two way trained for MEMS applications. *International Journal of Mechanics and Materials in Design* 4(2): 181–187.

Dickinson C A and Wen J T. 1998. Feedback control using shape memory alloy actuators. *Journal of Intelligent Material Systems and Structures* 9(4): 242–250.

Duerig T, Pelton A and Stoeckel D. 1999. An overview of nitinol medical applications. *Materials Science and Engineering: A* 273: 149–160.

Eggeler G, Khalil-Allafi J, Gollerthan S, Somsen C, Schmahl W and Sheptyakov D. 2005. On the effect of aging on martensitic transformation in Ni-rich NiTi shape memory alloys. *Smart Materials and Structures* 14: 186–191.

Elahinia M H, Koo J, Ahmadian M and Woolsey C. 2005. Backstepping control of a shape memory alloy actuated robotic arm. *Journal of Vibration and Control* 11(3): 407–429.

Elahinia M H, Ahmadian M and Ashrafiuon H. 2004a. Design of a kalman filter for rotary shape memory alloy actuators. *Smart Materials and Structures* 13(4): 691–697.

Elahinia M H, Esfahani E T and Wang S. 2010. Review of the state of the art of control of SMA systems. In Chen H R (ed): *Shape Memory Alloys: Manufacture, Properties and Applications*. Hauppauge: Nova Science Publishers, pp. 49–68.

Elahinia M H, Seigler T M, Leo D J and Ahmadian M. 2004b. Nonlinear stress-based control of a rotary SMA-actuated manipulator. *Journal of Intelligent Material Systems and Structures* 15(6): 495–508.

Elahinia M H. 2004. Effect of System Dynamics on Shape Memory Alloy Behavior and Control. Virginia Tech, Blacksburg, VA (PhD Dissertation).

Elahinia M H. and Ahmadian M. 2005. An enhanced SMA phenomenological model: II. The experimental study. *Smart Materials and Structures* 14(6): 1309.

Elahinia M H and Ashrafiuon H. 2002. Nonlinear control of a shape memory alloy actuated manipulator. *Journal of Vibration and Acoustics* 124(4): 566–575.

Elahinia M H, Hashemi M, Tabesh M and Bhaduri S B. 2012. Manufacturing and processing of NiTi implants: A review. *Progress in Materials Science* 57(5): 911–946.

Esfahani ET and Elahinia M H. 2010. Developing an adaptive controller for a shape memory alloy walking assistive device. *Journal of Vibration and Control* 16(13): 1897–1914.

Es-Souni M, Es-Souni M and Fischer-Brandies H. 2005. Assessing the biocompatibility of NiTi shape memory alloys used for medical applications. *Analytical and Bioanalytical Chemistry* 381(3): 557–567.

Frenzel J, Zhang Z, Neuking K and Eggeler G. 2004. High quality vacuum induction melting of small quantities of NiTi shape memory alloys in graphite crucibles. *Journal of Alloys and Compounds* 385(1): 214–223.

Frenzel J, George E P, Dlouhý A, Somsen C, Wagner M F X and Eggeler G. 2010. Influence of Ni on martensitic phase transformation in NiTi shape memory alloys. *Acta Materialia* 58: 3444–3458.

Fristov G S, van Humbeeck J and Koval J N. 2006. High temperature shape memory alloys—Problems and prospects. *Journal of Intelligent Material Systems and Structures* 17: 1041–1047.

Funakubo H. 1987. *Shape Memory Alloys*. New York: Gordon and Breach Science Publishers.

Gausemeier J, Echterhoff N and Wall M. 2012. Thinking ahead the future of additive manufacturing—Scenario-based matching of technology push and market pull. *RTejournal—Forum für Rapid Technologie* 9, digital publication, http://www.rtejournal.de/ausgabe9/3355.

Gédouin P-A, Delaleau E, Bourgeot J-M, Join C, Chirani S A and Calloch S. 2011. Experimental comparison of classical pid and model-free control: Position control of a shape memory alloy active spring. *Control Engineering Practice* 19(5): 433–441.

Gharaybeh M A and Burdea G C. 1994. Investigation of a shape memory alloy actuator for dextrous force-feedback masters. *Advanced Robotics* 9(3): 317–329.

Gibson I, Rosen D W and Stucker B. 2010. *Additive Manufacturing Technologies: Rapid Prototyping to Direct Digital Manufacturing*. New York: Springer Verlag.

Gorbet R B and Wang D W L. 1995. General stability criteria for a shape memory alloy position control system. In: *Proceedings of the International Conference On Robotics and Automation*, Nagoya, Japan, IEEE 3. pp. 2313–2319.

Haberland C. 2012. *Additive Verarbeitung von NiTi-Formgedächtniswerkstoffen mittels Selective Laser Melting*. Germany: Ruhr University Bochum; Aachen: Shaker Verlag GmbH.

Haga Y, Mizushima M, Matsunaga T and Esashi M. 2005. Medical and welfare applications of shape memory alloy microcoil actuators. *Smart Materials and Structures* 14(5): S266–S272.

Hartl D J and Lagoudas D C. 2007. Aerospace applications of shape memory alloys. *Journal of Aerospace Engineering* 221(4): 535–552.

Honma D, Miwa Y and Iguchi N. 1984. Micro robots and micro mechanisms using shape memory alloy. In *The Third Toyota Conference, Integrated Micro Motion Systems, Micro-machining, Control and Applications*, Nissan, Aichi, Japan.

Kanjwal K, Yeasting R, Maloney J D, Baptista C, Elsamaloty H, Sheikh M, Elahinia M H, Anderson W. 2011. Retro-cardiac esophageal mobility and deflection to prevent thermal injury during atrial fibrillation ablation: An anatomic feasibility study. *Journal of Interventional Cardiac Electrophysiology* 30(1): 45–53.

Khalil-Allafi J, Dloughý A and Eggeler G. 2002. Ni4Ti3-precipitation during aging of NiTi shape memory alloys and its influence on martensitic phase pransformations. *Acta Materialia* 50: 4255–4274.

Kilicarslan A, Grigoriadis K M and Song G. 2010. Nonlinear control of a shape memory alloy actuator via mu-synthesis. In: *Proceedings of Earth and Space 2010 - 12th International Conference on Engineering, Science, Construction, and Operations in Challenging Environments*, Honolulu, HI. ASCE.

Lagoudas D C. 2007. *Shape Memory Alloys: Modeling and Engineering Applications*. New York: Springer.

Lagoudas D C, Hartl D, Chemisky Y, Machado L and Popov P. 2012. Constitutive model for the numerical analysis of phase transformation in polycrystalline shape memory alloys. *International Journal of Plasticity* 32: 155–183.

Lagoudas D C, Rediniotis O K and Khan M M. 2000. Applications of shape memory alloys to bioengineering and biomedical technology. In: *Proceedings of the 4th International Workshop on Scattering Theory and Biomedical Applications*, Perdika, Greece, Oct. 1999. pp. 195–207.

Levitas V I and Ozsoy I B. 2009. Micromechanical modeling of stress-induced phase transformations. Part 1. Thermodynamics and kinetics of coupled interface propagation and reorientation. *International Journal of Plasticity* 25(2): 239–280.

Lexcellent C, Goo B C, Sun Q P and Bernardini J. 1996. Characterization, thermomechanical behaviour and micromechanical-based constitutive model of shape-memory Cu Zn Al single crystals. *Acta Materialia* 44(9): 3773–3780.

Li B-Y, Rong L-J and Li Y-Y. 1998. Porous NiTi alloy prepared from elemental powder sintering. *Journal of Materials Research* 13(10): 2847–2851.

Liang C. 1990. The Constitutive Modeling of Shape Memory Alloys. Virginia Tech, Blacksburg, VA (PhD Dissertation).

Liang C and Rogers C A. 1990. One-dimensional thermomechanical constitutive relations for shape memory materials. *Journal of Intelligent Material Systems and Structures* 1(2): 207–234.

Liu X, Wang Y, Yang D and Qi M. 2008. The effect of ageing treatment on shape-setting and superelasticity of a nitinol stent. *Materials Characterization* 59(4): 402–406.

Ma J, Karaman I and Noebe R D. 2010. High temperature shape memory alloys. *International Materials Reviews* 55(5): 257–315.

Mabe J H, Ruggeri R T, Rosenzweig E and Chin-Jye M Y. 2004. NiTinol performance characterization and rotary actuator design. In: *Proceedings of the SPIE 5388, Smart Structures and Materials 2004: Industrial and Commercial Applications of Smart Structures Technologies*, San Diego, CA, 95–109.

Mehrabi R and Kadkhodaei M. 2013. 3D phenomenological constitutive modeling of shape memory alloys based on microplane theory. *Smart Materials and Structures* 22(2): 1–11.

Meier H, Haberland C and Frenzel J. 2011. Structural and functional properties of NiTi shape memory alloys produced by selective laser melting. In Bartolo PJ et al. (eds): *Innovative Development in Virtual and Rapid Prototyping. Proceedings of the 5th International Conference on Advanced Research in Virtual and Rapid Prototyping*, Leiria, Portugal, London: Taylor & Francis, pp. 291–296.

Mirzaeifar R, Shakeri M, DesRoches R and Yavari A. 2011. A Semi-analytic Analysis of shape memory alloy thick-walled cylinders under internal pressure. *Archive of Applied Mechanics* 81(8): 1093–1116.

Moghaddam M M, Hadi A, Tohidi A and Elahinia M H. 2011. Design, modeling, and prototyping of a simple semi-modular biped actuated by shape memory alloys. *Journal of Intelligent Material Systems and Structures* 22(13): 1489–1499.

Morgan N B and Broadley M. 2004. Taking the art out of smart!-forming processes and durability issues for the application of NiTi shape memory alloys in medical devices. In *Medical Device Materials: Proceedings from the Materials & Processes for Medical Devices Conference 2003*, Anaheim, CA, 247. American Society for Metals.

Nevitt M V. 1960. Stabilization of certain Ti2Ni-type phases by oxygen. *Transactions of the Metallurgical Society of AIME* 218: 327–331.

Nespoli A, Besseghini S, Pittaccio S, Villa E and Viscuso S. 2010. The high potential of shape memory alloys in developing miniature mechanical devices: A review on shape memory alloy mini-actuators. *Sensors and Actuators A: Physical* 158(1): 149–160.

Ölander A. 1932a. The crystal structure of AuCd. *Zeitschrift für Kristallographie* 83 A: 145–148.

Ölander A. 1932b. An electrochemical investigation of solid Cadmium-Gold alloys. *Journal of the American Chemical Society* 54(10): 3819–3833.

Olier P, Barcelo F, Bechade J L, Brachet J C, Lefevre E and Guenin G. 1997. Effects of impurities content (oxygen, carbon, nitrogen) on microstructure and phase transformation temperatures of near equiatomic TiNi shape memory alloys. *Journal de Physique IV* 7(C5): 143–148.

Otubo J, Rigo O D, Neto C D M, Kaufman M J, and Mei P R. 2003. NiTi shape memory alloy ingot production by EBM. *Journal de Physique IV* 112: 813–820.

Qidwai M A and Lagoudas D C. 2000. Numerical implementation of a shape memory alloy thermomechanical constitutive model using return mapping algorithms. *International Journal for Numerical Methods in Engineering* 47(6): 1123–1168.

Raynaerts D and Van Brussel H. 1991. Development of a SMA high performance robotic actuator. In: *Advanced Robotics, Fifth International Conference On Robots in Unstructured Environments 1991*, Pisa, Italy, *ICAR*, 61–66. IEEE.

Reinoehl M, Bradley D, Bouthot R and Proft J. 2000. The influence of melt practice on final fatigue properties of superelastic NiTi wires. In *SMST-2000: Proceedings of the International Conference on Shape Memory and Superelastic Technologies. Pacific Grove, CA: International Organization on SMST*, 397–403.

Rofooei F R and Farhidzadeh A. 2011. Investigation on the seismic behavior of steel MRF with shape memory alloy equipped connections. *Procedia Engineering* 14: 3325–3330.

Saadat S, Salichs J, Noori M, Hou Z, Davoodi H, Bar-On I, Suzuki Y and Masuda A. 2002. An overview of vibration and seismic applications of NiTi shape memory alloy. *Smart Materials and Structures* 11(2): 218–229.

Saleeb A F, Padula S A and Kumar A. 2011. A Multi-axial, multimechanism based constitutive model for the comprehensive representation of the evolutionary response of SMAs under general thermomechanical loading conditions. *International Journal of Plasticity* 27(5): 655–687.

Shameli E, Alasty A and Salaarieh H. 2005. Stability analysis and nonlinear control of a miniature shape memory alloy actuator for precise applications. *Mechatronics* 15(4): 471–486.

Silva E P D. 2007. Beam shape feedback control by means of a shape memory actuator. *Materials & Design* 28(5): 1592–1596.

Song G, Chaudhry V and Batur C. 2003. Precision tracking control of shape memory alloy actuators using neural networks and a sliding-mode based robust controller. *Smart Materials and Structures* 12(2): 223–231.

Song G, Ma N and Li H-N. (2006). Applications of shape memory alloys in civil structures. *Engineering Structures* 28(9): 1266–1274.

Tabesh M. 2010. Finite Element Analysis of Shape Memory Alloy Biomedical Devices. University of Toledo, Toledo, OH (Master's Thesis).

Taheri Andani M. 2013. Constitutive modeling of superelastic shape memory alloys considering rate dependent non-Mises tension-torsion behavior. University of Toledo, Toledo, OH (Master's Thesis).

Taheri Andani M, Alipour A and Elahinia M H. In press. Coupled rate dependent superelastic behavior of SMA bars induced by combined axial-torsional loading; A semi-analytical modeling. *Journal of Intelligent Material Systems and Structures.*

Taheri Andani, M., A. Alipour, A. Eshghinejad, and M. Elahinia. In press. Modifying the torque-angle behavior of rotary shape memory alloy actuators through axial loading: A semi-analytical study of combined tension-torsion behavior. *Journal of Intelligent Material Systems and Structures.*

Weinert K and Petzoldt V. 2004. Machining of NiTi based shape memory alloys. *Materials Science and Engineering: A* 378(1): 180–184.

Van der Wijst M W M, Schreurs P J G and Veldpaus F E. 1997. Application of computed phase transformation power to control shape memory alloy actuators. *Smart Materials and Structures* 6(2): 190–198.

Williams E, Shaw G and Elahinia M H. 2010. Control of an automotive shape memory alloy mirror actuator. *Mechatronics* 20(5): 527–534.

Wu M H. 2002. Fabrication of Nitinol materials and components. *Materials Science Forum* 394: 285–292.

Yuan B, Chung C Y, Zhang X P, Zeng M Q and Zhu M. 2005. Control of porosity and superelasticity of porous NiTi shape memory alloys prepared by hot isostatic pressing. *Smart Materials and Structures* 14(5): 201–206.

13

Thermoelectric Materials and Generators: Research and Application

Vijay K. Varadan, Linfeng Chen, Jungmin Lee, Gyanesh N. Mathur, Hyun Jung Kim, and Sang H. Choi

CONTENTS

13.1 Introduction

As the world strives to meet a huge electricity demand, sustainable energy technologies are attracting significant attentions. Approximately 90% of the power in the world is generated by heat engines that use fossil fuel combustion as a heat source. These engines typically operate at 30–40% efficiency, and around 60% of energy produced by fossil fuel combustion is wasted in the form of heat (Yazawa and Shakouri, 2011; Vieira and Mota, 2009). Waste heat is an untapped major source for sustainable energy.

In recent years, due to the rapid advances in nanotechnologies, great achievements in the research on electrical energy production based on thermoelectricity have been made, and thermoelectric generators (TEGs) have been in waste heat recovery, solar energy harvesting, and deep space exploration. In this chapter, the latest advances in the research of thermoelectric materials, development of TEGs, and their applications in energy harvesting will be reviewed.

13.1.1 Thermoelectricity and Thermoelectric Generators

Thermoelectricity is a two-way process. It can refer either to the conversion of a temperature difference between two sides of a material into electrical energy, or to the conversion of an electric current through a material into a temperature difference between its two sides. The first part of the thermoelectric effect is often called the Seebeck effect. The reverse phenomenon is often called the Peltier effect. Based on the Peltier effect, an object can be heated or cooled by running an electric current through it, without combustion or moving parts. Usually, thermoelectric materials are solid-state conductors or semiconductors. Thermoelectric effects are caused by the charge carriers (either electrons or holes) within a thermoelectric material, diffusing from the hotter side to the cooler side.

This chapter concentrates on the direct conversion of thermal energy into electrical power based on the Seebeck effect. Thermoelectric power generation has the advantages of being maintenance free, silent in operation, involving no moving mechanical parts, and producing no greenhouse gas (Francioso et al., 2010).

13.1.1.1 Thermoelectric Parameters

Various thermoelectric parameters have been defined for thermoelectric materials and thermoelectric devices, and these parameters quantitatively describe the efficiency and capacity of a thermoelectric material or device to convert thermal energy into electric power.

1. *Thermoelectric parameters for thermoelectric materials*: For a thermoelectric material with temperature difference ΔT at its two sides, the produced voltage (ΔV) or thermoelectromotive force (EMF) due to the Seebeck effect is given by (Rowe, 2006b):

$$\Delta V = S\Delta T \tag{13.1}$$

   where S is the Seebeck coefficient. Equation 13.1 indicates that, under the same temperature difference, a thermoelectric material with a higher Seekbeck coefficient can produce a higher voltage. The power factor of a thermoelectric material is defined by

$$PF = S^2\sigma \tag{13.2}$$

   where σ is the electrical conductivity of the material.

   The thermoelectric conversion efficiency of a material is related to a dimensionless parameter, called figure of merit (ZT) defined by

$$ZT = \frac{PF}{\kappa}T = \frac{S^2\sigma}{\kappa} = \frac{S^2\sigma}{\kappa_L + \kappa_e}T \tag{13.3}$$

   where κ is the thermal conductivity of the material, and T is the absolute temperature of the material. The thermal conductivity ($\kappa = \kappa_L + \kappa_e$) is the sum of the lattice thermal conductivity (κ_L), known as phonon transmission, and the electronic thermal conductivity (κ_e). Equation 13.3 indicates that the thermal and electrical properties of a material affect its thermoelectric performances (Yazawa and Shakouri, 2011). In some cases, the figure of merit of a thermoelectric material is defined as

$$Z = \frac{PF}{\kappa} = \frac{S^2\sigma}{\kappa} \tag{13.4}$$

   In this definition, the figure of merit Z has a unit of K^{-1}.
2. *Thermoelectric parameters for thermoelectric devices*: Figure 13.1 shows a thermoelectric power converter with two types of thermoelectric elements (so-called thermolegs), a p-type semiconductor element and an n-type semiconductor element. In the heated n-type semiconductor element, the electrons in the

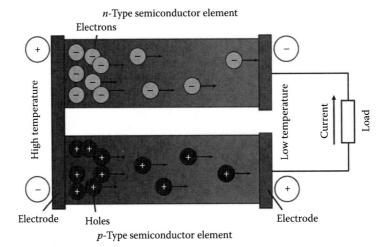

FIGURE 13.1
Mechanism of power converter in a thermoelectric module with π-type structure. (From Kawamoto, H, 2009. R&D trends in high efficiency thermoelectric conversion materials for waste heat recovery, *Science and Technology Trends*, 4, 54–69.)

high-temperature region are thermally energized to have high kinetic energy with increased potential, and they are transferred to the low-temperature region with lower potential. A thermal electromotive force is thus generated, and the high-temperature side reaches a high electric potential. On the other hand, in the p-type semiconductor element, the holes in the high-temperature region migrate to the low-temperature region, generating thermal electromotive force, and the low-temperature side achieves a high potential (Kawamoto, 2009). As these two semiconductor elements are electrically at their high-temperature ends, an output voltage is established between the two low-temperature ends.

Under a temperature ΔT, the output voltage ΔV of a TEG is given by

$$\Delta V = (S_p - S_n)\Delta T = S_{pn}\Delta T \tag{13.5}$$

where S_p and S_n are the Seebeck coefficients of the p-type element and n-type element respectively, and S_{pn} is the Seebeck coefficient of the element pair p–n. A thermoelectric converter is a heat engine. Like all heat engines it obeys the laws of thermodynamics. If we take the converter as an ideal generator in which there are no heat losses, its conversion efficiency is defined as the ratio of the electrical power delivered to the load to the heat absorbed by the converter (Kim et al., 2013; Rowe, 2006b).

The ideal thermoelectric conversion efficiency η_0 of a thermoelectric converter is given by

$$\eta_0 = \left[\frac{\sqrt{1 + ZT} - 1}{\sqrt{1 + ZT} + (T_c/T_h)} \right] \tag{13.6}$$

where T_c and T_h are the temperatures at the cold and hot ends, respectively. The maximum efficiency η_{TE} of a thermoelectric converter is given by

$$\eta_{TE} = \eta_c \cdot \eta_0 = \eta_c \left[\frac{\sqrt{1 + ZT} - 1}{\sqrt{1 + ZT} + (T_c / T_h)} \right] \tag{13.7}$$

where the Carnot efficiency η_c is given by

$$\eta_c = \frac{T_h - T_c}{T_h} \tag{13.8}$$

Figure 13.2 shows the conversion efficiency as a function of operating temperature and Carnot efficiency. A larger temperature difference corresponds to a larger Carnot efficiency. Figure 13.2 indicates that a thermoelectric convertor has higher conversion efficiency under a larger temperature difference, and under the same temperature difference, a thermoelectric convertor with a higher figure of merit has higher conversion efficiency. As a ballpark figure, a thermoelectric convertor fabricated from thermoelectric materials with an average figure-of-merit of $3 \times 10^{-3}\,K^{-1}$ would have a conversion efficiency around 20% when it operates under a temperature difference of 500 K (Rowe, 2006b). Practically the figure of merit 1 based on Si–Ge alloy that runs at 873°K hot and 644°K cold bases ($\eta_c = 26\%$) is equivalent to roughly 7% thermal efficiency after accounting all other losses and leaks through junction structures (Gross, 1961; Yazawa and Shakouri, 2012).

Figure 13.3 shows the relationship between the theoretical generating efficiency (conversion efficiency) of a thermoelectric convertor and the figure of merit of the materials used for fabricating the convertor, when the Carnot efficiency is 50%. Evidently, an increase of figure of merit (ZT) results in an increase of generating efficiency. For a thermoelectric converter made from thermoelectric materials with average figure of merit (ZT) around one, its theoretical generating efficiency is approximately 9% when the Carnot efficiency

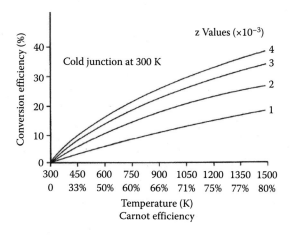

FIGURE 13.2

Conversion efficiency as a function of temperature and Carnot efficiency. (From Kawamoto, H, 2009. R&D trends in high efficiency thermoelectric conversion materials for waste heat recovery, *Science and Technology Trends*, 4, 54–69.)

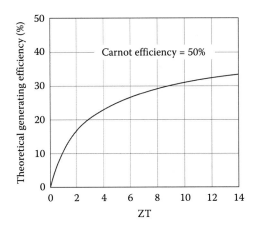

FIGURE 13.3

Relationship between theoretical generating efficiency and figure of merit. (From Kawamoto, H, 2009. R&D trends in high efficiency thermoelectric conversion materials for waste heat recovery, *Science and Technology Trends, 4*, 54–69.)

(η_c) is 50% (Kawamoto, 2009). If the materials have a figure of merit (ZT) around 2–3, the corresponding theoretical generating efficiency is around 15–20% (Kim et al., 2013).

13.1.1.2 Thermoelectric Generators

TEGs have been used to convert thermal energy into electrical energy in remote locations, such as in rural areas and on rockets where connection to a central power grid is not possible. They are also used for harnessing the heat wasted in cars and power plants, and heat produced by human body and computer chips.

In a thermoelectric power generator, to maximize the generated output voltage, several thermoelectric conversion units are connected electrically in series with each other and thermally in parallel to form a thermoelectric module (Francioso et al., 2010; Gou et al., 2010; Yazawa and Shakouri, 2011). As shown in Figure 13.4a, a general TEG with a load resistance RL connected is composed of several thermoelectric conversion units. Each unit consists of a pair of p- and n-type semiconductor legs which work between the high-and low-temperature heat reservoirs whose temperatures are T_H and T_L respectively. The heat flux between the generator and the two heat reservoirs are presented by Q_H and Q_L, which are the thermal energy the generator absorbs from the high-temperature reservoir and the thermal energy it releases to the low-temperature reservoir per unit time, respectively (Gou et al., 2010). The basic structure of a TEG is shown in Figure 13.4b.

A TEG, with n thermoelectric units connected electrically in series and thermally in parallel, is able to generate n times the output voltage of one thermoelectric unit. With optimal impedance matching, such a TEG can provide a maximum output electric power, P_{max}, given by (Gou et al., 2010)

$$P_{max} = \frac{(nS_{pn}\Delta T)^2}{4R_0} \tag{13.9}$$

where R_0 is the internal electrical resistance of the generator.

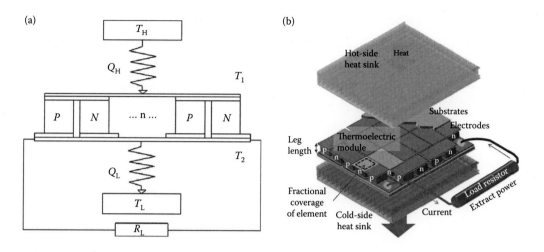

FIGURE 13.4
(a) Schematic diagram of multielement TEG (Gou et al., 2010), (b) the basic structure of a TEG. (From Yazawa, K, and Shakouri, A, 2011. *Environmental Science and Technology*, 45, 7548–7553.)

13.1.2 Choice and Optimization of Thermoelectric Materials

It can be found from Figures 13.2 and 13.3 that the power-generating efficiency of a TEG is closely related to the thermoelectric materials from which the TEG is fabricated. The choice and optimization of thermoelectric materials are essential for the development of high-efficiency TEGs.

13.1.2.1 Typical Thermoelectric Materials

Various types of thermoelectric materials preforming optimally at different temperature ranges have been used in the development of thermoelectric devices for different applications. According to their optimal temperature ranges, thermoelectric materials can be classified into low-temperature, intermediate-temperature, and high-temperature thermoelectric materials. Figure 13.5 shows the temperature dependence of the ZT values of typical thermoelectric materials (Kawamoto, 2009). The figure of merit of a thermoelectric material tends to increase with temperature and then decrease after reaching a maximum value. The ZT values of p-type and n-type Bi_2Te_3 and Zn_4Sb_3 reach their maximum values around 1.0–1.2 at the temperatures around 300–400 K, while a compound of $AgSbTe_2$/ GeTe shows a ZT value around 1.2 at 700 K and $Si_{0.2}Ge_{0.8}$ shows a ZT value of 0.7 at approximately 1100 K (Kim et al., 2012, 2013). Generally speaking, in the temperature region less than 500 K, Bi–Te-based compounds display high ZT. In the intermediate temperature region of 700–900 K, $AgSbTe_2$/GeTe and $CeFe_4CoSb_{12}$ are high ZT thermoelectric conversion materials, and in the high-temperature region above 900 K, the high ZT materials are $Si_{0.2}Ge_{0.8}$, $Bi_2Sr_2Co_2Oy$, and $Ca_3Co_4O_9$.

As shown in Figure 13.6, materials with high-thermoelectric figures of merit are typically heavily doped semiconductors (Shakouri, 2011). Insulators have poor electrical conductivity, σ, while metals have relatively low Seebeck coefficients. The thermal conductivity κ of a metal, which is dominated by energetic free electrons, is proportional to the electrical conductivity. In semiconductors, the thermal conductivity consists of the contributions from electrons (κ_e) and phonons (κ_L), and the majority contribution comes from phonons

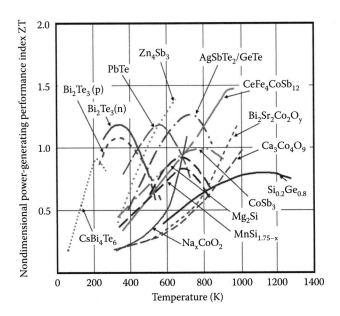

FIGURE 13.5
Temperature dependency on nondimensional power-generating performance index of main TE conversion materials. (From Kawamoto, H, 2009. *Science and Technology Trends*, 4, 54–69.)

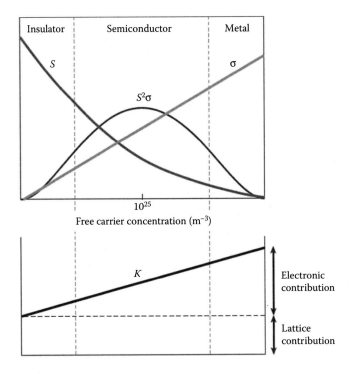

FIGURE 13.6
Trade-off between electrical conductivity (σ), Seebeck coefficient (S), and thermal conductivity (κ) that involves increasing the number of free carriers from insulators to metals. (From Shakouri, A, 2011. *Annual Review of Materials Research*, 41, 399–431.)

(Cahill, 1989). The phonon thermal conductivity can be reduced without causing too much reduction of the electrical conductivity, for example, by alloying or morphological realignment. The traditional TE cooling materials are alloys of Bi_2Te_3 with Sb_2Te_3 (such as $Bi_{0.5}Sb_{1.5}Te_3$; p type) and of Bi_2Te_3 with Bi_2Se_3 (such as $Bi_2Te_{2.7}Se_{0.3}$; n type), with a ZT at room temperature approximately equal to one. A typical power generation material is an alloy of PbTe (or SiGe), with a ZT around 1 (or 0.6) at 500°C (or 700°C).

13.1.2.2 Optimization of Thermoelectric Materials

High-performance TE materials require an ideal combination of the electrical and thermal properties. High electrical and low thermal conductivities produce high transport of electrons and low heat losses, respectively. One major challenge in the development of high-efficiency thermoelectric material is to achieve simultaneously high power factor ($S^2\sigma$) and low thermal conductivity ($\kappa = \kappa_L + \kappa_e$). For a thermoelectric material, its Seebeck coefficient, electrical conductivity, and thermal conductivity are not independent, since they are determined by the details of the electronic structure and scattering of charge carriers. The increase in electrical conductivity usually causes the increase of thermal conductivity at the same time. For semiconductors, κ_L is dominant in the total thermal conductivity (Slack and Hussain, 1991), and many efforts have been made to decrease their thermal conductivity for the improvement of their ZT values (Kim et al., 2012).

One strategy for reducing the lattice thermal conductivity of alloys is the introduction of scattering agencies, for example, alloy scatters. This strategy has employed the so-called "phonon glass electronic crystal" in which crystal structures containing weakly bound atoms or molecules that 'rattle' within an atomic cage should conduct heat like in a glass, but conducting electricity like in a crystal (Kim et al., 2013). Many of the recent high figures of merit materials similarly achieve a reduced lattice thermal conductivity through the disorder within unit cells. This disorder is achieved through interstitial sites, partial occupancies, or rattling atoms in addition to the disorder inherent in alloys. Another strategy is to use complex crystal structures to separate the electron crystal from the phonon glass. The exceptionally low thermal conductivity was attributed to the crystallinity of each layer combined with the randomness in the alignment between different layers. The impact of the superlattice structure on the thermal conductivity has been attributed to various effects, including modification of the phonon spectrum like bandgap formation and phonon localization, diffuse or specular scattering of phonons at interfaces due to the fact that the orientation of a layer lattice structure is different than the other layer and scattering of phonons at defects like dislocation induced lattice mismatch.

The Seebeck coefficient can be increased by the nanoparticle/matrix interface that creates a low electron energy filtering effect. The energy level bending at the interfaces between metallic nanoparticle and bulk TE materials causes low-energy electrons to be strongly scattered while allowing high-energy electrons unaffected for tunneling. This type of energy filtering precisely prescribes to increase the Seebeck coefficient of TE materials since the Seebeck coefficient depends on the energy derivative of the relaxation time at the Fermi energy. Another approach for the enhanced thermoelectric power factor is the carrier energy filtering. By introducing tall barriers in the conduction band (for n-type materials) or the valence band (for p-type materials), the higher-energy hot carriers can be selectively transmitted through the structure by filtering out the low-energy carrier. This can increase the Seebeck coefficient because its value depends on the thermal energy transported by the carriers (Kim et al., 2013).

13.1.2.3 Nanoscale Effects

By reducing its grain size, the thermal conductivity of a thermoelectric material can be decreased, and thus its ZT can be enhanced (He and Tritt, 2010). The concept of "nanostructured bulk" is often used in developing new thermoelectric materials: reduce the lattice thermal conductivity in a thermoelectric material by increasing phonon scattering over a large mean-free path range while maintaining good electrical properties. Theoretically, nanostructuring can introduce a large density of interfaces in which phonon scattering can occur without having to compromise carrier mobility values, due to the much shorter mean-free path of electrons versus phonons in heavily doped semiconductors. Theoretical modeling indicated that 90% of the thermal conductivity accumulation in Si is due to phonons that have a mean-free path greater than 20 nm (Borca-Tasciuc et al., 2001). Therefore, if the grain size were reduced to 20 nm, a 90% reduction in the lattice thermal conductivity could be achieved. The carrier mobility should not be affected as the electron mean-free path in Si was calculated to be only a few nanometers (Nolas et al., 2000). The guidelines for the nanoparticle dimensions vary for different material systems, but the critical feature sizes necessarily remain in the nanoscale (5–50 nm). To fully optimize the thermoelectric properties, composite structures with controlled distributions of nanoscale particles discrete in size and possibly in composition may be needed (Bux et al., 2010).

In the second part of this chapter, the fabrication and properties of typical thermoelectric nanomaterials will be discussed. TEGs will be discussed subsequently, followed by the discussion on TEG systems. After that, the applications of TEGs and TEG systems will be discussed. Concluding remarks will be given at the end of this chapter.

13.2 Thermoelectric Nanomaterials

For quite a long time, thermoelectrics have long been too inefficient to be cost-effective in most applications. The resurgence of interest in thermoelectrics began in the mid-1990s when theoretical studies suggested that thermoelectric efficiency could be greatly enhanced through nanostructural engineering, which led to experimental efforts to demonstrate the proof-of-principle and high-efficiency materials (Chen et al., 2003; Dresselhaus et al., 2007; Snyder and Toberer, 2008).

Many methods have been applied for producing nanomaterials, such as electrochemical synthesis (Hochbaum et al., 2008), chemical vapor deposition (Swihart, 2003), molecular beam epitaxy (Boukai et al., 2008), pulsed laser ablation (Swihart, 2003), and pyrolysis of volatile precursors (Reau et al., 2007). Although, these methods have been demonstrated to produce beautiful nanostructured materials of various geometries and complexities, they have inherent limitations that often include complex and expensive equipment, toxic precursors and scale-up challenges. Alternative methods for synthesizing large-scale quantities of nanostructured materials include solvothermal/hydrothermal methods, solution-based synthesis, self-assembly/nanoprecipitation techniques, and high-energy ball milling (Bux et al., 2010).

In the following, we discuss the synthesis and properties of four types of thermoelectric nanomaterials, including nanoparticles, nanorods and nanowires, superlattice structures, and complex nanomaterials. As most of their applications, thermoelectric nanomaterials are made into nanostructured bulks (nanobulks), the consolidation and densification of thermoelectric nanomaterials will be discussed at the end of this part.

13.2.1 Nanoparticles

Nanoparticles are often called zero-dimensional nanomaterials, whose three dimensions are quantum mechanically confined. In the following, Bi_2Se_3 nanoflakes and Bi_2Te_3 nanoparticles will be discussed.

13.2.1.1 Bi_2Se_3 Nanoflakes

Semiconductors having narrow band gap and high mobility carriers are best suited as thermoelectric materials. Bismuth selenide (Bi_2Se_3) is a V–VI semiconductor with a narrow band gap of about 0.3 eV (Lin et al., 2007). Among the various synthesis techniques employed for the formation of Bi_2Se_3 nanostructures, the solvothermal/hydrothermal process is attracting much interest due to the advantages of high yield, low synthesizing temperature, high purity, and high crystallinity (Kadel et al., 2011).

Kadel et al. synthesized Bi_2Se_3 nanoflakes via solvothermal route at different synthesis conditions using DMF as solvent (Kadel et al., 2011), and the surface morphology and crystal structure of the synthesized nanoflakes were shown in Figure 13.7. The calculated fringe separation is 0.311 nm, which corresponds to the d-spacing of (015) plane of rhombohedral Bi_2Se_3. Figure 13.7d shows the SAED spot pattern that is indexed to the corresponding lattice planes of rhombohedral Bi_2Se_3.

Figure 13.8 shows the temperature dependence of thermoelectric properties of synthesized Bi_2Se_3 (Kadel et al., 2011). The magnitude of the Seebeck coefficient for the sample at room temperature (300 K) is about 1.15×10^{-4} V/K, which is about two times as much as that for bulk Bi_2Se_3. The increase in the Seebeck coefficient is due to the quantum confinement of electrons induced by nanostructures. Figure 13.8b indicates that the

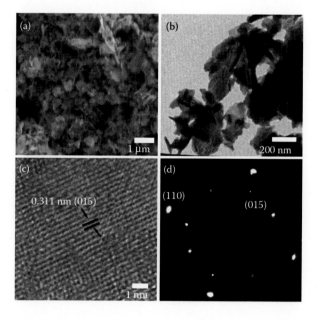

FIGURE 13.7
SEM and TEM images of the as-prepared Bi_2Se samples synthesized in DMF at 200°C for 24 h. (a) SEM image, (b) TEM image, (c) HRTEM image, and (d) SAED pattern. (From Kadel, K et al., 2011. *Nanoscale Research Letters*, 6, article no. 57.)

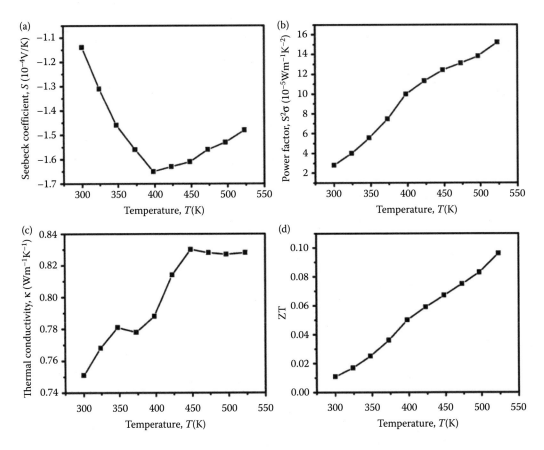

FIGURE 13.8
Temperature dependence of thermoelectric properties of sample prepared in DMF at 200°C for 24 h. (a) Seebeck coefficient (S) versus temperature (T), (b) power factor ($S^2\sigma$) versus temperature (T), (c) thermal conductivity (κ) versus temperature (T), and (d) figure of merit (ZT) versus temperature (T). (From Kadel, K et al., 2011. *Nanoscale Research Letters*, 6, article no. 57.)

power factor ($\sigma^2 s$) increases with the temperature. The increase in power factor with the temperature can be attributed to the increase in the electrical conductivity with temperature due to the semiconducting nature of the Bi_2Se_3 nanostructures. The maximum value of the power factor is 15.2×10^{-5} $Wm^{-1}K^{-2}$ at 523 K, and the room temperature value is 2.8×10^{-5} $Wm^{-1}K^{-2}$, which is comparable to the room temperature power factor value of about 7×10^{-5} $Wm^{-1}K^{-2}$ of Bi_2Se_3 nanoplates (Lin et al., 2007). The variation of thermal conductivity (κ) with temperature is shown in Figure 13.8c. The lowest value of the thermal conductivity, 0.751 $Wm^{-1}K^{-1}$, is recorded at room temperature, which is lower than that for the bulk Bi_2Se_3, 4 $Wm^{-1}K^{-1}$. The reduction in thermal conductivity of Bi_2Se_3 nanostructures is mainly due to the interface or boundary scattering of phonons in nanostructures. Figure 13.8d shows the plot of ZT versus temperature in the range of 300–523 K, indicating a nearly linear increase in ZT with the temperature. The maximum ZT value is 0.096 at 523 K, and the room temperature ZT value is 0.011. The as-prepared Bi_2Se_3 nanoflakes exhibit a higher Seebeck coefficient and a low thermal conductivity compared with the bulk counterpart at room temperature.

13.2.1.2 *Bi₂Te₃ Nanoparticles*

Nanoparticles have been predicted to show a strong scattering effect on phonons similar to that of atomic impurities or crystal boundaries. The effect was found to be inversely related to the nanoparticle diameter. Since phonon scattering depends on the size and shape of the nanoparticles, a good control over these parameters is essential in achieving further improvements in TE efficiencies. Several solution-based attempts to a more controlled synthesis of large amounts of small, crystalline bismuth telluride nanoparticles have been reported (Dirmyer et al., 2009; Lu et al., 2005; Purkayastha et al., 2006, 2008).

Scheele et al. synthesized sub-10-nm single-crystalline Bi_2Te_3 nanoparticles of narrow size distribution, as shown in Figure 13.9 (Scheele et al., 2009). To meet the key requirement for a sufficient electrical conductivity and thus a large power factor, they introduced a novel hydrazine hydrate-based ligand removal prior to the sintering of the nanoparticles to macroscopic pellets. This yielded an electrical conductivity that is virtually identical to typical n-type bulk samples. The total thermal conductivity of such nanoparticle pellets is by as much as one order of magnitude smaller than that of the bulk material. The power factor of 5 μW K^{-2} cm^{-1} is much higher than the samples purely made from solution grown Bi_2Te_3 nanoparticles. The transport properties of Bi_2Te_3 nanoparticles are shown in Figure 13.10 (Scheele et al., 2009).

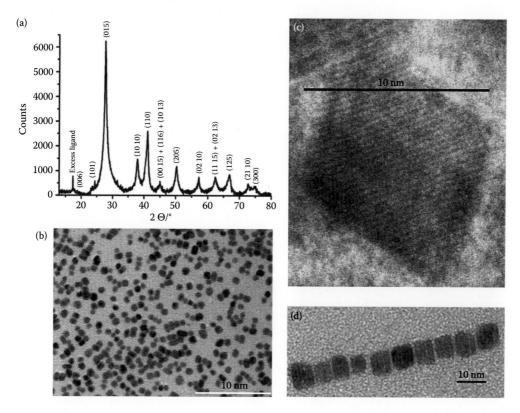

FIGURE 13.9
Bi_2Te_3 nanoparticles. (a) XRD; (b) TEM; (c, d) HRTEM of typical nanoparticles. (From Scheele, M et al., 2009. *Advanced Functional Materials*, 19, 3476–3483.)

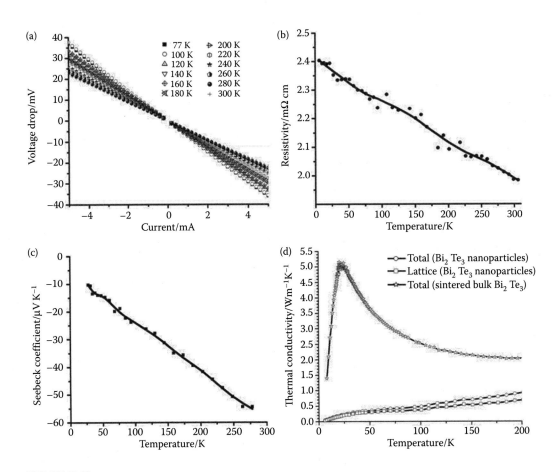

FIGURE 13.10
Transport properties of Bi_2Te_3 nanoparticles: (a) DC I/V-measurements, (b) AC resistivity, (c) thermopower, and (d) total and lattice thermal conductivity of a pellet of sintered nanoparticles. The thermal conductivity data includes a sintered bulk Bi_2Te_3 sample for comparison. (From Scheele, M et al., 2009. *Advanced Functional Materials*, 19, 3476–3483.)

13.2.2 Nanorods and Nanowires

Nanorods and nanowires are one-dimensional nanomaterials, whose two dimensions are quantum mechanically confined. Such materials exhibit strong shape anisotropy. In the following, we discuss Bi_2S_3 polycrystalline nanorods and Bi_2Te_3 ultrathin nanowires.

13.2.2.1 Bi_2S_3 Polycrystalline Nanorods

Bismuth sulfide (Bi_2S_3), a direct band gap material with Eg around 1.3 eV, attracts much attention in TE research as a potential material with low cost and low toxicity (Ge et al., 2012a). Chen et al. first reported the TE properties of Bi_2S_3, which has a high Seebeck coefficient (about 500 $\mu V\ K^{-1}$) and a low thermal conductivity (<1 $Wm^{-1}\ K^{-1}$), and got a peak ZT value of just 0.055 at 250 K due to the high electrical resistivity (Chen et al., 1997). The electrical resistivity of Bi_2S_3 is about two orders of magnitude higher than that of Bi_2Te_3 compounds. Element doping is an effective method to improve the carrier concentrations and reduce electrical resistivity, which was done in the polycrystalline Bi_2S_3 (Ge et al., 2012b). Controlling the microstructure and improving carrier mobility are another direction for enhancing the TE properties of polycrystalline Bi_2S_3. The fine grain size contributes to the low thermal conductivity and large Seebeck coefficient, but leads to an increase in the electrical resistivity. The coarse grain size contributes to low electrical resistivity, but causes the increase of thermal conductivity. Although many works have been done to optimize grain size for improving the TE properties of bulk and film materials, they just find a suitable grain size which could optimize TE properties, and could not fully utilize the characteristics of fine and coarse grains.

Ge et al. prepared polycrystalline Bi_2S_3 with a composite-like microstructure by spark plasma sintering (SPS) using the mixed precursor powders with Bi_2S_3 nanorods and mechanic alloying (MA)-derived Bi_2S_3 nanopowders (Ge et al., 2012a). Based on the fact that an electron selects its transport path but a phonon does not, they designed a composite like microstructure, as shown in Figure 13.11, to improve the TE transport properties of poly-crystalline Bi_2S_3. In this design, fine grains contribute to the low thermal conductivity, while strip-like grains with a rapid electron transport path contribute to the low electrical resistivity. The c-axis oriented single crystal nanorods with a good electron transport path along the c-axis direction were used as road materials, while the MA-treated nanopowders were used as matrix materials. Both of them were mixed in different weight fractions and consolidated by SPS technique, which is a rapid sintering process and could suppress the grain growth.

As shown in Figure 13.12, both HRTEM and SAED results indicate that the growth direction of the crystalline Bi_2S_3 nanorods is highly consistent with the crystallographic c-axis. Bi_2S_3 is featured with a layered structure that is easily cleaved in the plane parallel to the c-axis direction. The interplane parallel to the c-axis is considered as a good path for electron transport. Cantarero et al. proved that the carrier mobility of the Bi_2S_3 crystal along the c-axis direction is higher than that along the a-axis direction (Cantarero et al., 1987).

Figure 13.13 shows the temperature dependence of thermal conductivity, lattice thermal conductivity, and ZT value for Bi_2S_3 bulks with different nanorod contents (Ge et al., 2012a). The thermal conductivity (κ) of all the samples reduces with increasing the temperature due to the strong phonon scattering caused by the lattice thermal vibration. Due to the enhancement of the lattice thermal vibration with the increase of the temperature, lattice thermal conductivities (κ_L) will be reduced strongly, leading to a reduced κ. The κ values of the bulk-10 sample were from 0.79 to 0.59 W m^{-1} K^{-1} when the temperature was increased

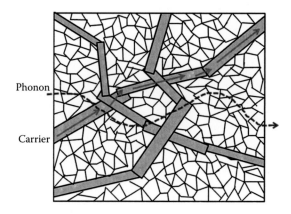

FIGURE 13.11
Design drawing of the composite-like microstructure in Bi_2S_3 thermoelectric polycrystal. (From Ge, ZH, Zhang, BP, and Li, JF, 2012a. *Journal of Materials Chemistry*, 22, 17589–17594.)

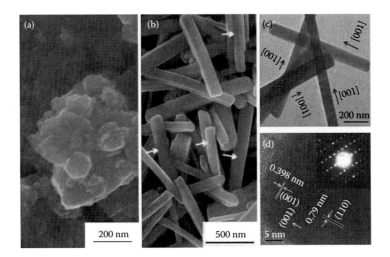

FIGURE 13.12
FESEM image of the Bi_2S_3 MA-powders (a) and Bi_2S_3 nanorods (b); TEM image (c), the corresponding SAED and HRTEM patterns (d); for the Bi_2S_3 nanorods. (From Ge, ZH, Zhang, BP, and Li, JF, 2012a. *Journal of Materials Chemistry*, 22, 17589–17594.)

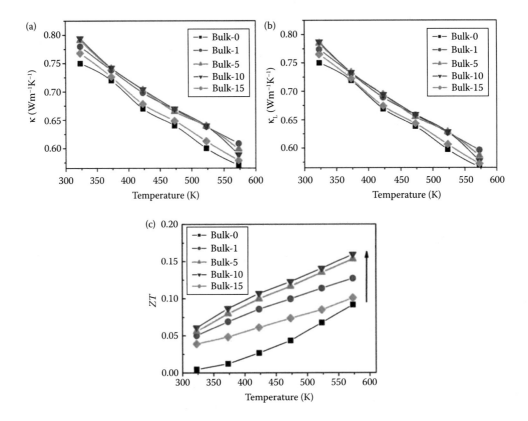

FIGURE 13.13
Temperature dependence of the thermal conductivity (a), lattice thermal conductivity (b) and ZT value (c) for Bi_2S_3 bulks with different nanorod contents. (From Ge, ZH, Zhang, BP, and Li, JF, 2012a. *Journal of Materials Chemistry*, 22, 17589–17594.)

from 323 to 573 K. The temperature dependence of the ZT value is shown in Figure 13.13c. The maximum ZT value of 0.16 was obtained for the bulk-10 sample at 573 K.

13.2.2.2 Bi₂Te₃ Ultrathin Nanowires

Nanowires are among the most promising routes to obtain higher ZT values through the sharp enhancement of electron density of states due to quantum confinement and the significant reduction of thermal conductivity due to increased surface/interface scattering of phonons (Hicks and Dresselhaus, 1993; Lin and Dresselhaus, 2003). There is a strong relationship between thermoelectric properties and nanowire diameter, and the ZT value could even be enhanced to higher than six if nanowires with diameters around 5 nm could be achieved (Zhang et al., 2012). Many effective routes to fabricate various ultrathin nanowires have been investigated, including noble metals (Huo et al., 2008), sulfides (Cademartiri et al., 2008), and oxides (Xi and Ye 2010).

Zhang et al. proposed a new approach to improve the ZT of n-type Bi_2Te_3 through a two-step synthesis of ultrathin n-type Bi_2Te_3 nanowires with an average diameter around 8 nm (Zhang et al., 2012). The TEM images and size distribution analyses of the synthesized Te-rich Bi_2Te_3 nanowires are shown in Figure 13.14. The simplicity, scalability, and extremely high yield of the nanowires with uniform diameter have made it possible to use SPS to consolidate nanowire powder into bulk pellets to test their thermoelectric performance.

Figure 13.15 shows the typical electrical and thermal properties of the nanowire bulk pellets after SPS [40]. The low electrical conductivity of nanowire composites is mainly

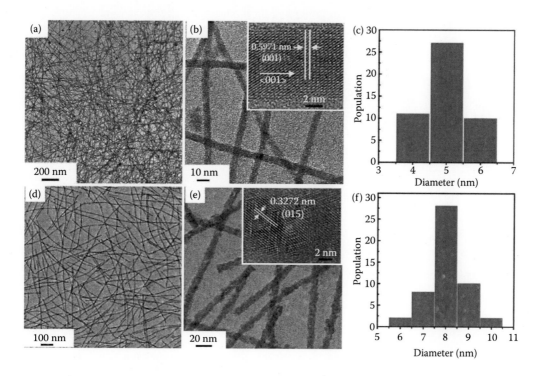

FIGURE 13.14
TEM images and size distribution analyses for (a – c) Te and (d – f) Te-rich Bi_2Te_3 nanowires. The insets in (b) and (e) are HRTEM images for Te and Bi2Te3 nanowires, respectively. (From Zhang, GQ et al., 2012. *Nano Letters*, 12, 56–60.)

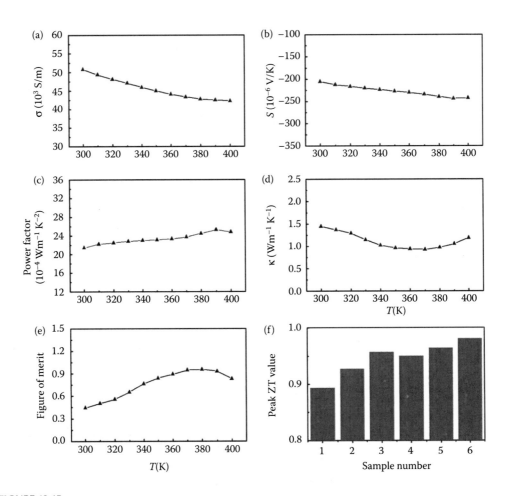

FIGURE 13.15
Thermoelectric property measurement of Bi_2Te_3 nanowire composites after SPS treatment: temperature dependence of (a) electrical conductivity; (b) Seebeck coefficient; (c) power factor; (d) thermal conductivity; (e) ZT calculation; and (f) typical peak ZT value distributions observed from multiple Bi_2Te_3 nanowire bulk pellets. (From Zhang, GQ et al., 2012. *Nano Letters*, 12, 56–60.)

due to the smaller diameter/grain size in ultrathin nanowires. The absolute value of the Seebeck coefficient gradually increases from 205 µV/K at 300 K to 245 µV/K at 400 K. A power factor ($S^2\sigma$, Figure 13.15c) of 21.4×10^{-4} $Wm^{-1}K^{-1}$ at room temperature is achieved, and it gradually increases to 25.2×10^{-4} $Wm^{-1}K^{-1}$ at 390 K mainly due to the enhancement of Seebeck coefficient along with increasing temperature. The thermal conductivity is 1.42 $Wm^{-1}K^{-1}$ at 300 K and decreases to 0.92 $Wm^{-1}K^{-1}$ at 370 K. After that, the thermal conductivity starts to increase and reaches 1.19 $Wm^{-1}K^{-1}$ at 400 K. The value of thermal conductivity of the nanowire bulk pellets is much lower than that of the best n-type commercial $Bi_2Te_{2.7}Se_{0.3}$ single crystals (around 1.65 $Wm^{-1}K^{-1}$). The ZT of the Bi_2Te_3 nanowire composites is shown in Figure 13.15e. The peak ZT value is around 0.96 at 380 K, corresponding to a 13% enhancement compared to that of the best n-type commercial $Bi_2Te_{2.7}Se_{0.3}$ single crystals (around 0.85). More significantly, the statistic distribution of ZT values (Figure 13.15f) measured from multiple nanowire bulk pellets is quite narrow (within 10%), which further proves the uniformity of the nanowires and provides a reliable and reproducible manufacture route for high-performance thermoelectric devices.

13.2.3 Superlattice Structures

In general, superlattice-integrated TE devices have potentially higher efficiency than those based on bulk materials due to quantum and classical size effects on electron and phonon transport. One of the dominant reasons for increasing ZT lies in the reduction of thermal conductivity. The reduction of thermal conductivity is mainly attributed to three mechanisms (Zhang et al., 2007): (1) increased scattering of the lattice vibrations due to the large number of interfaces, (2) increased diffusive reflection on the interface due to mismatches in material properties, and (3) wave interference that is created by the periodic nature of the structure.

Nanoscale thermoelectric materials, using phonon-blocking, electron-transmitting superlattices in thin-film form (Venkatasubramanian et al., 2001), using quantum-dot superlattices in thick-film form (Harman et al., 2002) and using nanoscale inclusions in bulk material form (Hsu et al., 2004) increasingly appear to be the route to achieving enhanced ZT in thermoelectric materials (Venkatasubramanian et al., 2006). In the following, two types of superlattices will be discussed, and they are InGaAs/InGaAsP superlattices and Bi_2Te_3/Sb_2Te_3 superlattice.

13.2.3.1 InGaAs/InGaAsP Superlattice

Zhang et al. studied InGaAs/InGaAsP superlattices because of the high-quality interfaces that can be achieved between the alternating layers (Zhang et al., 2012). A set of InGaAs/InGaAsP superlattices with different periodic lengths were fabricated by metal–organic chemical vapor deposition (MOCVD). The experimental results shown in Figure 13.16 indicate that the thermal conductivity decreases with the reduction of periodic length for comparatively long period superlattices and reaches a minimum at the periodic length of 20 nm; then, it increases when the periodic length becomes even shorter (Zhang et al., 2012).

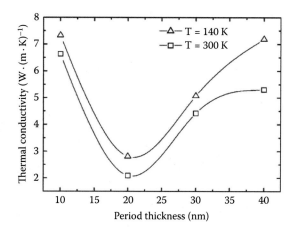

FIGURE 13.16
Thermal conductivity as a function of the periodic length for InGaAs/InGaAsP superlattice. The periodic length is 10, 20, 30, and 40 nm, corresponding to 5/5, 15/5, 15/15, and 30/10 structures, respectively. 5/5 means that the film thickness of the InGaAs layer is 5 nm, and the InGaAsP layer is 5 nm thick. (From Zhang, GQ et al., 2012. *Nano Letters*, 12, 56–60.)

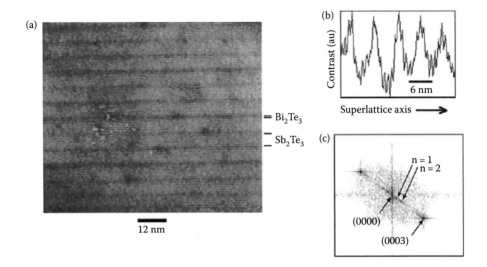

FIGURE 13.17

(a) TEM image of a 10 Å Bi_2Te_3/50 Å Sb_2Te_3 superlattice grown by MOCVD; (b) image contrast oscillations at 1 nm scale; and (c) Fourier transform of the image showing superlattice electron diffraction spots. (Venkatasubramanian, R et al., Energy harvesting for electronics with thermoelectric devices using nanoscale materials, *IEEE International Electron Devices Meeting*, 367–370 © 2007 IEEE.)

13.2.3.2 Bi_2Te_3/Sb_2Te_3 Superlattice

Figure 13.17 shows the nanoscale superlattices fabricated by MOCVD (Venkatasubramanian et al., 2007). *In situ* ellipsometry has been used to make a further control over nanometer-scale deposition. The MOCVD process can be scaled to multiwafer growth and large-area growth, similar to that for III–V semiconductor space photovoltaics and LEDs, for low-cost, volume production of modules. The fabrication of thin-film modules employs standard semiconductor device manufacturing tools such as photolithography, electroplating, wafer dicing, and pick-and-place tools. This allows scalability of the module fabrication, from simple modules that can provide a few milli Watts to multiconnected module-arrays that can provide tens of Watts.

13.2.4 Complex Nanomaterials

As discussed earlier, to maximize the thermoelectric figure of merit (ZT) of a material, it is necessary to optimize a variety of conflicting properties to achieve a large absolute value of the Seebeck coefficient, a high electrical conductivity, and a low thermal conductivity (Nolas et al., 2000). Complex nanomaterials provide spaces for such optimization. In the following, we will discuss flower-like dendritic PbTe and $Bi_{0.4}Te_3Sb_{1.6}$ porous thin film.

13.2.4.1 Flower-Like Dendritic PbTe

Flower-like dendritic PbTe microstructures were fabricated by a solvothermal method, and the possible growth mechanism was an oriented attachment accompanied by an Ostwald ripening process (Jin et al., 2012). Figure 13.18a shows the XRD pattern of the

FIGURE 13.18
Characterization of flower-like PbTe dendrites obtained at 180°C for 6 h in the presence of 0.4 g b-cyclodextrin: (a) XRD pattern, (b) low magnification FESEM image, (c) middle magnification FESEM image, and (d) high magnification FESEM image. (From Jin, RC et al., 2012. *CrystEngComm*, 14, 2327–2332.)

sample. Figure 13.18b not only reveals that the sample consists almost entirely of such dendritic superstructures with a diameter of 2–5 μm, but also shows that PbTe dendrites with a high yield and good uniformity can be easily obtained via this facile and easily controlled method. Figure 13.18c and d show that the well-defined branches are intercrossed with each other and have formed flowery structures.

The PbTe flower-like dendrites were pressed into a bar with a rectangular shape for the measurement of electrical conductivity and the Seebeck coefficient. As shown in Figure 13.19a, the electrical conductivities increase with increasing temperature, indicating a typical semiconductor characteristic. The corresponding Seebeck coefficient (S) in the temperature range 300–600 K reveals the p-type semiconductor behavior, as shown in Figure 13.19b. The S value of the PbTe flower-like dendrites is about 313.1 μVK^{-1} at room temperature, and reaches up to 349 μV K^{-1} at 400 K, higher than the 265 μV K^{-1} of the bulk sample, up to about 18% and 32%, respectively. The enhancement of the Seebeck coefficient may be attributed to the separation of higher energy electrons from lower energy electrons and the selective scattering of electrons (Jin et al., 2012).

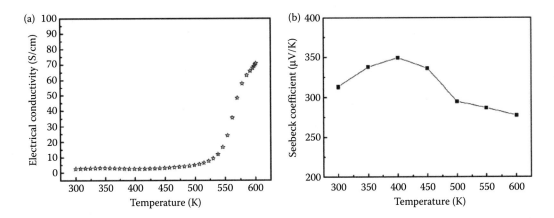

FIGURE 13.19
(a) Temperature dependence of the electrical conductivity and (b) the Seebeck coefficient for flower-like PbTe dendrites. (From Jin, RC et al., 2012. *CrystEngComm*, 14, 2327–2332.)

13.2.4.2 $Bi_{0.4}Te_3Sb_{1.6}$ Porous Thin Film

In general, incorporating structures with critical dimensions/spacings smaller than the phonon mean-free path should reduce the thermal conductivity without significantly affecting the electrical properties. Nanoporous structures should thus be a promising scalable thermoelectric nanostructured material with both good electrical properties and low thermal conductivity (Kashiwagi et al., 2011). A nanostructure containing isolated holes is necessary to decrease the thermal conductivity without significantly affecting the electrical conductivity, as already demonstrated for nanoporous Si (Yamamoto et al., 1998) and Al-doped ZnO containing nanovoids (Ohtaki et al., 2009).

Kashiwagi et al. proposed a new fabrication process for fabricating porous thin films of bismuth antimony telluride ($Bi_{0.4}Te_3Sb_{1.6}$) (Kashiwagi et al., 2011). A porous, thermoelectric thin film was fabricated by depositing a thin film of $Bi_{0.4}Te_3Sb_{1.6}$ on the porous alumina substrate. Flash evaporation was employed for depositing the thin film. Scanning electron microscope (SEM) images revealed that the thickness of the $Bi_{0.4}Te_3Sb_{1.6}$ film was 100 nm, while that of the porous alumina was 1 μm. The alumina substrate can be released after film deposition by chemical etching if required. The SEM images of the fabricated porous $Bi_{0.4}Te_3Sb_{1.6}$ thin film are shown in Figure 13.20. The average diameter of the holes was 20 nm and the average pitch of the hexagonally arranged holes was 50 nm. The sample has a packing density of 78%, and its electrical conductivity is about one-fifth of the bulk conductivity value.

It was found that the figure of merit of $Bi_{0.4}Te_3Sb_{1.6}$ was enhanced to 1.8 at room temperature (300 K) by the formation of a porous, thin film. The thermal conductivity was reduced without any major decrease in the electrical conductivity because of the hexagonal arrangement of the pores. The reduced thermal conductivity was consistent with a value calculated using a model for the full distribution of phonon mean-free paths (Kashiwagi et al., 2011).

13.2.5 Consolidation and Densification

Low-density materials typically have poor mechanical properties. In order for nanomaterials to be practical for an actual thermoelectric device, they must be consolidated into densified bulk pellets, which are often called bulk nanostructured materials, nanobulks,

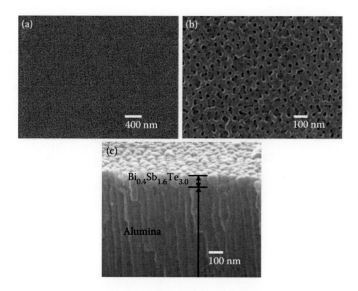

FIGURE 13.20

SEM images of a porous thin film of bismuth antimony telluride. (a) and (b) top view; (c) cross-sectional view. (From Kashiwagi, M et al., 2011. *Applied Physics Letters*, 98, 023114.)

or nanocomposites (Minnich et al., 2009). Numerous studies on porosity indicate that electrical properties can degrade by orders of magnitude when a sample is not at full theoretical density. Porosity can also affect thermal conductivity. A material with low density will inherently have a low thermal conductivity. Additionally, the pores can act as phonon scattering sites, thereby reducing the lattice thermal conductivity. Theoretically, the ideal nanostructured inclusion for reducing thermal conductivity is a nanoscale void, however, in order for a void to be an effective phonon scatterer it needs to be on the order of a few nanometers. Therefore, when examining the impact of nanostructuring on the thermoelectric properties, it is critical that materials that are being evaluated have similar high densities (Bux et al., 2010).

Three methods are widely used in the densification of nanostructured materials, including cold-pressing then sintering (Ur et al., 2007), hot uniaxial compaction (hot-pressing) (Savvides and Goldsmid, 1980), and SPS (Munir et al., 2006). A new technique, known as combustion driven compaction, can be used to sinter "softer" materials. In this process, a pressurized gas-based mixture undergoes a controlled combustion reaction, and the force due to the reaction is used to the drive upper and lower pistons to generate extreme pressures of up to 1000 metric tons (Nagarathnam et al., 2013). Unlike hot-pressing that provides extensive sintering at high temperatures and SPS that relies on very large currents to "fuse" the particle interfaces, this technique provides a different path to densification, with no thermal or electrical energy involved. This high-pressure compaction technique allows for materials to be compressed to very high densities without postsintering (Bux et al., 2010). One advantage of using this technique is that it could consolidate low bulk modulus nanostructured materials such as Bi_2Te_3 or PbTe to very high densities while maintaining their nanostructures. Though it is critical to compact samples into fully dense bulk materials, it is also critical to prevent grain growth. It is necessary to optimize temperature and pressure so that the nanoparticles consolidate, but they do not coalesce into large grains. This method has been successfully used in the densification of several materials (Bux et al.,

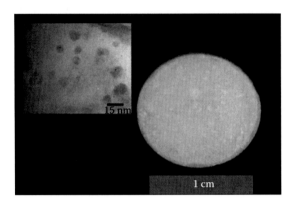

FIGURE 13.21
A fully dense nanostructured Si disk, the inset is a TEM image of an ion-milled pellet showing that nano-structured domains remain after high-temperature processing. (Modified from Bux, SK et al., 2009. *Advanced Functional Materials*, 19, 2445–2452.)

2009; Dresselhaus et al., 2007; Poudel et al., 2008). The TEM images shown in Figure 13.21 demonstrate that although there has been some grain growth, the nanostructured inclusions remain at the nanostructured grain boundaries (Bux et al., 2009).

13.3 Thermoelectric Generators

A TEGproduces a voltage when there is a temperature difference between its hot-side and cold-side based on the thermoelectric effect, and in this approach the thermal energy is converted directly into electric energy. It has no moving parts, and it is compact, quiet, highly reliable, and environmentally friendly (Liang et al., 2011). Most of the current commercially available TEG modules are hand assembled from bismuth telluride (Bi_2Te_3) dice. They display a large variation in efficiency and pose major challenges originating from the manual fabrication process and brittle nature of the materials.

In this part, after a brief introduction of the basic TEG structure, four types of new TEGs will be discussed, including MEMS TEGs (Kao et al., 2010), flexible TEGs, nanostructured TEGs, and printed TEGs.

13.3.1 Basic Structure of a Thermoelectric Generator

The basic structure of a TEG is shown schematically in Figure 13.4b. It mainly consists of a thermoelectric module, a heat source, and a heat sink. A thermoelectric module usually consists of more than 100 pieces of p–n thermoelectric conversion units, conductive tabs, and two ceramic substrates. The thermoelectric conversion units are configured so that they are connected electrically in series, but thermally in parallel. Ceramic substrates provide the platform for the thermocouples and the small conductive tabs that connect them. Heat source, thermoelectric conversion units, conductive tabs, ceramic substrates, and heat sink thus form a layered configuration. When the temperature of the heat source and the cold-side heat sink cause a temperature difference between the hot-side and cold-side of thermoelectric conversion units, an electric current will flow through an external load resistance because of the thermoelectric effect (Gou et al., 2010; Liang et al., 2011; Yazawa and Shakouri, 2011).

13.3.2 MEMS Thermoelectric Generators

For improving the power density per unit area in TEGs, different types of miniaturized TEGs have been developed and proposed based on in thin film and micromachining technologies (Huesgen et al., 2008; Jo et al., 2012; Leonov et al., 2011; Snyder et al., 2003). There are two general approaches for the miniaturization of TEGs (Dalola and Ferrari, 2011). One approach is to scale-down macroscopic cells. In this case, vertical pillars of thermoelectric materials are arranged and electrically connected in series and thermally sandwiched in parallel between two substrates. The micromachined thermogenerators are often fabricated by means of nonstandard techniques specifically developed to allow the fabrication of three-dimensional structures for the building blocks of p–n pairs, and postprocessing techniques may be needed to assemble the various elements that compose the microgenerators (Leonov et al., 2011; Snyder et al., 2003). The other approach is the adoption of membrane-based thermopiles, which are used for sensing applications. In this case the pillars of p–n pairs are substituted with planar elements (Huesgen et al., 2008; Jo et al., 2006).

To avoid the use of nonstandard micromachining technology, Dalola and Ferrari proposed a configuration of micromachined TEG based on readily available MEMS technologies (Dalola and Ferrari, 2011). Figure 13.22a shows the basic structure of a TEG based on a suspended membrane with a central hole. When the device is placed on a hot surface at temperature T_B, the heat Q flows into the device and local temperature differences are created. These internal temperature differences are converted into a voltage by means of the Seebeck effect. The planar thermoelectric units (thermocouples) are made up of thin layers of p-type polysilicon and a metal alloy of aluminium and copper. Figure 13.22b shows a 3D schematic of the micromachined TEG. The device consists of a square silicon membrane with sizes of 6×6 mm^2 and a thickness of 15 µm. The die dimensions are 7×7 mm^2. The silicon frame has a thickness of 450 µm. A square hole in the chip membrane, with size 500×500 µm^2, is used to improve the heat transfer between the thermogenerators and the ambient by means of the resulting airflow (Dalola and Ferrari, 2011).

The behavior of the TEG was analyzed by finite element simulations. Figure 13.23a shows the temperature distribution of the microgenerator with a uniform temperature at the bottom side $T_B = 80°C$, while the ambient temperature is 25°C. The micromachined TEG has been experimentally characterized as a function of the temperature T_B applied at

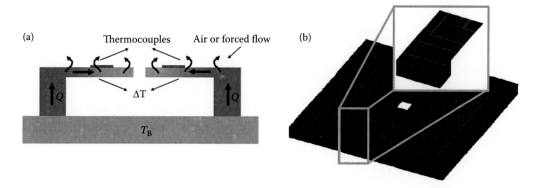

FIGURE 13.22
(See color insert.) Simplified cross section of the proposed device with suspended membrane and central hole (a). 3D schematic of the designed micromachined generator (b). (From Dalola, S, and Ferrari, V, 2011. *Procedia Engineering*, 25, 207–210.)

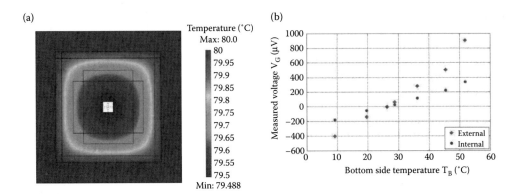

FIGURE 13.23
(See color insert.) (a) Simulated results of the temperature distribution for the microgenerator with a uniform temperature of the bottom side TB = 80°C and ambient temperature at 25°C. (b) Measured output voltage versus the applied bottom side temperature TB. (From Dalola, S, and Ferrari, V, 2011. *Procedia Engineering*, 25, 207–210.)

the chip bottom. The experimental results shown in Figure 13.23b indicate that, by placing the device on a hot surface, the internal and external thermocouples arrays will have an output voltage of about –180 and –400 μV, respectively, when the bottom side temperature TB is 10°C, and about 340 and 900 μV when TB is 50°C (Dalola and Ferrari, 2011).

13.3.3 Flexible Thermoelectric Generators

The operations of most of the wearable electronic devices are based on batteries. The transduction of ambient energy to electrical energy via energy harvesting technology can solve the problems caused by the capacities and working lives of batteries. The human body has abundant ambient energy sources in the form of vibrations, motions, and heat. Thermoelectric energy harvesting is advantageous because human body heat is steady, and the amount of energy released through metabolism is high. TEGs in direct contact with the human body can harvest this wasted energy (Jo et al., 2012).

However, most of the conventional TEGs have rigid substrates, and from the viewpoint of the curvature of the human body, rigid TEGs may be not suitable for harvesting the heat from a human body. On the other hand, flexible TEGs can transduce human body heat efficiently because they can be tightly attached to the skin, and could possibly make full use of the large area of the human body surface (Jo et al., 2012). Several methods for developing flexible TEGs have been reported (Itoigawa et al., 2005; Schwyter et al., 2008). In the following, we discuss two types of flexible TEGs: polydimethylsiloxane (PDMS)-based TEG and SU-8 based TEG.

13.3.3.1 PDMS-Based Thermoelectric Generator

Jo et al. proposed a flexible TEG comprising a PDMS substrate and thermoelectric materials (Jo et al., 2012). The use of the PDMS substrate provides flexibility to the TEG. The basic structure of the flexible planar TEG is shown in Figure 13.24. This TEG comprises a thick PDMS film that includes thermoelectric columns. In a planar TEG design, the substrate reduces the effective heat flow through the thermoelectric material. The use of a low thermal conductivity substrate such as PDMS can minimize such thermal losses. The

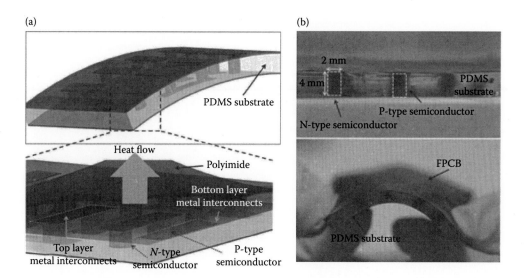

FIGURE 13.24
(a) Structure of a flexible TEG. (b) Image of fabricated TEG. (From Jo, SE et al., *Electronics Letters*, 48, 1015–1016.)

thermoelectric columns are inserted into the PDMS film, but the end points of the columns are exposed. The top sides of the p-type and n-type columns are connected to each other via metal interconnectors. The bottom sides of each couple are connected to the metal interconnectors beneath the substrate. As a result, the thermoelectric columns are connected in series. The top and bottom sides of the TEG are covered by polyimide films.

Figure 13.24b shows the image of a fabricated TEG. It had an output power of 2.1 μW when it was attached to the human body and the temperature difference between the human body and ambient air was 19 K (Jo et al., 2012).

13.3.3.2 SU-8-Based Thermoelectric Generator

Glatz et al. proposed an SU-8 based wafer level fabrication process for micro thermoelectric generators (μTEGs) for the application on nonplanar surfaces (Li et al., 2011). The generators are fabricated by subsequent electrochemical deposition (ECD) of Cu and Ni in a 190 μm-thick flexible polymer mold formed by photolithographic patterning of SU-8. Figure 13.25 shows the pictures of an SU-8-based TEG. Ninety-nine etch holes collocated across the device were used for releasing the device from the silicon support wafer in the fabrication process. The bright Cu-top-interconnect and the darker trough shining bottom interconnects electrically connect all the thermolegs in series. The diameter of the thermolegs and the etch holes is 210 μm. The TEG generated a power of 12.0 ± 1.1 nW/cm² for a ΔT of 0.12 K at the μTEG interface, which is equivalent to a thermoelectric power factor of 0.83 μWK⁻² cm⁻². Substituting SU-8 by another polymer may increase flexibility if necessary for the application of μTEGs on surfaces with high curvature.

13.3.4 Nanostructured Thermoelectric Generators

Low-dimensional thermoelectric materials have enhanced performances as compared with bulk counterparts. Silicon nanowires (SiNWs) have shown to be a much more efficient

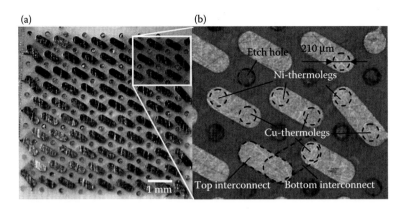

FIGURE 13.25
(See color insert.) Pictures of an SU-8-based TEG device. (a) Complete device with 90 Cu–Ni-thermocouples in a meander shape serial connection. (b) Detailed cut out view. (Li, Y et al., Chip-level thermoelectric power generators based on high-density silicon nanowire array prepared with top-down CMOS technology, *IEEE Electron Device Letters*, 32, 674–676 © 2011 IEEE.)

thermoelectric material than bulk silicon because of their lower thermal conductivity. While the σ and S values of 50-nm-diameter SiNWs do not deviate much from those of bulk silicon, the κ values of 1.6–25 W/mK (depending on the surface roughness) are clearly lower than that of bulk silicon (150 W/mK), resulting in an excellent ZT value of 0.6 at room temperature (Li et al., 2012). Therefore, SiNWs could be a promising candidate for use in applications such as microscale on-chip energy harvesting.

Li et al. proposed a high-density SiNW-based TEG prepared by a top–down CMOS-compatible technique (Li et al., 2011). The 5×5 mm TEG comprises of densely packed alternating n- and p-type SiNW bundles with each wire having a diameter of 80 nm and a height of 1 μm. Each bundle serving as an individual thermoelectric element, having 540×540 wires, was connected electrically in series and thermally in parallel. A schematic of the process flow is shown in Figure 13.26. The SiNW array was formed using a top–down approach: deep-ultraviolet (UV) lithography and dry reactive-ion etching. Specific groups of SiNWs were doped n- and p-type using ion implantation, and air gaps between the SiNWs were filled with silicon dioxide (SiO_2). The bottom and top electrodes were formed using a nickel silicidation process and aluminum metallization, respectively. The fabricated TEG demonstrates thermoelectric power generation with an open circuit voltage of 1.5 mV and a short circuit current of 3.79 μA with an estimated temperature gradient across the device of 0.12 K (Li et al., 2011). The generator can be further improved by the use of polyimide as a filler material to replace SiO_2. Polyimide, with a rated thermal conductivity value one order of magnitude lower than that of SiO_2, resulted in a larger measured thermal resistance when used as a filler material in a SiNW array (Li et al., 2012).

13.3.5 Printed Thermoelectric Generators

Printing technology has attractive advantages in the fabrication of electronic devices. First, it is a cost-effective processing due to its high fabrication speed and low material consumption. Second, as it is a low temperature process, it can be used for both flexible and rigid substrates. Third, it is well established, and is widely used in various areas of engineering and industries. Further, it is a direct patterning technology, and no additional etching is required.

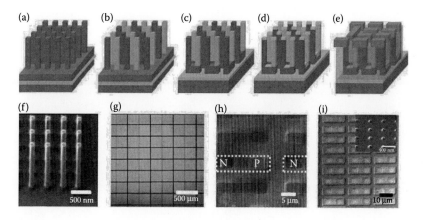

FIGURE 13.26
(**See color insert.**) Schematic of fabrication. (a) SiNW formation by dry etch. (b) Ion implantation and P/N element definition with each element consisting of hundreds of SiNW. (c) P/N couples formed by dry etch. (d) SiNW top and bottom silicidation while protecting the sidewall. (e) Dielectric deposition and etch back to expose only the tip of the SiNW and top metallization. (f) SEM images of pillar formation. (g) N and P implants can be seen clearly under microscope with a different shade. (h) SEM image of SiNW after N/P implant. (i) Metallization etch showing individual N/P couples. Inset shows the tips of the SiNW exposed after oxide etch which confirms the structure of the TEG. (Li, Y et al., Chip-level thermoelectric power generators based on high-density silicon nanowire array prepared with top-down CMOS technology, *IEEE Electron Device Letters*, 32, 674–676 © 2011 IEEE.)

As shown in Figure 13.27, printing techniques have been developed for the fabrication of TEGs (BMI, 2013; Wang et al., 2012). It is a two to three layer process, and the shape and size of the thermolegs are controllable. By rolling up the printed flexible substrate, a high-efficiency cylindrical TEG is formed. By integrating thermoelectric energy harvesting devices with electrochemical energy storage devices, for example, polymer lithium-ion batteries, integrated energy harvesting systems can be developed.

The roll-to-roll printing technology has obvious advantages of large area printing and high throughput (up to 60 m²/s). Figure 13.28 shows a roll-to-roll gravure printing system. A special process has been developed for the fabrication of flexible TEGs using roll-to-roll

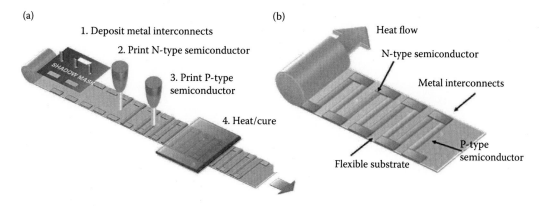

FIGURE 13.27
Schematic illustrations of printed TEGs. (a) Major steps in printing thermoelectric circuits. (b) Formation of a cylindrical printed TEG. (From BMI, http://bmi.berkeley.edu/thermo accessed on February 20th, 2013.)

FIGURE 13.28
A roll-to-roll gravure printing system at the University of Arkansas, Fayetteville.

printing techniques (Lee et al., 2012). This process involves the synthesis of thermoelectric inks, mainly consisting of thermoelectric fillers, binders, solvents, and additives. The thermoelectric properties of the printed circuits or devices are mainly determined by the properties of the thermoelectric fillers, and usually thermoelectric nanomaterials are used as the thermoelectric fillers. Meanwhile, appropriate binders, solvents, and additives should be used to ensure that the printed circuits have good physical properties so that they could be used in practical applications.

13.4 Thermoelectric Generator Systems

To meet up with the requirements for practical applications, TEG systems consisting of multiple TEGs may be needed. The TEGs in a TEG system may be connected in series and parallel. In some cases, hybrid systems, consisting of TEGs and other energy harvesting or storage devices, are needed.

The performances of a TEG system is more complicated than a single TEG, as it is hard to guarantee that every TEG in the system works in the same conditions (Liang et al., 2011). For example, when a TEG system is used in the waste heat recovery of industry furnaces, due to the different surface temperature of different furnaces, different TEGs in the system may have different hot-side temperatures. Therefore, the performances of a TEG systems consisting of TEGs connected in series or parallel is related to the working conditions of each TEG. Usually a management system is used to ensure that the TEG system have a stable output.

13.4.1 Cascaded Thermoelectric Generators

The output voltage of a cascaded TEG system is determined by the number of the generators in the system and the output voltage of each generator. Therefore, in a cascading approach, a TEG system with desired output voltage can be obtained by choosing suitable number of generators with suitable output voltages.

Another reason for the development of cascaded TEG systems is due to the intrinsic characteristic of thermoelectric materials. As the TE properties of materials vary considerably with temperature, it is not desirable, or is even impossible, to use the same material over a

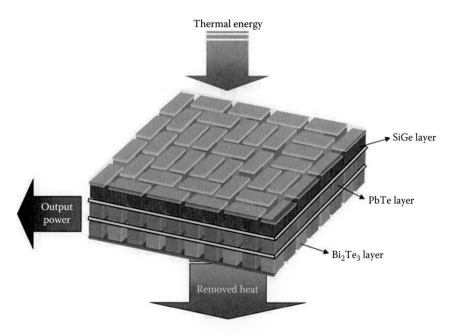

FIGURE 13.29
Schematic diagram of a cascaded TEG system. (From Choi, SH, Elliott, JR, and King, GC, 2006. Power budget analysis for high altitude airships, *Enabling Technologies and Design of Nonlethal Weapons*, edited by Glenn T. Shwaery, John G. Blitch, Carlton Land, Proceedings of SPIE, Vol. 6219, 62190C.)

wide temperature range. Consequently, segmented and cascaded TEGs are needed for high-efficiency energy conversion. Among these different TEG designs, the cascaded TEG design is more popular because of its simple structure and easy fabrication (Zhang et al., 2008). One of the most promising potential fields of application for cascaded TEGs is the energy recovery of high-temperature waste heat, which is an important issue in many industries. Figure 13.29 shows a schematic diagram of cascaded TEG for effective energy extraction and conversion from thermal energy (Choi et al., 2006). The cascade efficiency reveals the effective energy conversion by a cascaded regenerative cycle as shown in Figure 13.30. In a cascaded regenerative cycle, extracted thermal energy is used at the first layer of SiGe high-temperature TE material to generate electricity, then the rejected energy from the first layer is used at the second layer composed of lead-telluride (PbTe) mid-temperature TE material, and finally the energy left over from the second layer is used at the third layer (Bi$_2$Te$_3$). The cascade efficiency is tabulated in Table 13.1, as an example to show the effectiveness of cascaded regenerative cycle, in case when 30% of thermal efficiency was considered at each layer.

13.4.2 Parallel Thermoelectric Generators

The structure of a parallel TEG system is shown schematically in Figure 13.31 (Liang et al., 2011). It consists of multiple TE modules, heat sources, and heat sinks, but has only one load resistance. In such a system, only when all of the TE modules have the same inherent parameters and working conditions, the output voltage of the system equals to that of a single TE module, and the output current equals to the sum of all the TE modules. The inevitable electrical contact resistance among the TE modules increases the overall internal resistance of the system, which leads to the decease of the output power.

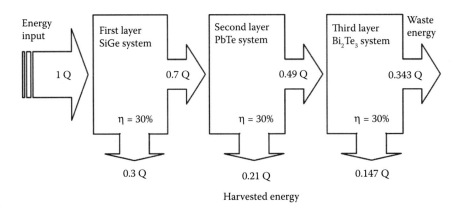

FIGURE 13.30
Cascaded efficiency of three-layer ATE systems in a tandem mode.

TABLE 13.1

Efficiencies of Cascaded TEG Systems

TE Tandem System	TE FoM ≥ 1.5 $\eta = 10\%$		TE FoM ≥ 3.5 T $\eta = 20\%$		TE FoM ≥ 4.5 $\eta = 30\%$		Solar Cells
	Loaded Energy, Q	η	Loaded Energy, Q	η	Loaded Energy, Q	η	η
First layer (Hi T)	1Q in	10%	1Q in	20%	1Q in	30%	30% (?) for membrane PV
	0.9Q out		0.8Q out		0.7Q out		
Second layer (Med T)	0.9 in	10%	0.8 in	20%	0.7 in	30%	
	0.81 Q out		0.64Q out		0.49Q out		
Third layer (Low T)	0.81 Q in	10%	0.64Q in	20%	0.49Q in	30%	
	0.729Q out		0.512Q out		0.343Q out		
Cascade efficiency	0.271Q Harvested	27%	0.488Q Harvested	48%	0.657Q Harvested	65%	

13.4.3 Hybrid Systems

To increase the overall energy efficiency and to develop integrated power systems, TEGs are often used in combination with other energy generation and storage devices, resulting in hybrid systems. In the following, we discuss photovoltaic-thermoelectric hybrid system and battery-TEG hybrid system.

13.4.3.1 Photovoltaic-Thermoelectric Hybrid System

The wavelength distribution of the solar radiation energy density is equivalent to that of a 6000 K black body: around 58% of the energy is in the visible and ultraviolet range (200–800 nm) and around 42% of the energy is in the infrared range (800–3000 nm). To improve the efficiency of harvesting solar energy, photovoltaic-thermoelectric (PV-TE) hybrid systems are proposed, with the photovoltaic cell and thermoelectric device covering the visible and ultraviolet spectrum, and the infrared spectrum respectively (He and Tritt, 2010; Kraemer et al., 2008; Wang et al., 2011).

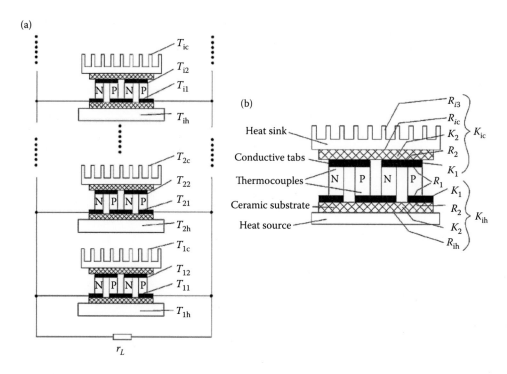

FIGURE 13.31
(a) Schematic diagram of a parallel TEG system. (b) Schematic diagram of thermal conductance and thermal resistance of every part of a TE module. (From Liang, GW, Zhou, JM, and Huang, XZ, 2011. *Applied Energy*, 88, 5193–5199.)

Two PV-TE hybrid system designs are extensively studied (Zhang et al., 2009). In one design, the solar cell is mounted on top of the concentrator which concentrates the solar radiation. In this configuration, the solar cell and the TE module work separately. The solar radiation that is not harvested by the solar cell is concentrated on the TE module, and then converted into electricity at an efficiency determined by the TE module. In the other design, the solar cell is placed directly on the TE module, or the TE module is affixed to the rear of the PV panel. In this configuration, the operation of the TE model will decrease the temperature of the PV panel, and thus improve the efficiency of the PV panel. Experimental results show that when the solar irradiance and the ambient temperature are 778 W/m² and 32°C respectively, the conversion efficiency would have an increase of 5.2% compared with the efficiency of a single PV panel (Zhang et al., 2009).

13.4.3.2 Battery-Thermoelectric Generator Hybrid System

In a battery-TEG hybrid system, the electricity produced by the TEG is stored in the battery. Various types of battery-TEG hybrid systems have been developed, and in these systems, converters are often used to boost-up the output voltage of the TE generators to charge the batteries (Vieira and Mota, 2009). Killander et al. developed a stove-top generator using two TE power modules. During the operating time, the output of the generator was about 10 W and supplied the battery with a net input from 1 to 5 W (Killander and Bass, 1996). Rahman et al. developed a TEG powered by butane gas, with a power output around 13.5 W, to charge a laptop computer battery (Mahmudur and Roger, 1995). Roth et al. developed a

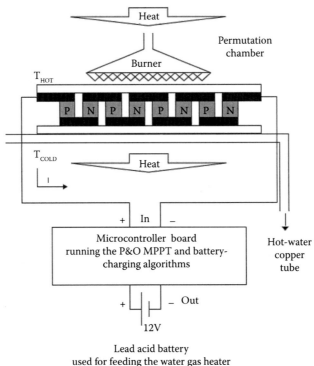

FIGURE 13.32
Block diagram of a battery charger system. (Vieira, JAB, and Mota, AM, Thermoelectric generator using water gas heater energy for battery charging, *2009 IEEE Control Applications CCA & Intelligent Control (ISIC),* 1–3, 1477–1482 © 2009 IEEE.)

photovoltaic/thermoelectric hybrid system as a power supply for a mobile telephone repeater, and this system can support a 50 W permanent load (Roth et al., 1997).

A TE generator is a nonlinear power source, and its output current/power strongly depends on the temperature between its two sides. To increase the system efficiency, it is important that TE generator operates in its maximum output power (MPP) system, therefore a maximum power point tracking (MPPT) algorithm is needed to optimize the transfer power from the TE generator to the battery. Figure 13.32 shows the block diagram of a battery-TEG hybrid system (Vieira and Mota, 2009). This system mainly consists of the three parts: a TE generator, an MPPT algorithm with a DC/DC converter, and a lead-acid battery. A Perturb and Observe MPPT algorithm was used in this system because of its simplicity, quick convergence, and low computational load.

13.5 Applications of Thermoelectric Generators and Thermoelectric Generator Systems

TEGs and thermoelectric generator systems have many advantages such as high reliability, low maintenance requirements, and long working life. They have been used in radioisotope thermal generators for deep space satellites (Bass and Allen, 1999) and remote power generation for unmanned systems (McNaughton, 1995). In recent years, due to the increase of the conversion efficiency and the decrease of the fabrication and installation costs, TEGs and TEG systems exhibited great application potentials in various engineering and industries, such as solar energy harvesting and waste heat recovery. As solar energy and waste heat are low-cost or no-cost energy resources, the low heat-to-electricity conversion efficiency is not a major problem, and the focus of concern is to get large power outputs with low capital costs (Liang et al., 2011).

In the following, we discuss the applications of TEGs and TEG systems in waste heat recovery, body heat energy harvesting, and communications.

13.5.1 Automotive Waste Heat Recovery

There are a large number of waste heats in our surroundings that cannot be recycled effectively by conventional methods. With ever-increasing demand on energy conservation, there is fast growing interest in the technologies to improve the fuel economy for hybrid electric vehicles (HEVs), especially the exhaust gas waste heat energy recovery. TEGs utilizing the heat of the exhaust gas from internal combustion engines mounted in vehicles, primarily automobiles, have been extensively investigated (Kusch et al. 2001; Matsubara, 2002). However, the requirements on such generators are quite complex and diverse. They must be light and compact, and can endure heavy vibration during transport. Further, TEGs for such applications must operate efficiently in different engine operating modes, and this entails additional difficulties (Anatychuk et al., 2011b).

The use of solar energy in HEVs is also proposed to promote on-board renewable energy and, hence, improve their fuel economy. As a result, thermoelectric-photovoltaic (TE-PV) hybrid energy sources have been developed for hybrid electric vehicles (HEVs) (Zhang et al., 2009). Figure 13.33 shows the configuration of a TE-PV hybrid energy source for HEV. The hybrid energy system mainly consists of TEGs, PV panels, a power conditioning system, a battery as the load, and an MPPT controller. In the operation of hybrid system, the MPPT controller measures the output voltages and currents of the TEG and PV array panel, respectively, and generates switching signals to the power conditioning system according to the MPPT algorithm.

13.5.2 Stationary Plant Waste Heat Recovery

For the TEGs that harvest the exhaust heat of stationary electric power plants, the requirements on weight, size, and vibration resistance are not so critical. In addition, the internal

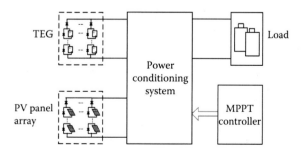

FIGURE 13.33
PV-TE hybrid system for HEV [80].

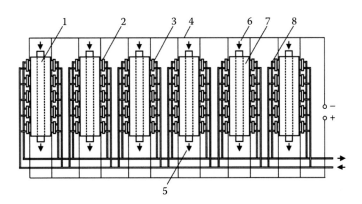

FIGURE 13.34
Schematic of the diesel plant TEG. 1, unit (primary generator); 2, thermoelectric module; 3, liquid heat exchanger; 4, generator electric circuit; 5, generator outflow gas; 6, generator inflow gas; 7, gas heat exchanger; 8, liquid circuit. (From Anatychuk, LI, Mykhailovsky, VY, and Strutynska, LT, 2011a. *Journal of Electronic Materials*, 40, 1119–1123; Anatychuk, LI, Rozver, YY, and Velichuk, DD, 2011b. *Journal of Electronic Materials*, 40, 1206–1208.)

combustion engines in such plants operate under sufficiently stable and optimal conditions. Therefore, the practical application of TEGs in stationary electric power plants appears more promising than those in vehicles (Anatychuk et al., 2011b).

Figure 13.34 schematically shows a diesel plant TEG. This system uses the exhaust heat of a diesel electric plant with electric power of 50 kW. The generator consists of six identical units that represent primary TEGs connected to each of the six diesel engine cylinders. Each primary generator comprises five sections with gas heat exchangers, thermoelectric modules, and liquid heat exchangers. The sections were optimized for the exhaust gas operating temperatures. The generator electric power was 2.1 kW, corresponding to about 4.4% of the diesel plant electric power (Anatychuk et al., 2011b).

13.5.3 Body Heat Energy Harvesting

Energy harvesting from the human body has been undergoing an interesting and quick development due to the technological availability of new electronic components and the advances of the various applications mainly for biomedical and social impacts on human beings daily life (Lay-Ekuakille et al., 2009).

13.5.3.1 Harvestable Body Heat

Like all the objects, the human body gains and loses heat by conduction, convection, and radiation. In most situations, these three effects operate together. To get a stable state, the human body absorbs heat from and emits heat to the environment. Meanwhile, the metabolic processes in the human body generate heat as well, similar to a heat-producing engine. Human beings and, more generally speaking, warm-blooded animals can serve as a heat source for TEGs attached to their skins (Lay-Ekuakille et al., 2009).

The Carnot efficiency puts an upper limit on how well the body heat can be recovered. Assuming normal body temperature and a relatively low room temperature (20°C), according to Equation 13.8, the Carnot efficiency is

$$\eta_c = \frac{T_{body} - T_{ambient}}{T_{body}} = \frac{310\,K - 293\,K}{310\,K} = 5.5\% \tag{13.10}$$

In a hot environment (27°C) the Carnot efficiency falls to 3.2%. This calculation provides an ideal value. The energy conversion efficiency of existent TEGs is much lower than the Carnot efficiency.

When a person is motionless, the total heat dissipation power is about 116 W. Using a Carnot engine to model the recoverable energy yields a power of 3.7–6.4 W. Evaporative heat loss accounts for about 25% of the total heat dissipation. This "insensible perspiration" includes water diffusing through the skin, sweat glands keeping the skin of the palms and soles pliable, and the expulsion of water-saturated air from the lungs. Therefore, the maximum power available drops to 2.8–4.8 W (Lay-Ekuakille et al., 2009).

13.5.3.2 Wearable Biometric Sensors

Continuous health monitoring is essential for recuperating patients and patients with chronic health conditions (Varadan and Chen, 2012). The interest of scientific community and healthcare providers for low-cost solutions for wearable biometric monitoring sensors and other energy autonomous biomedical devices has grown very fast during the last years. State-of-the-art monitoring nodes are typically powered by rechargeable batteries, with obvious drawbacks for continuous operation. To extend the lifetime of traditional batteries, intensive research is currently focused on the development of portable power generators which can harvest energy from environmental sources and convert it into electricity. Furthermore, the development of such microsystems on flexible substrates is the key point for the development of unobtrusive low cost wearable and ubiquitous integrated devices for healthcare and biometric monitoring (Francioso et al., 2010).

The conversion between thermal and electrical energy by thermoelectric effect is one of the simplest processes for the extraction of electrical energy from heat sources. By far the most widely used thermoelectric materials are Bi_2Te_3 and Sb_2Te_3 because of their high thermoelectric efficiency around the room temperatures and different deposition methods have been developed for such thermoelectric thin films, such as thermal co-evaporation (Zou et al., 2001), DC or RF co-sputtering (Shiozaki et al., 2006; Kandasamy et al., 2005), electrochemical deposition (Glatz et al., 2006), and flash evaporation (Foucaran, 1998).

Francioso et al. developed a low cost, energy autonomous, maintenance free, flexible, and wearable μTEG to power low consumption biometric monitoring electronics (Francioso et al., 2010). The high quality p-Sb_2Te_3 and n-Bi_2Te_3 thin films were deposited by using a co-sputtering technique with high level of process control. Kapton substrates were used because of their good thermal properties and superior temperature resistance, and moreover, the use of a flexible substrate is favorable for the development of wearable devices, such as integrated microsystems for healthcare and biometric monitoring.

Figure 13.35 shows a prototype with an array of 100 thin film thermocouples of Sb_2Te_3 and Bi_2Te_3, generates (Francioso et al., 2010). At a temperature difference of 40°C, it generates an open circuit output voltage of 430 mV and an electrical output power up to 32 nW with a matched load. In practical applications, the temperature difference is around, and under this condition, the device generates an open circuit output voltage of about 160 mV with an electrical output power up to 4.18 nW.

(a)

(b)

(c)

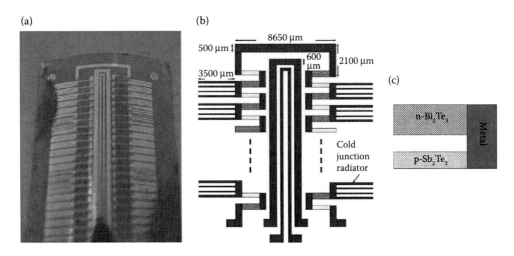

FIGURE 13.35
(a) Photograph of fabricated flexible μTEG on Kapton HN. (b) Schematic of flexible TEG. (c) A pair of p–n couple of the device, and parameters that specify the dimension of the device. (Francioso, L et al., Flexible thermoelectric generator for wearable biometric sensors, *IEEE Conference on Sensors,* 2010, 747–750 © 2010 IEEE.)

13.5.3.3 Biomedical Autonomous Devices

Lay-Ekuakille et al. demonstrated the application of TEGs to supply power to a biomedical hearing aid (Lay-Ekuakille et al., 2009). As the trunk of the human body usually has higher temperatures than the extremities, TEGs was placed on the trunk. Due to the temperature difference between the trunk and the environment, the TEGs produced an electric power, which was used to supply a biomedical hearing aid, as shown in Figure 13.36. The TEGs they used were MPG-D602. As one TEG could not provide sufficient power, an array of TEGs, together with a power conditioning circuit, was used to reach the necessary amount of voltage and current for the autonomous device.

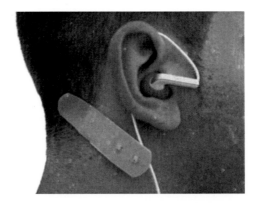

FIGURE 13.36
(See color insert.) Demonstration of TEGs for supplying a biomedical hearing aid. (From Lay-Ekuakille, A et al., 2009. *MeMeA 2009–International Workshop on Medical Measurements and Applications,* 1–4.)

13.5.4 Power Supply for Wireless Devices

Research has been made on the application of TEGs in powering wireless communication devices. In the following, we discuss self-contained TEG for cell phones and battery-free wireless sensors.

13.5.4.1 Self-Contained Thermoelectric Generator for Cell Phones

Anatychuk et al. developed a self-contained TEG, based on propane-butane catalytic combustion [96]. As shown in Figure 13.37a, the physical model contains a cylinder shaped catalytic heat source, hot and cold heat exchangers, thermoelectric modules, and an electric fan. The heat supply to the modules is made through the hot heat exchanger, which is in direct contact with the catalyst, and heat rejection is achieved by forced air delivery to the cold finned heat exchanger. In Figure 13.37a, Q is the thermal power generated by fuel combustion, Q_1 is the thermal power supplied from the hot heat sink to the thermopile, Q_2 is the thermal power rejected by the cold heat sink to the ambient, Q_3, Q_4, and Q_5 are the

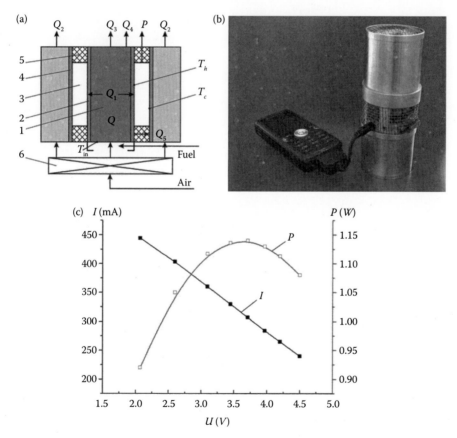

FIGURE 13.37
(a) Physical model of the thermogenerator: 1, catalytic heat source; 2, hot-side heat exchanger; 3, thermoelectric module; 4, cold-side heat exchanger; 5, insulation; 6, fan. (b) Thermogenerator with a catalytic heat source for charging cell phones. (c) Current–voltage characteristics (I) and electric power (P) of the thermogenerator for cell phones at $T_h = 250°C$ and $T_c = 60°C$. (From Anatychuk, LI, Mykhailovsky, VY, and Strutynska, LT, 2011a. *Journal of Electronic Materials*, 40, 1119–1123; Anatychuk, LI, Rozver, YY, and Velichuk, DD, 2011b. *Journal of Electronic Materials*, 40, 1206–1208.)

heat losses with combustion products to the ambient, from the hot heat exchanger and the catalyst due to radiation, and from the hot heat sink through the insulation due to thermal conductivity, respectively. P is the electrical power generated by the TEG.

As shown in Figure 13.37b, a TEG with a catalytic heat source was implemented for a self-contained power supply for low-power electronic equipment, in particular, cell phones (Anatychuk et al., 2011a). The generator has its own reservoir filled with propane–butane at the bottom and a fan for forced delivery of combustion air and generator cooling air. As shown in Figure 13.37c, the maximal power output of the generator for a matched load is 1.14 W at a voltage of 3.7 V. The TEG has an efficiency of about 2.2–2.4%. It will take four to five hours to fully charge a 900 mAh cell phone battery using this TEG.

13.5.4.2 Battery-Free Wireless Sensors

Typical power requirements for wireless sensors are in the range of 1–10 mW, with short peaks up to 100 mW. A heat flux driven generator may serve as a reliable source. Pasold et al. proposed a production-worthy concept for the fabrication of metal based (Ni–Cu) and semiconductor-based TEG devices (Pasold et al., 2011). As the TEGs were fabricated by PCB techniques in conjunction with electrochemical deposition, low-cost mass-production of TEGs is possible. The fabricated TEGs can be used for wireless sensing and monitoring applications.

A fully processed 5×5 cm TEG prototype is shown in Figure 13.38a, and the cross section of an individual Ni–Cu thermocouple is depicted in Figure 13.38b. An example of a typical load characteristic is shown in Figure 13.38c. For a fully packaged prototype consisting of 29.5 Ni/Cu couples per cm^2 at an effective area of 13 cm^2, at a temperature drop of 36 K across the packaged device, it generates an output power of 450 μW (around 27 $nWK^{-2}cm^{-2}$) (Pasold et al., 2011).

13.6 Concluding Remarks

The research on thermoelectricity-based energy recovery and harvesting can be generally categorized into two areas: development of thermoelectric materials with high figure of merit, and the development and application of TEGs. Due to the advances in nanotechnology, a number of high-performance thermoelectric nanomaterials have been developed, and some of them are commercially available. Many efforts are being made in developing low-cost and large-scale fabrication techniques for thermoelectric nanomaterials so that they could be practically used for engineering and industrial applications. In addition to the improvement on thermoelectric materials, thermal design and management of TEGs is equally important for improving the energy recovery and harvesting performance. TEGs may be connected in series or parallel, and they may also be used in combination with other energy conversion or storage devices, resulting in hybrid systems. TEGs is expected to be more widely used in various fields of industries, defense, and civilian life.

There are two major challenges in the research and application of thermoelectricity-based energy recovery and harvesting. One is the thermoelectricity at nanoscale, which is related to the further improvement of thermoelectric materials, and the other is cost-efficiency trade-off, which is related to the design, implementation, and operation of TEGs.

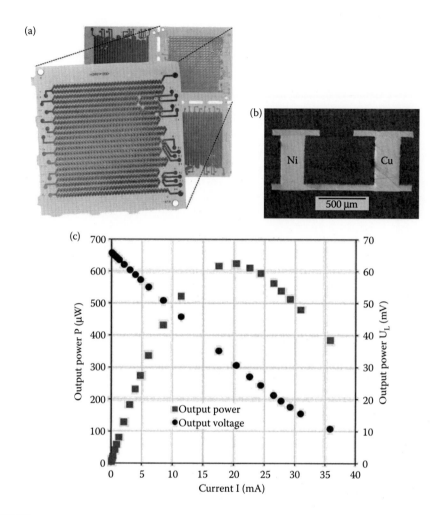

FIGURE 13.38

(a) Top view of a 5×5 cm TEG prototype cut from a 10×10 cm PCB substrate which is processed as a whole to give a total of 2×2 TEGs. (b) Cross-sectional view of an individual Ni–Cu thermocouple embedded in the FR4 material of the PCB. (c) Example of a typical load characteristic showing TEG output power and voltage as a function of current. (Pasold, G et al., Powering wireless sensors: Microtechnologybased large-area thermoelectric generator for mass applications, *IEEE Conference on Sensors, 2011*, 1293–1296 © 2011 IEEE.)

13.6.1 Thermoelectricity at Nanoscale

At the nanometer scale, thermoelectric materials exhibit properties that are quite different from those of their bulk counterparts. Numerous factors, such as composition, shape, size, and synthesis techniques and conditions, may affect the Seebeck coefficient, thermal conductivity, electric conductivity, and figure of merit of thermoelectric nanomaterials (Ohtaki et al., 2007). However, at this moment the understanding in this area is limited, though intensive investigations have been made on phonon scatterings at low-dimensional materials. Therefore, it is necessary to have a deeper understanding about the thermoelectricity at the nanometer scale, so that the properties of thermoelectric materials can be further improved. The research in this area should cover not only dispersed nanomaterials, but also nanobulks and inks.

13.6.2 Cost-Efficiency Trade-Off

In the real-world application of thermoelectricity for energy recovery and harvesting, a systematic discussion about the cost-efficiency trade-off, which is a cornerstone for any energy technology, is lacking (Yazawa and Shakouri, 2011), and there are different opinions about the prospects of this field (Vining, 2009). An analysis on the potential of thermoelectricity that focuses only on efficiency values is not complete. Though TEGs are not likely to replace the conventional combustion engines, they could play a big role in our society by direct conversion of heat into electricity, especially for waste heat recovery. As waste heats are low-cost or no-cost resources, and thermoelectricity-based energy recovery and harvesting has no pollution emissions, the low-efficiency problem is not the most important issue that we have to take into account (Yazawa and Shakouri, 2011). However, before the public can accept and adopt TEGs, great progress should be made in not only the improvement of the energy conversion efficiency, but also the decrease of capital cost, which is more important and urgent.

Acknowledgments

The authors thank Rama Venkatasubramanian, RTI International, Research Triangle Park, North Carolina; Eric Toberer, Colorado School of Mines, Golden, Colorado; Sabah K Bux, Jet Propulsion Laboratory (JPL), Pasadena, California; Jeff Sakamoto, Michigan State University, East Lansing, Michigan, for reviewing this chapter and for providing valuable technical comments and suggestions.

References

Anatychuk, LI, Mykhailovsky, VY, and Strutynska, LT, 2011a. Self-contained thermoelectric generator for cell phones, *Journal of Electronic Materials*, 40, 1119–1123.

Anatychuk, LI, Rozver, YY, and Velichuk, DD, 2011b. Thermoelectric generator for a stationary diesel plant, *Journal of Electronic Materials*, 40, 1206–1208.

Bass, JC, and Allen, DT, 1999. Milliwatt radioisotope power supply for space applications, *Proceedings of the IEEE 18th International Conference on Thermoelectrics*, 521–524.

BMI, http://bmi.berkeley.edu/thermo (accessed on February 20th, 2013).

Borca-Tasciuc, T, Achimov, D, Liu, WL, Chen, G, Ren, HW, Lin, CH, and Pei, SS, 2001. Thermal conductivity of InAs/AlSb superlattices, *Microscale Thermophysical Engineering*, 5, 225–231.

Boukai, AI, Bunimovich, Y, Tahir-Kheli, J, Yu, J K, Goddard, WA, and Heath, JR, 2008. Silicon nanowires as efficient thermoelectric materials, *Nature*, 451, 168–171.

Bux, SK, Blair, RG, Gogna, PK, Lee, H, Chen, G, Dresselhaus, MS, Kaner, RB, Fleurial, JP, 2009. Nanostructured bulk silicon as an effective thermoelectric material, *Advanced Functional Materials*, 19, 2445–2452.

Bux, SK, Fleurial, JP, Kaner, RB, 2010. Nanostructured materials for thermoelectric applications, *Chemical Communications*, 46, 8311–8324.

Cademartiri, L, Malakooti, R, O'Brien, PG, Migliori, A, Petrov, S, Kherani, NP, Ozin, GA, 2008. Large-scale synthesis of ultrathin Bi2S3 necklace nanowires, *Angewandte Chemie-International Edition*, 47, 3814–3817.

Cahill, DG, Fisher, HE, Klitsner, T, Swartz, ET, Poh, RO, 1989. Thermal conductivity of thin films: Measurements and understanding, *Journal of Vacuum Science and Technology, A*, 7(3), 1259.

Cantarero, A, Martinez-Pastor, J, and Segura, A, 1987. Transport-properties of bismuth sulfide single-crystals, *Physical Review B*, 35, 9586–9590.

Chen, BX, Uher, C, Iordanidis, L, and Kanatzidis, MG, 1997. Transport properties of Bi_2S_3 and the ternary bismuth sulfides $KBi_{6.33}S_{10}$ and $K_2Bi_8S_{13}$, *Chemistry of Materials*, 9, 1655–1658.

Chen, G, Dresselhaus, MS, Dresselhaus, G, Fleurial, JP, and Caillat, T, 2003. Recent developments in thermoelectric materials, *International Materials Reviews*, 48, 45–66.

Choi, SH, Elliott, JR, and King, GC, 2006. Power budget analysis for high altitude airships, *Enabling Technologies and Design of Nonlethal Weapons*, edited by Glenn T. Shwaery, John G. Blitch, Carlton Land, SPIE-The International Society for Optical Engineering, Bellingham, WA, Proceedings of SPIE, Vol. 6219, 62190C.

Dalola, S, and Ferrari, V, 2011. Design and fabrication of a novel MEMS thermoelectric generator, *Procedia Engineering*, 25, 207–210.

Dirmyer, MR, Martin, J, Nolas, GS, Sen, A, Badding, JV, 2009. Thermal and electrical conductivity of size-tuned bismuth telluride nanoparticles, *Small*, 5, 933–937.

Dresselhaus, MS, Chen, G, Tang, MY, Yang, RG, Lee, H, Wang, DZ, Ren, ZF, Fleurial, JP, and Gogna, P, 2007. New directions for low-dimensional thermoelectric materials, *Advanced Materials*, 19, 1043–1053.

Foucaran, A, 1998. Flash evaporated layers of $(Bi_2Te_3–Bi_2Se_3)(N)$ and $(Bi_2Te_3–Sb_2Te_3)(P)$, *Materials Science and Engineering B*, 52, 154–161.

Francioso, L, De Pascali, C, Farella, I, Martucci, C, Cretì, P, Siciliano, P, and Perrone, A, 2010. Flexible thermoelectric generator for wearable biometric sensors, *IEEE Conference on Sensors*, 2010, 747–750.

Ge, ZH, Zhang, BP, and Li, JF, 2012a. Microstructure composite-like Bi_2S_3 polycrystals with enhanced thermoelectric properties, *Journal of Materials Chemistry*, 22, 17589–17594.

Ge, ZH, Zhang, BP, Liu, Y, and Li, JF, 2012b. Nanostructured $Bi_{2-x}Cu_xS_3$ bulk materials with enhanced thermoelectric performance, *Physical Chemistry Chemical Physics*, 14, 4475–4481.

Glatz, W, Muntwyler, S, and Hierold, C, 2006. Optimization and fabrication of thick flexible polymer based microthermoelectric generator, *Sensors and Actuators A-Physical*, 132, 337–345.

Gou, XL, Xiao, H, and Yang, SW, 2010. Modeling, experimental study and optimization on low-temperature waste heat thermoelectric generator system, *Applied Energy*, 87, 3131–3136.

Gross, ETB, 1961. Efficiency of thermoelectric devices, *Amercian Journal of Physics*, 29, 729–731.

Harman, TC, Taylor, PJ, Walsh, MP, and LaForge, BE, 2002. Quantum dot superlattice thermoelectric materials and devices, *Science*, 297, 2229–2232.

He, J, and Tritt, TM, 2010. Chapter 3 Thermal-electrical energy conversion from the nanotechnology perspective, In *Nanotechnology for the Energy Challenge*. Edited by Javier Garcia-Martinez, Wiley-VCH, Weinheim.

Hicks, LD, and Dresselhaus, MS, 1993. Effect of quantum-well structures on the thermoelectric figure of merit, *Physical Review B*, 47, 12727–12731.

Hochbaum, AI, Chen, R, Delgado, RD, Liang, W, Garnett, EC, Najarian, M, Majumdar, A, and Yang, P, 2008. Enhanced thermoelectric performance of rough silicon nanowires, *Nature*, 451, 163–167.

Hsu, KF, Loo, S, Guo, F, Chen, W, Dyck, JS, Uher, C, Hogan, T, Polychroniadis, EK, and Kanatzidis, MG, 2004. Cubic $AgPb_mSbTe_{2+m}$: Bulk thermoelectric materials with high figure of merit, *Science*, 303, 818–821.

Huesgen, T, Woias, and P, Kockmann, N, 2008. Design and fabrication of MEMS thermoelectric generators with high temperature efficiency, *Sensors and Actuators*, 145–146, 423–429.

Huo, ZY, Tsung, CK, Huang, WY, Zhang, XF, and Yang, PD, 2008. Sub-two nanometer single crystal Au nanowires, *Nano Letters*, 8, 2041–2044.

Itoigawa, K, Ueno, H, Shiozaki, M, Toriyama, T, Sugiyama, S, 2005. Fabrication of flexible thermopile generator, *Journal of Micromechanics and Microengineering*, 15, S233–S238.

Jin, RC, Chen, G, Pei, J, Yan, CS, Zou, X, Deng MD, and Sun S, 2012. Facile solvothermal synthesis and growth mechanism of flower-like PbTe dendrites assisted by cyclodextrin *CrystEngComm*, 14, 2327–2332.

Jo, SE, Kim, MK, Kim, MS, and Kim, YJ, 2012. Flexible thermoelectric generator for human body heat energy harvesting, *Electronics Letters*, 48, 1015–1016.

Kadel, K, Kumari, L, Li, WZ, Huang, JY, Provencio, PP, 2011. Synthesis and thermoelectric properties of Bi_2Se_3 nanostructures, *Nanoscale Research Letters*, 6, article no. 57, 1–7.

Kandasamy, S, Pachoud, D, Holland, A, Kalantar Zadeh, K, Rosengarten, G, and Wlodarski, W, 2005, Thermoelectric properties of antimony telluride thin films deposited using RF magnetron sputtering, *East-West Journal of Mathematics*, 32, 459–464.

Kao, PH, Shih, PJ, Dai, CL, and Liu, MC, 2010. Fabrication and characterization of CMOS-MEMS thermoelectric micro generators, *Sensors*, 10, 1315–1325.

Kashiwagi, M, Hirata, S, Harada, K, Zheng, YQ, Miyazaki, K, Yahiro, M, and Adachi, C, 2011. Enhanced figure of merit of a porous thin film of bismuth antimony telluride, *Applied Physics Letters*, 98, 023114.

Kawamoto, H, 2009. R&D trends in high efficiency thermoelectric conversion materials for waste heat recovery, *Science and technology Trends*, 4, 54–69.

Killander, A, and Bass, JC, 1996. A stove-top generator for cold areas, *Proceedings of the IEEE 15th International Conference on Thermoelectrics*, 390–393.

Kim, HJ, Lee, J, Varadan, VK, and Choi, SH, 2013. Printed thermoelectric generator for hybrid tandem photovoltaic/thermoelectric device, *Recent Patents on Space Technology*, 3(1), 2013.

Kim, HJ, Park, Y, and Choi, SH, 2012. Thermoelectric materials design and performance, *Smart Nanosystems in Engineering and Medicine*, 1, 40–63.

Kraemer, D., Hu, L., Muto, A., Chen, X., Chen, G., and Chiesa, M., 2008. Photovoltaic-thermoelectric hybrid systems: A general optimization methodology, *Applied Physics Letters*, 92, 243503.

Kusch, AS, Bass, JS, Ghamaty, S, and Elsner, NB, 2001. Thermoelectric development at Hi-Z technology, *Proceedings of the IEEE 20th International Conference on Thermoelectrics*, 422–430.

Lay-Ekuakille, A, Vendramin, G, Trotta, A, and Mazzotta, G, 2009. Thermoelectric generator design based on power from body heat for biomedical autonomous devices, *MeMeA 2009–International Workshop on Medical Measurements and Applications*, 1–4.

Lee, J, Kim, HJ, Oh, S, Choi, SH, and Varadan, VK, 2012. Synthesis and characterization of thermoelectric ink for renewable energy applications, *Proceedings of SPIE*, 8344, 83440H.

Leonov, V, Fiorini, P, and Vullers, RJM, 2011. Theory and simulation of a thermally matched micromachined thermopile in a wearable energy harvester, *Microelectronics Journal*, 42, 579–584.

Li, Y, Buddharaju, K, Singh, N, and Lee, SJ, 2011. Chip-level thermoelectric power generators based on high-density silicon nanowire array prepared with top-down CMOS technology, *IEEE Electron Device Letters*, 32, 674–676.

Li, Y, Buddharaju, K, Singh, N, and Lee, SJ, 2012. Top-down silicon nanowire-based thermoelectric generator: Design and characterization, *Journal of Electronic Materials*, 41, 989–992.

Liang, GW, Zhou, JM, Huang, XZ, 2011. Analytical model of parallel thermoelectric generator, *Applied Energy*, 88, 5193–5199.

Lin, YF, Chang, HW, Lu, SY, and Liu, CW, 2007. Preparation, characterization, and electrophysical properties of nanostructured $BiPO_4$ and Bi_2Se_3 derived from a structurally characterized, single-source precursor $Bi[Se_2P(OiPr)_2]_3$, *Journal of Physical Chemistry C*, 111, 18538–18544.

Lin, YM, and Dresselhaus, MS, 2003. Thermoelectric properties of superlattice nanowires, *Physical Review B*, 68, 075304.

Lu, W, Ding, Y, Chen, Y, Wang, ZL, and Fang, J., 2005. Bismuth telluride hexagonal nanoplatelets and their two-step epitaxial growth, *Journal of the American Chemical Society*, 127, 10112–10116.

Mahmudur, R, and Roger, S, 1995. Thermoelectric power-generation for battery charging, *Proceedings of the IEEE Conference on Energy Management and Power Delivery*, 186–191.

Matsubara, K, 2002. Development of a high efficient thermoelectric stack for a waste exhaust heat recovery of vehicles, *Proceedings of the IEEE 21th International Conference on Thermoelectrics*, 418–423.

McNaughton, AG, 1995. Commercially available generators, in *CRC Handbook of Thermoelectrics*, Rowe, D. M., Ed., CRC Press, Boca Raton, FL.

Minnich, AJ, Dresselhaus, MS, Ren ZF, and Chen G, 2009. Bulk nanostructured thermoelectric materials: Current research and future prospects, *Energy & Environmental Science*, 2, 466–479.

Munir, ZA, Anselmi-Tamburini, U, and Ohyanagi, M, 2006. The effect of electric field and pressure on the synthesis and consolidation of materials: A review of the spark plasma sintering method, *Journal of Materials Science*, 41, 763–777.

Nagarathnam, K, Renner, A, Trostle, D, Kruczynaski, D, and Massey, D, 2013. Development of 1000-Ton Combustion Driven Compaction Press for Materials Development and Processing, http://www.utroninc.com/PowderMet2007-PaperPress.pdf (Accessed on February 20, 2013).

Nolas, GS, Kaeser, M, Littleton, RT, and Tritt, TM, 2000. High figure of merit in partially filled ytterbium skutterudite materials, *Applied Physics Letters*, 77, 1855–1857.

Ohtaki, M, Araki, K, and Yamamoto, K, 2009. High thermoelectric performance of dually doped ZnO ceramics, *Journal of Electronic Materials*, 38, 1234–1238.

Pasold, G, Etlin, P, Hahn, M, Muster, U, Nersessian, V, Bonfrate, D, Buser, R, Cucinelli, M, Gutsche, M, Kehl, M, Zach, N, and Hazelden, R, 2011. Powering wireless sensors: Microtechnology-based large-area thermoelectric generator for mass applications, *IEEE Conference on Sensors, 2011*, 1293–1296.

Poudel, B, Hao, Q, Ma, Y, Lan, YC, Minnich, A, Yu, B, Yan, X, Wang, DZ, Muto, A, Vashaee, D, Chen, XY, Liu, JM, Dresselhaus, MS, Chen, G, and Ren, ZF, 2008. High-thermoelectric performance of nanostructured bismuth antimony telluride bulk alloys, *Science*, 320, 634–638.

Purkayastha, A, Kim, S, Gandhi, DD, Ganesan, PG, Borca-Tasciuc, T, and Ramanath, G, 2006. Molecularly protected bismuth telluride nanoparticles: Microemulsion synthesis and thermoelectric transport properties. *Advanced Materials*, 18, 2958–2963.

Purkayastha, A, Yan, QY, Raghuveer, MS, Gandhi, DD, Li, HF, Liu, ZW, Ramanujan, RV, Borca-Tasciuc, T, and Ramanath, G, 2008. Surfactant-directed synthesis of branched bismuth telluride/sulfide core/shell nanorods, *Advanced Materials*, 20, 2679–2683.

Reau, A, Guizard, B, Mengeot, C, Boulanger, L, and Tenegal, F, 2007. Large scale production of nanoparticles by laser pyrolysis, *Materials Science Forum*, 1, 534–536.

Roth, W, Kugele, R, Steinhuser, A, Schulz, W, and Hille, G, 1997. Grid-independent power-supply for repeaters in mobile radio networks using photovoltaic/thermoelectric hybrid systems, *Proceedings of the IEEE 16th International Conference on Thermoelectrics*, 582–585.

Rowe, DM, 2006a. Thermoelectric waste heat recovery as a renewable energy source, *International Journal of Innovations in Energy Systems and Power*, 1, 13–23.

Rowe, DM, 2006b. *Thermoelectric Handbook: Micro to Nano*, Taylor & Francis Group, Boca Raton, FL.

Savvides, N, and Goldsmid, HJ, 1980. Hot-press sintering of GE-SI alloys, *Journal of Materials Science*, 15, 594–600.

Scheele, M, Oeschler, N, Meier, K, Kornowski, A, Klinke, C, and Weller, H, 2009. Synthesis and thermoelectric characterization of Bi_2Te_3 nanoparticles, *Advanced Functional Materials*, 19, 3476–3483.

Schwyter, E, Glatz, W, Durrer, L, and Hierold, C, 2008. Flexible micro thermoelectric generator based on electroplated $Bi_{2+x}Te_{3-x}$, *DTIP 2008: Symposium On Design, Test, Integration And Packaging of MEMS/MOEMS*, 46–48.

Shakouri, A, 2011. Recent developments in semiconductor thermoelectric physics and materials, *Annual Review of Materials Research*, 41, 399–431.

Shiozaki, M, Sugiyama, S, Watanabe, N, Ueno, H, and Itoigawa, K, 2006. Flexible thin-film BiTe thermopile for room temperature power generation, *Proceedings of the 19th IEEE International Conference on Micro Electro Mechanical Systems, MEMS 2006*, 946–949.

Slack, GA and Hussain, MA, 1991, The maximum possible conversion efficiency of silicon-germanium thermoelectric generators, *Journal of Applied Physics*, 70(5), 2694–2718.

Snyder, GJ, and Toberer, ES, 2008. Complex thermoelectric materials, *Nature Materials*, 7, 105–114.

Snyder, GJ, Lim, JR, Huang, CK, and Fleurial, JP, 2003. Thermoelectric microdevice fabricated by a MEMS-like electrochemical process, *Nature Materials*, 2, 528–531.

Swihart, MT, 2003. Vapor-phase synthesis of nanoparticles, *Current Opinion in Colloid & Interface Science*, 8, 127–133.

Ur, SC, Kim, IH, Nash, P, 2007. Thermoelectric properties of Zn_4Sb_3 processed by sintering of cold pressed compacts and hot pressing, *Journal of Materials Science*, 42, 2143–2149.

Varadan, VK, and Chen, LF, 2012. *Mobile Wearable Nano-Bio Health Monitoring Systems with Smartphone as Base Stations*, ASME Press, New York.

Venkatasubramanian, R, Siivola, E, Colpitts, T, and O'Quinn, B, 2001. Thin-film thermoelectric devices with high room-temperature figures of merit, *Nature*, 413, 597–602.

Venkatasubramanian, R, Watkins, C, Caylor, C, and Bulman, G, 2006. Microscale thermoelectric devices for energy harvesting and thermal management, *Proceedings of the Sixth International Workshop on Micro and Nanotechnology for Power Generation and Energy Conversion Applications*, 1–4.

Venkatasubramanian, R, Watkins, C, Stokes, D, Posthill J, and Caylor, C, 2007. Energy harvesting for electronics with thermoelectric devices using nanoscale materials, *IEEE International Electron Devices Meeting*, 367–370.

Vieira, JAB, and Mota, AM, 2009. Thermoelectric generator using water gas heater energy for battery charging, *2009 IEEE Control Applications CCA & Intelligent Control (ISIC)*, 1–3, 1477–1482.

Vining, CB, 2009. An inconvenient truth about thermoelectrics, *Nature Materials*, 8, 83–85.

Wang, N, Han, L, He, HC, Park, NH, and Koumoto, K, 2011. A novel high-performance photovoltaic-thermoelectric hybrid device, *Energy & Environmental Science*, 4, 3676–3679.

Wang, Z, Chen, A, Winslow, R, Madan, D, Juang, RC, Nill, M, Evans, JW, and Wright, PK, 2012. Integration of dispenser-printed ultra-low-voltage thermoelectric and energy storage devices, *Journal of Micromechanics and Microengineering*, 22, 094001.

Xi, GC, and Ye, JH, 2010. Ultrathin SnO_2 nanorods: Template- and surfactant-free solution phase synthesis, growth mechanism, optical, gas-sensing, and surface adsorption properties, *Inorganic Chemistry*, 49, 2302–2309.

Yamamoto, A, Takimoto, M, Ohta, T, Whitlow, L, Miki, K, Sakamoto, K, and Kamisako, K, 1998. Two dimensional quantum net of heavily doped porous silicon, *Proceedings of the IEEE 17th International Conference on Thermoelectrics*, 198–201.

Ohtaki, M, Hayashi, R, and Araki, K, 2007. Thermoelectric properties of sintered ZnO incorporating nanovid structure: Influence of the size and number density of nanovoids, *Proceedings of the IEEE 26th International Conference on Thermoelectrics*, 112–116.

Yazawa, K, and Shakouri, A, 2011. Cost-efficiency trade-off and the design of thermoelectric power generators, *Environmental Science and Technology*, 45, 7548–7553.

Yazawa, K, and Shakouri, A, 2012. Optimization of power and efficiency of thermoelectric devices with asymmetric thermal contacts, *Journal of Applied Physics*, 111, 024509-1~6.

Zhang, GQ, Kirk, B, Jauregui, LA, Yang, HR, Xu, XF, Chen, YP, and Wu, Y, 2012. Rational synthesis of ultrathin n-Type Bi2Te3 nanowires with enhanced thermoelectric properties, *Nano Letters*, 12, 56–60.

Zhang, L, Tosho, T, Okinaka, N, and Akiyama, T, 2008. Design of cascaded oxide thermoelectric generator, *Materials Transactions*, 49, 1675–1680.

Zhang, XD, Chau, KT, and Chan, CC, 2009. Design and implementation of a thermoelectric-photovoltaic hybrid energy source for hybrid electric vehicles, *World Electric Vehicle Journal*, 3, 1–11.

Zhang, YL, Chen, YF, Gong, CM, Yang, JK, Qian, RM, and Wang, YJ, 2007. Optimization of superlattice thermoelectric materials and microcoolers, *Journal of Microelectromechanical Systems*, 16, 1113–1119.

Zou, HL, Rowe, DM, and Min, G, 2001. Preparation and characterization of p-type Sb_2Te_3 and n-type Bi_2Te_3 thin films grown by coevaporation, *Journal of Vacuum Science & Technology A-Vacuum Surfaces and Films*, 19, 899–903.

14

High-Temperature Drilling Mechanisms

Yoseph Bar-Cohen, Xiaoqi Bao, Mircea Badescu, Stewart Sherrit, Kris Zacny,
Nishant Kumar, Thomas R. Shrout, and Shujun Zhang

CONTENTS

14.1 Introduction

There are many applications where it is necessary to penetrate rocks and/or the ground that are at very high temperature. These applications include drilling holes in very deep oil and gas reservoirs, drilling geothermal or enhanced geothermal wells, and exploration of hot planets such as Venus and Mercury. In geothermal drilling, for example, the temperature of formations can reach 300°C and more. Because of the relatively small market share, equipment manufactures do not have an incentive to develop geothermal-specific technologies or products. Most of the techniques or tools have been adopted from the oil and gas industry and do not work well past 300°C (GTP, 2010).

Since the authors have been focusing mostly on applications related to planetary exploration, the focus of this chapter is on drills and excavation systems that can be operated in temperatures up to 500°C found on the surface of Venus (Gershman and Wallace, 1999). In planetary explorations, drilling is required to acquire samples for *in situ* analysis or return to Earth (Bar-Cohen and Zacny, 2009; Zacny et al., 2008). Venus is one of the planets that has been specifically singled out as potential new Frontier-class mission in the latest NASA Decadal Survey study. This planet has a very hostile environment and, hence, its exploration has been avoided due to the lack of technologies able to withstand its high temperature. Venus' average surface temperature is about 460°C and it is mostly carbon dioxide atmosphere at approximately 90 bar pressure, which makes carbon dioxide supercritical. In fact, only very few spacecraft in the Soviet Venera and Vega series as well as the NASA Pioneer Venus 2 (Pioneer Venus Multiprobe) made it to the surface and survived long enough to perform surface measurements. These past missions were very simple, and apart from Venera 13 and 14, which had small drills, all other missions had passive instruments such as cameras, seismometers, or spectrometers.

Future exploration of Venus will require some means of collecting deeper samples and delivering them to onboard instruments. Unfortunately, the existing actuation technology cannot maintain functionality under these harsh conditions of Venus. Hence, this is one of the major obstacles to perform sampling, robotic manipulation, and other tasks that require the use of moving parts.

Generally, there are many issues that need to be considered when developing devices for operation at high temperature. These issues include material compatibility, chemical reactions, alloying, annealing, and diffusion characteristics that may affect the chemical and physical nature of various components. One of the key issues of concern is thermal expansion mismatch, which can be catastrophic to components that need to fit precisely inside a structure.

Actuators such as brush, brushless, and stepper motors require magnetic materials where some of the commercial units can be operated at approximately 200–300°C, although recently developed small electromagnetic actuators at Honeybee Robotics (Bar-Cohen and Zacny, 2009) and NASA JPL (Troy, 2011, private communication) were reported to operate at much higher temperatures (see Chapter 10). These actuators could be used to actuate drills, robotic arms, or enable mobility. Some motors designed for extraction of smoke and deadly toxic fumes during fire emergencies are available for operation at the range of several hundred degrees centigrade. However, they are usually large and have a lifetime of a couple of hours. The most recent 200 W switch reluctance motor (SRM) developed by Honeybee Robotics for high-temperature applications is relatively small (2 in. × 2 in.) and has been tested for over 20 h at 460°C without any degradation in its performance (see Section 14.2).

Piezoelectric actuators offer the potential of operating at high temperatures. In order to use such actuators at as high as 500°C, it is necessary to use materials with sufficiently high operation temperature that is outside the normal operating range for standard piezoelectric, ferroelectric, and ferromagnetic material-based actuators. The temperature limitation is dictated by the transition temperature (i.e., Curie temperature) where the material switches from ferro- to paramagnetic causing the loss of actuation capability. Studies of piezoelectric ceramics and single-crystal materials at JPL have shown the potential to provide the required performance characteristics (Sherrit et al., 2004). Advances in developing electromechanical materials, such as piezoelectric and electrostrictive, at Penn State University have enabled potential actuation capabilities that can be used to support such missions (Chapter 11).

High-temperature piezoelectric materials have the advantage that they can be attached to actuated structures; they do not require windings of electric wires and electrical or mechanical commutation. The piezoelectric actuated drill, which is described in this chapter, is based on the ultrasonic/sonic driller/corer (USDC) mechanism that was developed by the authors from JPL. The piezoelectric materials that were investigated include lithium niobate (LiNbO$_3$) (which has a Curie temperature of 1150°C), and bismuth titanate materials with various tungsten doping that were developed by the coauthors from Penn State University.

Besides drills that withstand high-temperature environments, there are drills that use very high temperatures as means of drilling. Such drills are also described in this chapter.

14.2 Geothermal and Oil Drilling

According to Bellarby (2009), 150°C (300°F) is considered high temperature in oil and gas drilling and it is attributed to the fact that temperatures above this level combined with very high pressure become a hostile environment. In geothermal and oil drilling, the temperature of the formations can reach over 300°C (Bellarby, 2009; GTP, 2010). To reach the great depths that are involved with the related drillings, fluids are used to enhance the penetration rate and the removal of the cuttings as well as controlling the drill head temperature (Jahn et al., 2008). Specifically, the main functions of drilling fluids include providing hydrostatic pressure to prevent formation fluids from entering into the well bore, help prevent well walls from collapsing, cooling and cleaning the drill bit during drilling, removing the drill cuttings from the borehole, and suspending the cuttings when the drilling process is either paused or the drilling assembly is brought in and out of the borehole. The drilling fluid passes through the bit nozzles and, while cooling, it cleans the bit face in order to have the bit cutters interact with fresh rock and avoid wasting energy on regrinding the cuttings. The specific type of drilling fluid that is used is selected such that it prevents formation damage and limits corrosion (Varnado, 1980). To provide seal at high temperatures, metal-to-metal is used rather than elastomers. Also, titanium is increasingly being used as durable construction material for the drilling components. There are three main categories of drilling fluids that are used, including (a) water-based muds (which can be dispersed and nondispersed); (b) nonaqueous muds, usually called oil-based mud; and (c) gaseous drilling fluid, in which a wide range of gases can be used.

There are two primary types of drill bits that are widely used in deep oil drilling: (a) fixed cutter and (b) roller cone. Fixed cutter bits consist of no individually moving components and the drilling takes place via rotation of the drill string. The cutters of the fixed cutter bits are generally made of either polycrystalline diamond compact (PDC) or grit hot-pressed inserts (GHI). The cutters in these bits are used to shear rock with a continuous scraping motion and two examples are shown in Figure 14.1: (a) one of the very early PDC type bits and (b) 20.3 cm (8 in.) diameter PDC drill bit. In contrast, the cutter of roller cone bits consists of rotating parts (Hughes, 1909; Jahn et al., 2008) and their cutters are made either with tungsten carbide inserts (TCI) or milled tooth (MT).

(b)

(a)

FIGURE 14.1
(See color insert.) Photographic view of diamond bits showing the cutters used to shear rock with a continuous scraping motion. (a) A very early type diamond bit (surface set diamond bit). (b) 8 in. diameter PDC (polycrystalline diamond compact) drill bit.

14.3 Rotary Drill Powered by High-Temperature Actuators

Conventional rotary or rotary percussive drills use motors to spin the bit or actuate percussive mechanisms and in turn advance the drill bit into rocks. However, for high-temperature applications, conventional motors will not be able to work.

For this reason, Honeybee Robotics has developed two types of high-temperature motors, an SRM and a brushless DC motor (Chapter 10). All the materials and components in both motors were selected based on the requirement to survive temperatures above a minimum of 460°C. A prototype SRM, which is about 2 in. in diameter and 2 in. in length, was made. After being tested noncontinuously for over 20 h at Venus-like conditions at 460°C temperature and mostly CO_2 gas environment, it remained fully functional with no degradation to motor torque or rpm. The prototype, controlled by a custom controller, has generated promising test data that is comparable with the Maxon RE-25 motor, which is used to actuate flight systems (Table 14.1). The SRM has stator windings as in a DC motor; however, the rotor has no magnets or coils attached. The rotor of the SRM motor becomes aligned as soon as the opposite poles of the stator become energized. In order to achieve a full rotation of the motor, the windings must be energized in the correct sequence (Chapter 10).

Since the SRM's phases are independent of one another, it has the following significant advantages—its capability to withstand higher temperatures benefiting from the absence of rotor windings and permanent magnets, it has simple and low-cost construction, and it has high fault tolerance capability. Even if one or more of the motor's phases fail, it will continue to operate, even though it will operate at a lower torque level. This is an important attribute for flight devices. Other advantages are that the torque-speed characteristics of SRMs can be tailored to the specific application requirements more easily during the design stage than in the case of induction and permanent magnet motors. The starting torque can be very high without the problem of creating excessive inrush current due to its higher self-inductance. The low rotor inertia and high torque/inertia ratio of the motor

TABLE 14.1

Characteristics of the Maxon RE-25 and the Honeybee Robotics HT Motor

Characteristics	Units	Maxon RE-25[a] Range at 25°C	Honeybee Robotics High-Temperature Motor Range at 460°C
Applied voltage	V	4.5–48	40
Maximum speed	rpm	5500	5000
No-load speed	rpm	4790–5500	1000–5000
No-load current	mA	7–80	1000–4000
Stall torque	mNm	119–144	100–150
Starting current	mA	1470–16,500	<5000
Maximum continuous torque	mNm	12–29.3	20–40
Maximum continuous current	mA	322–1500	<4000

[a] "Maxon DC Motor, RE-25 mm," http://www.maxonmotor.com/docsx/Download/catalog_2005/Pdf/05_077_e.pdf

allow for fast dynamic response over a very wide operating speed range. Also, extremely high speeds are possible with SRMs, though operating speeds of less than 10,000 rpm are well within the SRM's capabilities.

The high-temperature drill that was developed for testing the mechanism is a simplified version of the Champollion/ST4 Sample Acquisition and Transfer Mechanism (SATM) that can operate with low thrust reaction in low-gravity environment (Zacny et al., 2008). The current best estimates for the Venus drill system requirements, along with a comparison to the SATM drill, are given in Table 14.2.

The prototype drill did not include sample acquisitions since the goal of the tests was to determine drilling data and prove the feasibility of drilling at high temperature. The drill system included two degrees of freedom for (1) rotation of the drill bit and auger, typically called the rotary axis, and (2) linear motion of the bit and auger into and out of a rock or other surface, typically called the z-axis.

TABLE 14.2

Comparison of the Venus Drill with SATM Drill System

	Venus Drill	SATM
Maximum drill depth	30 cm	120 cm
Bit diameter	1.3 cm	1.3 cm
Volume envelope (approximately)	12 cm × 24 cm × 69 cm	15 cm × 40 cm × 150 cm
Mass	3–10 kg	9 kg
Degrees of freedom	2–5	5
Penetration rate	>1.0 cm/min	1.0 cm/min
Preload required	>80 N	>130 N
Surface/rock design metric	Thermally hydrated nonvesicular basalt	Limestone/chalk
Power consumption	TBD	25 Wh
Sample state	Locally mixed powder/regolith	Locally mixed powder
Sample volume	>1 cm^3	<0.1 cm^3
Associated missions	New Frontiers Venus *in situ* Explorer; Venus Surface Sample Return	ST4/Champollion Comet Sample Return

Two high-temperature, switch reluctant motors were integrated into the system to actuate the auger axis and the *z*-axis. Both were driven by the custom controller using sensorless control algorithm. Both motors were geared down at the output shaft to accommodate a relatively low auger speed and *z*-axis rate. The auger speed and *z*-axis rate could be adjustable using controls software.

The final volume of the drill was 18 cm × 11 cm × 48 cm with drill stroke of up to 25 cm. The drill included off-the-shelf 1.27 cm (0.5 in.) diameter carbide drill bit. Other components of the drill, including gears, bearings, and bushings, were all selected based on the requirement to survive temperatures above a minimum of 460°C, at Earth's atmosphere.

A test chamber was set up to reach 460°C (and higher) with a constant CO_2 purge to roughly simulate Venus surface conditions (Figure 14.2). The constant CO_2 flow into the chamber provided a decrease in O_2 content from about 20% to about 5% by volume. Two furnaces, capable of operating at temperatures between 200°C (392°F) and 1100°C (2012°F) were set up with a ceramic extension. Two thermocouples attached to the drill were also connected to a thermal control box to automatically turn off the furnace once the set temperature was reached. For easy observation during the test, three openings were cut in the top plate of the oven and covered with clear high-temperature quartz glasses.

The drilling system was tested for more than 15 h in Venus-like conditions, 460°C temperature, and mostly CO_2 gas environment. The drill successfully completed three tests by drilling into chalk up to 22 cm deep in approximately 20 min. Thus, the average penetration rate was 1 cm/min. The average power was 40–50 W at 400 rpm, the auger motor current was 0.8 A and auger motor voltage ~28 V. Therefore, the energy required to drill to 20 cm was ~17 Wh. Owing to lack of high-temperature load cells, no weight-on-bit data was acquired. Figure 14.3 shows power and depth versus drilling time for one of the tests. Note that the rate of penetration was getting lower as the drill was approaching the 22 cm depth (this can be identified by the reduction in the gradient of the depth versus time curve). This penetration rate decrease was also accompanied by an increase in drilling power, suggesting that cuttings were not being conveyed out of the hole in an efficient and effective manner. In fact, when the drill initially penetrated the rock, the power was only ~20 W. This was the actual drilling power, while the remaining power increase was due to conveyance of cuttings out of the borehole and parasitic losses from drill auger rubbing against the side wall.

(a)   (b)

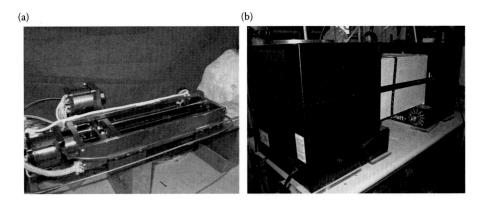

FIGURE 14.2
(**See color insert.**) (a) High-temperature drill with two SR motors before the drill test. The drill was integrated and tested outside of the chamber first. Once the algorithm was confirmed, the drill was set up in the chamber to test at 460°C. (b) The drill was placed inside an oven at 460°C.

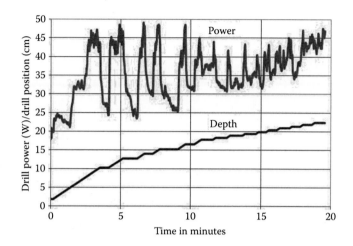

FIGURE 14.3
Typical drilling data with a high-temperature drill.

14.4 The Ultrasonic/Sonic Driller/Corer

Addressing the need for drilling at low-gravity conditions using low preload, the authors from JPL and engineers from Cybersonics, Inc. jointly developed the USDC (Bar-Cohen et al., 2001, 2007; Bao et al. 2003). This mechanism was developed to support the NASA exploration missions with the objective to search for existing or past life in the universe and it allows for probing and sampling of rocks, ice, and soil. It was designed to produce both core and powdered cuttings, operate as a sounder to emit elastic waves, and serve as a platform for sensors (Bar-Cohen and Zacny, 2009). The requirement for low axial load allows for its operation from lightweight robots and rovers. The USDC is a mechanism of penetration that is driven by piezoelectric stacks where an intermediate free-mass converts high-frequency vibrations to low-frequency hammering (Figure 14.4).

The USDC consists of three key components: actuator, free-mass, and bit (Figure 14.4) (Bao et al., 2003). The actuator operates as an ultrasonic vibration mechanism that imparts energy to the free-mass and the free-mass in turn impacts the bit, producing stress impulses. These impulses fracture rock when its ultimate strength is exceeded. The piezoelectric stack actuator consists of a piezoelectric stack, a backing, and horn and a prestress bolt. The prestress bolt maintains the piezoelectric stack in compression between the backing and the horn base. The backing layer is also used for forward power delivery and the horn is used for amplification of the induced displacement. In the basic configuration, the actuator is driven at a resonance frequency of about 20 kHz. Using software or hardware control, the drive electronics maintains tuning of the actuator to ensure maximum electric current input. This tuning is required since there are several factors that affect the resonance frequency, including the action of the drilled medium reducing the Q (the Q is mechanical energy stored divided by the energy loss in each cycle) of the resonator and slightly shifting the frequency. In addition, the USDC requires attention to the impacts that cause time variations in the current signal. Its effect is minimized using various control algorithms, including hill climbing, extremum seeking, and others (Aldrich et al., 2008). Unlike typical ultrasonic drills where the bit is acoustically coupled to the horn, in

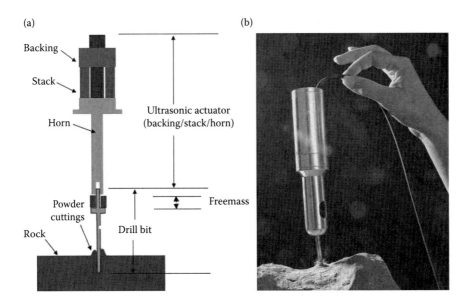

FIGURE 14.4
(See color insert.) A schematic cross-section view (a) of the USDC and a photo showing its ability to core with minimum axial force (b).

the USDC, the actuator drives a free-mass that converts ultrasonic impacts to hammering at sonic frequencies.

Following the development of the USDC, numerous novel designs were conceived and disclosed in NASA New Technology Reports and patents (e.g., Aldrich et al., 2008; Badescu et al., 2006a,b, 2011a,b,c; Bao et al., 2004, 2010b; Bar-Cohen et al., 1999, 2001, 2002, 2003, 2005, 2008, 2010; Bar-Cohen and Sherrit, 2003a,b; Blacke et al., 2003; Chang et al., 2004; Dolgin et al., 2001a, 2001b; Sherrit et al., 2001, 2002, 2003, 2005, 2006, 2008, 2009, 2010ba,b). These include the Ultrasonic/Sonic Rock Abrasion Tool (URAT), the Ultrasonic/Sonic Gopher, and the Auto-Gopher for deep drilling (Bar-Cohen et al., 2012), the Lab-on-a-drill, and many others (Bar-Cohen and Zacny, 2009). Further, it was demonstrated to drill ice and various rocks, including granite, diorite, basalt, and limestone. In this configuration, where the bit vibrates longitudinally and is not designed to rotate, sensors (e.g., thermocouple and fiberoptics) were integrated into the bit to examine the borehole during drilling. Another benefit that nonrotating bits enabled has been the ability to produce nonround shape cores. While developing the analytical capability to predict and optimize its performance, efforts were made to enhance its ability to drill at higher power and high speed.

14.4.1 HT Piezo-Ceramic Actuators

Increasingly, new piezoelectric materials with a high Curie temperature are being developed (Chapter 11). These materials are added to the pool of known high-temperature piezo-ceramic materials as well as the single-crystal $LiNbO_3$, which is known to have a high Curie temperature higher than 1100°C. To determine the aging characteristics of $LiNbO_3$ with Pd–Au and Pt electrodes at 500°C, an isothermal test was done for 1000 h (Bar-Cohen et al., 2012) and, to the level of the measurement error, no appreciable change in properties was observed demonstrating its high stability. The results shown in Figure 14.5 for this aging test were plotted for the capacitance and planar coupling as a function of time.

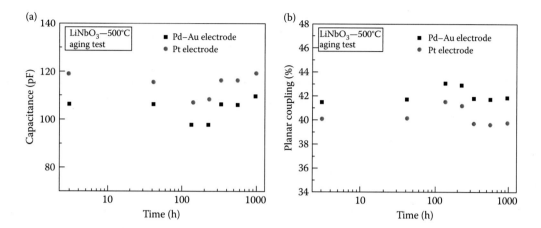

FIGURE 14.5
(a) Capacitance and (b) planar coupling as a function of time at 500°C for LiNbO$_3$.

Since the electrodes need to sustain the high-temperature environment too, three different types were subjected to 500°C:

1. Sputtered platinum film
2. Gold–palladium thick film
3. Sputtered gold film

Evaluating the results of the electrodes exposure to 500°C showed that the sputtered gold electrodes sustained severe degradation while the Pd–Au and Pt electrodes operated quite effectively. Based on these results and the fabrication ease, the electrodes made of sputtered platinum films were chosen for the actuators.

For the purpose of making piezoelectric stacks for driving high-temperature percussive drills, a series of bismuth titanate with various composition levels of titanate and dopants (W, Fe, Ca, Sr, and Mn) were made. Disks were made and tested for high temperature (HT) performance and the data were tabulated for the various samples (Table 14.3). For this purpose, an Agilent 4294A precision impedance analyzer and a d$_{33}$ meter were used to measure the electrical characteristics at room and high temperatures of up to 500°C. The test results have shown significant capability compared to the previously known materials where the thickness coupling coefficient at 500°C was about 15–20% of that at room temperature. Following these successful tests, efforts were made to optimize the response using dopant additive of tungsten into the powder mixture. Effective mixtures were identified and HT piezoelectric rings were made using hot isostatic pressing to ensure the production of robust low-porosity ceramic compositions. The material with a tungsten (W) dopant was selected because of its relatively high Q, low loss, high Curie temperature (T$_c$), high resistivity, and high d$_{33}$. Specifically, bismuth titanate rings were made with diameters of 25.4 and 38.1 mm (1.0 and 1.5 in.) and, in parallel, LiNbO$_3$ rings with 25.4 and 50.8 mm (1.0 and 2.0 in.) diameter were used.

In Table 14.3, T$_c$ is the Curie temperature, K is the relative dielectric constant, loss is an imaginary part of the K over the real part, d$_{33}$ is the piezoelectric "d" constant in poling axis, which is in thickness direction of the sample disks, Q is the mechanical quality factor, subscript p means the radial mode of thin disks, k is the electromechanical coupling coefficient, and resistivity is in unit of ohm·cm. Generally, high piezoelectric constant, high

TABLE 14.3

Electromechanical Measured Properties of $LiNbO_3$ and Bismuth Titanate with Various Doping Contents

Material	T_C (°C)	K	Loss (%)	d_{33} (pC/N)	Q_p	k_p (%)	500°C K	500°C Loss	500°C Q_p	500°C Resistivity
Modified $Bi_4Ti_3O_{12}$	666	118	0.5	16 (16)[a]	3000	3.7	300	41%	200	7.4×10^6
$Bi_{3.887}Ti_{2.866}W_{0.146}O_{12} - Fe_2O_3$	~620	154.6	1	14 (13)	2900	3.3	590	62%	50	1.5×10^6
$Bi_{3.9}Ti_{2.85}W_{0.15}O_{12}-Fe_2O_3$	~620	156.8	1	11.5 (11)	2000	3.1	670	67%	46	1×10^6
$Sr_{0.8}Ca_{0.2}Bi_4Ti_4O_{15}-Fe_2O_3$	595	143.7	0.4	12 (11.5)	5600	2.9	461	42%	360	1.9×10^6
$Sr_{0.6}Ca_{0.4}Bi_4Ti_4O_{15}-Fe_2O_3$	644	146.8	0.25	8 (8)	5800	2.6	463	100%	120	2.9×10^5
$Bi_{3.93}Ti_{2.9}W_{0.1}O_{12}-MnO_2$	657	158	0.5	17 (11)	3700	4.3	421	44%	45	3.6×10^6
$Bi_{3.96}Ti_{2.9}W_{0.1}O_{12}-MnO_2$	~650	145	0.3	18 (12)	3900	4.3	370	40%	40	5.6×10^6
W doped $Bi_4Ti_3O_{12}$	637	165.7	1.5	17 (15.5)	1800	3.4	309	16%	1000	$~5 \times 10^6$
$LiNbO_3$ (36° Y-cut)	1150	62	0.5	40 (40)	1500	46	104	6%	500	3.8×10^6

[a] The values in parentheses are the data obtained after 500°C.

Note: The unit for resistivity is ohm.cm.

mechanical factor, high dielectric constant, low loss, high resistivity, and high electromechanical coupling coefficient are preferred for actuator applications.

14.4.2 Materials for Fabricating the Drill

There are many materials that can be used to produce devices for operation at high temperatures. The making of drills requires the material to sustain impacts challenging the durability of the devices (Chapters 4 and 7). Several materials were used to produce piezoelectric actuated drills that were found effective at room temperatures, including CPM-3 V with hardness (HRC 59) that is a tool steel made by the crucible particle metallurgy process, designed to provide maximum resistance to breakage and chipping in a highly wear-resistant steel. However, the bits that were made of crucible hardened CPM-3 V and exposed to 500°C have degraded during drilling. As a replacement for the drill bit material, tungsten carbide WA-2 was used and this grade of carbide is considered one of the hardest commercially available materials having hardness of HRA 92. To make sure that the actuator can be operated at 500°C, the horn was fabricated of titanium; the stress bolt and backing of stainless steel, while the Belleville washers were made of Inconel. This selection allowed increasing the thermal stability and thermal expansion matching, while maintaining the prestress of the piezoelectric stack as the actuator is heated.

For testing the performance of the high-temperature drills, the following samples were used: limestone, basalt, and bricks. Their estimated range of the unconsolidated compressive strength at room temperature is basalt 100–300 MPa, limestone 30–250 MPa, and bricks 10–100 MPa (http://www.stanford.edu/~ tyzhu/Documents/Some%20Useful%20 Numbers.pdf). Based on a study by Schultz (1993), basalt rock that had unconsolidated compressive strength of 262 MPa at room temperature was measured at 450°C to have a strength of 210 MPa. Generally, large variability was found between the different rocks that were drilled as well as along the depth while drilling. To drill samples with more uniform properties at high temperatures, clay bricks were used to perform the drilling tests.

14.4.3 Modeling and Analysis of Transducers of the HT USDC

The piezoelectric transducer that drives the USDC is the key to its operation and, in order to maximize the drilling capability using piezoelectric materials with relatively low performance, it is essential to optimize its design. For this purpose, finite element (FE), equivalent circuit models, and impedance spectrum measurements were used to predict, analyze, and characterize the performances of various configurations of the transducers.

ANSYS FE package, which is capable of dealing with piezoelectric materials, is used to evaluate and optimize various transducer designs. With reasonable accuracy, properties data of the involved materials are needed to perform the FE analysis. However, many materials data, especially those of piezoelectric materials at high temperature and under large signal excitation, are not available. The FE modeling was first performed using the available materials data at room temperature and then for verification, the corresponding equivalent circuit was calculated with lumped components (Figure 14.6) (Bao et al., 2003). The verification was performed by impedance spectrum measurement at room temperature and the details of the impedance measurement are described in the following paragraph. Then, the impedance data of the transducer were measured at high temperature

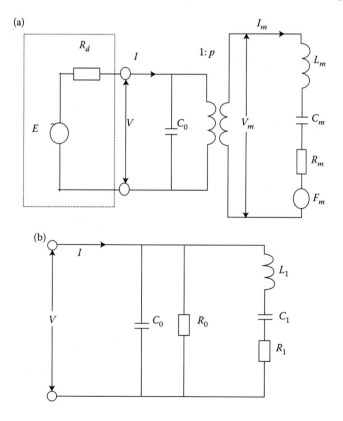

FIGURE 14.6
Schematic diagram of the equivalent circuit of the transducer around resonance. (a) The form of the electromechanical equivalent circuit. The C_0 is the static capacitance of the transducer and the resistor R_0 represents the electric loss. The subscript m denotes equivalent mechanic components, that is, L_m is equivalent mass, C_m is equivalent spring, and so on. All the components can be determined by FE. (b) The form of electric equivalent circuit, which is a simplified version of the form (a) with zero mechanical force F_m. The L_1, C_1, and R_1 form a dynamic branch, which represents the resonance of the transducer.

to obtain the ratios of changes of the components of the circuit. For various designs of transducers having similar configurations, the corresponding equivalent circuits calculated by FE were modified using the changing ratios determined with the first fabricated transducer to estimate the performances at high temperature.

Precision impedance analyzer Agilent 4294a was used to measure the impedance spectrum and characterize the fabricated transducer. To clearly identify the resonance frequency, the complex admittance Y, that is, one over impedance, was measured. A typical curve of the transducer can be seen in Figure 14.34. The resonance frequency, bandwidth (or mechanical Q), and other parameters could be determined from the measured curve. A more efficient method to estimate the characteristics of a resonant transducer is to determine its equivalent circuit with lumped components. It can be done by curve fitting of the measured impedance and is a built-in function of the Agilent 4294a if the R0 is large enough to be neglected. The form of equivalent circuit is shown in Figure 14.6b, where C0 and R0 are the capacitance and loss of the piezoelectric elements and the C1, L1, and R1 branches are corresponding to the mechanical vibration. The performances of the transducer (e.g., input power, electric loss, mechanic Q, and bandwidth at frequency around the resonance) can be calculated easily from the circuit. Furthermore, vibration displacement or velocity can be computed if the normalized displacement distribution (mode shape) is available according to the relationship between Figure 14.6a and b. The mode shape can be computed by the FE model.

A transducer design that consists of a piezoelectric stack, horn, backing block, and prestress bolt, developed for room temperature operation, was used as a starting configuration (Bao et al., 2003) (see Figure 14.7). The piezoelectric stack was made of four

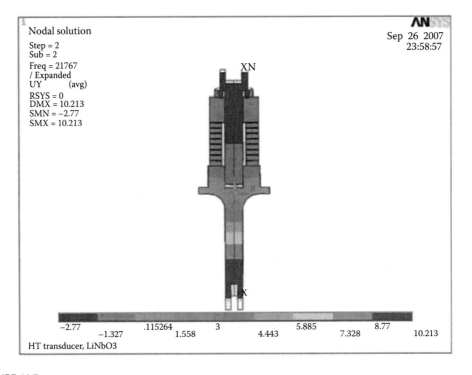

FIGURE 14.7
(See color insert.) The first longitudinal mode of a LiNbO$_3$ shape with resonance frequency at 21.767 kHz. The color scale shows the displacement in the vertical direction.

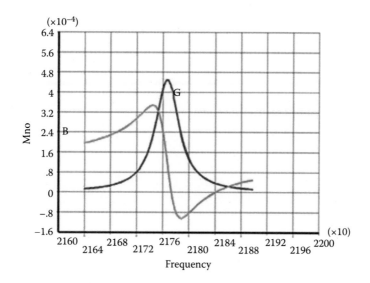

FIGURE 14.8
The input admittance response around the first resonance frequency. G is the real part (conductance) and B is the imaginary part. The mechanical Q is set at 500.

pieces of 0.2 in. thick PZT-8. Using eight pieces of 0.1 in. LiNbO$_3$ single crystal to replace the admittance response calculated by the FE model with material properties at room temperature was determined for frequencies around the first resonance (see Figure 14.8). The calculated component values of the equivalent circuit are listed in Table 14.4 labeled as LN $8 \times 1.0 \times 0.1$ together with those for PZT-8 marked as PZT-8 $4 \times 1.0 \times 0.2$. The FE results suggest that using LiNbO$_3$ as a transducer material has a resonance frequency and mode shape similar to PZT-8 (Figure 14.9). However, the R1 is much higher and it resulted in much lower actuation power and tip vibration velocity under the same voltage. The fabricated transducer was characterized from room temperature to 500 °C. The results were shown in Figures 14.10 through 14.13. The parameters of equivalent circuit at room temperature and 500 °C are listed in Table 14.4 marked as *LN exper $8 \times 1.0 \times 0.1$* under the corresponding temperature labels. In general, the experimental characterization at room temperature confirmed the FE results. However, the R1 increased more at 500°C. Increasing the driving voltage partially compensates for the disadvantage of the high input resistance, which is ultimately limited by electric breakdown. Further, the high loss at high temperatures (see Table 14.4) may result in quick overheating when operating the drill without duty cycling. To address this issue, the possible development of a transducer using more (16) pieces of LiNbO$_3$ or larger diameter (1.5 in.) elements were investigated. Figure 14.14 shows the structures and FE meshes of possible two designs. The corresponding equivalent circuit parameters and predicted performances by the FE at room temperature are listed in Table 14.4 labeled as *LN $16 \times 1.0 \times 0.1$* and *LN $8 \times 1.5 \times 0.1$*, respectively. To estimate the performances at 500°C, the parameters of the equivalent circuits (C0, R0, L1, C1, and R1) were modified by multiplying the same ratios found for the *LN $8 \times 1.0 \times 0.1$* transducer and listed in Table 14.4.

The estimated results show the use of longer or larger diameter LiNbO$_3$ transducers should improve performances although, at 500°C, they are still poorer by a factor of 5–6 than the reference transducer of PZT-8 as judged by the tip velocity under the same voltage. A longer and thicker design was examined as a way to improve the performance

TABLE 14.4

Estimated and Measured Characters and Performances of the LiNbO$_3$ Actuator Designs Shown in Figure 14.14 at Room Temperature and 500°C and the Data for PZT-8 as Baseline

	Room Temperature					500°C			
	PZT-8 4 × 1.0 × 0.2	LN 8 × 1.0 × 0.1	LN 16 × 1.0x0.1	LN 8 × 1.5X0.1	LN exper 8 × 1.0 × 0.1	LN exper 8 × 1.0 × 0.1	LN 8 × 1.0 × 0.1	LN 16 × 1.0 × 0.1	LN 8 × 1.5 × 0.1
R1 (ohm)	116	513	115	147	544.067	2710	2610	570	732
C1 (F)	1.09E − 10	1 55E − 11	8.30E − 11	5 66E − 11	1 41E − 11	1.66E − 11	1 82E − 11	9.82E − 11	6.67E − 11
L1 (H)	S.30E − 01	3.33E + 00	8.62E − 01	9,61E − 01	3.28569	3.96593	4.01E + 00	1.03E − 00	1.16E − 00
C0 (F)	2.78E − 09	7.39E − 10	1.43E − 09	1.64E − 09	7.61E − 10	1.18E − 09	1.17E − 09	2.27E − 09	2.60E − 09
Fr (Hz)	20908	22182	18816	21575	23366	19629	18634.56	15806.866	18124.636
Fa (Hz)	21316	22413	19356	21944	23585	19766	18779.17	16144.905	18355.629
Ct (F)	2.88E − 09	7 55E − 10	1 51E − 09	1 70E − 09	7.65E − 10	1.20E − 09	1.18E − 09	2.369E − M	2.666E − 09
Eff. Coupling	0.1947	0.1432	0.2346	0.1826	0.1358	0.1176	0−1239	0.2036	0.1581
Norm D_tip	10.569	10.604	7.4	6.997	10.213	10.213	10.604	7.4	6.997
Om	00	887	887	887	887	180	180	180	180
Voltage (V)	10	10	10	10	10	10	10	10	10
Pm (W)	0.43	0.10	0.44	0.34	0.09	0.02	0.02	0.09	0.07
Velo_tip (m/s)	0.66	0.37	0.60	0.47	0.34	0.08	0.08	0.13	0.10

Note: The rows R1–C0 are the components of equivalent circuit of Figure 14.6b; R0 is omitted because it is large and negligible; Fr is the resonant frequency; Fa is the antiresonant frequency; Ct is the capacitance at low frequency; Eff. Coupling is the effective electromechanical coupling coefficient; Norm D tip is the normalized tip displacement; Qm is the mechanical quality factor; Voltage is the voltage set to calculate input power and tip velocity; Pm is the input power under the Voltage at resonance; Velo_tip is the tip velocity under the Voltage at resonance.

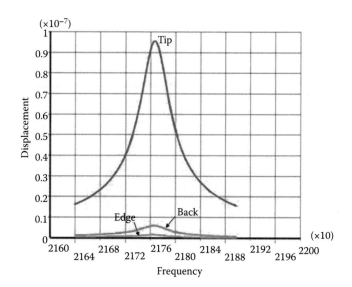

FIGURE 14.9
The displacement distribution as a function of frequency for $LiNbO_3$ at the tip of the horn, as well as the back of the transducer and the plane of the mounting frame.

further. Considering the larger diameter transducer having thick horn tip would be able to drive heavier free-mass and create larger impacts at the same velocity, a transducer of 2 in. diameter was fabricated. The details are presented in Section 14.4.5.

Although the Agilent 4294A impedance analyzer has a build-in function to extract the parameters of the equivalent circuit, but it could not function correctly with the piezoelectric materials having relatively high electric loss since the equivalent circuit model used by the analyzer neglects the dielectric loss. The dielectric loss of the piezoelectric ceramics bismuth titanate at high temperature is quite high. A computer program was developed to extract the parameters for the case.

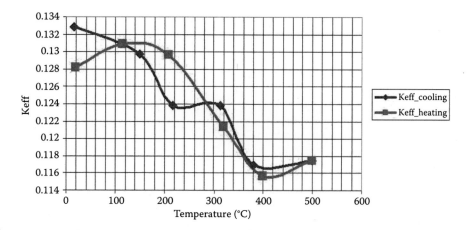

FIGURE 14.10
The effective electromechanical coupling coefficient as a function of temperature. This coefficient is defined as square root of mechanical energy over total (mechanical and electrical) energy of the oscillating transducer at the resonance. This coefficient directly affects the power capacity and efficiency of the transducer.

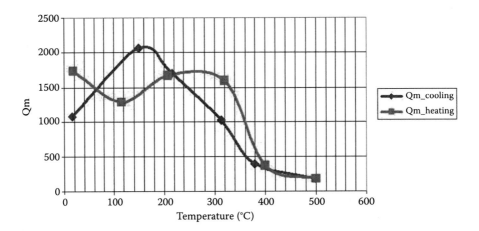

FIGURE 14.11
The mechanical Q of the transducer as a function of temperature. The Q is mechanical energy stored divided by the energy loss in each cycle. A low Q means a low efficiency of the transducer.

In additions, the Agilent 4294a can measure the impedance of the transducers at room or high temperature with small signal (≤1 V) only. To characterize the transducer performances under large voltage excitation, we constructed a homemade setup to perform impedance measurements under large signals. The setup consists of a function generator generating a linear frequency modulated sine signal, an amplifier amplifying the signal to a higher voltage close to the operating voltage level, and a digital oscilloscope recording the applied voltage and the voltage across a 1 Ω resistor that was connected to the

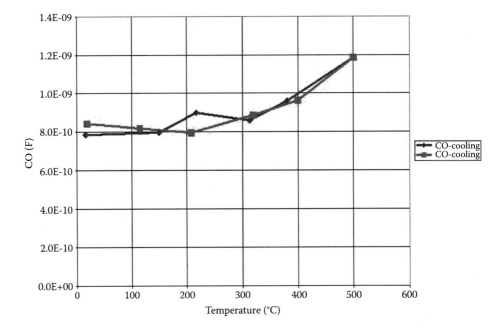

FIGURE 14.12
The capacitance C_0 of the transducer as a function of temperature. A 50% increase of the capacitance was observed.

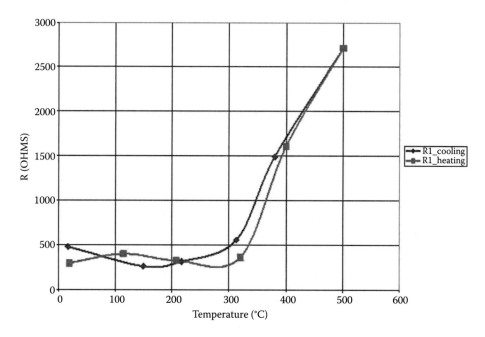

FIGURE 14.13

The resistance R_1 of the transducer as a function of temperature. The increase of the resistance at temperature above 300°C implies higher input voltage will be required to excite the transducer at the same power levels.

transducer in series. The voltage across the 1 Ω resistance was numerically equal to the current through the transducer. The impedance at the scanned frequency range was determined from the recorded voltage and current data. The large signal level was of the order of hundreds of volts. The scans can be done at room temperature and high temperature of 500°C.

An example of the experimental data obtained by the measurement of a high voltage scan and the corresponding admittance of the equivalent circuit with the extracted parameters are presented in Figure 14.15. The agreement between the raw data and fitted,

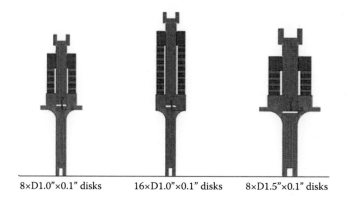

8×D1.0"×0.1" disks   16×D1.0"×0.1" disks   8×D1.5"×0.1" disks

FIGURE 14.14

Using 2.54 mm (0.1 in.) thick piezoelectric disks, the response of a 25.4 mm (1 in.) diameter actuator stacks with 16 disks were examined analytically and compared. Also, the performance of 8 disks with a 25.4 mm (1.0 in.) and 28.1 mm (1.5 in.) diameters were compared.

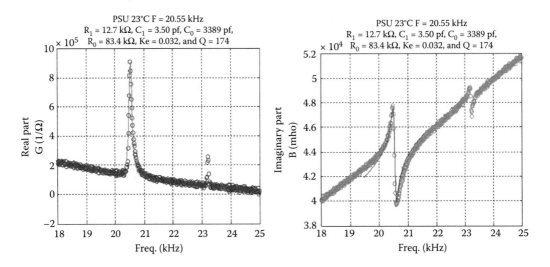

FIGURE 14.15
Example of large signal impedance scans: the complex admittance and parameters extracting. Circles—experimental data; line—fitted curve with the equivalent circuit parameters shown in title.

calculated from the extracted circuit in the frequency range around the resonance, is an indicator of the accuracy of the model.

The data in Table 14.5 is a summary of the impedance analysis results for bismuth titanate transducers made by ferroperm (PZ46) (Tichý et al., 2010) and by Penn State University (PSU). The tests were done at low and high voltages as well as at 23°C and 500°C. In the table, the resonance frequency Fr, static capacitance C_0 and its loss resistance

TABLE 14.5

Summary of Impedance Analysis for Transducers Made of Bismuth Titanate

Transducer	PZ46				PSU			
Temperature	23°C		500°C		23°C		500°C	
Voltage level	0.5 V	1100 Vp	0.5 V	850 Vp	0.5 V	1000 Vp	0.5 V	570 V p
Fr (Hz)	22996.88	22678.2	2080 5.25	20480.57	20830	20555	19384.1	19082.851
C0 (F)	1.84E − 09	2.17E − 09	2.75E − 09	3.07E − 09	3.46E − 09	3.39E − 09	4.20E − 09	5.70E − 09
R0 (ohm)	1.00E + 07	187318	48780.488	52423	1.00E + 07	83400	15038	8319
C1 (F)	2.40E − 12	4.03E − 12	3.62E − 12	4.58E − 12	4.15E − 12	3.50E − 12	4.82E − 12	3.79E − 12
R1 (ohm)	2359	24391	2968	20012	1446	12699	1152	7900
Qm	1224.32	71.38	711.29	84.83	1272.06	174.21	1479.70	278.86
K dielectric	116.18	136.99	173.72	193.92	216.05	211.57	262.19	355.75
Effec. Coup. Ke	0.0360	0.0430	0.0362	0.0386	0.0346	0.0321	0.0338	0.0258
Transformer M/E (n)	0.2237	0.2861	0.2488	0.2753	0.2668	0.2416	0.2673	0.2333
Efficiency M/E	1.000	0.885	0.943	0.724	1.000	0.868	0.929	0.513
|Z| at Fr (ohm)	1998	3210	1971	2490	1209	2237	938	1376
Vtipat applied V (m/s)		1.6552		1.6198		3.4221		3.2473

R_0, dynamic capacitance C_1, and resistance R_1 are the extracted parameters of the equivalent circuit. The dynamic inductance L_1 can be determined by the resonance Fr and C_1. The Qm is the mechanical Q value and K is the dielectric constant of the material. K_{eff} is the effective coupling factor defined as the square root of mechanical energy over the total (mechanical and electrical) energy of the oscillating transducer at the resonance frequency (see Figure 14.10). The "Vtip at applied V(voltage)" was estimated by using the equivalent circuit and the normalized mode shape computed by the FE model (Bao et al., 2010a).

Both PZ46 and the PSU transducers worked at 500°C with increased capacitance and higher electric loss compared with room temperature results. Under high voltage, both have higher electric and mechanical loss (lower Qm) compared with low voltage values. At high temperature and high voltage, the PSU transducer exhibited better overall performance than the PZ46. It is easier to reach higher tip velocity with the PSU material than using the PZ46 material as the "Vtip data at applied V" in the table shows.

14.4.4 Testbed Setup

In order to test developed samplers at high temperatures, a testbed chamber was needed. The chamber has to provide a controlled elevated temperature environment and allow placement of the drill and samples inside the chamber and perform tests while controlling the preload externally. The JPL authors tested their drill at temperatures as high as 500°C and tracked the drilling rate as a function of various test parameters. Specifically, the testbed was designed to allow for placing the drill inside the chamber with the bit pushed against the test rock while controlling the preload of the drill. The testbed consisted of a commercial horizontal tube shape furnace (made by Carbolite, UK) with customized insulation side caps with center hole (see the sketch in Figure 14.16). This hole was used in the initial studies for inserting the bit where the drill was kept outside the chamber (Figure 14.16) and later the drill was inserted completely inside the chamber while the cable was inserted through the hole. To control the load on bit and feed rate, a pushrod was used that was controlled by a pneumatic cylinder (behind the USDC in the photo of Figure 14.16). The preload controller was driven by air from an air cylinder with 3/4 in. bore diameter (Model Norgren RLC09A), which was mounted on a frame behind the support fixture. The pressure in the cylinder and the pushing force were controlled by Omega Electropneumatic Transducer CS-2250. A dial indicator with resolution of 0.0254 mm (0.001 in.) was used

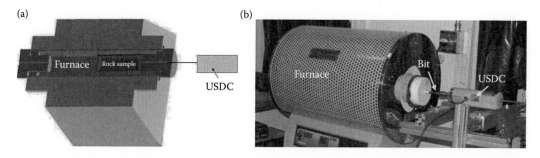

FIGURE 14.16
(a) Schematic cross-section and (b) photographic views of the furnace of the developed HT testbed. For the initial testing, rock samples were placed inside the chamber and were drilled at various temperatures while the USDC was placed outside and the preload was controlled pneumatically.

to measure the forward distance of the shaft of the cylinder. In this configuration, Santa Barbara limestone was drilled at room temperature (RT) reaching a depth of 22.3 mm in 23 min, while at 500°C, a depth of slightly greater than 2 mm was reached. The bit was made of crucible hardened CPM-3 V with hardness HRC 59. Degradation of the cutting teeth of the bit was observed after drilling at 500°C, necessitating the use of a more durable material.

To test the USDC with the HT piezoelectric stacks inside the chamber, a fixture was constructed as shown schematically and photographically in Figure 14.17. This fixture allows sliding the USDC while drilling as well as securing the position of the drilled rocks along the bit path. The USDC was pushed from outside the chamber using the controlled preload system.

The driving power was provided by a power amplifier ENI1104LA with an input from Tektronix AFG 3022 Dual Channel Function Generator. The latter was programmed to generate a sine wave with controlled duty cycle (e.g., on for 1 s per 2 s cycles). A custom-made transformer with a voltage ratio factor of up to 1:5 was placed between the amplifier and the sampler to improve the impedance match to the ultrasonic transducer. A 1 ohm resistor was connected to the sampler input in series as a current–voltage converter for the monitoring oscilloscope. The driving voltage and current were monitored and recorded by a digital oscilloscope (Model Tektronix TDS5054B-NV). The input electric power was calculated in real time by programming the oscilloscope. The instrumented chamber that makes up our testbed is shown photographically in Figure 14.18.

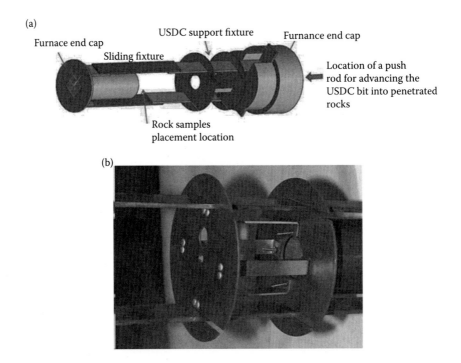

FIGURE 14.17
(a) Schematic and (b) photographic views of the support fixture for testing the USDC and its drilling performance inside the HT chamber. The photograph shows a close-up of the USDC mount.

FIGURE 14.18
The high-temperature testbed setup included the instrumented chamber, the support of the sampler (with the sampler and sample), the pneumatic pushing setup, as well as the drive and measurement instruments.

14.4.5 HT Piezoelectric Actuated Drills

14.4.5.1 Basic Designs of the HT Piezo-Actuated Drills

The components of a USDC that can operate at high temperatures were produced to support the testing of the developed piezoelectric stacks. The key components of the USDC are the actuator, free-mass, and bit. A schematic and a cross-section diagram of the USDC as well as the actuator are shown in Figures 14.19 and 14.20, respectively. To make sure that the actuator can operate at the required temperature of 500°C, the horn was fabricated of titanium and the stress bolt and backing of stainless steel, whereas the Belleville washers were made of Inconel to increase thermal stability and maintain the piezoelectric stack prestress as the actuator is heated. The use of titanium allows for producing a low mass device that has high mechanical strength. The $LiNbO_3$ or piezoelectric ceramic disks were mounted into the actuator and held in compression by the stress bolt to prevent fracture of the disks during operation. In order to design the actuator and determine its resonance frequency and input impedance, it was analytically modeled using both thermal and electromechanical analyses. Using the HT chamber that is shown in Figure 14.18, the impendence characteristics of the actuator at RT and 500°C were measured.

High-temperature
ultrasonic transducer

Bit

FIGURE 14.19
Schematic view of the high temperature USDC.

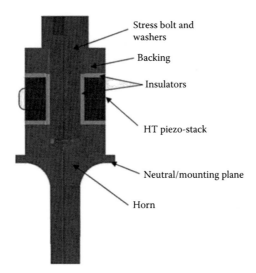

FIGURE 14.20

A cross section view of the high temperature USDC actuator.

TABLE 14.6

Baseline Transducers That Were Made of 1 in. Diameter Piezoelectric Rings

Transducer #	Material	Number of Rings	Outer Diameter (mm)	Inner Diameter (mm)	Thickness (mm)
1	LiNbO$_3$	10	25.4	12.7	2.0
2	PZ46	10	25.0	13.0	2.0
3	PSU BT	10	25.4	12.7	2.1

TABLE 14.7

Summary of the Number and Dimensions of the Rings

Xducer (mm)	Piezo Material	Number of Rings	Out Diameter (mm)	Inner Diameter (mm)	Thickness (mm)	Horn Tip Diameter (mm)
38.1 BT	PSU BT	10	38.1	12.7	2.0	9.5
50.8 LN	LiNbO$_3$	4	48.3	25.4	5.0	17.8

In the initial tests, three transducers were made using LiNbO$_3$, PZ46, and BT ceramic rings. The summary of the number and dimensions of the rings are listed in Table 14.6; the baseline transducers that were made of 1 in. diameter piezoelectric rings.

The test results showed that more powerful HT transducers were needed to obtain a better performance. The analytical results were used to guide the design of two larger diameter transducers: 38.1 mm (1.5 in.) diameter bismuth titanate and a 48.3 mm (2 in.) diameter LiNbO$_3$. The summary of the dimensions and number of the rings that were used are listed in Table 14.7.

14.4.5.2 2.5 cm (1 in.) Diameter Samplers

Samplers with 2.5 cm (1 in.) actuators using three different piezoelectric materials were tested in both room temperature and 500°C drilling brick. The dimensions of the related drills

TABLE 14.8

Summary of Drilling Test of 1 in. Sampler

	23°C				500°C			
	V rms	A rms	Minutes	Drill Rate (mm/mm)	V rms	A rms	Minutes	Drill Rate (mm/mm)
PZ46	1320	0.4	12	1.94	830	0.37	10	0.02
PSU	900	0.35	7	1.1	460	0.3	4	0.14
LiNbO$_3$	216	0.054	4	3	580	0.041	2	1.27

and stacks are given in Table 14.8. The drilling was stable at room temperature. However, at high temperature, the drilling started with a lower rate and dropped quickly, especially for the LiNbO$_3$ sampler. A summary of the maximum drilling rates and driving conditions are given in Figure 14.21. The LiNbO$_3$ sampler initial drilling lasted for 2 min only.

For the 500°C tests, a transformer with primary/secondary ratio of 5 was used to increase the voltage that was applied to the actuator of the samplers. The highest drilling rate was obtained by the LiNbO$_3$ sampler. The drilling rate was ~1.4 mm/min during the first minute under a drive voltage of ~580 V rms and decreased quickly in the following runs as shown in Figure 14.21. Electric current surges were encountered in the following consecutive runs. It is believed that electric arcs through the cracks in the broken crystals were the source of the current surges (Ryan, 2011). After the drilling tests postdrilling measurements of the electrical properties of the actuator showed that the resistance between the electrodes was greatly reduced even at room temperature.

The samplers that were driven by the PZ46 and the bismuth titanate actuators worked at 500°C and were found to be relatively robust, where the bismuth titanate actuator showed better performance than the PZ46. Only a few cracks in the ceramic rings were found after the tests. However, the achieved drilling rates were lower than with the LiNbO$_3$ actuator. A photograph of the actuator after being exposed to the 500°C environment is

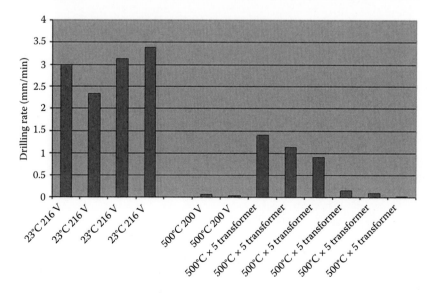

FIGURE 14.21
Drilling test results of 1 in. sampler using lithium niobate crystals.

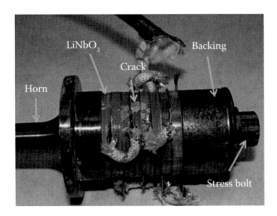

FIGURE 14.22
Photographic view of the first HT USDC actuator breadboard. The piezoelectric stack was made of $LiNbO_3$ and was tested at 500°C.

shown in Figure 14.22. While tightening the stress bolt, a malfunction in the electrical meter led to overstressing the crystals, which apparently fractured some of the piezoelectric disks. However, the test results were encouraging since the resonance characteristics of the $LiNbO_3$ transducer indicated good response in spite of the fracture of the individual disks. To improve on this issue, the disks were manufactured with beveled edges to minimize stresses on the corners and softer electrodes were used to avoid stress risers. In addition, more stringent assembly procedure was initiated. Another issue that was identified is the electric shorting to the stress bolt from the electrodes through the mica film electrical insulation. To prevent applying transverse stresses on the mica insulation and possibly damaging the film and causing a short, great efforts were made to assure the alignment of the electrodes. Also, a thicker layer of mica was used to increase its insulation capability.

14.4.5.3 3.81 cm (1.5 in.) Diameter Bismuth Titanate Samplers

After assembling of the 3.81 cm (1.5 in.) diameter bismuth titanate, the sampler was tested drilling basalt at room temperature. The bit used was made of tungsten carbide with six teeth. The preload for the free-mass was equal to 15 N. The sampler was driven by 900 Vrms with a duty cycle of 0.5 s per second. The current was around 1 A and power was ~77 W (see Figure 14.23). The results are presented in Figures 14.24 and 14.25. The penetration rate at the first 0.5 min was ~0.15 mm/min. However, it dropped in the following drilling experiment. At 5 min, a rotation of the bit slowly by hand was initiated and the drilling rate was found to be stable at ~0.24 mm/min. When applying percussive drilling, tooth-prints are formed that slows the drilling rate because of the increased contact surface area. The application of rotation eliminates the formed tooth-print at the bottom of the hole and it helped increasing the drilling efficiency.

The 3.81 cm (1.5 in.) diameter bismuth titanate-driven USDC-based sampler was also tested at 460°C drilling a brick. The bit assembly consisted of two teeth and with a center bar extending to the outside of the furnace but without self-rotation mechanism. The bit was manually rotated at a low speed of about 4–5 rpm. The weight-on-bit (WOB) was 20 N and the preload for free-mass was 17 N. The sampler was driven in a duty cycle of 0.5 s per second. The bit was much easier to rotate when the sample was active

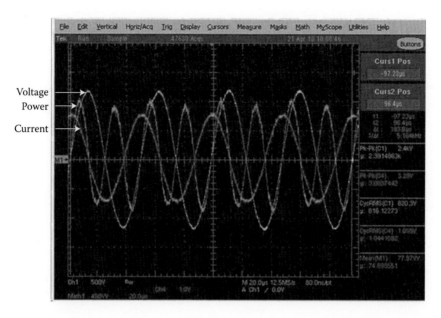

FIGURE 14.23
A sample of monitoring screen of the digital oscilloscope showing the waveforms of driving voltage, current, and power in the basal drilling test.

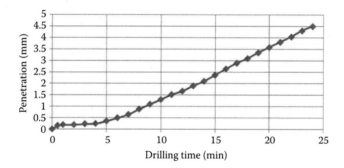

FIGURE 14.24
Penetration versus time for drilling basalt with the 3.81 cm (1.5 in.) diameter bismuth titanate samplers.

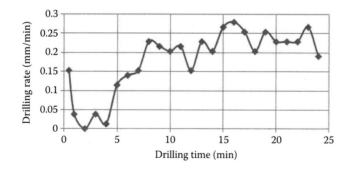

FIGURE 14.25
Drilling rate versus time for drilling basalt with the 3.81 cm (1.5 in.) diameter bismuth titanate samplers.

FIGURE 14.26
(See color insert.) 1.5 in. bismuth titanate sampler on fixture after test.

than inactive. The torque needed for the rotation was estimated to be less than 0.02 Nm (0.2 in.-lbs). The close-up photos of the drill and the sample are shown in Figures 14.26 and 14.27.

The test results are presented graphically in Figures 14.28 and 14.29. The brick sample was 26 mm in thickness. A starting hole of about 1 mm deep was drilled initially at room temperature and the drilling was continued at high temperature. A sample of the screen of the monitoring oscilloscope is shown in Figure 14.30. The driving frequency was 20.9–21.0 kHz, averaged voltage was 400 Vrms, and averaged current was 0.69 A. The averaged power was 26 W. The power factor was low, approximately ~0.094. The sampler drilled through the sample in 21 active minutes before reaching the other end of the brick sample. The average drilling rate was 1.2 mm/min. The decrease of the drilling rate with time or depth was clearly seen. A room temperature test after the high-temperature drilling showed a reasonable rate compared to previous tests. Besides a decrease in efficiency of

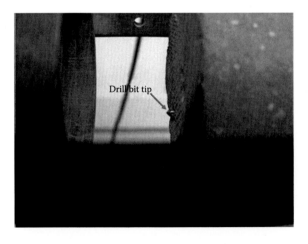

FIGURE 14.27
The tip of the bit was seen in the other end of the brick.

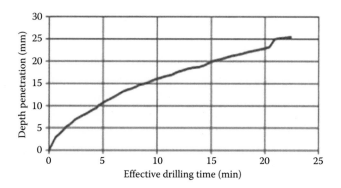

FIGURE 14.28
Penetration versus time at 460°C.

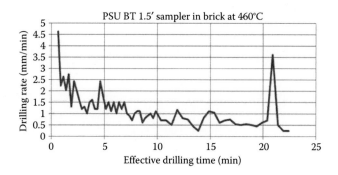

FIGURE 14.29
Drilling rate versus time at 460°C.

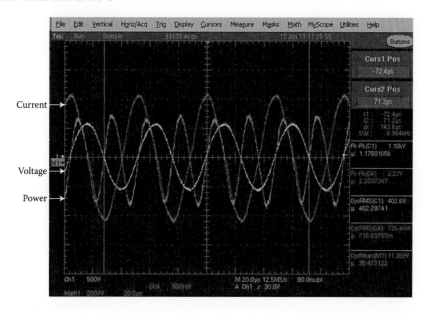

FIGURE 14.30
A sample of monitoring screen of the digital oscilloscope showing the waveforms of driving voltage, current, and power in the brick drilling test at 460°C.

cuttings removal, the misalignment, which can result in friction between the bit and the hole, might be the major cause of the decrease of the drilling rate.

14.4.5.4 4.83 cm (2 in.) Diameter LiNbO₃ Samplers

Two samplers with 4.83 cm (2 in.) diameter $LiNbO_3$ actuators were assembled consecutively.

The first (#1) sampler: This sampler was manually wrapped using a fiberglass wire where the wrapping was not tight and the resulting prestress was estimated to be low. It was equipped with a helical adapter but without a central bar. At room temperature, the bit rotated but the rotation was not stable. Because no central bar was used that extended the end out of the furnace, it was not possible to determine if it rotated in the 500°C furnace or not. The following tests were then performed:

- At room temperature—Drilled basalt sample at a rate of 1.65 mm/min
- At 500°C—The drilling results in the basalt for the sampler were
  - 1st minute 0.91 mm, driven with 22.44 kHz, 465 V_{rms}, 0.11 A, 46.5 W (power factor 0.91), WOB 13.75 N
  - 2nd minute 0.13 mm, driven with 22.44 kHz, 466 V_{rms}, 0.10 A, 44.3 W (power factor 0.91), WOB 13.75 N

The large drop of the drilling rate might be due to the lack of using rotation, the increase of the cross section of the bit near the tip, or material property changes at high temperature. After cooling, significant cracking of the crystals was observed.

The second (#2) sampler: This sampler was built after the above tests. The stack was wrapped using an improved wrapping method, and two de-poled PZT rings added to provide further protection of the $LiNbO_3$ stack from cracking. The latest version of the self-rotating bit assembly was used. Here, the self-rotating bit worked but stopped frequently. To keep the bit rotating, the rotation was done manually. The tests that were conducted are listed below:

- At room temperature, the 4.83 cm (2 in.) LN #2 rotary/hammering sampler drilled at a rate of 7–8 mm/min in a brick sample.
- The test results at high temperature (460°C) are given in Figures 14.31 and 14.32. The driving frequency, voltage (Figure 14.33), and weight-on-bit were adjusted during the test in order to search for the best operating parameters.

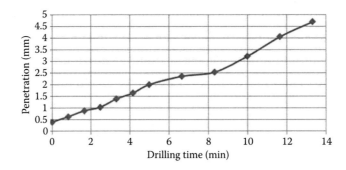

FIGURE 14.31
The penetration depth as a function of drilling time for sampler #2 at 460°C.

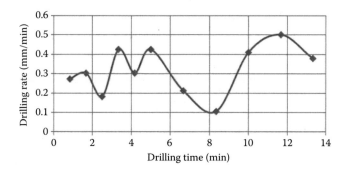

FIGURE 14.32
The penetration rate as a function of drilling time for sampler #2 at 460°C.

- In the last 5 min, the driving frequency was 19.9 kHz, the average power was 78 W, the weight-on-bit was 18.7 N, the preload for free-mass was 40 N, and the average drilling rate was 0.43 mm/min.
- The test ended due to the fact that the power amplifier shut off. Tests have shown that the shapes of the impedance curves of the transducer at room temperature were similar, but the resonance frequency was lower than what was measured before the test (Figure 14.34). Drilling tests at room temperature showed large degraded performance.

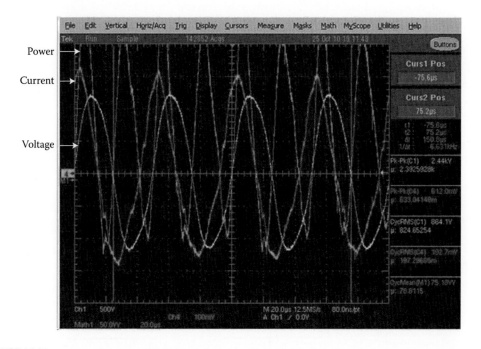

FIGURE 14.33
A sample of monitoring screen of the digital oscilloscope showing the waveforms of driving voltage, current, and power for sampler #2.

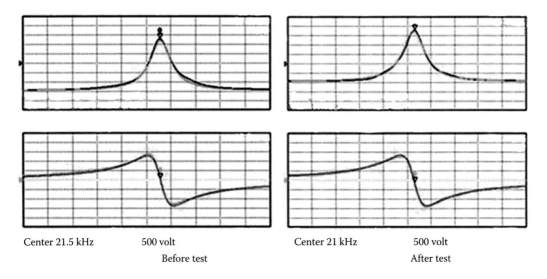

Center 21.5 kHz	500 volt	Center 21 kHz	500 volt
	Before test		After test

FIGURE 14.34

The impendence curves for sampler #2 before and after the test showing relatively similar results but a 0.6 kHz shift down in the resonance frequency.

- After the fiberglass wire was unwrapped, the two crystals at both ends of the stack were found cracked. It was determined that sampler #2 was a better transducer than sampler #1 when one looked at the overall performance.

14.4.5.5 Special Technical Measures for the Operation at HT

To address the challenges that are involved with the operation of samplers that are actuated by piezoelectric stacks at high temperatures, several special technical measures were initiated and applied to the samplers' design and fabrication. Some of them were based on the lessons learned.

1. *Stabilization of the axial prestress*: The piezoelectric elements are held under compression by a stress steel bolt. The coefficient of thermal expansion (CTE) of the piezoelectric elements is much lower than that of steel. While the transducer is assembled at room temperature, it is expected that the prestress applied to the elements is lost at high temperature. To stabilize the compressive stress, Belleville spring washers made of Inconel were used. These washers are spring shaped and also known as coned-disk springs, conical spring washers, disk springs, or cupped spring washers.

2. *Machinable ceramic for electric insulation*: At the beginning of the task, a mica sheet was used as an insulation tube between the steel bolt and electrodes. Tests have shown that the mica is easily damaged during the assembling and electrically fractured at the high-temperature tests when high voltage is applied. Therefore, machinable alumina was selected to make the isolation tube.

3. *Application of radial prestress*: Piezoelectric materials, especially the $LiNbO_3$ crystal, are brittle and have much lower CTE than the metals that are used to produce the sampler. Cracks were developed after the operation at high temperatures. To prevent cracking, the $LiNbO_3$ crystals were wrapped by fiberglass wire (Figure

FIGURE 14.35
Winding the fiberglass fiber around the actuator.

14.35) in order to provide radial compression stress on the crystal and prevent their cracking. The fiberglass wrapping parameters for 2 in. LN sampler are given in Table 14.9 and the induced radial prestress was calculated as 23.4 MPa.

4. *Low CTE caps for the crystal stack*: To reduce the thermal stress in the crystal and avoid cracks, depoled PZT rings were used at both ends of the $LiNbO_3$ stack. These caps are intended to prevent direct contact between the $LiNbO_3$ and metallic parts like the horn base and backing reducing radial stress on the crystals.

5. *Bit assembly for independent control of free-mass preload and WOB*: The analytical results showed that the high-temperature transducer would have a lower performance than a room-temperature transducer because of the lower performances of the available materials at these temperatures. Optimization of the operation condition is important for the HT samplers. The design using a center bar plus a spring for the drill bit assembly was applied. This design allows independent control of the preload provided to the free-mass and the WOB. It also provides more choices to operate the sampler and increase the drilling rate.

6. *Self-rotary bit*: Tests have showed that even slow rotation of the bit significantly helps to increase the drilling rate of percussive drill. To employ rotation via the vibratory actuation of the USDC, a self-rotating bit design was developed (Figure 14.36). A helical set of slots was introduced to the bit assembly and this structure partially converts the longitudinal impacts to a twisting deformation and bit rotation. This mechanism rotated the bit slowly in the tests of the 4.83 cm (2 in.) diameter $LiNbO_3$ sampler. However, the observed rotation was not sufficiently stable particularly at high temperatures.

TABLE 14.9

Fiberglass Wire Wrapping Parameters for the 2 in. $LiNbO_3$ Sampler

Tension	# Layers	# Strands per Layer	# Strands per Crystal	Stress (MPa)
250 N	7	2	14	23.4

(a)

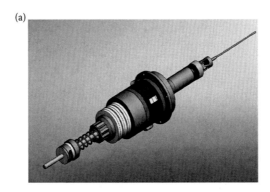

(b)

FIGURE 14.36
(See color insert.) The components of the rotary-hammering sampler. (a) General schematic view and (b) schematic view of the components.

14.5 Thermal Drilling

Heating of rocks is another method of penetrating the subsurface. There are two principal methods of such thermal drilling: thermal-spalling and thermal-melting. The difference between these two methods is the mechanism of breaking rocks that is determined by their applied temperature range. The thermal-spalling occurs at lower temperatures of about 400–600°C and it uses the resulting thermal expansion and mismatch; while thermal-melting uses vaporization at temperatures that that are, typically, in the range of 1100–2200°C (Maurer, 1980).

14.5.1 Thermal-Spalling

Heating a heterogeneous rock causes thermal stresses due to mismatch in the thermal expansion of its constituents and the grains within it structure (Just, 1963). These stresses cause flaking of rock fragments due to the fracture and degradation of the rock. The thermal gradients produced in the rock determine the effectiveness of the fracturing process and the use of its fracturing capability is limited since not all rocks are sufficiently heterogeneous to sustain spalling. Generally, thermal-spalling is a natural process of rock breakage and it is also known as "exfoliation." For example, in the desert, rock surfaces are heated and expand during the day and contract during the night due to cooling (Figure 14.37). The expansion and contraction creates small cracks that grow with time until entire layers of rock break off. The process is accelerated in the presence of water and at below freezing temperatures due to the increase in the expansion coefficient of water ice. Heating a rock nonuniformly may also cause development of internal stresses and cracks. Laser or plasma can be used as forms of heating (Xu et al., 2003) and the degraded rock material can be removed by such techniques as water jet erosion.

FIGURE 14.37
Granite dome exfoliation due to the cyclic increase and decrease in the temperature of the surface layers. (From Wikimedia Commons, http://en.wikipedia.org/wiki/File:GeologicalExfoliationOfGraniteRock.jpg.)

14.5.2 Melting and Vaporization

Subjecting a rock to very high temperatures can cause its melting followed by vaporization. This method could use a high-intensity laser beam to form controlled-shape holes (Ready, 1997). The laser does not necessarily have to be close to a material. For example, the laser on the NASA's Curiosity rover can be used to vaporize rocks from up to 7 m distance. The laser is part of a laser-induced breakdown spectroscopy (LIBS) instrument called Chemistry and Camera (ChemCam) analyzer. Once the laser strikes a rock creating plasma, the resulting emission spectra is captured by a remote microimager (RMI) to determine the composition of the rock. This method is destructive since a small hole is left behind. If the laser continues to strike the same area in a rock, one can learn about the rock stratigraphy since the hole gets deeper with every laser shot. An example of holes produced by the laser in a martian rock at a distance of 3.5 m (11.5 ft) from the Curiosity rover is shown in Figure 14.38. The diameter of the circular field of view is about 3.1 in. (7.9 cm).

FIGURE 14.38
Laser holes produced by the Chemistry and Camera instrument aboard NASA's Curiosity Mars rover. From left to right: The photos are showing before and after the laser interrogation of the rock, respectively. (NASA/JPL-Caltech/LANL/CNES/IRAP http://photojournal.jpl.nasa.gov/catalog/PIA16075.)

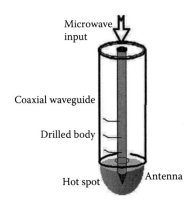

FIGURE 14.39
Schematic illustration of a microwave-drill. (Courtesy of Eli Jerby, Tel Aviv University, Ramat Aviv, Israel.)

Melting rocks requires very high power. However, in ice, the power requirement is lower. In the north and the south poles, this method is used to drill ice and make holes in ice sheets and ice shelves, allowing reaching penetration rates as fast as 1 km in 10 h. The use of ice melting drills has also been proposed by NASA for exploration missions to Europa (one of the icy moons of Jupiter), and the water–ice cap at the north and the south poles of Mars (Smith et al., 2006; Rapp, 2007).

Instead of using direct heat, melting could also be achieved using microwaves (Maurer, 1968). Researchers at Tel Aviv University, Israel (Jerby et al., 2002) have been developing this approach; they used a coaxial near-field radiator driven by a conventional microwave source (Figure 14.39). The center electrode of a coaxial radiator acts as the bit and it drills by softening and then melting the penetrated medium forming a hole. The drill bit serves as an antenna that focuses the microwave energy onto a small spot below the surface of the drilled material. Melting of any nonconductive material along the path of the microwave is possible. The microwave drilling increases the temperature of the drilled formation to as high as 1500°C. In concrete drilling tests, a 2 mm diameter bit drilled a 2 cm deep hole in about a minute.

An example of a 12 mm diameter and 10 cm deep hole in concrete is shown in Figure 14.40 (note that drilled cuttings were removed by mechanical means). The benefits of using this method include the elimination of the need to rotate the drill bit; it does not produce dust or noise. The latter advantage is important for such applications as construction in urban areas or for some military applications. However, this method is not ideal for *in situ* planetary exploration due to its high power requirement that can be in the range of hundreds of watts and higher and the thermal and physical/chemical alteration or destruction of the drilled material and the material surrounding the hole.

14.6 Summary and Conclusions

This chapter presented mechanisms of drilling at high-temperature environments. These mechanisms include the use of high temperature motors and piezoelectric actuators that drive the drills.

FIGURE 14.40
A 12 mm diameter, 10 cm deep hole in a concrete slab produced by a microwave drill. (Courtesy of Eli Jerby, Tel Aviv University, Ramat Aviv, Israel.)

Two SRM developed for high-temperature applications by Honeybee Robotics were used for actuating a conventional drill. These two degrees of freedom were required for rotating auger and the bit, and for advancing the bit into the rock. Tests conducted at 460°C have shown that this conventional approach to drilling is feasible. In particular, a drilling depth of 20 cm was reached in approximately 20 min with a maximum power of 45 W.

The second approach actuators are based on piezoelectric materials and electromagnetic mechanism. These were specially designed to operate at temperatures as high as 500°C. The greatest accomplishment of making a drill that is operated by piezoelectric materials was reached using the bismuth titanate actuator material (made by Penn State University). This sampler drilled a brick sample to a depth of 25 mm in 21 accumulated minutes (accounting for the use of duty cycling). The developed sampler is based on the USDC and it was demonstrated that a modified breadboard sampler could be operated at the Venus ambient temperature. The drill was designed with a novel method of operating the USDC as a rotary-hammer where the rotation is induced by the vibration of the piezoelectric actuator. The drill was actuated by piezoelectric materials that have a higher Curie temperature than the temperature on the surface of Venus. Key benefits of the developed sampler is its ability to create fine powdered cuttings that are ideal for x-ray diffraction and x-ray fluoroscopy analysis and many other analytical instruments. The high pressure on Venus was not considered since it is not expected to affect the operation of the USDC.

Six different USDC breadboards were produced, including 2.54 cm (1.0 in.) diameter $LiNbO_3$, PZ46 (commercially made by Ferroperm), bismuth titanate (made by Penn State University) in a rotary-hammering configuration; 3.81 cm (1.5 in.) bismuth titanate; 4.83 cm (2.0 in.) diameter $LiNbO_3$; and 4.83 cm (2.0 in.) diameter $LiNbO_3$ in a rotary-hammering configuration. Isothermal tests of the $LiNbO_3$ piezoelectric disks at 500°C were made and showed high stability with no measureable change in properties after 1000 h of exposure. Another accomplishment includes the finding that the bismuth titanate with various dopings of tungsten shows a thickness coupling coefficient of about 15–20% at 500°C comparable to the room temperature value.

To simplify the analysis of the high-temperature USDC actuator and designing effective samplers, equivalent circuit models were developed for the piezoelectric rings as lumped circuit components. To test the operation of the drill, a high-temperature testbed was developed, capable of operating significantly beyond the ambient temperature on Venus. Brick samples were used in this study since they showed more uniform properties at high temperatures than the natural rocks that were initially tested, including basalt and limestone.

Many significant challenging issues had to be overcome, including cracking when using $LiNbO_3$ crystal as an actuator for the USDC-based sampler. Based on the lessons learned, making a sampler using a bismuth titanate actuator that performs consistently will require improvements to the self-rotating bit mechanism, and the power of the actuator will need to be increased to a level that allows it to effectively drill hard rocks.

In addition to various drills that operate at high temperatures, there are also drills that use high temperatures to penetrate rocks. These drills use two primary methods of high-temperature rock destruction: thermal-spalling and thermal-melting. The difference between these two methods is the mechanism of breaking rocks that is determined by their applied temperature range. The thermal-spalling occurs at lower temperatures of about 400–600°C and it uses the resulting thermal expansion and mismatch, while thermal-melting uses vaporization at temperatures that are, typically, in the range of 1100–2200°C. Thermal-spalling takes advantage of the difference in coefficients of thermal expansion of constitutive minerals. As the temperature is increased, some minerals expand more than others creating high stresses at the boundaries between the two minerals. These stresses can be so high as to break the bond between the minerals and hence fracture the rock. Only a range of rocks that contain such minerals with different coefficient of thermal expansion will be suitable for thermal-spalling. Thermal-melting requires the generation of high enough temperatures to melt the rock. In both cases, some means of rock removal out of the hole is still required.

Acknowledgments

Some of the research reported in this chapter was conducted at the Jet Propulsion Laboratory (JPL), California Institute of Technology, funded by the NASA Planetary Instrument Definition and Development Program (PIDDP), under a contract with the National Aeronautics and Space Administration (NASA). The authors thank Timothy J. Szwarc, Department of Aeronautics and Astronautics, Stanford University, California, and George Cooper, University of California at Berkeley, for reviewing this chapter and for providing valuable technical comments and suggestions. The authors also express their thanks to R. Peter Dillon, JPL/Caltech, for assisting in obtaining photos from the Mars's Curiosity Rover Mission.

References

Aldrich J., Y. Bar-Cohen, S. Sherrit, M. Badescu, X. Bao, and J. Scott, Percussive Augmenter of Rotary Drills (PARoD) for Operating as a Rotary-Hammer Drill, NTR Docket No. 46550, 2008.

Badescu M., S. Sherrit, A. Olorunsola, J. Aldrich, X. Bao, Y. Bar-Cohen, Z. Chang et al. Ultrasonic/ sonic Gopher for subsurface ice and brine sampling: Analysis and fabrication challenges, and testing results, *Proceedings of the SPIE Smart Structures and Materials Symposium*, Paper #6171–07, San Diego, CA, 2006a.

Badescu M., S. Sherrit, Y. Bar-Cohen, X. Bao, and S. Kassab, Ultrasonic/Sonic Rotary-Hammer Drill (USRoHD), U.S. Patent No. 7,740,088, June 22, 2010. NASA New Technology Report (NTR) No. 44765, 2006b.

Badescu M., D. B. Bickler, S. Sherrit, Y. Bar-Cohen, X. Bao, and N. H. Hudson, Rolling tooth core break-off and retention mechanism. NTR Docket No. 47354, Submitted on (October 19, 2009a). *NASA Tech Briefs*, 35(6), 2011a, 50–51.

Badescu M., S. Sherrit, Y. Bar-Cohen, X. Bao, and P. G. Backes, Scoring dawg—core break-off and retention mechanism, NTR Docket No. 47355, Submitted on (October 19, 2009b). *NASA Tech Briefs*, 35(6), 2011b, 48–49.

Badescu M., S. Sherrit, Y. Bar-Cohen, X. Bao, and R. A. Lindemann, Praying mantis—bending core break-off and retention mechanism, NTR Docket No. 47356, Submitted on (October 19, 2009c). *NASA Tech Briefs*, 35(6), 2011c, 49–50.

Bao X., Y. Bar-Cohen, Z. Chang, B. P. Dolgin, S. Sherrit, D. S. Pal, S. Du, and T. Peterson, Modeling and computer simulation of ultrasonic/sonic driller/corer (USDC), *IEEE Transaction on Ultrasonics, Ferroelectrics and Frequency Control (UFFC)*, 50(9), 2003, 1147–1160.

Bao X., J. Scott, Y. Bar-Cohen, S. Sherrit, S. Widholm, M. Badescu, T. Shrout, and B. Jones, Ultrasonic/ sonic drill for high temperature application, SPIE Sensors and Smart Structures Technologies for Civil, Mechanical, and Aerospace Systems Smart Structures and Materials Symposium, Paper 7647–116, San Diego, CA, March 8–11, 2010a.

Bao X., S. Sherrit, M. Badescu, Y. Bar-Cohen, S. Askins, and P. Ostlund, Free-mass and interface configurations of hammering mechanisms, Patent was filled on October 27, 2011, NTR Docket No. 47780, 2010b.

Bao X., Y. Bar-Cohen, Z. Chang, S. Sherrit and R. Stark, Ultrasonic/Sonic Impacting Penetrator (USIP), NASA NTR No. 41666, 2004.

Bar-Cohen Y. and K. Zacny (Eds.), *Drilling in Extreme Environments—Penetration and Sampling on Earth and Other Planets*, Wiley—VCH, Hoboken, NJ, ISBN-10: 3527408525, ISBN-13: 9783527408528, 2009, pp. 1–827.

Bar-Cohen Y., S. Sherrit, B. Dolgin, T. Peterson, D. Pal and J. Kroh, Smart-ultrasonic/sonic driller/ corer, U.S. Patent No. 6,863,136, March 8, 2005, NASA NTR No. 20856, 1999.

Bar-Cohen, Y., S. Sherrit, B. P. Dolgin, N. Bridges, X. Bao, Z. Chang, A. Yen et al. Ultrasonic/sonic driller/corer (USDC) as a sampler for planetary exploration, *Proceedings of the IEEE Aerospace Conference on the Topic of "Missions, Systems, and Instruments for In Situ Sensing*, (Session 2.05), March 2001.

Bar-Cohen Y., J. Randolph, C. Ritz, G. Cook, X. Bao, and S. Sherrit, Sample preparation, acquisition, handling and delivery (SPAHD) system using the ultrasonic/sonic driller/corer (USDC) with Interchangeable Bits," NTR Docket No. 30640, May, 2002.

Bar-Cohen Y., and S. Sherrit, Self-Mountable and Extractable Ultrasonic/Sonic Anchor (U/S-Anchor), NASA NTR No. 40827, 2003a.

Bar-Cohen Y., S. Sherrit and J. L. Herz Ultrasonic/Sonic Jackhammer (USJ), NASA New Technology Report (NTR), Docket No. 40771, 2003, Patent was submitted on January 31, 2007 (Serial No. 11/700,575).

Bar-Cohen Y., and S. Sherrit, Thermocouple-on-the-bit a real time sensor of the hardness of drilled objects, NASA NTR No. 40132, 2003b.

Bar-Cohen Y., S. Sherrit, B. Dolgin, X. Bao and S. Askin, Ultrasonic/Sonic Mechanism of Deep Drilling (USMOD), U.S. Patent No. 6,968,910, 2005.

Bar-Cohen Y., M. Badescu, and S. Sherrit, Rapid Rotary-Percussive Auto-Gopher for deep subsurface penetration and sampling, NASA NTR No. 45949, 2008.

Bar-Cohen Y., M. Badescu, and S. Sherrit, Acquisition and retaining granular samples via rotating coring bit, NASA NTR No. 47606, 2010.

Bar-Cohen Y., M. Badescu, S. Sherrit, K. Zacny, G. Paulsen, L. Beegle, and Xiaoqi Bao, Deep drilling and sampling via the wireline auto-gopher driven by piezoelectric percussive actuator and EM rotary motor, *Proceedings of the SPIE Smart Structures and Materials/NDE Symposium*, San Diego, CA, March 12–15, 2012.

Bellarby J., *Well Completion Design, Vol. 56 (Developments in Petroleum Science)*, ISBN-10: 0444532102, ISBN-13: 978–0444532107, Elsevier Science, New York, NY, 2009, pp. 639–642.

Blake D. F., P. Sarrazin, S. J. Chipera, D. L. Bish, D. T. Vaniman, Y. Bar-Cohen, S. Sherrit, S. Collins, B. Boyer, C. Bryson, and J. King, Definitive mineralogical analysis of martian rocks and soil using the CHEMIN XRD/XRF instrument and the USDC sampler, *Proceedings of the Sixth International Conference on Mars*, held at Caltech, Pasadena, CA, July 20–25, 2003.

Chang Z., S. Sherrit, X. Bao, and Y. Bar-Cohen, Design and analysis of ultrasonic horn for USDC (Ultrasonic/Sonic Driller/Corer), *Proceedings of the SPIE Smart Structures and Materials Symposium, Paper #5387-58*, San Diego, CA, March 15–18, 2004.

Dolgin B., S. Sherrit, Y. Bar-Cohen, R. Rainen, S. Askins and D. Sigel, D. Bickler et al., Ultrasonic Rock Abrasion Tool (URAT), NASA NTR No. 30403, 2001.

Gershman R., and R. A. Wallace, Technology needs of future planetary missions, *Acta Astronautica*, 45, Nos. 4–9, 329–335, 1999.

Geothermal History Reports (GTP), A History of Geothermal Energy Research and Development in the United States, 1976–2006, Volume 2: Drilling, 2010.

Honeybee, High temperature motor specification sheet, downloaded January 2011 http://www.honeybeerobotics.com/images/stories/pdf/HTM_Data_Sheet.pdf.

Hughes H., X D Drill, US Patent No. 930,758, 1909.

Jahn F., M. Cook, and M. Graham, *Hydrocarbon Exploration and Production, Development in Petroleum Science Series*, 2nd Edition, ISBN-10: 0444532366, ISBN-13: 978-0444532367, Elsevier Science, New York, NY, 2008, pp. 1–456.

Jerby E., V. Dikhtyar, O. Aktushev, and U. Grosglick, The microwave drill, *Science*, 298(5593), 2002, 587–589.

Just G. D., The Jet piercing process, *Quarry Managers' Journal, Institute of Quarrying Transactions*, June 1963, pp. 219–26.

Maurer W. C., *Novel Drilling Techniques*, ISBN-10: 0080036155, ISBN-13: 978-0080036151, Pergamon Press, New York, NY, 1968, pp. 1–114.

Maurer W. C., *Advanced Drilling Techniques*, ISBN-10: 0878141170, ISBN-13: 978–0878141173, Petroleum Pub. Co., Tulsa, OK, 1980, pp. 1–698.

Rapp D., *Human Missions to Mars: Enabling Technologies for Exploring the Red Planet*, Springer/Praxis Publishing, ISBN: 3-540-72938-9, Appendix C, Water on Mars. 2007.

Ready J. F., *Industrial Applications of Lasers*, ISBN-10: 0125839618, ISBN-13: 978-0125839617, Academic Press, New York, 2nd Edition, 1997.

Ryan H. M. (Ed.), *High Voltage Engineering Testing*, 3rd Edition, The Institution of Engineering and Technology, Charity, England, ISBN-10: 1849192634, ISBN-13: 978-1849192637, 2011, pp. 1–750.

Schultz R. A., Brittle strength of basaltic rock masses with applications to Venus, *Journal of Geophysical Research*, doi: 10.1029/93JE00691, 98(E6), 1993, 10,883.

Sherrit S., S. A. Askins, M. Gradziel, B. P. Dolgin, Y. Bar-Cohen, X. Bao, and Z. Cheng, Novel ultrasonic horns for power ultrasonics, *NASA Tech Briefs*, 27(4), 2003, 54–55, NASA NTR No. 30489, 2001.

Sherrit S., Y. Bar-Cohen, B. Dolgin, X. Bao, and Z. Chang, Ultrasonic Crusher for Crushing, Milling, and Powdering, NASA NTR No. 30682, 2002.

Sherrit S., Y. Bar-Cohen, X. Bao, Z. Chang, D. Blake and C. Bryson, Ultrasonic/Sonic Rock Powdering Sampler and Delivery Tool, NASA NTR No. 40564, 2003.

Sherrit S., X. Bao, Y. Bar-Cohen, and Z. Chang, Resonance analysis of high temperature piezoelectric materials for actuation and sensing, *SPIE Smart Structures and Materials Symposium*, Paper #5388-34, San Diego, CA, March 15–18, 2004.

Sherrit S., M. Badescu, Y. Bar-Cohen, Z. Chang, and X. Bao, Portable Rapid and Quiet Drill (PRAQD), Patent disclosure submitted on Feb. 2006. U.S. Patent No. 7,824,247, November 4, 2010. NASA NTR No. 42131, 2005.

Sherrit S., M. Badescu, and Y. Bar-Cohen, Miniature Low-Mass Drill Actuated by Flextensional Piezo-Stack (DAFPiS) NASA NTR No. 45857, 2008.

Sherrit S., Y. Bar-Cohen, M. Badescu, X. Bao, Z Chang, C. Jones, and J. Aldrich, Compact Non-Pneumatic Powder Sampler (NPPS), NASA NTR No. 43614, 2006.

Sherrit S., X. Bao, M. Badescu, and Y. Bar-Cohen, Single Piezo-Actuator Rotary-Hammering (SPaRH) Drill, NASA NTR No. 47216, 2009.

Sherrit S., X Bao, M. Badescu, Y. Bar-Cohen, and P. Allen, Monolithic Flexure Pre-Stressed Ultrasonic Horns, A Provisional Patent Application 61/362,164 was filed on July 8, 2010. NASA NTR No. 47610, 2010a.

Sherrit S., X. Bao, M. Badescu, Y. Bar-Cohen, P. Ostlund, P. Allen, and D. Geiyer, Planar Rotary Piezoelectric Motor using Ultrasonic Horns, A Provisional Patent was filed on July 7, 2011, NASA NTR No. 47813, 2010b.

Smith M., G. Cardell, R. Kowalczyk, and M. H. Hecht, The chronos thermal drill and sample handling technology, *The 4th International Conference on Mars Polar Science and Exploration*, Davos, Switzerland, Abstract 8095, October 2–6, 2006.

Tichý J., J. Erhart, E. Kittinger, and J. Prívratská, *Fundamentals of Piezoelectric Sensorics: Mechanical, Dielectric, and Thermodynamical Properties of Piezoelectric Materials*, Springer, Berlin, Germany, ISBN-10: 3540439668, ISBN-13: 978-3540439660, 2010, p. 160.

Varnado S. G. (ed), Geothermal Drilling and Completion Technology Development Program, Annual Progress Report, October 1979–September 1980, SNL Report SAND80-2179, SNL National Laboratories, 1980.

Xu Z., C. B. Reed and G. Konercki, B. C. Gahan, R.A. Parker, S. Batarseh, R. M. Graves, H. Figueroa, and N. Skinner, Specific energy for pulsed laser rock drilling, *Journal of Laser Application*, 15(1), doi: 10.2351/1.1536641, 2003, 25–30.

Zacny, K., Y. Bar-Cohen, M. Brennan, G. Briggs, G. Cooper, K. Davis, B. Dolgin, D. Glaser, B. Glass, S. Gorevan, J. Guerrero, C. McKay, G. Paulsen, S. Stanley, and C. Stoker, Drilling systems for extraterrestrial subsurface exploration, *Astrobiology Journal*, 8(3), doi: 10.1089/ast.2007.0179, 2008, 665–706.

Internet Resources

Nano materials http://www.dtic.mil/cgi-bin/GetTRDoc?AD=ADA476834

Solutions for drilling high-temperature, high-pressure wells http://msdssearch.dow.com/PublishedLiteratureDOWCOM/dh_0516/0901b803805167cc.pdf?filepath=oilandgas/pdfs/noreg/812-00020.pdf&fromPage=GetDoc

The USDC development at JPL http://ndeaa.jpl.nasa.gov/nasa-nde/usdc/usdc.htm

15

High-Temperature Electronics

Zhenxian Liang

CONTENTS

15.1 Introduction

Electronics deals with groups of electronic circuits and components which are designed to build up complex measurement, control or power-supplying functional systems, including signal detection and processing, data storage and display, computation, power distribution, and many others. An exceptionally broad range of novel technologies have been invented and developed, such as electric power conversion/generation and transmission, electronic signal and information processing, telecommunication, automation, and robotics. These, in turn, have made it possible for people to manufacture a wide array of products, from consumer to industrial, and defense. Indeed, it can be said that emergence of electronic industry is one of the most significant revolutions of the twentieth century.

The information processing of electronics systems can be recognized to accomplish certain algorithms through manipulation of voltage and electric current such as rectification, amplification, oscillation, switching of electronic signals, as well as through conversion of information form factors by optoelectronic, thermoelectric, electromechanical and electromagnetic effects, and so on. All these operations in electronics systems are performed through the electric circuit, which is a network of a large range of devices and components such as resistors, capacitors, inductors, diodes, transistors, and even integrated circuits (ICs), the building blocks comprised of large number of components.

Electronic systems are constructed by a hierarchical packaging scheme, in which four electrical interconnection levels are employed, they are chip to package, component on board, subassembly of multiple boards, and system assembly levels. The packaging functions include mechanical support and thermal management in addition to the electrical connection and functional integration. The large range of materials, such as polymers, ceramics, and metals have been employed to realize these packaging functions.

It is well known that the electronics system is an integration of multiple, diverse materials with specially designed functionality based on their properties. Usually, these electrically functional properties and their mechanical and chemical properties change to certain extent with temperature. In most cases, they may lose their functionalities before structural damage. Thus, the criteria for the lowest and highest temperature limits of an electronic system are based on functionality and structural integrity. The normal operating temperature range of today's electronic systems, comprised mostly of silicon (Si) semiconductors and their auxiliary components, is usually defined as from −55 to 125°C (McCluskey et al., 1997; National Research Council, 1995). As rapid expansion of electronics systems' application scope, the operating temperature beyond the limit is frequently encountered. All the electronics systems with operation temperature higher than 125°C are regarded as high-temperature electronics (HTE) systems. The requirement to operate electronics systems at high temperature brings tremendous challenges to existing electronics technologies. This chapter deals with the mechanisms of temperature limitation and the technologies developed to extend the operation temperature range of the advanced electronic systems.

15.2 Operating Temperature of Electronics Systems

Generally speaking, the operating temperature of electronic components can be elevated by both hot environment and self heating. It is well-known that any electronic components have an effective or equivalent electric resistance, large or small. When the electric current flows through these components, the electric power will be lost according to $P_{loss} = IV$ or I^2R, where P_{loss} is electric power loss in the component, I is electric current, V is voltage drop on the component, and R is the value of the equivalent resistance. This power loss will be absorbed by these components and generate heat in them. This self heating plays two harmful roles for electronics systems. One is the reduction of electronic processing efficiency, especially for power electronics systems, and the other effect is to raise the component temperature. The heat generated in these components must be removed promptly to prevent the temperature rise over the limit that would cause immediate failure, long-term degradation. The technique for heat dissipation (or removal) in electronics systems is usually called thermal management (electronic cooling), which utilizes convection, conduction, and radiation mechanisms to transfer the thermal energy from the components to the ambient. The electrothermal relationship of a component in a system can be expressed as

$$T_j = T_a + \theta_{ja} \cdot P_{loss} \tag{15.1}$$

where, T_j is the maximum operating temperature of the core element in a component; T_a is the ambient temperature of an electronic subsystem, which is usually the baseline temperature of the cooling system; θ_{ja} is thermal resistance, which is defined by the heat transfer capability of materials and construction of thermal path from hot spots to the ambient. θ_{ja} represents the thermal characteristics of both the package and thermal management system. Clearly, no matter how good the cooling system applied is, T_j will be higher than the base temperature T_a. In many modern electronics systems, high ambient temperature is often met. For example, in exploration and monitoring of petroleum and geothermal wells, many electronics sensors are sent into deep holes, where the temperature may reach 200–300°C (Veneruso, 1979). In commercial and military aircrafts, electronic interface circuitry can be co-located with monitoring and control transducers for engines, actuators, and other system elements; many of these locations involve a hot environment, possibly greater than 300°C (Mehdi and Brockschmidt, 2006). In the automotive industries, the trend for extensive use of electronics in vehicles is extending to electrification of most subsystems from electronic engine management, control of traction and braking, to electric driven pumps and even electric traction system. As high-density packaging of components for automobiles is desired, most of these electronics units are mounted as close to the engine compartment as possible, as well as in locations such as inside the gearbox or other transmission components. This requires electronic boards and components to perform normally at the operating temperature of the engine or the gearbox. Moreover, smaller engine compartments impose tighter restrictions on heat sink sizes, and changes to vehicle aerodynamics deliver less cooling airflow to the radiator, so that the underhood temperature can reach 150°C and some locations near the engine can reach 200°C. The use of power electronics for motor drives in hybrid and electric cars that deal with electric power as high as 100 kW is also pushing temperature requirements for automotive-qualified electronic components, which are now moving up from 150°C maximum to 200°C (Johnson et al., 2004). Other applications with extreme temperature environments include satellite, spacecraft, and the space station, whose temperature regulation or cooling is difficult. Thus, the high-temperature components with greater reliability and stable performance are highly required.

15.3 Thermal Management and Challenges

The general purpose of thermal management in electronics systems is to ensure the component's operation temperature T_j is lower than its ratings. The conventional techniques for heat removal employ forced fluid convection heat exchange mechanism, such as in forced air cooling and liquid cooling systems. The forced air or liquid cooling provides higher efficiency, in other words, lowers the thermal resistance of the thermal management system. The electronics cooling systems are usually comprised of multistage coolant circulation systems, in which the heat is finally dissipated into the surrounding environment. The base temperature T_a, in Equation 15.1, is usually determined by the ambient temperature of the final stage. In high-temperature applications, electronic systems must be either remotely located from the high-temperature region or actively cooled with air or liquid coolant pumped in from elsewhere. However, these distant mount schemes of cooling systems bring additional overhead that can negatively offset the desired benefits of the electronics relative to overall system operation. The additional overhead, in the form of longer wires, more connectors, and/or cooling system plumbing, can add undesired size and weight to the system, as well as increase complexity, parts count, and corresponding increased potential for failures.

It is envisioned that mounting electronic control and processing modules close to sensors and actuators is a promising solution to overcome all these disadvantages and improve the cost-effectiveness of electronic systems in electromechanical applications. For example, integrating the power electronics drive into the electric machine, mostly used in the electric drive vehicles, can eliminate heavy connector cables, share the cooling loop, and save space and weight. On the other hand, as expressed in Equation 15.1, higher operation temperatures can also help increase the electrical processing capability if the system is with the same thermal management efficiency. Thus, it will reduce the usage of component materials and in turn, reduce the system cost and volume.

In summary, with the growing demand for high-temperature electronics, there is a push to raise the operating temperature of various electronic components beyond the conventional limitations. However, there are two main challenges to HTE. First, as discussed above, all electronic components are made based on certain physical mechanisms or their electric functions, which are also tightly affected by the operating temperature. For example, some functionality will be poor or lost completely at high temperature. Second, the electronic components and systems are made of multiple hybrid materials in their core elements and packaging structures. It is well-known that the average temperature, temperature distribution, and time variation within the components are among the driving forces for degradation and failure of electronic components and systems. Additionally, the thermal management system must not just ensure operation temperature below the component's withstand capability, but also needs to be powerful enough to handle possible thermal instability such as thermal escalation in components with temperature increase. So far, there is a lack of a good solution for high-efficiency thermal management of HTE when considering their efficiency, weight, volume, cost, and so on.

15.4 High-Temperature Semiconductor Devices

15.4.1 Temperature Effects of Semiconductors

Semiconductors are the most important electronics materials and have been used to manufacture numerous kinds of electronics devices, including various transistors, diodes, and even ICs. In fact, the advancement of the modern electronic industry relies intensively on the development of semiconductor technologies. A semiconductor can be either of a single element, such as silicon (Si), germanium (Ge), or a compound such as gallium arsenide (GaAs), indium phosphide (InP) and cadmium telluride (CdTe), or an alloy such as silicon–germanium ($Si_xGe_{(1-x)}$), aluminum gallium arsenide ($Al_xGa_{(1-x)}As$) and many others, where x is the fraction of the particular element and it ranges from 0 to 1. Among them, Si is a dominant material nowadays due to its abundance, safety, and abundant electronic functions.

As important building materials of electronic components, in addition to their thermal and mechanical properties, the electrical properties of semiconductor materials have been intensively characterized. The fundamental parameters include energy bandgap (E_g), Fermi level (E_f), effective energy band density of states (N_c, N_v), intrinsic carrier concentration n_i, mobility (μn, μp) of carriers, electron and hole, electron saturation drift velocity v_{sat}, carrier lifetime (τ_{HL}, τ_{sc}), and breakdown electric field strength (E_c), and so on. The definition and details of these parameters can be found in handbooks, for example in Yu and Cardona (2010). The physical properties are strongly dependent on their chemical composition and crystalline structure.

The basic electronic processing functions are realized based on the specially designed semiconductor device structures including stack of p-type and n-type layers (PN junction), metal–semiconductor contact (MS junction), metal-dielectric-semiconductor (MOS junction), and so on. The combination of all these junctions will build up the various devices such as diode (PN, Schottky), transistors (pnp, npn), MOS field effect transistor (MOSFET), junction field effect transistor (JFET), insulated gate bipolar transistor (IGBT), thyristor (npnp), and so on. The set of parameters to represent the electronic performance of the devices has been well established, including forward voltage drop V_{ce}, equivalent on-resistance R_{on}, current gain α or transconductance g_m, breakdown voltage V_{ces}, reverse leakage current I_r, switching speed or operation frequency f, and so on, respectively for different bipolar and field effect devices (Sze and Kwok, 2007).

It has been well-known that the electrical properties of semiconductor are greatly dependent on temperature (Wondrak, 1999). Table 15.1 summarizes the temperature dependence

TABLE 15.1

Temperature Dependence of Si Electrical Properties and Device Parameters

Serial Number Electrical Parameter	Formula	Parameter Value @25°C (T_0)	Parameter Value @125°C	Parameter Value @200°C
1. Energy bandgap	$E_g = 1.18\text{-}9 \times 10^{-5}(T_j)\text{-}3 \times 10^{-7}(T_j)^2$	1.12 eV	1.09 eV	1.06 eV
2. Intrinsic carrier concentration	$n_i = (N_cN_v)^{1/2} \cdot \exp(-E_g/2kT_j)$ $= 1.042 \times 10^{15}(T_j{}^2\exp(-6884/T_j)$	$1.01 \times 10^{10} \times cm^{-3}$	$5.59 \times 10^{12} \times cm^{-3}$	$1.19 \times 10^{14} \times cm^{-3}$
3. Electron mobility	$\mu_n = 1500(T_0/T_j)^{2.5}$	$1500\ cm^2/V \cdot s$	$739.92\ cm^2/V \cdot s$	$475.51\ cm^2/V \cdot s$
4. Electron mobility near surface	$\mu_{ns} = 400(T_0/T_j)^{2.5}$	$400\ cm^2/V \cdot s$	$197.31\ cm^2/V \cdot s$	$126.80\ cm^2/V \cdot s$
5. Hole mobility	$\mu_p = 450(v/T_j)^{2.5}$	$450\ cm^2/V \cdot s$	$221.98\ cm^2/V \cdot s$	$142.65\ cm^2/V \cdot s$
6. Electron diffusion coefficient	$D_n(T_j) = \mu_nkT_j/q$ $= 2.0187 \times 10^5(T_j)^{-1.5}$	$38.8572\ cm^2/s$	$25.23\ cm^2/s$	$19.50\ cm^2/s$
7. Hole diffusion coefficient	$D_p(T_j) = \mu pkT_j/q$ $= 6.0561 \times 10^4(T_j)^{-1.5}$	$11.66\ cm^2/s$	$7.57\ cm^2/s$	$5.85\ cm^2/s$
8. Ambipolar diffusion coefficient	$D_a(T_j) = 2D_n(T_j)D_p(T_j)/$ $(D_n(T_j) + D_p(T_j))$ $= 9.3171 \times 10^4(T_j)^{-1.5}$	$17.93\ cm^2/s$	$11.65\ cm^2/s$	$8.9972\ cm^2/s$
9. Ambipolar (high-level) lifetime	$\tau_{HL}(T_j) = \tau_{HL0}(T_j/T_0)^{1.5}$ $= 1.9245 \times 10^{-4}(T_j)^{1.5} \tau_{HL0}$	$0.9\ \mu s$	$1.385\ \mu s$	$1.793\ \mu s$
10. Built-in potential	$V_{bi} = (kT/q) \ln(N_AN_D/n_i{}^2)$ $= V0\text{-}2.1\ mV/(T_j\text{-}T_0)$	$700\ mV$	$490\ mV$	$332.5\ mV$
11. PN junction reverse current	$J_r = a\ n_i{}^2 + b\ n_i/\tau_{sc}$ $= J_{ro}.2^{(T-T0)/11}$	$1\ \mu A$	$545\ \mu A$	$61,147\ \mu A$
12. PN junction breakdown voltage	$V_{br} = V_{br0}(1 + 0.001/K)$	$1000\ V$	$1100\ V$	$1175\ V$
13. Threshold voltage	$V_{TH}(T_j) = V(T_0)\text{-}0.009(T_j\text{-}T_0)$	$4\ V$	$3.1\ V$	$2.43\ V$
14. Thermal conductivity	$\kappa = \kappa_0(T_0/T)^{1.3}$	$1.56\ W/cm \cdot K$	$1.07\ W/cm \cdot K$	$0.86\ W/cm \cdot K$

of important Si electrical properties, and most used device performance parameters, the relationship of the parameters variation with temperature, including the typical values of the parameters at 25°C, 125°C, and 200°C, respectively.

The sensitivity of temperature dependence of each parameter is characterized with a temperature coefficient. For example, the breakdown voltage of the Si devices with PN junction increases with temperature rise so it has a positive temperature coefficient, while the voltage drop of a bipolar device like diode's built-in potential, mostly has a negative temperature coefficient, which means it will decrease with the temperature increase. The detailed temperature characteristics of various semiconductors and their devices have been extensively investigated and can be found in many handbooks.

15.4.2 Thermally Induced Failures

The change of most semiconductor parameters with temperature brings mild effects on the device's operation performance. However, some of them cause significant variation and may have severe consequences. A noticeable one among them is the intrinsic carrier concentration n_i, which is the density of electric charge carriers (electrons and holes) inherently generated in the semiconductor due to thermal excitation. It also follows the relationship $n = p = n_i$, where n and p are the electron and hole concentration, respectively. As we know, in semiconductor devices, the majority carrier concentration n or p is designed to be generated extrinsically by introducing "impurity" elements (donors and acceptors) to form n-type and p-type conduction regions through doping processes. The concentrations of electron and hole carriers are greatly determined by the fixed dopant density, mostly from 10^{13} to 10^{19} cm^{-3}, which is much greater than n_i, ($n_i = 10^{10}$ cm^{-3} at room temperature). However, the intrinsic carrier density n_i increases exponentially with temperature as described with the equation shown in Table 15.1. Figure 15.1 shows the graph of the Si intrinsic carrier concentration increase with temperature. It can be seen that at very high temperatures, the thermal generation would become the dominant mechanism, surpassing

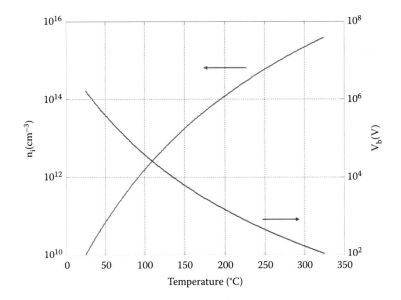

FIGURE 15.1
(**See color insert.**) Intrinsic carrier concentration n_i and breakdown voltage versus temperature in Si.

the dopant, for the creation of carriers. Thus, the doped regions also change from a specific type p-type (*p»n*) or n-type (*n»p*) to a compensated type (*n = p*), leading to the abrupt change of device performance and even functionality, for example, disappearance of the carrier barriers (built-in potential) in bipolar devices and loss of gate control on carrier channel in field effects devices. To set an ultimate temperature limit of a semiconductor device, an intrinsic temperature T_{int} is defined at which intrinsic carrier concentration n_i equals the background doping such as N_D. Furthermore, the voltage blocking capability, for example, for a PN junction, is determined by the background doping concentration N_D (Sze and Kwok, 2007). So it is easy to derive the relationship between the breakdown voltage and T_{int}, as drawn in Figure 15.1. Higher voltage devices, such as power semiconductor ones, will have lower intrinsic temperature, and low voltage ones like most ICs will have higher characteristic temperature.

In some cases, the variation of devices' performance parameters with temperature is beneficial to device operation, for example, the voltage drop (built-in potential) of Si rectifier diode decreases at elevated temperature, leading to lower power loss. However, some of these shifts may cause catastrophic failures if an irreversible electrical–thermal interaction process happens. For instance, the leakage (reverse) current of a PN junction diode is well-known to have a large positive temperature coefficient, due to strong dependence on the intrinsic carrier concentration n_i, as listed in Table 15.1. The complete leakage current density J_r expression is as follows:

$$J_r = qn_i \left[\frac{n_i}{N_D} \sqrt{\frac{kT\mu_p}{q\tau_L}} + \frac{W}{2\tau_{sc}} \right] \tag{15.2}$$

where carrier mobility and lifetimes in the equation are also functions of temperature. A relationship of power loss and leakage current versus temperature of a power PiN diode is illustrated in Figure 15.2, in which the power loss is the product of leakage current and

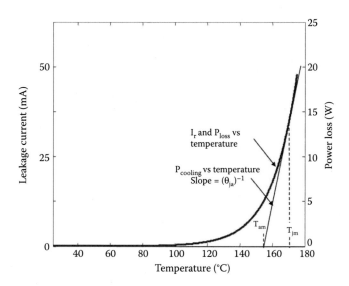

FIGURE 15.2
Power loss induced by leakage current I_r in a Si diode versus temperature and maximum operation temperatures under stable conditions.

a fixed reverse biased voltage. The curve characterizes that the leakage current increases tremendously with temperature. In fact, there is an empirical estimation that the leakage current will double in magnitude with temperature rise of every 9–11 degrees (Obreja, 2000). The power loss from this leakage current will be large enough to heat up the device itself. As the temperature further increases, the device will get hotter, which again leads to larger leakage current. If this thermal power increase cannot be instantly removed by the thermal management system, the device will be inevitably destroyed. This is a typical thermal instability phenomenon that can happen to electronic components, called thermal runaway (Sinkevitch and Vashchenko, 2008).

To assess the thermal stability of an electronic component or system, the cooling efficiency of the thermal management system must be considered simultaneously with the temperature dependency of electric power loss. If *Ploss* is the power loss or heat power density and *Pcool* is the power density that can be maximally drawn out via the thermal management, the criterion for potential thermal runaway can be expressed as

$$\frac{\partial Ploss}{\partial T} \geq \frac{\partial Pcool}{\partial T} \tag{15.3}$$

The left term represents the temperature coefficient of power loss or thermal power, while the right term is related to the cooling efficiency, which is the reciprocal of the package's thermal resistance, θ_{ja}^{-1}. The equation indicates that thermal stability cannot be maintained if the cooling capability is insufficient. This relationship is greatly dependent on the operating temperature because both power loss and cooling power are functions of temperature. Usually the cooling efficiency (thermal resistance) does not change much with the temperature, while the power loss in devices will increase significantly as shown in the example.

In Figure 15.2, a cooling line is also drawn on the power loss graph. The cooling line is with a slope of 0.9 W/°C (cooling efficiency), corresponding to the thermal resistance of 1.1°C/W. The line was drawn in such way that there is a point of tangency of the power loss curve and cooling line. Thus, the maximum safe operation temperature for a device can be defined. The intersection point of the cooling line with the x-axis defines the maximum base (or ambient) temperature T_{am}, where the point of tangency defines the maximum junction temperature, T_{jm}. It can also be concluded that smaller thermal resistance permits higher temperature operation of a device.

There are also other mechanisms in extreme operation conditions that generate strong feedback of heat generation with temperature such as latch up, short circuit, avalanche, second breakdown, and so on. The electrothermal dynamic balance must be characterized individually to determine their limitations to the device maximum temperatures.

Since the beginning of semiconductor technology, imperfection in the wafers and devices has been a well-known issue, which not only comes from the interface microstructure fluctuation and composition transition of multiple materials in a semiconductor device, but also from the defects and contaminations in crystalline semiconductor regions. The existence of these imperfections worsens the performance of the devices; also, they are closely associated with the root causes of the device failures. During device operation, interacting with other stresses such as high electric voltage and current, the elevated temperature is a major driving force to accelerate the device degradation and shorten lifetime. The Arrhenius model is widely accepted to simulate the behavior of component degradation with regard to temperature. If degradation is modeled as a "chemical" reaction, the chemical reaction rate R can be expressed as in an Arrhenius equation:

$$R = Ae^{(-Ea/kT)} \tag{15.4}$$

TABLE 15.2

Main Failure Mechanisms and Activation Energy Values (Examples)

Failure Mode	Failure Mechanism	Activation Energy (eV)
Metal trace failure (open, short, corrosion)	Al metal electromigration	0.4–1.2
	Al metal stress migration	0.5–1.4
	Au–Al alloy growth	0.85–1.1
	Cu metal electromigration	0.8–1.0
	Al corrosion	0.6–1.2
Oxide film breakdown (leakage current increase, short circuit)	Oxide film breakdown	0.3–0.9
h_{FE} degradation	Ion movement acceleration due to moisture	0.8
Parameter value fluctuation	Degradation by NBTI	0.5 and up
	Na ion drive in SiO_2	1.0–1.4
	Slow trapping of Si–SiO$_2$ interface	1.0
Increased leakage current	Inversion layer formation	0.8–1.0

where Ea is activation energy (eV), k is Boltzmann's constant, T is absolute temperature (K), and A is a constant. Ea quantifies the minimum amount of energy needed to allow a certain chemical reaction to occur. The chemical reaction here indicates the failure begins to take place. Table 15.2 provides the activation energy values for typical failure mechanisms within conventional Si devices (JEDEC, 2010), which is tightly associated with the materials and manufacture processing. Based on Equation 15.4, the failure modes with smaller Ea will occur easily. However, the temperature will accelerate the chemical reaction and thus make degradation quickly. For any given reaction (degradation or failure mechanism) obeying the Arrhenius equation, the product of the reaction rate R times the elapsed reaction time t is a constant (C), that is, $R \times t = C$. If t_f represents the time to failure, then $t_f = C/R$, thus:

$$t_f = C/(Ae^{(-Ea/kT)})$$

$$= B\left(e^{(Ea/kT)}\right)$$

$$Ln(t_f) = Ln(B) + Ea/T \tag{15.5}$$

where A, B, or C includes possible other factors but is constant regarding to temperature. It can be clearly seen that the lifetime t_f of a semiconductor device drastically decreases with elevated operation temperature. Therefore, an operating temperature limitation is also introduced based on reliability (lifetime) requirements.

15.5 High-Temperature Semiconductor Technologies

A semiconductor device includes not only semiconductor crystals but also other materials such as dielectric and metals, playing roles of electric conduction and isolation. Figure 15.3a shows the structure of a widely used MOSFET cell in ICs, in which the silicon dioxide

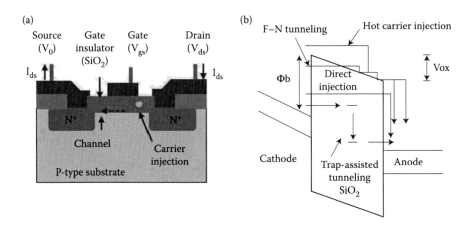

FIGURE 15.3
(a) Cross-sectional view of a MOSFET cell used in ICs; (b) Schematic of carrier injections into gate insulator (SiO$_2$).

(SiO$_2$), grown directly from Si substrate, couples the electric charge on the gate to control the current going through the channel. The SiO$_2$ layer forms two interfaces with adjacent materials, Si substrate and gate electrode. During device operation, the undesired electrons transportation may happen at these interfaces. The well-studied mechanisms include Fowler-Nordheim (F-N) injection, direct injection, and hot carrier injection (HCI) (DiMaria et al., 1993; Yang and Saraswat, 2000), as schematically illustrated in Figure 15.3b. These carrier injections to the dielectric from the substrate or gate (metal or polysilicon) will cause damage and trapped charges in the oxides or at interfaces. The presence of such mobile carriers in the oxides triggers numerous physical damage processes that can drastically change the device characteristics over prolonged periods. The accumulation of damage can eventually cause the circuit to fail as key parameters, such as the threshold voltage shifts. It has been demonstrated that the injection and damages are the root cause of the MOSFET failure modes such as time-dependent dielectric breakdown (TDDB). The different injection is related to energy gain of the carriers for them to tunnel through the barrier created by the energy band discontinuity at the semiconductor/oxide interface, and bending of the bands in the oxide. The effects of the electrical stresses and temperature on electron injection were well investigated (Bravaix et al., 1999; Yang and Saraswat, 2000). At higher temperature the electrons gain more thermal energy and also experience stronger scattering from atoms in oxide during their transportation. The higher energy electrons cause more severe damage in dielectrics because they will easily go through the barrier between Si and SiO$_2$ (Schroeder and Avellán, 2003).

The junctions formed by metal–semiconductor contacts are widely used in electron devices, such as Schottky barrier diodes and MOSFETs. The electrical characteristics of these junctions electric current under different bias can be generally described as

$$J\_forward = Js(exp(qVa/kT)-1)$$

$$J\_reverse = Js = AT^2exp(-q\Phi_B/kT) \tag{15.6}$$

where *Va* is applied voltage, and Φ_B is the electron energy barrier between metal and semiconductor. The current is determined mainly by the emission or injection of electrons going through this barrier. The detailed temperature dependences of all parameters are extensively investigated (Sharma, 2010). It is easy to determine that the temperature coefficient of leakage current increases exponentially with the effective potential barrier height. Thus, to increase device high-temperature capability, the barrier height should be increased through tailoring the energy band by applying new material combinations. In Yeo et al. (2003) various dielectrics replacing SiO_2 were discussed to suppress the leakage current under electric stress on gate. The effects of metal or alloys for different metal-Si contacts on the barrier height were discussed in Mönch (1994). All these measures enhance the ruggedness of electron devices operating at elevated temperatures.

In general, different type of parasitic devices may co-exist in a designed main semiconductor structure. Among them, the undesired PN junctions lie in a MOSFET as depicted in Figure 15.4a. When it operates, a positive voltage will be applied between the drain and source electrodes. The current will go through the channel under the control of the gate signal. When it is turned to off-state, the drain-source current comprises of channel leakage current and the substrate leakage current of the n-type drain to p-type body substrate, which is a parasitic PN junction diode. This leakage current increases with temperature exponentially, as described in Equation 15.2, while the on-state current decreases significantly with the temperature increase. The most effective method to eliminate this failure regard to temperature is to remove the parasitic mechanisms or cut off the leakage current paths. The widely accepted semiconductor on insulator (SOI) structure is one of these technologies.

As shown in Figure 15.4b, an embedded SiO_2 layer is built into the MOSFET, replacing partial p-type body substrate. Thus, it removes the junction area associated with the horizontal bottom layer under each n-type drain of the MOSFET. The presence of the embedded insulator layers also reduces the source to drain channel leakage that physically occurs in the p-type regions of conventional MOSFETs. This technology enables MOSFET circuits to successfully operate at temperatures higher than 200°C, which is beyond the operating realm of conventional bulk silicon MOSFETs. The SOI concept has also been used in many other device structures and circuits, such as bipolar transistors and brought also the benefit to control the undesired leakage current inside the device structure (Cristoloveanu, 2000). The mass production of uniform high-quality silicon and high-quality insulator has been a promising technology. Significant technology and reliability breakthroughs have successfully promoted the applications for HTE. The SOI approach is now well commercialized for a variety of high-temperature electronics.

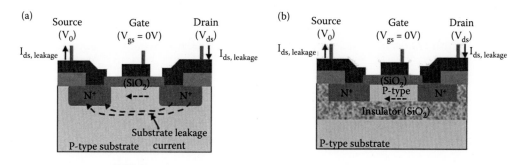

FIGURE 15.4
(a) Substrate leakage current of parasitic PN junction in MOSFET device, and (b) semiconductor on insulator (SOI) structure of MOSFET device.

TABLE 15.3

Electrical Properties of WBG Semiconductors at Room Temperature (25°C)

Property	Si	GaAs	SiC	GaN	Diamond
Bandgap, Eg (ev)	1.12	1.43	3.26	3.45	5.45
Breakdown electric field E_c (kV/cm)	300	400	2200	2000	10,000
Intrinsic carrier concentration n_i (cm^{-3})	9.65E9	1.8E6	1.6E-6	1E-7	1E-27
Electron mobility μ_n (cm^2/V · s)	1,500	8,500	500–1000	1250	2200
Hole mobility μ_p (cm^2/V · s)	600	400	100–115	850	850
Dielectric constant ε_r	11.9	13.1	10.1	9	5.5
Thermal conductivity κ (W/cm · K)	1.5	0.46	4.9	1.3	22
Saturated electron drift velocity v_{sat} (10^7 cm/s)	1	1	2	2.2	2.7

The power supplies or power electronics subsystems are necessary parts of the HTE systems. Unlike in microelectronics, the devices and components used in power electronics operate at higher voltages and current levels, leading to higher power dissipation themselves. To realize the high voltage blocking capability, the doping concentration is usually low and, thus, limits its critical temperature to even lower than its low-voltage microelectronics counterparts. To overcome this limitation, one class of semiconductor materials, featuring wide energy bandgap (WBG), have been explored to manufacture high power, high-temperature semiconductor switches (Neudeck et al., 2002). Among them, the most practical ones are silicon carbide (SiC) and gallium nitride (GaN). Table 15.3 lists the physical properties of these WBG semiconductors, as well as their comparison to silicon. It can be seen that the electron energy bandgap of WBG semiconductors are 3–5 times of that of Si. The electric breakdown strength and the intrinsic carrier concentration are strongly dependent on the bandgap of semiconductor crystals. It leads to a seven times higher breakdown electric field E_c and more than 15 orders lower intrinsic carrier concentration of WBGs at room temperature. The intrinsic temperature T_{int}, as discussed above, is a characteristic temperature determined by device voltage rating, in turn, by the dopant concentration. In Wondrak (1999) the intrinsic temperature versus breakdown voltage for all WBGs and Si has been presented. Taking 1000 V device as an example, the T_{int} is 200°C, 600°C, 700–1300°C, and 1400°C for Si, SiCs, and GaN, respectively. It can be seen that T_{int} of all WBG semiconductors are higher than 600°C. This will result in the leakage current reduction by tens of order of magnitude, pushing the operating temperature of WBG devices tremendously high.

However, as shown in the previous discussion, the injection density of electrons into oxidation is exponentially dependent on the barrier height. The reported conduction-band offset values between SiO$_2$ and 4H-SiC vary between 2.45 and 1.92 eV, which is 0.65–1.18 eV lower than that for SiO$_2$ on silicon (Gurfinkel et al., 2008). Considering its interface states and Fermi level in the bandgap, the effective barrier height of electron injection for SiC MOS structures at room temperature is 2.7–2.9 eV, while it is 3.15 eV for Si MOS structure. The lower barrier height and band offset cause the SiC MOS structure to have higher F-N injection. The energy band bending and these parameters are strong functions of temperature, resulting in tremendous F-N tunneling increase with temperature in SiC MOS structure (Hadjadj et al. 2002). Since the temperature is a critical factor to control the magnitude of electron tunneling, the safe operating temperature of SiC with MOS structure devices has been derated to much lower than its intrinsic temperature (Agarwal et al., 1997).

In Schottky-based devices, the reverse leakage currents are dominated by the Schottky barrier height between metal and SiC structure, which is in the 0.7–1.2 eV range (Crofton and Sriram, 1996). The Schottky leakage current increases with temperature, similar to that of silicon PN junction-based devices. The leakage current in a Schottky diode is exponentially dependent on both the metal–semiconductor barrier height and the operating temperature. Since SiC-powered Schottky diodes are expected to compete with Si PIN diodes, they must have a comparable forward on-state voltage drop. The most commonly used Schottky metals (e.g., titanium and nickel) have barrier height values close to Si bandgap value of 1.1 eV. The high-temperature blocking performance of these Schottky diodes will also behave similar to the Si counterpart. So the maximum temperature will be determined by thermal stability, which is also dependent on cooling capability.

Although these limits exist in current SiC devices, the combination of necessary material properties can meet the demanding high-temperature and high-power application requirements. After special design, the new junction structures with low reverse leakage at junction temperature as high as 600°C enables power-device operation at higher ambient temperature. Superior power switching properties of wide bandgap devices are also present at high-temperature ambient. Therefore, if the remaining technical challenges can be overcome, wide bandgap semiconductors are likely to play a critical role in realizing high-power electronics beyond the capability of silicon at all temperature (Singh, 2006).

15.6 High-Temperature Limits of Passive Components

Electronics components may be categorized as passive, active, or electromechanical types. A passive component may be either one that consumes (but does not produce) energy (thermodynamic passivity), or a component that is incapable of power gain (incremental passivity). Active components rely on a source of energy and usually can inject power into a circuit; the semiconductor devices discussed above are active components. Electromechanical components can carry out electrical operations by using moving parts or by using electrical connections. In this section, we will discuss the high-temperature performance and limitations of two mostly used types of passive electronics components.

15.6.1 Temperature Characteristics of Capacitor Materials

A capacitor is a two-terminal electrical component used to store energy in an electric field. The forms of practical capacitors vary widely, but all contain at least two electrical conductors separated by a dielectric material. The voltage of a capacitor depends on the amount of electric charge it holds. To vary its voltage, a capacitor must be charged or discharged. The breakdown voltage of a capacitor is the maximum voltage that it can withstand.

A dielectric material is an electrical insulator that can be polarized by an applied electric field. When a dielectric is placed in an electric field, electric chargers do not flow through the material as they do in a conductor or semiconductor, but only slightly shift from their average equilibrium positions, causing dielectric polarization. Because of dielectric polarization, an internal electric field is created, which reduces the overall field within the dielectric itself and realizes charge storage. Dielectric materials can be solids, liquids, or gases. In addition, a high vacuum can also be a useful, lossless dielectric even though its relative dielectric constant is only unity.

Several solid dielectrics are commonly used, including paper, plastic, glass, mica, and ceramic materials. They exhibit different properties.

Paper was used extensively in older capacitors and offers relatively high voltage performance. However, it is susceptible to water absorption, and has been largely replaced by plastic film capacitors. Plastics offer better stability and aging performance, which makes them useful in timer circuits. Ceramic capacitors are generally small, cheap, and useful for high-frequency applications, although their capacitance varies strongly with voltage and they age poorly. They are broadly categorized as class 1 dielectrics, which have predictable variation of capacitance with temperature or class 2 dielectrics, which can operate at higher voltage. Glass and mica capacitors are extremely reliable, stable, and tolerant to high temperatures and voltages, but are too expensive for most mainstream applications.

Electrolytic capacitors use an aluminum or tantalum plate with an oxide dielectric layer. The second electrode is a liquid electrolyte, connected to the circuit by another foil plate. Electrolytic capacitors offer very high capacitance, but suffer from poor tolerances, high instability, and gradual loss of capacitance especially when subjected to heat and high leakage current. Tantalum capacitors offer better frequency and temperature characteristics than aluminum ones, but higher dielectric absorption and leakage.

The characteristics of a capacitor usually includes its nominal capacitance, dielectric strength, leakage current, equivalent series resistance (ESR), dissipation factor (DF), tan δ (Kaiser, 2011), and so on. In order to maximize the charge that a capacitor can hold the dielectric material needs to have as high permittivity as possible, while also having as high dielectric strength as possible. In general, different types of dielectrics and capacitor show distinctive temperature behaviors.

The dielectric's capacitance variation with temperature, determines the working temperature range of a capacitor based on a defined tolerance. The temperature coefficient of a capacitor is the maximum change in its capacitance over a specified temperature range. The temperature coefficient of a capacitor is generally expressed linearly as parts per million per degree centigrade (PPM/°C), or as a percent change over a particular range of temperatures. Some capacitors increase their value as the temperature rises, giving them a temperature coefficient that is expressed as a positive "P". Some capacitors decrease their value as the temperature rises, giving them a temperature coefficient that is expressed as a negative "N." For example, "P100" is +100 ppm/°C, or "N200" is –200 ppm/°C etc. However, some capacitors do not change their value and remain constant over a certain temperature range; such capacitors have a zero temperature coefficient or "NPO." These types of capacitors such as mica or polyester ones are generally referred to as class 1 capacitors.

Capacitance variation is caused by a change in the dielectric constant (permittivity) and an expansion or shrinking of the dielectric material/electrodes itself with temperature. Some dielectrics in these capacitors are strong temperature-sensitive materials. They have a phase transition temperature called Curie temperature, T_c; above it the ferroelectric dielectrics change to be paraelectric, resulting in dielectric constant change tremendously.

The dielectric strength level decreases as the temperature increases. This is due to the chemical activity of the dielectric material, which causes a change in the physical or electrical properties of the capacitor. The rated DC voltage U_R is the maximum DC voltage, or peak value of pulse voltage, or the sum of an applied DC voltage and the peak value of a superimposed AC voltage applied continuously to a capacitor at any temperature between the upper category temperature T_{RC} and the rated temperature T_R. The breakdown voltage U_R of a film capacitor decreases with temperature increase. When using film capacitors at temperatures between the rated temperature and the upper category temperature, only a temperature-derated category voltage U_C is allowed. Some manufacturers may have quite different derating curves for different types of capacitors. Figure 15.5 shows a generic curve of the capacitor voltage derating versus temperature. It can be seen that the rated voltage U_R is valid until the upper rated temperature T_R. After that, the allowed voltage of the capacitor decreases with increasing temperature. The minimum breakdown voltage, related to T_{RC}, is reduced to U_{RC}.

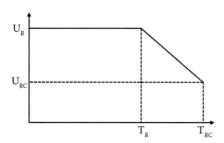

FIGURE 15.5
Temperature derating of maximum voltage of a capacitor.

An AC voltage will cause an AC current (with an applied DC bias this is also called "ripple current"), with cyclic charging and discharging of the capacitor causing oscillating motion of the electric dipoles in the dielectric. This results in dielectric losses, which are related to the equivalent series resistance (ESR) of capacitors, and the self-heating causes capacitor temperature rise. The maximum alternating voltage at a given frequency, which may be applied continuously to a capacitor (up to the rated temperature), is defined as the rated AC voltage $U_{R\,AC}$. Rated AC voltages usually are specified at 50 Hz (or 60 Hz for the US).

The rated AC voltage is generally calculated so that an internal temperature rise of 8–10°K sets the allowed limit for film capacitors. Because these losses are frequency dependent, the curves for derating the AC voltages are applied at higher frequencies.

The DF or tan δ, is a complex function involved with the "inefficiency" of the capacitor. It may vary either up or down with increased temperature, depending upon the dielectric material.

15.6.2 Thermally Induced Failures of Capacitors

Changes in temperature around the capacitor affect the value of the capacitor's parameters because of changes in the dielectric properties, as discussed above. If the air or surrounding temperature becomes too hot, the parameter value of the capacitor may change so much that it affects the correct operation of the circuit. In addition, catastrophic failures will occur or be accelerated if a capacitor operates continuously at high temperature.

Capacitors are subject to two classic failure modes: opens and shorts. Included in these categories are intermittent opens, shorts, and high resistance shorts. In addition to these failures, capacitors may fail due to capacitance drift, instability, high dissipation factor, or low insulation resistance with temperature increase.

Open capacitors usually occur as a result of overstress in an application. For instance, operation of capacitors at high AC current levels can cause a localized heating at the end terminations. The localized heating is caused by high resistive losses. Usually capacitors have poor thermal management. Continued operation of the capacitor can result in additional heating, and eventual failure. For example, the multilayer ceramic capacitor (MLCC) is constructed of alternate layers of silver/palladium (Ag/Pd) alloy, with a CTE of around 20 ppm/°C, and ceramic with a CTE of 10–12 ppm/°C. When this composite structure is heated during operation, the electrodes tend to force the capacitor apart. This tendency is made worse by Ag/Pd being a much better conductor of heat (>400 W/m.K) than ceramic (4–5 W/m.K), so that a thermal gradient will exist across the ceramic layer. This may result in increased leakage or shorts, as well as capacitance loss due to oxide vacancy migration, which is enhanced at high temperature (Freiman and Pohanka, 1989).

Transient heat energy may be generated through avalanche breakdown with high voltage or energy surges. The extremely high thermal energy can cause the failure by evaporating the connection between the metallization and the end contact.

For electrolytic capacitors and especially aluminum electrolytic capacitors, at temperatures greater than 85°C, the liquids within the electrolyte can be lost to evaporation, and the body of the capacitor, especially the small sizes, may become deformed due to the internal pressure and leak outright. As the temperature increases, the internal pressure inside the capacitor increases. If the internal pressure becomes great enough, it can even cause a breach in the capacitor, causing leakage of impregnation fluid or moisture susceptibility.

The capacitance of certain capacitors decreases as the component ages. This is caused by degradation of the dielectric. The most significant aging factors are the type of dielectric and ambient operating and storage temperatures. If the device is operating at or below its maximum rated conditions, most dielectric materials gradually deteriorate with time and temperature to the point of eventual failure. Most of the common dielectric materials undergo a slow aging process by which they become brittle and are more susceptible to cracking. The higher the temperature, the more the process is accelerated. The service life of a capacitor must be taken into consideration.

Lifetime is defined as the period during which a specified failure rate is not exceeded under given operating conditions and under specified failure criteria. Lifetime is continuously confirmed by accelerated sample tests at the upper category temperature. For electrolyte capacitor, at temperatures higher than 40°C, for every temperature rise of 10°C the acceleration factor is assumed to halve the lifetime at the same failure rate. In principle, the mechanism of failure is the loss of electrolyte. The degree of electrolyte loss (diffusion through the sealing elements) depends on the time, the electrolytic vapor pressure, the individual interaction of electrolytic solvent with the sealing materials, and geometric factors. The standard methods have been established for identification of the failure rates of capacitors, such as MIL-HDBK-217 (Military Handbook, 1991). Based on these, large amounts of experimental data demonstrating the temperature effects on lifetime have been obtained. In general, the temperature dependence of all capacitors failure rate is also described by Arrhenius Equation 15.4; for example, the effect of temperature on ceramic capacitor reliability is explained by a similar equation (Liu and Sampson, 2011).

15.6.3 Temperature Characteristics of Magnetic Materials

In electronic circuits there is another type of passive component that uses magnetism of materials. These devices include inductor, coil, choke, transformer, magnetic sensors, and many others. Specifically, these devices operate based on electromagnetic effects, which are realized by a basic structure of magnetic core plus an electrically conductive coil or winding. The magnetic material is a kind of substance that offers strong magnetism, characterized with a high permeability. They have the ability to retain a magnetic field when magnetized. A magnet is used to confine and guide magnetic fields in electrical, electromechanical, and magnetic devices. It is made of ferromagnetic metals and alloys, or ferrimagnetic compounds. The high permeability, relative to the surrounding air, causes the magnetic field lines to be concentrated in the core material. The magnetic field is often created by a coil of wires around the core that carries a current. The presence of the core can increase the magnetic field of a coil by a factor of several thousand over what it would be without the core.

The materials and form factors to make the magnets include laminated iron or Si–Fe alloys, powdered cores made of carbonyl iron, and ferrites consisting of ceramic materials with iron

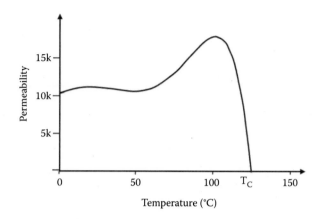

FIGURE 15.6
The temperature dependence of permeability of ferrite magnetic material.

oxide (Fe_2O_3) as their principal constituent. The characteristic temperatures of a magnetic component are determined by several performance parameters. Among them, the Curie temperature (T_c) of the magnet is the temperature at which it will be completely demagnetized (Chung, 2010). Figure 15.6 shows a typical characterization of magnetic materials with temperature. The permeability is at its maximum just below the T_c; above it the core permeability sharply disappears. The range of existing magnet's T_c is from 125 to 500°C, depending on the magnetic materials. Ferrites usually offer low Curie temperatures.

Below the Curie temperature, the magnetic properties of most magnets also show strong dependence on the temperature. For example, the permeability shifts with temperature variation in all types of magnetic cores and even with nonlinear temperature coefficient. This may require special adjustment in composition of the alloys for ensuring the temperature stability of the component.

In some magnets, the saturation flux density (Bsat) declines as temperature goes up. In many applications, to keep the losses low, cores are not operated near saturation. The Bsat reduction limits the components' operating range.

In most cases, some cores' maximum continuous operating temperature is limited by the case or coating on magnets and winding wires. Cores are placed in cases in order to protect the strain-sensitive magnetic material from physical stresses. Also, coatings and cases serve to prevent windings from being damaged at the sharp corners on bare cores under mechanical vibration. Most coating materials are polymer type and with an operation temperature below 150°C. Typical temperature ratings for cases, coatings, wire, and other components are far below the alloys' temperature limitations.

15.6.4 Thermally Induced Failure of Magnetic Components

It is important to note that heating a magnet to T_c can cause structural or mechanical damage, depending on the process used, heating and cooling rates, and the environment that the magnet is exposed to during the temperature cycle. For example, ferrites can easily crack if exposed to thermal shock. The material becomes vulnerable when it is changing by 4–8°C or more per minute. An increase or decrease of 2 or 3°C per minute is generally safe. Ferrites are poor thermal conductors, which is a benefit in relation to thermal shock. A short exposure to high temperature (such as a soldering operation) is often too brief to cause core heating or cracking. Likewise, after cores have been assembled into wound units, and possibly taped, varnished, potted, and attached to other components, the thermal shock seen by the core can be quite different from the nominal changes in temperature due to manufacturing, testing, or operation (de Graaf et al., 1995).

The core temperature can be elevated by internal heat from core loss and winding loss. The core loss of magnets becomes usually larger at elevated temperature. This positive feedback, also, may cause thermal runaway if the cooling condition is insufficient of a magnetic device, similar to the phenomena of power loss from leakage current in semiconductor devices discussed in Section 15.4.2.

The failure rate of a magnetic device is strongly affected by temperature. Specifically, it is dependent on how close the operating hot spot temperature is to the maximum rated temperature of its insulation materials.

15.7 High-Temperature Passive Materials

The wide variety of applications and increased demands challenged the electronics industry to make new products that are able to operate in exceedingly higher temperatures for longer periods of time. Various dielectrics have been intensively developed and modified, including ceramic, tantalum, plastic film, mica, solid aluminum, glass, silicon, diamond-like carbon (DLC), and aluminum oxynitride. Ceramic, tantalum, plastic film, glass, and DLC capacitors have established reliability in the high-temperature arena, whereas solid aluminum, silicon, and aluminum oxynitride are emerging technologies.

The commercially available capacitors rated for use at more than 200°C in power electronics systems are summarized in Wang et al. (2013). The typical three types of high-temperature capacitor are: ceramic, tantalum, and mica ones. Generally, the tantalum capacitor covers a high-capacitance range with a very low voltage rating. Conversely, the mica capacitor covers a low-capacitance range with a high voltage rating. The ceramic capacitor fills the gap between the tantalum capacitor and mica capacitor. The high-temperature film capacitors are manufactured from polyimide (PI) and polytetrafluoroethylene (PTFE) respectively. Capacitors wound with these dielectrics are most often standard paper-foil construction because the materials are only available in thick films with large thickness variations for any one casting. Although this technology is stable and readily available, the capacitor is bulky and costly. DLC and polycrystalline diamond (PCD) capacitors have the potential of achieving size and weight reductions compared to present polycarbonate (PC) capacitors, but the manufacturing technology limits their availability for large value, high voltage capacitors. Additionally, since DLC and PCD are deposited dielectric material, on a supporting base metal, the size, weight, and performance of the capacitor is dependent upon the base metal selected. Fluorene isophthalate terephalate or fluorene polyester (FPE) is seen as the only feasible near term option to 250°C capacitors. Other films have temperature limitations and/or would result in much larger and expensive capacitors due to less-favorable electrical characteristics. FPE dielectric constant is 3.4 versus PC at 3.0 for the same film thickness and FPE breakdown voltage is 12 kV/mil compared to 8 kV/mil for PC. FPE offers superior properties to manufacture high-temperature capacitors.

The high-temperature magnetic components rated above 200°C are also summarized in Wang et al. (2013). For applications requiring low-saturation flux density and high-operation frequency, most publications utilize ferrite magnetic core. For high-saturation flux density applications, such as a boost filter or an electromagnetic interference (EMI) filter in a high-power device, a nanocrystalline soft magnetic core could be a good candidate. High-temperature low-permeability NiZn ferrite toroid cores (4C65) are used in the gate drive transformer design, etc.

15.8 Summary

Modern electronics feature the operation temperature range of –55–125°C. More and more applications of electronics systems, such as in aerospace, automotive, and well-logging systems require electronics operation at much higher temperatures beyond the high end temperature. The development of new HTE components is the key to achieving cost-effectiveness of electronics implementation in such electromechanical systems. The limits to operation of electronic components come from their physical property and thermomechanical behavior dependence on the temperature. Thus, the breakthrough of electronics technology for high-temperature applications aims at the innovation in changing materials physical and chemical properties, structure optimization, and processing development of devices and components. So far, numerous electronics technologies have been successfully developed to enable the operation temperature beyond 200°C while achieving high reliability, safety, and economical operations.

Acknowledgments

The author thanks Leon M. Tolbert, University of Tennessee, Knoxville, Tennessee; Michael Pecht, University of Maryland, College Park, Maryland; and Fred (Fei) Wang, University of Tennessee, Knoxville, Tennessee, for reviewing this chapter and for providing valuable technical comments and suggestions.

References

Agarwal, A. K., Seshadri, S., and. Rowland, L. B., Temperature dependence of Fowler–Nordheim current in 6H- and 4H-SiC MOS capacitors, *IEEE Electron Device Letters*, 18(12), 592–594, 1997.

Bravaix, A., Goguenheim, D., Revil, N., Vincent, E., Varrot, M., and Mortini, P., Analysis of high temperature effects on performances and hot-carrier degradation in DC/AC stressed 0.35 μm n-MOSFETs, *Microelectronics Reliability*, 39(1), 35–44, 1999.

Chung, D. D. L., *Functional Materials: Electrical, Dielectric, Electromagnetic, Optical and Magnetic Applications*, Singapore: World Scientific Publishing Company, 2010.

Cristoloveanu, S., Silicon on insulator technology, in *The VLSI Handbook*, W.-K. Chen, Ed. Boca Raton, FL: CRC, Electrical Engineering Handbook Series, ch. 4, pp. 4.1–4.15, 2000.

Crofton, J., and Sriram, S., Reverse leakage current calculations for SiC Schottky contacts, *IEEE Trans on Electron Device*, 43(12), 2305–2307, 1996.

de Graaf, M., Dortmans, L., and Shpilman, A., Mechanical reliability of ferrite cores used in inductive components, *Proceedings of the Electrical Electronics Insulation Conference, 1995, and Electrical Manufacturing & Coil Winding Conference*, Page(s):485–488, 1995.

DiMaria, D. J. et al., Impact ionization, trap creation, degradation, and breakdown in silicon dioxide films on silicon, *Journal of Applied Physics*, 73, 3367–84, 1993.

Freiman, S. W., and Pohanka, R. C, Review of mechanically related failures of ceramic capacitors and capacitor materials, *Journal of the American Ceramic Society*, 72(12), 2258–2263, 1989.

Gurfinkel, M., Horst, J. C., Suehle, J. S., Bernstein, J. B., Shapira, Y., Matocha, K. S., Dunne, G., and Beaupre, R. A., Time-dependent dielectric breakdown of 4H-SiC/SiO$_2$ MOS capacitors, *IEEE Transactions on Device and Materials Reliability*, 8(4), 635–641, 2008.

Hadjadj, A., Simonetti, O. Maurel, T., Salace, G., and Petit, C., Si–SiO2 barrier height and its temperature dependence in metal-oxide-semiconductor structures with ultrathin gate oxide, *Applied Physics Letters*, 80, 3334, 2002.

JEDEC Solid State Technology Association, Failure Mechanisms and Models for Semiconductor Devices, JEDEC PUBLICATION-JEP122F, Nov. 2010.

Johnson, R. W., Evans, J. L., Jacobsen, P., Thompson, J. R., and Christopher, M., The changing automotive environment: High-temperature electronics, *IEEE Trans. on Electronics Packaging Manufacture*, 27(3), 164–176, 2004.

Kaiser, C. J., *The Capacitor Handbook*, C J Publishing, Olathe, KS, 2011.

Liu David (Donhang), and Sampson, M. J., Reliability Evaluation of Base-Metal-Electrode Multilayer Ceramic Capacitors for Potential Space Applications, CARTS USA 2011, March 28–31, 2011.

McCluskey, F. P., Grzybowski, R., and Podlesak, T., *High Temperature Electronics*, CRC Press, Inc. Boca Raton, FL, 1997.

Mehdi, A. E. I., and Brockschmidt, Karimi K. J, A case for high temperature electronics for aerospace, *IMAPS International Conference on High Temperature Electronics* (HiTEC), 2006.

Military Handbook, Reliability Prediction of Electronic Equipment, MIL-HDBK-217F, 1991.

Mönch, W., Metal-semiconductor contacts: Electronic properties, *Surface Science*, 299–300(1), 928–944, 1994.

National Research Council, Appendix a: Silicon as a high-temperature material, *Materials for High-Temperature Semiconductor Devices*, Washington, DC: The National Academies Press, 1995.

Neudeck, P. G., Okojie, R. S., and Chen, L.-Y., High-temperature electronics—A role for wide bandgap semiconductors? *Proceedings of the IEEE*, 90(6), 1065–1076, 2002.

Obreja, Vasile V. N., On the leakage current of present-day manufactured semiconductor junctions, *Solid-State Electronics*, 44(1), 49–57, 2000.

Schroeder, D. and Avellán, A., Physical explanation of the barrier height temperature dependence in metal-oxide-semiconductor leakage current models, *Applied Physics Letters*, 82, 4510, 2003.

Sharma, Rajinder, Temperature dependence of I-V characteristics of Au/n-Si schottky barrier diode, *Journal of Electron Devices*, 8, 286–292, 2010.

Singh, R., Reliability and performance limitations in SiC power devices, *Microelectronics Reliability*, 46, 713–730, 2006.

Sinkevitch, V. F., Vashchenko, V. A., *Physical Limitations of Semiconductor Devices*, US: Springer, 2008.

Sze, S. M., Ng, K. K., *Physics of Semiconductor Devices*, John Wiley & Sons, Inc. Hoboken, NJ, 2007.

Veneruso, A. F., High temperature electronics for geothermal energy, *Circuits & Systems Magazine* 1(3), 11–17, 1979.

Wang, R., Boroyevich, D., Ning P., Wang, Z., Wang, F., Mattavelli, P., Ngo, K. D. T., and Rajashekara, K., A high-temperature SiC three-phase AC–DC converter design for >100°C ambient temperature, *IEEE Trans. on Power Electronics*, 28(1), 555, 2013.

Wondrak, W. W., Physical limits and lifetime limitations of semiconductor devices at high temperatures, *Microelectronics Reliability* 39, 1113–1120, 1999.

Yang, T.-C., Saraswat, K. C., Effect of physical stress on the degradation of thin SiO$_2$ films under electrical stress, *IEEE Trans. on Electron Devices*, 47(4), 746–755, 2000.

Yeo, Y.-C., King, T.-J., and Hu, C., MOSFET gate leakage modeling and selection guide for alternative gate dielectrics based on leakage considerations, *IEEE Trans. on Electron Devices*, 50(4), 1027, 2003.

Yu, P. Y. and Manuel Cardona, *Fundamentals of Semiconductors: Physics and Materials Properties*, Springer, New York, 2010.

16

Ultra-High-Temperature Ultrasonic Sensor Design Challenges

Matthew M. Kropf

CONTENTS

16.1 Systems for Ultra-High-Temperature Sensors

16.1.1 Applications: Sensors in Extreme Environments; Geo-Technical, Space, Nuclear Reactors, Gas Turbines, and So On

The application base for sensors in high temperature, harsh environments is expanding rapidly. This growth is partly due to the pace of development of inexpensive electronics making the integration of self-diagnostic sensor systems practical to the point of pervasive adoption. The growth in the application base for high-temperature sensors is also being driven by the performance optimization and useful life extension in critical infrastructure, systems, and operations. Energy exploration and production is a category with several applications for high-temperature sensors.

For example, the use of natural gas combustion turbines is growing rapidly in the United States partly due to the boon in natural gas production from unconventional geologic reservoirs and partly due to the diversification away from traditionally configured coal-fired electrical power plants. With more of the electrical grid being supplied through gas turbines, competition is driving manufacturers to maximize performance and extend the life of products to gain market favor. In the case of gas turbines, both performance optimization and reliability can be addressed through the use of *in situ* sensors monitoring the active and critical components of the system. This requires small sensors capable of handling temperatures, pressures, and combustion by-products all while maintaining attachment to turbine blades rotating upwards of 10,000 rpms. The performance enhancements afforded through the apt use of live data from embedded sensors ultimately improve the efficiency of the machine, while similar sensors monitor critical components ready to provide early failure warning to minimize downtime while maintaining safety.

Another example of energy-related use of high-temperature sensors exists in the nuclear power industry. From making critical measurements in the operating core of a nuclear reactor to the eventual handling and storage of the radioactive by-products, sensors capable of operating at high temperature and in harsh environments contribute to the safety and efficiency of nuclear power generation. The nuclear application of high-temperature sensors is a significant challenge due to the limitation of housing and wiring materials suitable for radiation exposure and the additional environmental factors affecting sensor performance.

In addition to energy applications, there is a growing demand for high-temperature sensors to enable in situ measurements facilitating process control. For example, during the production of carbon–carbon composites, temperatures reach above 800°C. During this phase of the process, the composite material is vulnerable to developing cracks at the fiber–matrix interface that result in delamination. This process can be quantified through the use of ultrasonic and acoustic emission techniques (see Yen and Tittmann 1995). These measurement techniques have been designed for high-temperature deployment demonstrating process control abilities (Tittmann and Yen, 2008).

The focus of this chapter addresses design aspects of mechanical-based sensors for high (100 < 600°C) and ultra-high (>1000°C) environments. Specific examples will focus on the deployment of ultrasonic sensors and measurements, with reference to universal considerations for sensor packaging for ultra-high-temperature environments. The sensor configuration assumed in general is a user-controlled active element capable of generating and receiving ultrasonic mechanical waves through a waveguide, representing the object of inspection. For this reason, the chapter will focus on the electro-mechanical transducers and mechanical wave propagation. For a review of high-temperature sensors focused on optical-based sensors (see e.g., Barrera et al., 2012).

16.1.2 Materials and Environmental Limitations

Nonthermal environmental constraints on material choices for sensor systems are not limited to radiation from nuclear reactor cores or from celestial origins in space, for example material constraint could result from chemical compatibility in high-temperature material processing or geotechnical applications, or by weight limitations in aerospace applications. The application to nuclear power reactors creates even more extreme environments by delivering other environmental factors in conjunction with the formidable challenges of high and even ultra-high temperatures. In such environments, limitations on material selection extend beyond the thermal stability of the sensor and packaging to include the radioactive susceptibility and potential radio-chemical transmutations that could occur under neutron influence. This complication adds further complexity to sensor design by limiting the selection of materials to a narrow set and leaving little room for design choices based on optimal material properties.

16.2 Sensor Material Selection for High- and Ultra-High-Temperature Applications

High-temperature ultrasonic applications is not a new concept (see e.g., Tittmann and Aslan, 1999). However, research into sensing elements for ultra-high-temperature applications has continued to produce new options for ultrasonic measurements in extreme environments. The electromechanical coupling factor is a quantity describing how well a transducer converts electrical currents into mechanical strains. Measuring this over a range of temperature provides a practical measurement of a transducer's temperature performance. The material's ability to transduce depends on the crystalline molecular structure and its associated electromagnetic field. Practically, these are polarized electric fields in the case of piezoelectric (ferroelectric) materials and permanent magnetism in the case of magnetostrictive (ferromagnetic) materials. The Curie temperature describes the temperature at which this electric (or magnetic) field is sufficiently weakened. In this way, the Curie temperature represents the point at which most transducers fail to operate. Beyond the Curie temperature, the phase transition, or glass transition, temperature and melting temperature mark the absolute end of transducer viability. Table 16.1 shows the Curie temperatures of leading ultra-high-temperature piezoelectric materials.

16.2.1 Application Specific Environmental Limitations on Material Selection

Certain high-temperature environments can pose material compatibility issues based on factors peripheral to thermal energy. Specifically, reactivity between the sensor components and the sensor's environment can lead to the selection of materials that would

TABLE 16.1

High-Temperature Performance Characteristics of Ultrasonic Piezoelectric Transducers

Formula/Cut	T_c (°C)	Source
$Bi_4Ti_2O_{12}$	685	(Cummins, 1968)
$BiTi_3NbO_3$	940	(Subbarao, 1962)
$(1-x)BiScO_3$-$xPbTiO_3$	~450	(Eitel et al., 2001)
$LiNbO_3$	1200	(Baba et al., 2010; Lynnworth, 1999)
$La_2Ti_2O_7$	1300	(Nanamatsu et al., 1974)
AlN (Z)	2800[a]	(Parks et al., 2013)
LN (Y-36°)	1150	(Parks et al., 2013)
YCOB (XYlw -15°/45°)	1500[a]	(Parks et al., 2013)
PZT	350	(Kazys et al., 2005)
$Bi_4Ti_2O_{12}$	650	(Kazys et al., 2005)
(GaP)4	970[b]	(Kazys et al., 2005)
$LiNbO_3$ (36° Y-cut)	1210	(Kazys et al., 2005)
AlN	2200[a]	(Kazys et al., 2005)
Single-crystal ordered langasite	900[c]	(Zhang et al., 2009)

[a] Melt temperature.
[b] Phase transition temperature.
[c] Demonstrated.

otherwise be outperformed by other candidate materials. For example, in a nuclear reactor environment, the radioactive damage or decay to sensor housing materials such as nickel and iron-based alloys could produce long-lived radioactive isotopes creating high-level waste. This limits the selection for sensor housings to materials that would otherwise not provide the ideal mechanical performance.

A less extreme, but just as critical, example is the use of sensors in chemically reactive environments like high-temperature gases in geotechnical or natural gas turbine applications. In these cases, the thermal energy present creates an energetic tendency toward oxidation, coking, and other high-temperature chemical reactions. In this situation, ideal materials can still be chosen based on sensor performance provided they are sufficiently insulated from the reactive environment by way of coating or containment. Such a protective coating would be subject to the thermal expansion coefficient mismatch phenomena described in the case of spray-on transducers.

The application-specific limitations on sensor material selection may result in an increased reliance on temperature compensation techniques, as the optimal transducer material may not be suitable. However, ignoring environmental incompatibilities in the selection of sensor materials would obscure sensor measurements with competing physical–chemical effects.

16.2.2 Useful Life Limitations in Extreme Environments

The useful life a sensor system is limited primarily by the component's thermally weakest component. In the category of structural materials, for housings and waveguides, the lifetime limitation is related to the thermal degradation of the material through mechanical fatigue. For sensor housing components, this limitation in lifetime is realized by the compromised structural integrity of the sensor components. In this situation, due to the sudden and irreversible nature of the damage, the lifetime is marked by catastrophic failure where the sensor ceases to operate.

In other cases, the useful lifetime of the sensor could be marked by a deviation in performance that places the measured signals outside of the predicted and compensated parameters used in design. This form of lifetime limitation must be determined absolutely through experiment prior to sensor deployment in order to eliminate the possibility of false readings. This can occur based on the gradual nature of a failure due to sensor performance degradation. At some point, unknown to the user, the degradation phenomena will outweigh the effect of the measured phenomena on the sensor producing spurious data.

Both aspects of limited useful life indicate the critical step of reliability testing of high-temperature sensors. Reliability tests at expected operating environments to the point of failure must be conducted repeatedly, before the useful life of a sensor can be determined. In addition, tests beyond the operating environment should be made to determine the effects of thermal excursions on the useful life of the sensor element. For example, with piezoelectric sensors, the Curie temperature predicts the temperature where the transducer will no longer operate. Depending on the material, excursions near and above the Curie temperature may present irrecoverable damage to the sensing element. In this way, a reliability test over expected temperatures that are sufficiently below the Curie temperature will result in useful lifetime determinations significantly longer than a reliability test that includes even a few temperature excursions near the Curie temperature. Such tests are important in order to determine a means to detect the evidence of operating conditions extending beyond design parameters. Without such means, the sensors readings could be incorrectly interpreted, potentially exacerbating the cause of the system departing from design parameters.

16.3 Thermal Mechanisms for Consideration in Extreme Environments

The formal descriptions of thermal degradation mechanisms have been provided; thermal phenomena are selected in this section to highlight the practical design consideration for packaging ultrasonic sensors. Ultimately the sensor packaging is responsible for maintaining the mechanical contact with the waveguide and the electrical connection to the user. So, provided a viable sensing element, the design of the sensor packaging becomes the most important aspect to achieving accuracy and reliability of the measurement.

16.3.1 Thermal Expansion

16.3.1.1 Thermal Expansion of Sensor Package

Thermal expansion can present a complex problem to the design of sensors for ultra-high temperatures. In the simplest of terms, as the materials being used to connect, attach, and hold together the sensor assembly rise in temperature, they expand. From this general notion, maintaining the structural integrity of the assembly and the continuity through electrical leads over a broad range of temperatures is clearly a design challenge.

This notion is compounded by the fact that the various materials of the sensor assembly potentially have different coefficients of thermal expansion and that the different aspect ratios of the assembly parts both result in diverging rates of thermal expansion (Figure 16.1). Additionally, transient fluctuations in temperature will induce stresses even in a structure composed of materials with matched thermal expansion coefficients.

Furthermore, for ultrasonic sensors, the compression of the piezoelectric elements and the force coupling the sensors to the measurement substrate has direct implications on the quantitative amplitude measurements made by the sensors. This compression and coupling can be achieved in one of two ways: adhesive coupling of thin films and physical pressure of piezoelectric stacks.

In another case, thin film transducers, the piezoelectric element is adhered or deposited on to the waveguide to be inspected. This application can include surface acoustic wave (SAW) such as in Zhao and Tittmann (2010). The thermal expansion mismatch from this configuration is a result of the difference in thermal expansion coefficients of the sensor material and the substrate. Often, the example involves a metal substrate with a higher coefficient of thermal expansion than the ceramic piezoelectric deposited as the thin film transducer. In this case, the sensor will begin to delaminate or fracture when the thermally

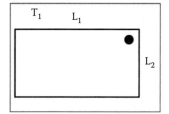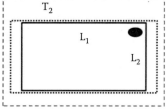

FIGURE 16.1

Illustration of nonuniform thermal expansion based on designed geometric aspect ratio, where the longer initial dimension ($L_1 > L_2$) manifests a nonuniform expansion of the hole in the upper right from the T_1 to the elevated temperature T_2.

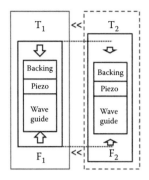

FIGURE 16.2
Illustration of the thermal expansion mismatch due to geometric aspect ratio of a piezoelectric sensor housing reducing the compression force from F1 in the initial case to F2 in the thermally expanded case resulting in signal loss at elevated temperatures.

induced strain exceeds the mechanical strength of the ceramic at the elevated temperature. In either case—delamination or fracture—the result is a loss of signal transmission from the sensor. In practical terms, the mismatch between the transducer's and substrate's thermal expansion creates a limitation to the working temperature range of the sensor system, which can be more restrictive than the expected range based on the sensor's thermal degradation characteristics. For example, the Curie temperature of a piezoelectric sensor may be significantly higher than the temperature at which the thermal expansion mismatch effectively destroys the sensing element. High temperatures can also affect the adhesion between sensor and substrate within the working temperatures defined by the thermal expansion mismatch, particularly in cases of long-term elevated temperatures and/or multiple temperature cycles (Searfass et al. 2010).

The alternative method of coupling ultrasonic transducers by physical pressure also takes careful design consideration to prevent nonuniform thermal expansion from limiting the functional temperature range of a sensor to below the range dictated by the physical properties of the sensor medium. To illustrate this problem, Figure 16.2 shows a cylindrical piezoelectric stack representing an ultrasonic transducer. In conventionally designed ultrasonic sensors, the backing material, piezoelectric element, electrodes, and working faces are compressed and sealed in a cylindrical housing. Without compensation for the mismatch in thermal expansion due to the long dimension of the sensor, the housing's elongation would exceed the thermal expansion of the piezoelectric and backing components resulting in weakening of the mechanical coupling of components and reduced effectiveness of the sensor. Furthermore, considering a waveguide that is coupled within the transducer's housing, this loss of compressive force would result in a loss of mechanical coupling.

16.3.1.2 Thermal Expansion of Sensing Element and Waveguide

In addition to the mechanical and electrical contacts being challenged by thermal expansion, the thermal expansion of the sensor element and waveguide materials alters the measurement frequency. Ultrasonic sensors generally operate at a mechanical resonance to transmit waves to the waveguide for inspection. Consider the simple case of a single element longitudinal bulk acoustic wave (BAW) sensor, the operating frequency is related to the thickness of the piezoelectric disc as indicated by the following equation; where n is the wave number, f_r is the resonant frequency, d is the thickness of piezoelectric disc, and c is the wave velocity in the transducer.

$$\frac{n}{f_r} = \frac{2d}{c}$$

Accordingly, the frequency output of the sensor will change as depicted in Figure 16.3 when the temperature is elevated from an initial low temperature (T1) to a much higher temperature (T2). The figure also depicts a frequency broadening and increase in loss due to increased scattering from mechanical degradation phenomena typically coincident with the frequency shift due to the thermal expansion of the element along its resonant dimension at ultra-high temperatures. Note that the loss is represented in decibels (dB), with 0 dB loss representing ideal signal transmission and values below 0 dB representing increasing loss in signal transmission.

The alteration of resonant response of ultrasonic transducers due to thermal expansion is not confined to single element, BAW sensors. Surface acoustic wave (SAW) and thin-film ultrasonic transducers are also sensitive to dimensional change due to thermal expansion. For SAWs, the thermal expansion increases the physical spacing of inter-digital finger pairs resulting in decreasing the resonant and, thus, operating frequency of the sensor. Thin-film ultrasonic sensors can be used to generate various wave modes, and are similarly affected by length dilations along the thickness mode. However, these and other guided wave sensor-based measurements can be inherently more sensitive to frequency fluctuations due to thermal expansion. In guided wave sensors, wave propagation itself is limited to specific frequencies (Rose, 1999). In these cases, the effect of a change in frequency generation due to thermal expansion is not limited to time of flight resolution as in the case of the single element, BAW sensor. In guided wave applications, the frequency dictates the wave mode meaning the ultrasound generated would not predictably propagate outside of a known frequency regime.

Compensation for a guided wave mode ultrasound has been demonstrated in refractory wire waveguides at ultra-high temperatures (Kropf et al., 2005). For guided wave modes, high thermal expansion compensation involves both the change in physical dimension of the waveguide and the mechanical property alteration. Each factor will determine the wave speed and the displacement profiles of a traveling wave. In ultra-high-temperature environments the displacement profile of the propagating wave mode can affect its sensitivity

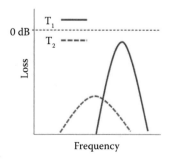

FIGURE 16.3
(See color insert.) This illustration depicts the frequency shift resulting from the thermal expansion along the resonant dimension of a single element piezoelectric transducer and increased scattering due to material degradation. Where the difference in temperatures T_1 and T_2 can be characterized by: $T_2 \gg T_1$.

to degradation phenomena. Specifically, wave modes can be selected with displacement profiles with maximum displacements at a controlled distance from the surface, creating mode selections that can either avoid or accurately measure corrosion on the surface of a waveguide.

16.3.2 Mechanical Degradation Mechanisms

For ultrasonic sensors and actuators, the action of the transducer is reliant on the organization of its crystalline structure. For piezoelectrics, this organization creates the polarized domains that cause the material to elongate or contract when subjected to an electric field. Similarly, magnetostrictive transducers rely on established ferromagnetic domains that deform when aligning physical domains to an external magnetic field. In both cases, either the polarized domains in the ferroelectric crystals or the magnetic domains aligned in ferromagnetic grain boundaries, the function of the sensor is derived from resultant electromagnetic properties that emerge from specific molecular structure and periodicity. The ultrasonic sensor is limited by the material's Curie temperature. Above this temperature, the thermal energy is sufficient in the material to prevent the alignment of the respective electromagnetic domains, preventing it from functioning as a sensor.

In addition to the ultrasonic sensor material degradation, the degradation and transformation of the waveguide dramatically affects measurement accuracy. For instance, corrosion is a degradation mechanism that is isolated to the surface of a material. This type of degradation can affect wave propagation by attenuating signals traveling through the waveguide or even providing an alternative path for mechanical wave to travel. While the first scenario is detrimental to signal quality, the latter can lead to spurious measurements. The degraded material, in this example a layer of corrosion, has a different wave velocity, and thus will affect time-of-flight measurements. A general mechanical degradation in crystalline materials will occur as thermal energy stimulates lattice dislocation migration to form micro-cracks.

16.4 Design for Thermal Expansion Compensation

There are two categories of design paradigms for sensors in high-temperature environments. The first is designing to maintain the structural integrity of the sensor's critical operating components. The second design consideration is the method of thermal compensation in the sensor measurement protocol. Both aspects of high-temperature design are critical to the performance of sensors deployed at high temperatures.

16.4.1 Maintaining Structural Integrity and Environmental Seals

As described in preceding sections, the effect of thermal expansion has been shown to cause a loss of electrical connections, mechanical contacts between sensors and subject, and compression, which can degrade a sensor's performance to the point of failure. One design technique to overcome thermal expansion mismatches is to incorporate a passive thermal expansion compensator, such as a pre-stressed spring element into the transducer housing. Hardened, stainless-steel alloys can be used for temperatures in the range of

1300°C. Deployment of this technique was shown to increase the longevity of ultrasonic transducers from less than 8 h, to greater than 40 h (Parks et al., 2010).

Another technique used to minimize the concern of environmental seals at ultra-high temperatures is to locate the sensing element remotely and use a single robust waveguide to direct the ultrasonic waves into and away from the component in the high-temperature environment. Using magnetostrictive transducers coupled to refractory metal waveguides, ultrasonic guided wave inspections in >1000°C were made at 30 m distance from the high-temperature environment (Kropf et al., 2006). This approach effectively avoids the issue of mechanical housing, coupling, and environmental seals in the high-temperature environment.

16.4.2 Compensating for Thermal Mechanisms

In an earlier section, both the effect of thermal expansion and thermally induced degradation was shown to change the performance of sensing elements, specifically in the case of ultrasonic sensors. Realistically, these changes in physical properties cannot be avoided entirely through sensor design. The result is a sensor where the active transduction element remains functional throughout the desired temperature range. However, while the sensor element is functioning, its signal response is still affected by thermal mechanisms in the sensor and subject body. Accordingly, thermal compensation remains a key aspect to high-temperature design, particularly in sensors reliant on wave propagation (Stepanova et al., 2010). In general, there are two methods of temperature compensation: physically integrating a control measurement into the sensor and digitally processing signals based on thermal models.

Introducing a physical control measurement into a sensor system has the advantage of providing a physical temperature correlation in the actual sensor environments. This avoids the reliance on an external temperature measurement with a sensor susceptible to similar failings. This physical compensation technique (Figure 16.4) was demonstrated in a precision ultrasonic position indicator reported in Pedrick and Tittmann (2004). In this work, the position of a metal target is tracked by monitoring changes in the time of flight of

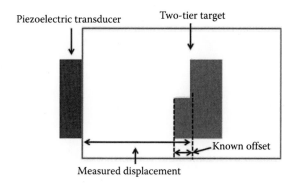

FIGURE 16.4
Depicts the introduction of a physical compensation technique creates a secondary measurement that is correlated to a known dimension enabling a temperature compensated ultrasonic velocity that can be used to determine the position of the target. (Pedrick M., and B. R. Tittmann, Ultrasonic micrometer position indicator with temperature compensation, *Proceeds of the Ultrasonics Symposium, IEEE-UFFC 50th Anniv. Joint Symp.* 2, 1199–1202. © 2004 IEEE.)

round-trip ultrasonic waves. To accurately calculate the change in distance of a round trip by the change in time of flight relies on a known ultrasonic velocity. The ultrasonic velocity is variable within a specified material over a range of temperatures. Therefore, fluctuations in temperature would otherwise be read as changes in position if not adequately compensated. To achieve this compensation, the displacement target was designed with a two-tiered surface that created two reflections from the target's surface with a known separation. With this addition, each reflected signal indicating the position of the target contained a smaller set of signals representative of the ultrasonic wave propagation over a known distance. With one known distance associated with the smaller reflection signals, one can calculate the actual velocity for each measurement. This temperature-adjusted velocity of sound is then used to calculate the change in position of the target (Pedrick and Tittmann, 2004).

Digitally processing signals from high-temperature environments based on theoretical models and experimental data sets can also be implemented to achieve temperature compensated sensor measurements. One example was demonstrated in Kropf et al. (2005), in which ultrasonic-guided wave modes were modeled to account for thermal expansion and physical property changes of the waveguide. In this work, theoretical and experimentally determined thermal expansion, density, and elastic modulus alterations over a range of temperatures were used to predict the guided wave mode velocities. These theoretical and experimental determinations were then confirmed by measurement up to 1000°C.

Practically, a degree of both design techniques for temperature compensation are used to ensure the proper correlation between measurement and effect at high temperatures. In general, temperature-compensated systems relying on either or both design elements are subjected to calibration runs at known temperatures and environmental conditions within the performance expectation of the sensor. By controlling the measurement parameter and potentially interfering environmental aspects before deployment, the performance of the designed system is verified.

16.5 Future Approaches for the Integration of Packaging and Sensor Electronics

With the capabilities provided by nanotechnology, the future of packaging sensor electronics for high temperature will trend toward integrated approaches. For example, the use of multi-functional housings (see, e.g., Rhimi, 2012), creates sensor packaging options that can serve to counteract the effects of thermal expansion in an automated and integrated way. Furthermore, the category of smart materials will begin to replace the deployment of dedicated sensors in favor of materials with the ability to both sense and act. In this way, the future of high-temperature sensors will terminate at the point of high-temperature smart materials. These high-temperature smart materials will serve both the primary purpose of a structural component to a larger machine and the secondary purpose of monitoring and maintaining its performance.

Despite this future, one must acknowledge the vast amount of existing systems and infrastructure bound to be in operation for decades to come. This leaves ample room for research and innovation in the area of retrofitted, external sensor applications for ultra-high-temperature environments. For these applications, the enhanced signal processing capabilities afforded through the progress in micro-computing will open the door for enhanced temperature compensated measurement capabilities for sensors. Such capabilities can be manifested in sensor designs featuring self-inspection of critical packaging features and access to reference databases of physical state performance behaviors.

Acknowledgments

The author thanks the editors for their support and patience in shaping this chapter. The author also thanks his colleagues: Carl W. Chang, NASA Glenn Research Center, Cleveland, Ohio; Liang-Yu Chen, NASA Glenn Research Center, Cleveland, Ohio; Michael Iten, Marmota Engineering AG, Zürich, Switzerland; David Parks, Idaho National Laboratory, Idaho Falls, Idaho; Bernhard R. Tittmann, Pennsylvania State University, University Park, Pennsylvania; for their useful inputs and helpful suggestions.

References

Baba A., C. T. Searfass, and B. R. Tittmann, High temperature ultrasonic transducer up to 1000 °C using lithium niobate single crystal, *Applied Physics Letters*, 97(23), 2010, 232901.

Barrera D., V. P. Finazzi, J. Villatoro, S. Sales, and V. Pruneri, Packaged optical sensors based on regenerated fiber Bragg gratings for high temperature applications, *IEEE Sensors Journal*, 12(1), 2012, 107–12.

Cummins, S. E., Electrical and optical properties of ferroelectric $Bi_4Ti_3O_{12}$ single crystals, *Journal of Applied Physics*, 39(5), 1968, 2268.

Eitel R. E., C. A. Randall, T. R. Shrout, P. W. Rehrig, W. Hackenberger, and S.-E. Park, New high temperature morphotropic phase boundary piezoelectrics based on Bi(Me)O3–PbTiO3 ceramics, *Japanese Journal of Applied Physics*, 40, 2001, 5999–6002.

Kazys R., A. Voleisis, R. Sliteris, L. Mazeika, R. Van Nieuwenhove, P. Kupschus, and H. A. Abderrahim, High temperature ultrasonic transducers for imaging and measurements in a liquid Pb/Bi eutectic alloy, *IEEE Transactions on Ultrasonics, Ferroelectrics, and Frequency Control*, 52(4), 2005, 525–537.

Kropf, M. M., Pedrick, M, and Tittmann, B.R., Remote high temperature thermometry using ultrasonic guided waves in thin wires, *AIP Conf. Proc.* 820, 2006, 1570.

Kropf M., M. Pedrick, and B. R. Tittmann, Remote sensing using ultrasonic guided waves in thin wires, *Journal of American Society for Nondestructive Testing*, Fall Conference and Quality Testing, 2005, ISBN: 1-57117-138-X.

Lynnworth L., High temperature flow measurement with wetted and clamp-on ultrasonic sensors, *Sensors*, 16(10), 1999, 36–52.

Nanamatsu S., M. Kimura, K. Doi, S. Matsushita, and N. Yamada, A new ferroelectric: La2Ti2o7, *Ferroelectrics*, 8(1), 1974, 511–513.

Parks D. A., M. M. Kropf, and B. R. Tittmann, Aluminum nitride as a high temperature transducer. *Review of Progress in Quantitative Nondestructive Evaluation*, edited by D. O. Thompson and D. E. Chimenti, 29A, 2010, 1029–1034.

Parks, D. A., S. Zhang, and B. R. Tittmann, High temperature (>500°C) ultrasonic transducers: An experimental comparison among three candidate piezoelectric materials, *IEEE Transactions UFFC*, 60(5), 2013, 1010–1015.

Pedrick M., and B. R. Tittmann, Ultrasonic micrometer position indicator with temperature compensation, *Proceeds of the Ultrasonics Symposium, IEEE-UFFC 50th Anniv. Joint Symp.* 2, 2004, 1199–1202.

Rhimi, M., and N. Lajnef, Passive temperature compensation in piezoelectric vibrators using shape memory alloy–induced axial loading, *Journal of Intelligent Material Systems and Structures*, 23.15, 2012, 1759–1770.

Rose J. L., *Ultrasonic Waves in Solid Media*, Cambridge University Press, Cambridge, United Kingdom, 1999.

Searfass C. T., A. Baba, B. R. Tittmann, and D. Agrawal, Fabrication and testing of microwave sintered sol–gel spray-on bismuth titanate-lithium niobate based piezoelectric composite for use as a high temperature (>500°C) ultrasonic transducer, *Review of Progress in Quantitative Nondestructive Evaluation*, edited by D. O. Thompson and D. E. Chimenti, 29A, 2010, 1035–1042.

Stepanova L. N., K. V. Kanifadin, and I. S. Ramazanov, The influence of temperature on the characteristics of piezoelectric transducers and errors of localization of acoustic-emission signals, *Russian Journal of Nondestructive Testing*, 46(5), 2010, 377–385.

Subbarao E. C., A family of ferroelectric bismuth compounds. *Journal of Physics and Chemistry of Solids*, 23, 1962, 665–676.

Tittmann B. R., and C. E. Yen, Acoustic emission technique for monitoring the pyrolysis of composites for process control, *Ultrasonics*, 48(6–7), 2008, 621–630.

Tittmann, B. R. and M. Aslan, Ultrasonic sensors for high temperature applications, *Japanese Journal of Applied Physics* 38, 1999, 3011–3013.

Yen C. E., and B. R. Tittmann, Fiber–matrix interface study of carbon–carbon composites using ultrasonics and acoustic microscopy, *Composites Engineering*, 5(6), 1995, 649–661.

Zhang, Shujun et al. Characterization of high temperature piezoelectric crystals with an ordered langasite structure, *Journal of Applied Physics*, 105.11, 2009, 114107–114107.

Zhao, X., and B. R. Tittmann, High-temperature surface acoustic wave transducer, *NASA Tech Briefs LEW-18547-1*, 34(10), 2010, 26.

17

High-Temperature Materials and Mechanisms: Applications and Challenges

Yoseph Bar-Cohen

CONTENTS

17.1 Introduction

An important tool in processing materials is to subject them to high temperatures and to use the resulting change in properties, which includes softening or melting, and changes in chemical and material phase. Thus, it is feasible to perform desired modifications in the material shape and characteristics. Other applications of high temperatures include heating, power harvesting, and other energy-related tasks. As covered in the various chapters of this book, there are many examples of operations that are implemented when materials are subjected to high temperatures, including forming and casting. To operate at extremely high temperatures, there is a need for devices, tools, structures, and mechanisms that are used to operate at such conditions. High-temperature materials have been used by humans as early as the primitives have started using fire and heat. The materials that are used have grown in capability and sophistication from as simple as rocks to the enormous selection that is available today (Bar-Cohen, 2003; Spear et al., 2006). The pool of high temperature (HT) materials that are available nowadays includes metal alloys such as stainless steels, superalloys, and refractory metals as well as ceramics, composites, and nano-carbon-based materials. The application of high-temperature materials has been expanded from food preparation and heating to fabrication processes and applications that require sustaining enormously high temperatures for extended periods of time. Examples of applications of high-temperature materials include aircraft jet engines, industrial gas turbines, hypersonic aircraft structures, power generators and nuclear reactors, space entry heat shields, as well as furnaces, ducting, electronics, and lighting devices (Wuchina et al., 2007). The requirements for materials and the material selection are dictated by the service operation conditions, as well as the ability of the selected materials to sustain the exposure over extended periods while being subjected to forces, pressures, corrosive conditions, and torques (in spikes, cycles, or constant levels). As the operating temperatures increase, there are fewer material choices that are available. The range of applications where high temperatures are used is quite large and some were covered in the previous chapters of this book. The focus of this chapter is on providing a brief review of some of the major and/or critical applications.

17.2 Operating Underground and at Great Earth Depths

The temperature of Earth rises with the increase in depth and the rate of the temperature rise in the upper part of the crust layer is approximately 20–25°C per kilometer but the rise gradient becomes smaller as the depth increases, reaching as much as 400–500°C at the boundary with the underlying upper mantle. A tabulated list of the temperatures at the various layers of Earth is given in Table 17.1 covering the crust, mantle, and core. Most applications that involve penetration of Earth's surface are limited to the upper part of the crust and they involve energy harvesting, mining, oil and gas extraction, and other geological requirements.

17.2.1 Geothermal Energy

Earth's underground crust and mantle are an enormous source of resources, including minerals and energy (EERE Report, 2010). As described in Table 17.1, the various layers of Earth from the upper mantle and further are quite hot. Volcanic emission of molten lava (Figure 17.1) illustrates the significant level of heat that is contained inside our planet and this heat offers enormous potential for generating clean energy. One-fifth of the geothermal energy of Earth's crust has been produced from the original formation of our planet

TABLE 17.1

Temperature at the Various Layers of Earth

Layer/Depth	Temperature (°C)
Crust—0–100 km thick	Up to about 500
Upper mantle—down to 1000 km	~500–900
Mantle (dense, hot layer of semisolid rock)—1000–2900 km	~1200
Outer core—2900–5100 km (liquid/magma)	~000
Inner core—5100–6378 km (solid)	>5000

Source: Based on the USGS website "Inside the Earth" http://pubs.usgs.gov/gip/dynamic/inside.html.

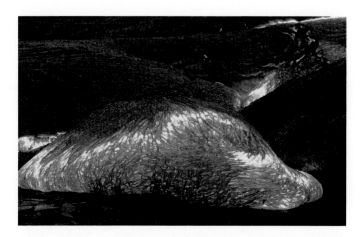

FIGURE 17.1
Molten lava shows the extremely high temperature to which rocks are subjected to at deep layers of Earth. (From Wikimedia Commons, http://en.wikipedia.org/wiki/File:Pahoehoe_toe.jpg.)

while the other 80% results from ongoing radioactive decay of minerals (Turcotte and Schubert, 2002). The temperature at the core of Earth is over 5000°C and, as a result of the heat and the enormous pressure (particularly in the outer core), the material is in a molten form known as magma. The heat from the core of Earth is conducted outward to the upper layers, heating rocks and water in the vicinity. The geothermal energy that can be reached has temperatures as high as 350°C. Geothermal energy provides a cost-effective, reliable, sustainable, and environmentally friendly source of energy for electrical power generation. Since ancient Roman times, naturally heated water in the form of hot springs have provided a source of warm water for bathing. The use of geothermal energy for generating electricity is now common in many countries (Figure 17.2).

FIGURE 17.2
Nesjavellir Geothermal Power Station in Iceland and steam that is emitted as part of power harvesting in this facility. (From Wikimedia Commons, http://en.wikipedia.org/wiki/Geothermal_energy.)

Geothermally heated water is directly used in district heating systems for both home and industrial consumers. The use of geothermal energy to produce electricity and as a source of hot water in district heating systems has reached the level of production and power of many megawatts. Even though there is an emission of some greenhouse gases that are trapped deep within Earth and are released from geothermal wells, geothermal energy is an Earth-friendly source. The quantities of the released gases are significantly lower than the level per energy unit that is produced by fossil fuels. The major challenge to using geothermal energy in greater levels is the cost of deep drilling in areas of hot rocks and it is far more difficult than the use of steam for energy harvesting. The higher cost is the result of the difficulties in placing and controlling the curing time of the cement used to create the borehole casing at high temperatures. Another cause of difficulty in drilling geothermal formations is that often they contain highly corrosive fluids.

17.2.2 Oil and Gas Exploration

The pressure of natural gas and water causes upward flow of petroleum and natural gas through Earth's crust. Oil and gas seep along fault lines and cracks and tend to accumulate in areas where they are trapped and form reservoirs of these resources. Through geological and seismic studies, as well as oil exploration, companies are continually seeking locations of such reservoirs. The generally accepted model for the formation of fossil fuel has been that it resulted from the compression of organic materials at great depths of Earth where carbon bonds in decomposing organic materials have broken down. Generally, higher temperatures create more natural gas than oil and, therefore, such gases are located at greater depths. In search of oil, gas, or other minerals, there are two drilling categories (Bar-Cohen and Zacny, 2009):

Downhole drilling—This drilling refers to penetrating the surface of Earth downward to a certain depth.

Horizontal drilling—This type of drilling starts as a downhole category and, at a selected depth, is followed by drilling horizontally in parallel to the surface of Earth. This category covers the trenchless method of installing infrastructure pipes for gas, oil, power, sewer, telecommunications, and water.

The capability of the high-power electronics currently available for drilling at great depths is quite limited. The functions they need to provide include monitoring the drilling operation, the health of the drilling system, as well as guiding the drill head toward the geological target. Initial exploration of sites involve various measurements of the characteristics of the formation in order to determine the presence and location of fluids, and if there are sufficient quantities of hydrocarbons that can be extracted. Once the completion (referring to the process of finishing a well in order to make it ready to produce oil or natural gas) and production phases of oil extraction start, measurements are made of the temperature, pressure, and fluids flow rates. The process of oil exploration and extraction is very costly and critically dependent on the quality of the equipment and mechanisms that are used.

As drilling for oil and gas is deepened, the temperature and the pressure are increasing and they pose greater challenges to the required tools and mechanisms that can sustain the related harsh conditions. At today's oil and gas exploration depths, the drilling tools can be subjected to as high as 230°C and pressure as high as 25,000 psi. The conditions that are encountered at increasingly growing depth of drilling, which can be 10 km and more, are requiring advances in the capacities and technologies that are being used. These include high-strength metal alloys, sensors, electronics, mechanisms, and *in situ* power generation and storage.

17.2.3 Underground Mining

Underground mining is subjected to the increased temperature levels of Earth's crust as listed in Table 17.1. When operating underground, besides the need to use equipment that can sustain the temperature and pressure conditions, there is even greater concern to the miners' health and safety. Generally, personnel who are working in areas with high temperatures may suffer heat illness and even death. The heat causes loss of concentration, reduces the miners' productivity, and may lead to accidents due to human errors. The effects are even more serious when there is also exposure to relatively high humidity inhibiting the body from the ability to cool off by the evaporation of sweat. According to the Institute of Occupational Medicine, temperatures as high as 30°C can pose great health concerns and require appropriate measures to prevent risk to the operators in mines.

17.2.4 Underground Steam Pipes

District heating systems carry steam or hot water from central generating stations under the streets of cities in various locations in the world to provide heat, hot water, and cooling (via centrifugal or absorption chillers) to buildings, businesses, and facilities. This section focuses on district steam and Figure 17.3 shows a typical steam street infrastructure and equipment that is used. The delivered steam is used by buildings, businesses, and facilities for cleaning, disinfection, sterilization, food preparation, and humidification. Such

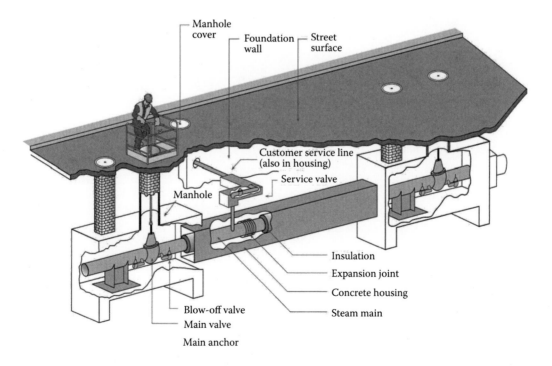

FIGURE 17.3
Steam distribution system. (Courtesy of Con Edison, NY.)

applications can be found at dry cleaners, laundries, hospitals, restaurants, and museums. In 1882, the New York Steam Company began providing steam in lower Manhattan. Today, Consolidated Edison Company of New York (known as Con Edison) owns and operates this system, which has grown to become the largest steam system in the world. About 26 billion pounds of steam flow through its system every year at speeds that can reach over 100 miles per hour. The pipes are as hot as 200°C (390°F) or more and the pressure can be as high as 400 psig. The steam vapor that is sometimes observed rising from manholes in Manhattan is generally caused by external water (rain, leaking water pipe, etc.) being boiled by contact with the steam pipe in the specific manhole rather than resulting from a leak.

Monitoring the steam system is important to assuring the safety of its operation. Safety concerns arise when excessive condensation accumulates in the steam pipe. Steam traps are installed at strategic locations along the pipe run to remove condensed water from inside the steam pipes. Excessive condensation can cause the potential for water hammer that may lead to damaged steam piping and equipment. Water hammer is caused when steam condensate accumulates and is trapped in certain configurations and locations of the steam pipes. There are two types of water hammer:

- Steam that rapidly flows over condensed water causes ripples in the water and builds up turbulence that may result in the formation of a pressure slug. The water slug travels at high velocities and can cause damage when it hits a pipe elbow or configuration of piping equipment that is a flow obstacle.

- Excessive condensate accumulation can subcool below the saturation temperature and, if such subcooled condensate is disturbed by the steam flow, it could envelop a large steam bubble. Under certain conditions, the bubble collapses within the water, resulting in an enormous pressure spike. Such a water hammer effect is more catastrophic in terms of potential damage and it is more critical to prevent.

To minimize the potential for water hammer conditions, there is a need for a system that can monitor the level of condensed water inside the steam pipes. For this purpose, jointly with members of the authors group at JPL/Caltech/NASA, they are developing a monitoring system that can measure the height of the condensed water through the pipe wall (Chapter 8). The system is being developed with an ultrasonic probe that can sustain the harsh environment of the steam pipe system at temperature as high as 250°C (480°F). Also, the drive electronics that consists of pulser/receiver, amplifier, and microprocessor are developed to sustain 75°C (167°F) and humidity as high as 85% and are designed to be placed away from the pipe but inside the manhole. A pulse-echo ultrasonic method was demonstrated to provide the required capability and detailed description of the method is given in Chapter 8 and in Bar-Cohen et al. (2010).

17.3 Space Exploration of Hot Planets in the Solar System

NASA is increasingly launching exploration missions to planets in the solar system that have harsh ambient conditions including very high temperatures. Among the hot planets in the solar system, one can list Mercury (Figure 17.4) and Venus (Figure 17.5). Mercury is only slightly larger than Earth's Moon; it has very little atmosphere to stop impacts from

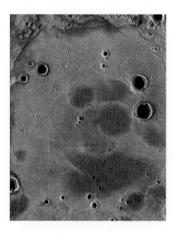

FIGURE 17.4
Topographic image from the Mercury Laser Altimeter (MLA) that was used to colorize a mosaic image of Goethe basin, located in Mercury's northern region. (NASA/Johns Hopkins University Applied Physics Laboratory/ Carnegie Institution of Washington. http://photojournal.jpl.nasa.gov/catalog/PIA15536.)

meteorites and it is covered with many craters. Its dayside is superheated by the sun to as high as 427°C, but due to the thin atmosphere at night, the temperature drops significantly to as low as –173°C, where ice may even exist in its craters.

Venus is a neighbor of Earth and is closer to the Sun having an ambient temperature of about 460°C. Like Earth, its surface shows that the geological processes and climate are driven by feedback between the atmosphere, the surface, and the interior. Over the last 50 years, various missions have been launched to Venus and currently it is considered an objective for further exploration of its surface and atmosphere in order to help understand the origin of Earth, and perhaps other bodies in the solar system and beyond. Efforts are being made to help answer the question "if Venus was ever like Earth and will someday Earth become like Venus?" Also, it may determine the factors that make a rocky body evolve to a warm wet world like Earth, or to a sulfurous, desiccated, extremely hot planet like Venus. The past exploration missions by the Soviet Union (Venera and Vega) and the

FIGURE 17.5
(**See color insert.**) A synthetic image of the volcano Maat Mons on the surface of Venus that was created from Magellan orbital radar data. (From Wikimedia Commons, http://en.wikipedia.org/wiki/File:Venus_-_3D_ Perspective_View_of_Maat_Mons.jpg.)

United States did not bring enough answers to these questions and a number of recent *in situ* exploration mission proposals have been made in recent years including the Surface and Atmosphere Geochemical Explorer (SAGE) and Venus *In Situ* Explorer (VISE).

Technology limitations of operating at high temperature and pressure on or near the surface of Venus have been dictating the scope of all the *in situ* exploration missions that were launched or planned to be launched. Various techniques have been used to slow the destructive effect of heating the landers, including strong thermal insulation and using phase-change heat sink materials that maintain the inside of the lander at the phase-transition temperature. Over the short duration of prior missions (up to 2 h), various tasks and functions have been executed, including taking photos, making various measurements, and drilling. A synthetic image of the volcano Maat Mons on the Venus surface that was created from Magellan orbital radar data is shown in Figure 17.5. Improvements in technology are allowing the planning of missions with longer duration period. The proposed NASA mission SAGE is seeking to have a lander that would reach the surface of Venus and survive for 2–3 h. The surface will be analyzed remotely by a suite of instruments that include cameras, spectrometers, meteorology package, as well as instruments to determine the mineralogy and surface texture. By trenching the surface and the subsurface exposed formation would be examined by these instruments.

17.4 Commercial and Military Applications to Superfast Flights

Air flight at super high speeds, which include supersonic and beyond, causes significant temperature rise, particularly at the leading edges and the control surfaces of the flying vehicle (Anderson, 2006). At such speeds, there are physical changes in the airflow that involve viscous dissipation of heat leading to nonequilibrium characteristics of vibrational excitation, dissociation, and ionization of molecules; these changes result in convective and radiation heating of the vehicle. The higher the temperature to which an aircraft structure is subjected, the greater the challenges to testing the materials and structures at the related service conditions. Generally, the flying speed of combat planes is in the supersonic regime of above 1.0–2.5 Mach. The regime of speed that is 5–10 Mach is known as hypersonic and it represents the latest challenge in developing aerodynamic systems for military aircraft (Figure 17.6). The next regimes are known as high-hypersonics and they cover 10–25 Mach and it is followed by the reentry speed. The Space Shuttle used to operate in the reentry speed regime and these speed regimes cover the increasing number of private space aircraft currently being developed. For operating at the speeds of reentry into Earth's atmosphere, heat shields that are based on materials such as carbon/carbon were developed, as described in Chapters 1 and 3 (Beck et al., 2010).

17.5 Electronics

Driving and controlling high-temperature mechanisms require the use of electronic systems. In many industries, particularly if the temperature is outside the operating range of available electronics, the systems are either passively or actively cooled. There are reasons

FIGURE 17.6
The jet-propelled aircraft hypersonic NASA X-43, which is an unmanned experimental hypersonic aircraft. (From Wikimedia Commons, http://en.wikipedia.org/wiki/File:X-43A.jpg.)

for which one would want to operate directly at the high temperatures of the applicable system and these may include reducing the cost of the operation or increasing the system reliability. Operating at high ambient temperatures poses challenges to the capability of existing electronic systems and requires advances in materials science, design techniques, methods and means of interconnecting, packaging, and methods of qualifying the related hardware (Neudeck et al., 2002; McCluskey et al., 1997). Some of the issues related to fabrication and operation of high-temperature electronics include the melting (softening) point of the solders that are used and for this purpose there are choices of soft solders that melt at about 180°C and hard solders that melt at as high as 400°C. Many applications that require high-temperature electronics were covered in this chapter and throughout this book, including Chapter 16. For applications such as geothermal energy harvesting and others, efforts are underway to develop robust sensors and telemetry electronics that can operate at as high as 300°C (Neudeck et al., 2002).

For active devices and passive components, the high-temperature limit of electronics is mostly determined by the interconnections and the related packaging (Zhang et al., 2010, 2011; Fang et al., 2011). The practical upper limit of commercially available circuits is about 300°C (see, e.g., http://www.ssec.honeywell.com/hightemp/). Discrete semiconductor devices have been reported to operate at temperatures as high as about 650°C for silicon carbide (SiC) MOSFET devices and +700°C for a diamond Schottky diode. Integrated circuits based on Si and GaAs have been reported to operate at as high as +400 °C to 500°C. Silicon-based integrated circuits have been reported to operate at +300°C for 1000 h or longer. Also, transistors were reported to work at as high as +50°C to 400°C (Kirschman, 1998).

17.5.1 Military Electronics

Electrical and mechanical devices that are used in military and aerospace applications are required to sustain temperatures that are higher than commercial or industrial devices. For example, commercial-grade resistors need to sustain 70°C while the industrial-grade resistors are required to sustain 85°C. In contrast, military-grade resistors are required to sustain 125°C and sometimes even 175°C. The grading of components is intended to assure that the devices are able to sustain the environmental conditions at which they are used. For the assurance of the service life of the military devices, military standards define the requirements for the design and testing limits and an example of a related document is MIL-STD-810.

17.5.2 Automobile Electronics

The temperature under the hood of an automobile can be as high as 125°C (Figure 17.7). The thermostat plays a critical role in assuring the reliable operation of automobile engines and their various parts, including the control electronics (Blalock et al., 2008; Johnson et al., 2004). The exposure duration and maximum temperature to which the automobile electronics is subjected depends on the specific type of vehicle that is used and the location in it (Table 17.2). Increasingly, efforts are made to push the durability of the automotive electronics to higher temperatures, but the increase is limited by cost constraints and reliability requirements. The increased efforts to replace mechanical and hydraulic systems with electromechanical ones as well as the development of hybrid cars are pushing these technologies toward higher-performance power electronics. These applications necessitate operation temperatures that are as high as 175–200°C.

17.5.3 Aerospace Electronics

As in automobiles, the hottest parts of an aircraft are the engine and the exhaust system. In addition, the structural areas facing the direction of the flight are subjected to higher temperatures during flight, particularly, the leading edges of high-speed airplanes and aerospace systems. Components that are in close proximity to the engine, including the electronics, are subjected to ambient temperatures of up to 200°C. Also, depending on the aircraft speed, the leading edges can be subjected to hundreds of degrees centigrade. Even though the electronics can be cooled, there are advantages to operating at the ambient conditions, including reduction of cost, lowering of the required hardware weight, and reducing the probability of failure by minimizing the number of parts. Also, aircraft systems

FIGURE 17.7
Under the hood of an automobile engine, the temperature is as high as 125°C and the automobile electronics technology is increasingly being pushed toward operation at 200°C.

TABLE 17.2

Typical Automotive Maximum Temperature Ranges

Automobile Part	Temperature Range
On the engine and in the transmission	150–200°C
On the wheel sensors	150–250°C
Cylinder	200–300°C
Exhaust	Ambient 300°C and can reach as high as 850°C

are increasingly designed with minimal wiring and this is achieved by placing electronics close to the actuators and this requires operation at the related high ambient-temperature environment of certain parts and components.

17.6 Summary/Conclusions

The availability of materials that can be effectively used at high temperatures allows pushing the limits of the potential applications that can be considered. Materials that are available and suitable for these applications were described in this book and they include metals, ceramics, and polymers, both monolithic and in composites forms. Also, many applications were described and discussed in this book, including aircraft and aerospace structures, space exploration, geophysics, and energy harvesting. Further, high-temperature processes were described, including reduction, reuse, recycling, and recovering of metals, rare earth, and other elements. The applications and mechanisms that can be considered are dictated by the limitations of magnets and actuators that are generally in the range of 350°C. In recent years, with the introduction of piezoelectric materials with higher Curie temperature and efficiency as well as such magnets, the temperature limit has been raised to the range of 500°C. Moreover, limitations on the temperature of applicable mechanisms and devices are dictated by the capability of the available electronics.

In conclusion, high-temperature materials are increasingly being needed for a growing number of applications, including automotive and aerospace industries as well as power generation and space exploration. As new materials, including nanomaterials, are developed, the ability of mechanisms and structures to withstand and perform at high temperatures is becoming easier to realize.

Acknowledgments

Some of the research reported in this chapter was conducted at the Jet Propulsion Laboratory (JPL), California Institute of Technology, under a contract with the National Aeronautics and Space Administration (NASA). The author thanks Jeff Hall, Jet Propulsion Laboratory/Caltech, Pasadena, California, and Robert Cormia, Foothill College, Los Altos Hills, California, for reviewing this chapter and for providing valuable technical comments and suggestions. The author also thanks Edward Ecock, Josephine Aromando, and

Dowlatram Somrah, for their inputs, comments, and helpful suggestions regarding the health monitoring of the height of condensed water in steam pipes. The author also thanks Arsham Dingizian and Linda Y. Del Castillo, JPL/Caltech, for their helpful comments regarding high-temperature electronics.

References

EERE Report, A History of Geothermal Energy Research and Development in the United States—Drilling 1976–2006, http://www.eere.energy.gov/, EERE Information Center, Energy Efficiency & Renewable Energy, US Dept. of Energy, September 2010.

Anderson J., *Hypersonic and High-Temperature Gas Dynamics*, 2nd Edition, ISBN 10: 1-56347-780-7, ISBN-13: 978-1-56347-780-5, AIAA Education Series, The American Institute of Aeronautics and Astronautics (AIAA), Reston, VA, 2006, pp. 1–813.

Bar-Cohen Y., High Temperature Technologies for Sample Acquisition and In-Situ Analysis, JPL Technical Report, Document No. D-31090, September 8, 2003.

Bar-Cohen Y. and K. Zacny (eds.), *Drilling in Extreme Environments—Penetration and Sampling on Earth and Other Planets*, Wiley-VCH, Hoboken, NJ, ISBN-10: 3527408525, ISBN-13: 9783527408528, 2009, pp. 1–827.

Bar-Cohen Y., S. Lih, M. Badescu, X. Bao, S. Sherrit, S. Widholm and J. Blosiu, In-service monitoring of steam pipe systems at high temperatures. SPIE Health Monitoring of Structural and Biological Systems IV Conference, Smart Structures and Materials Symposium, Paper 7650-26, San Diego, CA, March 8–11, 2010.

Beck R. A. S., H. H. Hwang, M. J. Wright, D. M. Driver, and E. M. Slimko, The evolution of the Mars Science Laboratory heatshield, 7th International Planetary Probe Workshop, Barcelona, Spain, June 16, 2010.

Blalock B., C Huque, L. Tolbert, M. Su, S. Islam, and R. Vijayaraghavan, Silicon-on-insulator based high temperature electronics for automotive applications, 2008 IEEE International Symposium on Industrial Electronics, 2008, pp. 2538–2543.

Fang K., R. Zhang, R. W. Johnson, E. Andarawis and A. Vert, Thin film multichip packaging for high temperature digital electronics, 2011 High Temperature Electronics Network (HiTEN), Oxford, UK, July 18–20, 2011.

Johnson R. W., J. L. Evans, P. Jacobsen, J. R. Thompson, and M. Christopher, The changing automotive environment: High-temperature electronics. *IEEE Transactions on Electronics Packaging Manufacturing*, Vol. 27, No. 3, (July 2004) pp. 164–176. http://ieeexplore.ieee.org/stamp/stamp.jsp?tp = &arnumber = 1393072&userType = inst.

Kirschman R., *High-Temperature Electronics*, ISBN-10: 0780334779, ISBN-13: 978-0780334779, Wiley-IEEE Press, Hoboken, NJ, 1998, 912 pages.

McCluskey F. P., R. Grzybowski, and T. Podlesak, *High Temperature Electronics*, ISBN-10: 0849396239, ISBN-13: 978-0849396236, CRC Press, New York, NY, 1997, pp. 1–352.

Neudeck P. G., R. S. Okojie, and L.-Y. Chen, High-temperature electronics—A role for wide bandgap semiconductors, *Proceedings of the IEEE*, Vol. 90, No. 6, June 2002, 1065–1076.

Spear K. E., S. Visco, E. J. Wuchina, and E. D. Wachsman, High temperature materials, *Interface Journal, the Electrochemical Society Conference* (Spring 2006), Vol. 15, No. 1, pp. 48–51.

Turcotte, D. L., G., Schubert, *Geodynamics*, 2nd Edition, ISBN-10: 0521666244, ISBN-13: 978-0521666244, Cambridge University Press, Cambridge, England, UK, ISBN 978-0-521-66624-4 2002, pp. 1–472.

Wuchina E., E. Opila, M. Opeka, W. Fahrenholtz, and I. Talmy, UHTCs: Ultra-high temperature ceramic materials for extreme environment applications, Proceedings of the Winter 2007 Electrochemical Society Interface, 2007, pp. 30–36.

Zhang R., R. W. Johnson, D. Shaddock, T. Zhang and V. Tilak, Characterization of thick film technology for 300°C packaging, 2010 High Temperature Electronics Conference, Albuquerque, NM, May 11–13, 2010.

Zhang T., D. Shaddock, A, Vert, R. Zhang and R. W. Johnson, Characterization of LTCC-thick film technology for 300°C packaging, 2011 HiTEN, Oxford, UK, July 18–20, 2011.

Internet Resources

High Temperature Quartz Oscillators Meet Oil and Gas Exploration Challenges http://www.ecnmag.com/articles/2010/01/high-temperature-quartz-oscillators-meet-oil-and-gas-exploration-challenges

High-Temperature Electronics Pose Design and Reliability Challenges http://www.analog.com/library/analogDialogue/archives/46-04/high_temp_electronics.html

How were fossil fuels formed? http://wiki.answers.com/Q/How_were_fossil_fuels_formed

NASA Solar System Exploration—Mercury: http://solarsystem.nasa.gov/planets/profile.cfm?Object=Mercury&Display=OverviewLong

NASA Solar System Exploration—Venus: http://solarsystem.nasa.gov/planets/profile.cfm?Object=Venus

The packaging challenges of high temperature electronics http://ge.geglobalresearch.com/blog/the-packaging-challenges-of-high-temperature-electronics/

Temperature Range http://www.national.com/AU/design/courses/213/nsi03/02nsi03.htm

The Temperature Ratings of Electronic Parts http://www.electronics-cooling.com/2004/02/the-temperature-ratings-of-electronic-parts/

Index

For Product Safety Concerns and Information please contact our EU representative GPSR@taylorandfrancis.com Taylor & Francis Verlag GmbH, Kaufingerstraße 24, 80331 München, Germany

Printed and bound by CPI Group (UK) Ltd, Croydon, CR0 4YY
01/05/2025
01858616-0004